Methods in Cell Biology

VOLUME 60

The Zebrafish: Genetics and Genomics

Series Editors

Leslie Wilson
Department of Biological Sciences
University of California, Santa Barbara
Santa Barbara, California

Paul Matsudaira
Whitehead Institute for Biomedical Research and
Department of Biology
Massachusetts Institute of Technology
Cambridge, Massachusetts

Methods in Cell Biology

Prepared under the Auspices of the American Society for Cell Biology

VOLUME 60

The Zebrafish: Genetics and Genomics

Edited by

H. William Detrich, III

Department of Biology
Northeastern University
Boston, Massachusetts

Monte Westerfield

Institute of Neuroscience
University of Oregon
Eugene, Oregon

Leonard I. Zon

Howard Hughes Medical Institute
Children's Hospital
Boston, Massachusetts

ACADEMIC PRESS

San Diego London Boston New York Sydney Tokyo Toronto

Cover photograph (paperback edition only): The *leopard* mutation (upper fish) was the first to be identified in *Danio rerio*, the first zebrafish mutation to be mapped, and it identifies Linkage Group 1. Courtesy of Steve Johnson, Jerry Gleason, and John Postlethwait.

This book is printed on acid-free paper. ♾

Academic Press
a division of Harcourt Brace & Company
525 B Street, Suite 1900, San Diego, California 92101-4495, USA
http://www.apnet.com

Academic Press
24-28 Oval Road, London NW1 7DX, UK
http://www.hbuk.co.uk/ap/

International Standard Book Number: 0-12-544162-2 (hb)
International Standard Book Number: 0-12-212172-4 (pb)

PRINTED IN THE UNITED STATES OF AMERICA
98 99 00 01 02 03 EB 9 8 7 6 5 4 3 2 1

These volumes are dedicated to George Streisinger, whose insight, generosity, and encouragement inspired and nurtured a vibrant field of research.

CONTENTS

CONTRIBUTORS

Numbers in parentheses indicate the pages on which the authors' contributions begin.

Chris T. Amemiya (235), Center for Human Genetics, Boston University School of Medicine, Boston, Massachusetts 02118-2394

Angel Amores (149, 323), Institute of Neuroscience, University of Oregon, Eugene, Oregon 97403-1254

Adam Amsterdam (87), Center for Cancer Research and Department of Biology, Massachusetts Institute of Technology, Cambridge, Massachusetts 02139

Christine E. Beattie (71), Neurobiotechnology Center and Department of Pharmacology, Ohio State University, Columbus, Ohio 43210

David R. Beier (185), Genetics Division, Brigham and Women's Hospital, Harvard Medical School, Boston, Massachusetts 02115

Mario Chevrette (303), Department of Surgery, Urology Division, McGill University and Montreal General Hospital Research Institute, Montreal, Quebec, Canada

Ricky Critcher (287), Department of Genetics, University of Cambridge, Cambridge CB2 3EH, United Kingdom

H. William Detrich, III (361), Department of Biology, Northeastern University, Boston, Massachusetts 02115

Eckehard Doerry (339), Computer and Information Science Department, University of Oregon, Eugene, Oregon 97403-1254

Sarah A. Douglas (339), Computer and Information Science Department, University of Oregon, Eugene, Oregon 97403-1254

Pat Edwards (373), Institute of Neuroscience, University of Oregon, Eugene, Oregon 97403-1254

Fredericus J. M. van Eeden (21), Wellcome/CRC Institute, Cambridge CB2 1QR, United Kingdom

Judith S. Eisen (71), Institute of Neuroscience, University of Oregon, Eugene, Oregon 97403-1254

Marc Ekker (303), Department of Medicine and Department of Cellular and Molecular Medicine, Ottawa Hospital Loeb Research Institute, University of Ottawa, Ottawa, Ontario K1Y 4E9, Canada

Mark C. Fishman (235), Cardiovascular Research Center, Massachusetts General Hospital, Charlestown, Massachusetts 02129; and Department of Medicine, Harvard Medical School, Cambridge, Massachusetts

Dorothee Foernzler (185), Genetics Division, Brigham and Women's Hospital, Harvard Medical School, Boston, Massachusetts 02115

Allan Force (149), Institute of Neuroscience, University of Oregon, Eugene, Oregon 97403-1254

Michael A. Gates (165), Developmental Genetics Program, Skirball Institute of Biomolecular Medicine, New York University Medical Center, New York City, New York 10016

Zhiyuan Gong (213), Department of Biological Sciences, National University of Singapore, Singapore 119260

Michael Granato (21), University of Pennsylvania, Philadelphia, Pennsylvania 19104-6058

Perry B. Hackett (99), Department of Genetics and Cell Biology, Institute of Human Genetics, University of Minnesota, St. Paul, Minnesota 55108-1095

Pascal Haffter (21), Max-Planck-Institüt für Entwicklungsbiologie, 72070 Tubingen, Germany

Paul D. Henion (71), Neurobiotechnology Center and Department of Cell Biology and Neuroanatomy, Ohio State University, Columbus, Ohio 43210

Nancy Hopkins (87), Center for Cancer Research and Department of Biology, Massachusetts Institute of Technology, Cambridge, Massachusetts 02139

Zoltan Ivics (99), Division of Molecular Biology, The Netherlands Cancer Institute, Amsterdam 1066CX, The Netherlands

Zsuzsanna Izsvak (99), Division of Molecular Biology, The Netherlands Cancer Institute, Amsterdam 1066CX, The Netherlands

Jason R. Jessen (133), Institute of Molecular Medicine and Genetics, Medical College of Georgia, Augusta, Georgia 30912

Stephen L. Johnson (357), Department of Genetics, Washington University Medical School, St. Louis, Missouri 63110-1010

Lucille Joly (303), Department of Medicine and Department of Cellular and Molecular Medicine, Ottawa Civic Hospital, Loeb Institute for Medical Research, University of Ottawa, Ottawa, Ontario K1Y 4E9, Canada

Donald A. Kane (361), Department of Biology, University of Rochester, Rochester, New York 14627

Arthur E. Kirkpatrick (339), Computer and Information Science Department, University of Oregon, Eugene, Oregon 97403-1254

Cheni Kwok (287), Genetic Technologies, SmithKline Beecham Pharmaceuticals (NFSPCN), Harlow CM19 5AW, United Kingdom

Eric C. Liao (181), Division of Health Sciences and Technology, Massachusetts Institute of Technology, Cambridge, Massachusetts 02139

Shuo Lin (133), Institute of Molecular Medicine and Genetics, Medical College of Georgia, Augusta, Georgia 30912

Anming Meng (133), Institute of Molecular Medicine and Genetics, Medical College of Georgia, Augusta, Georgia 30912

Jorg Ödenthal (21), University of Sheffield, Sheffield S10 2TN, United Kingdom

Francisco Pelegri (1), Max-Planck-Institüt für Entwicklungsbiologie, 72076 Tübingen, Germany

John H. Postlethwait (149, 165, 323), Institute of Neuroscience, University of Oregon, Eugene, Oregon 97403-1254

David W. Raible (71), Department of Biological Structure, University of Washington, Seattle, Washington 98195

David G. Ransom (195, 365), Children's Hospital, Boston, Massachusetts 02115

Alexander F. Schier (259), Department of Cell Biology, Developmental Genetics Program, Skirball Institute of Biomolecular Medicine, New York University School of Medicine, New York City, New York 10016

Karin Schmitt (287), Millenium Pharmaceuticals, Cambridge, Massachusetts 02139

Stefan Schulte-Merker (1), Max-Planck-Institüt für Entwicklungsbiologie, 72076 Tübingen, Germany

Gary A. Silverman (235), Department of Pediatrics, Harvard Medical School, Children's Hospital, Boston, Massachusetts 02115

William S. Talbot (259), Department of Cell Biology, Developmental Genetics Program, Skirball Institute of Biomolecular Medicine, New York University School of Medicine, New York City, New York 10016

Patricia Tellis (303), Department of Surgery, Urology Division, McGill University and Montreal General Hospital Research Institute, Montreal, Quebec, Canada

Charline Walker (43), Institute of Neuroscience, University of Oregon, Eugene, Oregon 97403-1254

Monte Westerfield (339), Institute of Neuroscience, University of Oregon, Eugene, Oregon 97403-1254

Yi-Lin Yan (149, 165), Institute of Neuroscience, University of Oregon, Eugene, Oregon 97403-1254

Fengchun Ye (303), Department of Surgery, Urology Division, McGill University and Montreal General Hospital Research Institute, Montreal, Quebec, Canada

Tao P. Zhong (235), Cardiovascular Institute, Massachusetts General Hospital East, Charlestown, Massachusetts 02129-2060

Leonard I. Zon (181, 195, 235, 357, 361, 365), Howard Hughes Medical Institute, Children's Hospital, Boston, Massachusetts 02115

PREFACE

The design of this two-volume set of *Methods in Cell Biology* devoted to the zebrafish, *Danio rerio,* was stimulated by the converging recognition among developmental biologists and geneticists that this organism may be the ideal model system for genetic analysis of vertebrate development. Teleosts, such as the killifish (*Fundulus heteroclitus*), medaka (*Oryzias latipes*), and zebrafish, have been used since the late 1800s as excellent systems for studying vertebrate embryogenesis. In the mid-1970s, George Streisinger of the University of Oregon recognized that the embryological advantages of the zebrafish (external fertilization and development, optical clarity of the embryo, ease of manipulation of the embryo, etc.) complemented its genetic potential (short generation time, ease of mutagenesis, large embryonic clutches per mating). Streisinger and his Oregon colleagues pioneered many of the genetic technologies now available for use in the zebrafish, and their enthusiasm and generosity encouraged many other laboratories to adopt the zebrafish for molecular–genetic analyses of vertebrate development. Indeed, their efforts directly stimulated the emergence of a large, cooperative, international zebrafish research community and the proliferation of technical advances in the zebrafish system. These volumes are a testament to the effort and spirit of the community of zebrafish researchers, many of whom have contributed chapters.

Our goal in these volumes is to provide a unified resource that describes the state of the art in the molecular, cellular, embryological, genetic, and genomic methods available for analysis of zebrafish embryogenesis. We have deliberately designed the volumes to complement the extensive protocols described in *The Zebrafish Book* by Monte Westerfield, to which many of our contributors make reference. The first volume, *Biology of the Zebrafish,* is divided into five sections. The first section introduces the zebrafish and explores its embryology. The second presents techniques for cell culture and for production of uniparental embryos. Strategies and methods for analyzing gene expression and function are described in the third section, and section four emphasizes cellular techniques for studying embryonic development. Finally, the volume concludes with a section devoted to methods used to explore the development of important organ systems. The second volume, *Genetics and Genomics,* includes two major sections, the first covering strategies and techniques for mutational screens and transgenesis and the second presenting methods for mapping mutations, cloning the causative genes, and using zebrafish databases. The Appendixes summarize practical, but difficult to find, information on zebrafish strains, centromeric markers, etc. To facilitate the sharing of technical information subsequent to publication of these volumes, we have also included a listing of zebrafish websites.

We trust that the methods presented here will benefit experienced zebrafish biologists as well as attract newcomers to the field. We particularly hope that the attractive optical advantages of the zebrafish embryo will entice cell biologists to work on the system. (Of the more than 2600 abstracts presented at the 1997 Annual Meeting of the American Society for Cell Biology, only 7 contained "zebrafish" as a keyword in their titles!) To the uninitiated we stress, naturally, the advantages and utility of the zebrafish as a genetic system for vertebrate developmental biology, but we suspect that they, like us, will also be attracted by the profound beauty of the embryo and of the developmental processes that it so elegantly reveals.

We express our gratitude to each of the contributors who worked so diligently to provide timely, up-to-date descriptions of the many methodologies presented in the two volumes. We also thank the series editors, Leslie Wilson and Paul Matsudaira, and the staff of Academic Press, especially Jasna Markovac and Jennifer Wrenn, for their help, patience, and encouragement as we developed these volumes.

H. William Detrich, III
Monte Westerfield
Leonard I. Zon

CHAPTER 1

A Gynogenesis-Based Screen for Maternal-Effect Genes in the Zebrafish, *Danio rerio*

Francisco Pelegri and Stefan Schulte-Merker
Max-Planck Institut für Entwicklungsbiologie
72076 Tübingen, Germany

I. Introduction

A. Maternal Factors in Zebrafish and Other Organisms

Maternal factors are expected to be involved in cell-fate decisions and basic cellular functions prior to, and possibly after, the activation of the zygotic genome at the midblastula transition (Signoret and Lefresne, 1971; Gerhart, 1980; Kane

METHODS IN CELL BIOLOGY, VOL. 60

0091-679X/99 $30.00

and Kimmel, 1993). Maternally derived factors in the oocyte cytoplasm have been shown to govern a variety of embryonic processes, for example, the cleavage orientation of snails (Freeman and Lundelius, 1982), cell divisions and the establishment of axes and germline in *Drosophila* (St. Johnston and Nüsslein-Volhard, 1992), early cell lineage decisions in *C. elegans* (Kemphues and Strome, 1997; Schnabel and Priess, 1997), and mesoderm induction in amphibians (Heasman, 1997). Much attention has been paid to the elucidation of the earliest events during embryogenesis, and particularly in *Drosophila* and *C. elegans,* where genetic screens for maternal-effect genes have been carried out, this analysis has been spectacularly successful and informative.

In vertebrates, analysis of maternal factors governing development relies much more on embryological and overexpression studies rather than on genetic approaches. In the zebrafish, *Danio rerio,* the role of maternal determinants is still largely unknown, and our current knowledge in many instances mirrors parallel advances in amphibians. In this section, we briefly discuss several cases of particular relevance to maternal factors in zebrafish, such as dorsoventral specification, mesoderm induction, and germ cell determination.

In *Xenopus,* the dorsal side is specified very early in development through cortical rotation, where movements upon fertilization between cortex and inner core lead to the repositioning of vegetally localized determinants toward the future dorsal side (reviewed in Gerhart *et al.,* 1989; see also Sakai, 1996; Kageura, 1997). Similar specification processes also occur in teleosts, possibly through transport of vegetally located components to the equatorial region along cortical microtubules (Oppenheimer, 1936; Jesuthasan and Strähle, 1996). The identity of the dorsal-determining signal(s) is unknown, although these are presumably, at least in part, responsible for the stabilization of β-catenin, which can be observed to accumulate in the nuclei of the dorsal side of vegetal cells in *Xenopus* and the yolk syncitial layer in the zebrafish (Schneider *et al.,* 1996; Larabell *et al.,* 1997; Jesuthasan and Strähle, 1996). These two structures are known to be axis-inducing centers in both amphibians (Nieuwkoop, 1969) and teleosts, including the zebrafish (Long, 1983; Mizuno *et al.,* 1996). Much experimental evidence has accumulated to describe the processes involved (reviewed in Slack, 1994; Moon *et al.,* 1997); however, the molecular identity of many of the components responsible for eliciting dorsal-specific gene activation remains unknown.

Similarly, mesoderm induction in *Xenopus* has been studied intensively and suggested to occur during the early cleavage stages (Jones and Woodland, 1987), when signals from the vegetal blastomeres induce neighboring blastomeres in the equatorial region of the blastula to form mesoderm. Although a whole range of secreted factors has been implicated in this process (Slack, 1994), we still do not know which factor(s) is the real *in vivo* inducer. In the Medaka fish, interference with the signaling pathway of the secreted TGF-β family-member activin has suggested a role for maternal activin signaling in the formation of mesoderm (Wittbrodt and Rosa, 1994), but the specificity of reagents used in such studies are always a caveat (see Schulte-Merker *et al.,* 1994).

Another example is the determination of the germ plasm. Yoon *et al.* (1997) have shown that the RNA for the zebrafish gene *vasa* localizes to the cleavage planes of two- and four-cell stage embryos and later segregates into four cells, which are likely to be the primordial germ cells. As pointed out by Yoon *et al.*, this striking localization fits very well with previous studies in *Xenopus,* in which the electron-dense granulofibrillar components associated with the germ plasm also segregate during the early cleavage to four-germ-cell primordia (Whittington and Dixon, 1975). The presence of similar electron-dense material has been reported in zebrafish oocytes (Selman *et al.,* 1993), although it is not yet known whether *vasa* RNA or protein colocalizes with it.

Much remains to be known about these and other roles of maternal factors in early processes in the zebrafish. Other patterning processes and cell-fate determinations are expected to depend on maternal products, for example, the determination of the left–right axis (Hyatt *et al.,* 1996). In addition, specific products may be required for basic cellular processes such as the stepwise maturation of the oocyte within an oocyte–follicle complex (Kessel *et al.,* 1985, 1988; Selman *et al.,* 1993, 1994), fertilization and egg activation (Hart and Collins, 1991; Hart *et al.,* 1992), and the rapid early embryonic cleavages before zygotic gene activation (Kane and Kimmel, 1993).

B. Rationale of the Approach

Dominant negative and antisense technology approaches have proven useful in understanding the function of candidate genes in *Xenopus* (see, for example, Moon *et al.,* 1997; Wylie *et al.,* 1996) and in zebrafish (Hammerschmidt *et al.,* Chapter 7, Vol. 59; Barabino *et al.,* 1997). Nevertheless, these approaches are largely limited by our knowledge of candidate genes and can be undermined by cross-specificity or residual function, which often leave in doubt the identity of the actual endogenous factor (Schulte-Merker *et al.,* 1994). The use of forward genetics, on the other hand, is unparalleled as an unbiased approach to identifying genes required for a particular process.

A small number of zebrafish genes with maternal-effects have already been found. For example, *janus* affects the adhesion of the early blastula cells, leading to a split blastoderm and the formation of two axes (Abdelilah *et al.,* 1994; Abdelilah and Driever, 1997), *yobo* is required for anteroposterior extension of the axis (Fig. 1A, B, see color plate; Odenthal *et al.,* 1996), and *somitabun* (Fig. 1C, D, see color plate; Mullins *et al.,* 1996) plays a role in dorsoventral patterning. The uncovering of these few mutants was fortuitous, however, and a systematic search for maternal-effect genes seems the only way of addressing by genetic means the multitude of biological processes outlined earlier. A genetic screen for maternal-effect genes has not yet been attempted in a vertebrate. In the following, we outline the rationale and methodology to carry out such a screen in the zebrafish, which is suited for mutant screens as a result of its unique features such as small size, optical clarity of the embryo, large offspring numbers, and the applicability of a variety of genetic tricks.

In principle, it is possible to carry out a screen for maternal-effect mutations in the zebrafish utilizing a scheme involving natural crosses similar to that used for the large-scale zygotic screens carried out in Tübingen and Boston (Haffter *et al.,* 1996; Driever *et al.,* 1996). One more generation would be required, however, in order to test putative homozygous F_3 females for maternal-effect mutations (Fig. 2A, see color plate). This poses a considerable problem in the zebrafish because, as a result of the lack of balancer chromosomes, crosses are essentially blind, and therefore every generation in a genetic screen generates an exponentially increasing number of crosses. A scheme based solely on natural crosses would therefore be very labor-intensive and require enormous amounts of tank space. Fortunately, the zebrafish has the capacity to produce gynogenetic progeny under experimental conditions (Streisinger *et al.,* 1981). Artificially induced gynogenesis involves diploidization of the maternal haploid genome, producing viable offspring with solely a maternal genetic contribution. In a basic gynogenesis-based scheme (Fig. 2B, see color plate), mutations are induced in the germline of parental (P) males by exposing them to the point-mutagen *N*-ethyl-*N*-nitrosourea (ENU) (van Eedin *et al.,* Chapter 2, this volume). P males are then mated to produce F_1 progeny heterozygous for the induced mutations. Eggs are stripped from the F_1 females, and gynogenesis is induced. This allows newly induced mutations to become homozygous in up to 50% of the gynogenetic F_2 generation (Section II.A). Finally, F_2 adult females are screened for maternal effects by crossing them to wild-type males and testing their progeny for embryonic phenotypes. The incorporation of gynogenesis into a genetic scheme for maternal-effect mutations allows the direct production of homozygotes for induced mutations from a single heterozygous F_1 carrier, bypassing one generation in comparison to a scheme based solely on natural crosses (compare Figs. 2A, B) and therefore greatly reduces the time, labor, and space required. Clearly the incorporation of gynogenesis would greatly simplify a maternal-effect genetic screen; however, gynogenesis itself is only efficiently induced under specific conditions. Here we describe the methodology that we have used to optimize this procedure for its incorporation into a maternal-effect screen, as well as preliminary results from a pilot screen.

II. Optimization of the Gynogenetic Method

In a scheme for a gynogenesis-based maternal-effect screen, the main goal is the efficient production of fertile gynogenetic F_2 females, which are homozygous for newly induced mutations. A number of issues are important to optimize this procedure. First, a suitable method of gynogenesis needs to be selected. Second, lines amenable to gynogenetic procedures need to be found and selected. Third, an appropriate mutagenesis dosage needs to be chosen to induce a reasonably high rate of mutations while allowing the production of viable homozygous adult mutants. We discuss each of these particular points.

A. Choice of Gynogenetic Technique

There are two main techniques for the artificial induction of gynogenesis in the zebrafish: early pressure (EP) and heat shock (HS) (Streisinger *et al.,* 1981). In both methods, eggs are first artificially fertilized with sperm whose genetic material has been inactivated by UV irradiation. In the absence of further treatment, these eggs would develop into haploid embryos which are inviable. Both EP and HS lead to the diploidization of the genetic content of the egg, thus producing viable, diploid embryos.

In EP, diploidization is induced by the application of hydrostatic pressure between minutes 1.33 and 6 after egg activation (Fig. 3A, see color plate, Section V.C.6). This treatment inhibits completion of the second meiotic division and the expulsion of the 2° polar body, resulting in a diploid egg. On the other hand, HS inhibits cytokinesis of the first mitotic division of haploid embryos by applying a heat pulse during minutes 13 to 15 after egg activation (Fig. 3B, see color plate, Section V.C.5), transforming haploid embryos into diploid ones. Hydrostatic pressure, applied late, has also been used as an alternate method to inhibit the first mitosis, although it has been found to be less effective and more cumbersome than HS (Streisinger *et al.,* 1981). In theory, HS is more efficient than EP in the direct induction of homozygosity and therefore might be the technique of choice in a maternal-effect screen. This is because HS-derived progeny are homozygous at every single locus; therefore, 50% of HS-derived F_2 progeny are homozygous for a mutation present in heterozygous form in the F_1 mother. On the other hand, as a result of recombination during meiosis, EP leads to a variable degree of homozygosity ranging from 50% toward 0% depending, respectively, on whether loci are linked to the centromere or are distally located. Thus, HS would, in principle, provide the highest possible yield of homozygous mutant adults for all loci regardless of their chromosomal location. Moreover, the expectation of a fixed percentage of mutant individuals would aid in the assessment of newly identified phenotypes.

In spite of these obvious theoretical advantages of HS over EP, in practice EP is superior to HS as a gynogenetic method for a number of reasons. First, HS is about twofold less efficient than EP in inducing diploid, viable gynogenotes (Table I, see also Streisinger *et al.,* 1981), presumably because of a greater intrinsic ease of inhibiting the meiotic cell division rather than the mitotic one. Moreover, EP-derived adults, probably due to their higher heterozygosity, show viability and fertility rates that combined are about fourfold higher than those in HS-derived clutches (Table I). Thus, the final yield of fertile adult gynogenotes derived from EP is about eightfold higher than the derived from HS.

Higher levels of heterozygosity in EP-derived gynogenotes are beneficial for additional reasons. First, under mutagenic conditions, the yield of HS-

Table I
Comparison of Heat Shock- and Early Pressure-Induced Gynogenesis (gol-mix line)

	Heat shock		Early pressure	
Viability at d5 (viable[a]/fertilized eggs)	0.09	n = 3590	0.21	n = 4368
Fraction clutches with >6 viable d5 fish	0.41	n = 29	0.93	n = 29
Clutch size (viable d5 fish/clutch)	10	n = 29	37	n = 29
Adult viability (viable at 3 mo./d5 viable)	0.53	n = 324	0.66	n = 218
Fertility (fertile adults/total adults)	0.23	n = 13	0.65[b]	n = 226

[a] Viable at day 5 are defined as fish that can inflate their swim bladders.
[b] Value from F_2 descendants of P males mutagenized with 2 mM.

gynogenetic clutches is expected to be further reduced by a factor of 0.5 per induced zygotic recessive-lethal mutation, whereas EP-derived clutches are expected to be reduced by a factor of only 0.23 (Section II.C). Second, the increased heterozygosity of EP gynogenotes improves the odds of recovering newly identified mutations. This occurs because the overall fraction of *fertile* siblings that carry a given mutation, as a result of a decrease in the fraction of the (sterile) homozygous mutant females, is greater in EP-derived clutches than in HS-derived ones (Section IV.B). The main drawback of the higher heterozygosity of EP gynogenotes is that it leads to an intrinsic bias against the identification of distally located mutations. Measurements of the frequency of homozygosity (F_m) of random zygotic mutations after EP-induced diploidization range from 0.50 to 0.04, with an average value of 0.23 (16 loci; Streisinger *et al.,* 1986; Neuhauss, 1996). With the assumption that maternal genes are similarly distributed throughout the chromosomes, these data suggest that the majority of these genes are sufficiently centromere-linked to be identified through an EP-based screen. Thus, we chose EP over HS as a gynogenetic method for our screen, although it is possible that HS may become applicable in the future with the use of highly selected lines.

B. Selection of Lines Amenable to Gynogenetic Methods

The majority of lines we have examined, including lines recently derived from the wild, tend to produce low yields of fertile gynogenotes (unpublished observations). Selection of appropriate lines is therefore very important for an efficient gynogenetic-based maternal screen.

1. Selection of Lines with a High Yield of Gynogenotes

The experimental induction of gynogenesis relies on the manipulation of *in vitro* fertilized eggs at very early stages (see following discussion). Therefore, it is necessary that females should, as a first requirement, readily yield oocytes when manually stripped. Different fish strains differ greatly in their ability to be manually stripped (not shown). The capacity to be stripped of eggs is distinct from being fertile and to successfully mate under standard "natural" crosses in the laboratory (Eaton and Farley, 1974; our observations). This may be related to the fact that, under natural conditions, release of mature oocytes from their follicles into the ovarian lumen requires hormonal stimulation (Selman *et al.,* 1994), which may normally be triggered by vigorous chasing by the males (Eaton and Farley, 1974). Lines that can be most easily stripped appear to be those which have been propagated by artificial fertilization methods, which also involve stripping of females, such as those derived from the AB Oregon line (Streisinger *et al.,* 1981). In contrast, lines from the wild or laboratory lines which have been propagated mostly by natural crosses tend not to be easily stripped. We have found that the gol-mix line, a hybrid line with both AB and Tübingen genetic backgrounds, is robust, and that its females can be easily stripped of eggs as well as produce a high yield of gynogenotes. Thus, we chose to continue our selections and genetic screen schemes with this starting population. In addition, the fact that this line is marked with the recessive pigment marker *golden* allows the detection of unwanted products of incompletely inactivated sperm (isolated from *golden*$^+$ males) after the EP procedure.

2. Selection for Lethal/Sterile-Free Genetic Backgrounds

Another preferable characteristic of a genetic background is the absence of preexisting mutations, whether maternal or zygotic. The use of lines free of preexisting mutations is important to diminish unwanted background lethality and false positives in any kind of screen, whether based on gynogenetic techniques or natural crosses. The gynogenetic method of HS induces homozygosity at every single locus (Section II.A) and is thus particularly effective at selecting in one single generation for fish which lack background mutations. After growing a large number of HS-derived gynogenetic clutches from our substrate line gol-mix, we selected for adult fish free of lethal or sterile mutations. From our starting gol-mix population, we generated two lines, golFL-1 and golFL-2, from four different HS-derived individuals. These two lines were combined to create golFL-3 (golFL-2 was 100% male and could not be propagated as a pure stock). Similar strategies had been used previously to select for such lethal/sterile-free strains, which can be further propagated through EP to generate clonal lines (Streisinger *et al.,* 1981). Selection of lines through HS and EP can also lead to stocks of higher viability under gynogenetic conditions, presumably by the reduction of background detrimental alleles (Streisinger, 1981).

3. Selection for Favorable Sex Ratios under Gynogenetic Conditions

Sex determination in fishes varies from systems with sex-determining chromosomes to multifactor autosomal ones, and in some cases sex has been shown to be influenced by external factors (reviewed in Chan and Yeung, 1983). The mechanism of sex determination in zebrafish, although poorly understood, appears to fall in the latter category, lacking a single sex chromosome and being sensitive to growing conditions. Most gynogenetic clutches, after grown to adulthood, exhibit sex ratios that are strongly biased toward maleness (86–88% males, Figs. 4A, B). The phenomenon of sex bias in gynogenetic clutches is likely related to the tendency of zebrafish and other teleosts to develop into males under suboptimal conditions, for example, in overcrowded conditions or in subviable genetic backgrounds (see Chan and Yeung, 1983; our observations). Presumably, gynogenetic clutches, because of their high degree of inbreeding, also have a suboptimal genetic background that under normal circumstances produces males. Nevertheless, a small fraction of the gynogenetic clutches (5–10%) are composed of at least 50% females.

The observation of rare gynogenetic clutches with a high female to male ratio shows that it is possible to select for genetic backgrounds which produce a high proportion of females even under gynogenetic conditions. In fact, the gynogenetic procedure itself may act as a selection for female-rich genetic backgrounds, as exemplified by the fact that one out of two lethal/sterile-free lines that we produced consists of mostly females (92% females, Fig. 4C). In this line, golFL-1, there exists a small fraction of males, which can mate with wild-type females but which tend to produce unfertilized eggs. Nevertheless, treatment of this line with testosterone for the first 14 days of development leads to the production of larger percentages of fertile males (72% males, Fig. 4D), and thus allows the production of males both for mutagenesis and for the propagation of the line through natural crosses.

4. A Hybrid/Inbred Approach

If, on one hand, selection of lines can increase the frequency of certain desired traits, such selection also leads to inbreeding which often causes a reduction in overall robustness and fertility (Thorgard, 1983). Thus, the best lines for gynogenetic-based maternal screens might be hybrids between gynogenetically selected lethal/sterile-free lines.

C. Choice of Mutagenesis Dose

In the F_2 gynogenetic generation, homozygosity for mutations in essential zygotic genes will lead to a decreased survival of gynogenotes. For example, the

---

Fig. 4 Sex ratio in HS- and EP-derived clutches from gol females. The fraction of gynogenetic clutches is plotted against the fraction of females present in them. HS-derived (A) and EP-derived (B) clutches from the gol-mix strain. Note the low number of clutches with a high fraction of females. (C) EP-derived clutches from the golFL-1 line show much higher fractions of females. (D) Clutches as in (C) but with subsequent exposure to testosterone result in higher proportions of males.

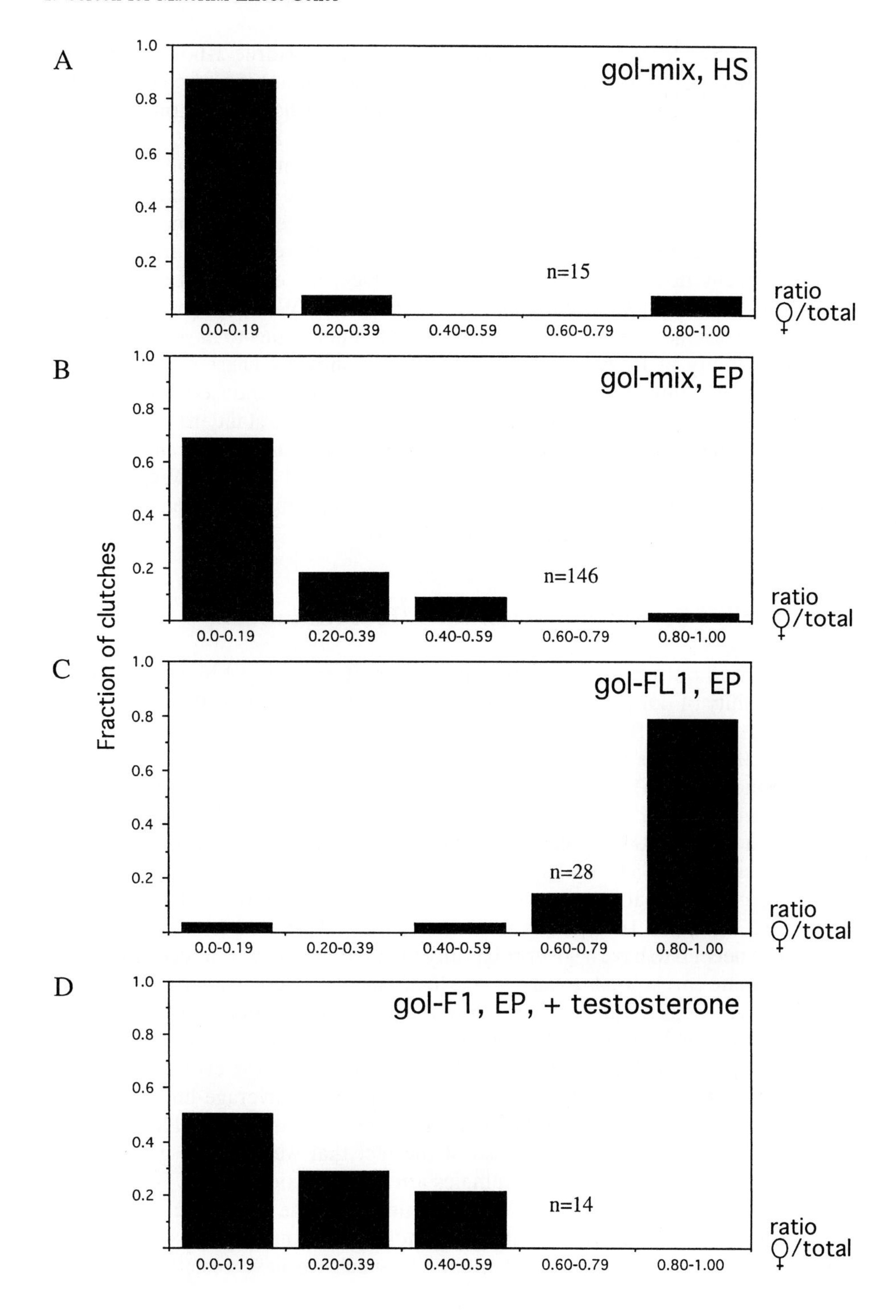
A
1.0
0.8
0.6
0.4
0.2
gol-mix, HS
n=15
0.0-0.19
0.20-0.39
0.40-0.59
0.60-0.79
0.80-1.00
ratio
♀/total
B
gol-mix, EP
n=146
C
gol-FL1, EP
n=28
D
gol-F1, EP, + testosterone
n=14
Fraction of clutches

mutagenic dosage used in large-scale zygotic screens (three 1-hour treatments with 3m*M* ENU; van Eedin *et al.,* Chapter 2, this volume) is expected to induce about one embryonic lethal and one larval lethal per haploid genome (Mullins *et al.,* 1994; Solnica-Krezel *et al.,* 1994; Haffter *et al.,* 1996). This implies that under this mutagenic condition, only 59% of what would be otherwise viable EP-derived gynogenotes (25% using HS) would survive to adulthood. Thus, we reduced the ENU dosage in our maternal-effect screen experiments. Similar reductions in the strength of mutagenic treatments were adopted for maternal-effect screens in *Drosophila* and *C. elegans* (see, for example, Lehmann and Nüsslein-Volhard, 1986 and Kemphues *et al.,* 1988). We have observed that a mutagenic dosage of 3×1 hr 2m*M* ENU treatments begins to have a mild effect on the viability of F_2 gynogenetic clutches (not shown). These conditions lead to a mutagenic rate, as assayed by the frequency of newly induced *albino* alleles, estimated to be about one-third of the rate induced by the standard (3m*M* ENU) treatment (data not shown; Mullins *et al.,* 1994; Solnica-Krezel *et al.,* 1994), or about 0.3–0.4 embryonic lethal mutations per haploid genome. We chose for our screen the ENU concentration of 2m*M* as a compromise between a moderate mutagenic rate and a practical level of viability.

III. A Pilot Screen for Maternal-Effect Mutations

We carried out a small-scale pilot run of a maternal screen in order to assess the feasibility of our procedure. In this screen, we mutagenized P males of a gol-mix line (the line golFL-1 line was not yet available) with three 1 hour treatments with 2m*M* ENU (as in van Eedin *et al.,* Chapter 2, this volume, except for the ENU concentration). A total of 402 egg clutches stripped from F_1 females were treated with EP. Of the resulting clutches, we grew those with at least 6 viable d5 fish, a total of 181 clutches (45% of EP-treated clutches). From these, we grew an estimated 3542 adults, of which 432 were females. Of these females, 281 (65%) produced healthy eggs which were tested for maternal effects. With an F_m of 0.23 (Section II.A), one can estimate that 9 F_2 tested females per clutch would be needed to have a 90% probability of detecting a newly induced mutation. This estimate corresponds to an average locus and varies greatly depending on the centromere–locus frequency: in order to reach a similar frequency of detection, 4 or 44 tested females per clutch would be necessary, respectively, for centromere-linked (F_m: 0.5) and distal (F_m: 0.05) loci. Considering these estimates, the 281 F_2 females we tested correspond to 24 genomes for average-linked loci (33 genomes for centromere-linked loci or 6 genomes for distal loci). The efficiency of this pilot screen was low because of the fact that we used the gol-mix line, which produces a low fraction of females after EP (Section II.B.3). Presently we are carrying out additional rounds of screening using the female-rich line golFL-1, which has increased our yield of adult gynogenetic females. In this pilot screen, we identified one maternal-effect mutation which we have recently recovered and are beginning to characterize. Embryos from homozygous mutant females can be recognized at the one-cell stage because of an increased turbidity of the cytoplasm. Later, during the blastula and gastrula stages, cell–cell adhesion and convergence extension appear to be reduced, and the embryos lyse by 18–36 hours. More detailed studies on this mutant will be presented elsewhere.

IV. Recovery of Mutations

Once an F_2 female is identified as exhibiting a maternal-effect phenotype, the mutation needs to be recovered. The observed maternal-effect phenotypes are expected to be caused by maternal homozygosity for recessive mutations because dominant mutations present in the F_1 females are unlikely to be propagated in the F_2 gynogenetic clutches. Because homozygosity of the mutation in F_2 gynogenetically derived females leads to the inviability of the F_3 embryos, a genetic scheme has to allow the recovery of the mutations through genetically related individuals. A mutation can be recovered by three means.

A. Recovery through Parental F_1 Females

F_1 females which produce the F_2 EP-derived gynogenotes to be tested for maternal effects are heterozygous carriers of the tested mutations. Thus, F_1 females can be stored separately until their clutches are tested for maternal-effect phenotypes. We have found that after the 3–5 months that are required to grow up and test the gynogenetic clutches, the majority of the separated F_1 females are still alive and fertile and, thus, can be used to recover the mutation.

B. Recovery through Gynogenetic F_2 Siblings

The mutation can also be recovered from the F_2 gynogenetic siblings. In EP-derived clutches, the frequency of heterozygotes and homozygotes for a given mutation varies depending on the centromere–locus distance. For centromere-linked loci (F_m close to 0.5), 50% of the siblings are homozygous for the mutation. As the centromere–locus distance increases, the fraction of homozygous siblings (F_m) decreases, but the fraction of heterozygous siblings increases two times as rapidly. For a distal mutation of $F_m = 0.05$, for example, 5% of the F_2 siblings are homozygous mutant, whereas 90% are heterozygous carriers. Thus, the overall frequency of F_2 carrier siblings (homozygous or heterozygous) varies from 50% for centromere-linked loci to percentages approaching 100% for distal loci. Therefore, the recovery of mutations through F_2 siblings can also be an efficient strategy, which we have successfully used to recover the maternal-effect mutation found in the pilot screen. Outcrossing mutations through sibling F_2 males is preferable, when possible, over their outcrossing through females. This is because females homozygous for a maternal-effect mutation are sterile, whereas homozygous males should be fertile unless the mutated gene also affects male fertility. This is particularly important in the case of centromere-linked loci (F_m toward 0.5), in which the proportion of carriers that are homozygotes is greater. In large F_2 EP-derived clutches, F_m can be estimated by the proportion of screened F_2 females that exhibit the maternal-effect mutant phenotype.

C. Recovery through Rare F_3 Survivors

In cases of incompletely penetrant phenotypes, mutations may also be able to be recovered through rare survivors within the F_3 clutch that exhibits the maternal-effect phenotype. The presence of such "escapers" may be the result of variability in the phenotype caused by residual function of the mutated gene or some degree of redundancy in the affected pathways. Escapers are expected to be heterozygous carriers for the mutation, and the mutation can be propagated through incrossing of a family derived from them. Of the preceding options, the schemes in A and C are the most effective because they use individuals that are known to be carriers for the mutation to be recovered.

V. Solutions, Materials, and Protocols for Gynogenetic Methods (adapted from Westerfield, 1993)

A. Solutions

MESAB stock solution: 0.2% ethyl-*m*-aminobenzoate methanesulfonate. Adjust to pH 7.0 with 1*M* Tris pH 9.0. Keep at 4°C.

MESAB working solution: 7 ml stock solution/100 ml fish water.

Hank's Solutions: Stock solutions 1, 2, 4, and 5 and Premix can be stored at 4°C. Stock Solution 6 is prepared fresh and added to the premix to form the final Hank's solution.
Solution 1: 8.0 g NaCl, 0.4 g KCl in 100 ml double-distilled (dd) H_2O.
Solution 2: 0.358 g Na_2HPO_4 Anhydrous, 0.60 g KH_2PO_4 in 100 ml dd H_2O.
Solution 4: 0.72 g $CaCl_2$ in 50 ml dd H_2O.
Solution 5: 1.23 g $MgSO_4 \cdot 7H_2O$ in 50 ml ddH $_2O$.
Hank's Premix: Combine the following in order:
10.0 ml Solution 1
1.0 ml Solution 2
1.0 ml Solution 4
86.0 ml dd H_2O
1.0 ml Solution 5
Solution 6 (prepare fresh): 0.35 g $NaHCO_3$ in 10 ml dd H_2O
Hank's (Final): 990 μl Hank's Premix and 10 μl Solution 6.

E2 saline (used specially during the testosterone treatment because of its higher buffering properties): 15mM NaCl, 0.5mM KCl, 1mM $CaCl_2$, 1mM $MgSO_4$, 0.15mM KH_2PO_4, 0.05mM Na_2HPO_4, and 0.7 mM $NaHCO_3$.

E3 saline (a simpler version of E2 saline used for routine embryo raising): 5mM NaCl, 0.17mM KCl, 0.33mM $CaCl_2$, 0.33mM $MgSO_4$, 10^{-5}% Methylene Blue.

Testosterone stock: 150 mg 17-α-methyl testosterone in 50 ml absolute ethanol. Store in aliquots at −20°C.

Testosterone working solution: While stirring, add 10 μl of stock solution per 600 ml of (a) E2 saline, for babies before day 6, or (b) fish water supplemented with 3 g/l Red Sea Salt (Read Sea Fish pHarm, Israel), for larvae between days 6 and 15. In our hands, E2 (instead of E3) and Red Sea Salt in the fish water improve the survival of testosterone-treated larvae. Stir 10 minutes.

B. Other Materials

UV lamp: Sylvania 18-in., 15-W germicidal lamp.

French Press Cell, 40 ml (SLM-Aminco)

French Pressure Cell Press (SLM-Aminco) or Hydraulic Laboratory Press (Fisher)

Heat Shock baskets: Make by cutting off the bottom of Beckman Ultraclear centrifuge tubes and heat-sealing a fine wire mesh to the bottom edge of the tube.

EP vials: Disposable glass scintillation vials with plastic caps (3.2 cm height and 2.2 cm diameter, Wheaton), or similar vials. The plastic caps are perforated several times with a needle to better allow exposure to the hydrostatic pressure. Only two vials of this type can fit at once in a pressure cell. In order to fit four vials in one cell, we have custom-built shorter plastic vials (1.8 cm height, including cap, 2.5 cm diameter, 0.3 mm wall thickness), which fit the plastic caps from the scintillation vials.

C. Protocols

1. Sperm Collection (adapted from D. Ransom)

A sperm solution can be made with testes dissected from 10 males for each 1 ml of Hank's solution. Keep the isolated testes and Hank's solution on ice. Shear the testes with a small spatula and by pipetting up and down with a 1000-μl Pipetteman. Allow debris to settle and transfer supernatant to a new tube. Sperm solution on ice is effective for about 2 hours. Sperm still remaining inside the sheared, settled testes can be further collected by adding 300 μl of fresh Hank's and letting the mixture rest for 1/2 hour or longer. For more details, see Ransom (Appendix 3, this volume).

2. UV-Inactivation of Sperm

Transfer the sperm solution to a watch glass. Avoid pieces of debris; they may shield sperm from the UV light. Place the watchglass on ice at a distance of 30 cm (12 in.) directly under the UV lamp. Irradiate for 2.5 minutes with gentle stirring every 30 seconds. Transfer to a new Eppendorf tube with a clean pipette tip. UV-treated sperm solution on ice is effective for about 2 hour.

3. Stripping of Eggs

Our observations suggest that females are more amenable to manual stripping if removed from their tank and placed in a clean tank (1–10 females per 2-l tank) the evening before stripping. Best stripping and egg clutch quality are obtained during the first 4 hours after the start of the light cycle of the first day after the separation of the females. The presence of males together with the separated females does not significantly affect the ability of gol-mix females to be stripped (our observations), although it may have an effect when working with other fish lines (Eaton and Farley, 1974).

Anesthetize females in MESAB working solution until they reduce their gill movements (2–4 minutes, MESAB solution may have to be boosted through time with more stock solution in 0.5–1 ml increments). Overexposure to MESAB will impede the recovery of the female, and fish should be placed in fresh water if they are not going to be used within 1 or 2 minutes after they stop their movements.

With the aid of a spoon, rinse a female in fish water and place her on several paper towels to remove excess moisture.

Place the female on the bottom half of a Petri plate. With a soft tissue, dry further the anal fin area. Excess water may prematurely activate the eggs.

Slightly moisten the index fingers of both hands (dry hands will stick to the skin of the fish). With one finger support the back of the female and with the other gently press her belly. Females which can be stripped will release their eggs upon gentle pressure. Healthy eggs have a translucent, yellowish appearance. Separate the eggs from the female with a small, dry spatula. Females can be placed separately in boxes and identifying tags can be attached to the box with the female and the corresponding egg clutch. If necessary, clutches can wait for several minutes before being activated. In this case, we cover the clutches with the Petri plate lid to reduce drying of the clutch. Fertilization can occur after even longer delays (in our hands, up to 6 minutes), although not in a consistent manner. Egg activation can be delayed for periods of 1.5 hour or more with ovarian fluid from the rainbow trout or coho salmon (Corley-Smith *et al.*), or with Hank's saline buffer supplemented with 0.5% bovine serum albumin (BSA) (Sakai *et al.,* 1997), although we have not tested these methods in combination with gynogenesis.

4. *In Vitro* Fertilization

Add 25 μl of untreated or UV-irradiated sperm to the egg clutch. Mix the sperm and eggs by moving the pipette tip without lifting it from the Petri plate (to minimize damage to the eggs). If desired, proceed at this point to Heat Shock or Early Pressure Protocols. If not, add 1 ml of E3 saline to activate the eggs, and, after 1 minute, fill the Petri plate with E3. Incubate at 27–29°C.

5. Heat Shock

After *in vitro* fertilization with UV-treated sperm, add 1 ml of E3 saline to activate the eggs and start the timer.

Add more E3 after 30 seconds. Transfer the eggs to a heat shock basket. Immerse the basket in a water bath with stirring and E3 saline at 28.5°C.

At 13.0 minutes, blot briefly the bottom of the basket onto a stack of paper towels and transfer the basket to a water bath with stirring and E3 saline at 41.4°C.

At 15.0 minutes, blot briefly the bottom of the basket and transfer the basket back to the 28.5°C E3 bath.

Allow the embryos to rest for about 45 minutes and transfer to a Petri plate. Allow embryos to develop in a 27–29°C incubator (see note in Section V.C. 6).

6. Early Pressure

In order to maximize the number of clutches produced, we work on cycles in which we include up to four clutches in separate vials within the Pressure Cell. For this, we typically anesthetize 6–12 females. Once four healthy-looking clutches are obtained, the females that have not yet been stripped of eggs are transferred to fresh fish water until they completely recuperate. It works well to begin to anesthesize females for the next EP cycle at around minute four within a current cycle.

After mixing eggs with UV-treated sperm (see *In Vitro* Fertilization), activate up to four clutches simultaneously by adding 1 ml of E3 saline to each clutch and start the timer (at least two people are required to timely manipulate four clutches).

After 12 sec, add more E3. A squirt to the side of the Petri plate will make the fertilized eggs collect in the middle of the plate.

With a plastic pipette, transfer the fertilized eggs to an EP vial. Fill the vial with E3 and cap it with the perforated plastic lid. Avoid large air bubbles. Place the vials inside the pressure cell, ensuring that no air remains trapped inside it. Record the relative position of the clutch within the pressure cell by placing the tags in the corresponding order on a dry surface. Fill the pressure cell with E3 and close it allowing excess E3 to be released from the side valve. Close the side valve without overtightening. Insert entire assembly on the French Press apparatus, and apply pressure to 8000 lb/in.^{2} by time 1 minute, 20 seconds after activation. For different strains and/or presses, different pressure values may be optimal (see Gestl *et al.,* 1997).

At 6.0 min., release the pressure and remove the pressure cell from the French Press apparatus. Maintaining the relative order of the vials, remove the vials from the pressure cell, dry them with a towel, and label them with their corresponding number tags. Place the vial in a 27–29°C incubator.

After all EP cycles have been completed, allow the embryos to rest in the vial for at least 45 minutes but no more than 4 hours. Transfer embryos with their corresponding tags to Petri plates. Let embryos develop in a 27–29°C incubator.

Note: Because of the large amount of embryonic lethality induced by the HS and EP procedures, we incubate the embryos at a low density of 80 embryos maximum per 94 mm Petri plate (this is particularly important for the first 24 hours of development).

7. Testosterone Treatment

Before embryos reach 24 hours of development, remove the chorions from the embryos, remove as much E3 as possible, and replace with testosterone/E2 working solution.

Each consecutive day, replace half of the testosterone/E2 with fresh testosterone/E2.

On day 6, transfer the embryos to mouse cages with 1 l of testosterone solution in fish water supplemented with 3 g/l Coral Reef Salt. Start feeding as normally. Continue replacing half of the solution every day by carefully aspirating the solution and refilling with fresh testosterone solution.

On day 15, remove the testosterone by aspirating most of the solution and refilling with fresh fish water. Rinse again by repeating this procedure. Embryos can now be connected to the water system.

VI. Conclusions

We describe background work that has enabled us to implement a gynogenesis-based genetic screen for maternal-effect mutations in the zebrafish. Gynogenesis, by allowing the production of viable descendants from a mother in the absence of paternally derived genetic material, leads to the direct homozygosis of mutations present in a heterozygous individual. This in turn allows us to greatly simplify a screen for recessive maternal-effect mutations. We have tested the main methods to induce gynogenesis in the fish and conclude that inhibition of the second meiotic division through early hydrostatic pressure (EP) can be used as a gynogenetic method applicable to a genetic screen for maternal-effect mutations. We find that fish strains can be selected for specific traits that improve the applicability of the technique, such as the ability to be stripped of eggs and to produce a high yield of viable adult progeny with a favorable sex ratio, as well as the absence of lethal/sterile mutations. Further improvement of HS or EP derived lines, as described here, or the combination of such lines in order to increase the hybrid strength of the background may continue to improve the efficiency of future rounds of screening.

In principle, gynogenesis is applicable to any juvenile or adult trait. Indeed, in our screens, we frequently observe adult phenotypes, such as abnormal pigmentation or skeletal morphology, which in some cases we have shown to be caused by recessive mutations (our unpublished results). This suggests that gynogenesis-based screens should be applicable to any robust juvenile or adult trait. Adult traits that exhibit quantitative variation, on the other hand, may be less amenable for such type of screen because of the relatively low number of fish in a gynogenetic clutch. Female fertility (i.e., absence of maternal-effects in its progeny) is a particularly robust "adult" trait because each gynogenetic female can produce hundreds of eggs with a relatively low background of abnormalities.

With the methods outlined in this report, a pilot screen for maternal-effect mutations has been carried out, in which we have isolated and recently recovered one maternal-effect mutation. We are currently carrying out more rounds of screening with an improved, gynogenetically selected line. These efforts should lead to the identification of additional mutations in maternal-effect genes and eventually the genetic analysis of early development in a vertebrate.

Acknowledgments

We are grateful to Hans-Martin Maischein and Catrin Seydler for their excellent and persistent technical support, Cosima Fabian for her help during the early stages of this work, Don Kane for demonstrating the gynogenetic procedures, and Karin Rossnagel for providing us with photographs of *yobo* embryos. We also thank Christiane Nüsslein-Volhard and Darren Gilmour for critically reading this manuscript. We are specially grateful to Christiane Nüsslein-Volhard for support and advice throughout this project. F.P. was supported by a long term EMBO fellowship. S.S.M. was supported by a grant from the DFG (SFB 1635).

References

Abdelilah, S., and Driever, W. (1997). Pattern formation in *janus*-mutant zebrafish embryos. *Dev. Biol.* **184,** 70–84.

Abdelilah, S., Solnica-Krezel, L., Stainier, D. Y., and Driever, W. (1994). Implications for dorsoventral axis determination from the zebrafish mutations *janus. Nature* **370,** 468–471.

Barabino, S. M. L., Spada, F., Cotelli, F., and Boncinelli, E. (1997). Inactivation of the zebrafish homologue of *Chx10* by antisense oligonucleotides causes eye malformations similar to the ocular retardation phenotype. *Mech. Dev.* ***63,*** 133–143.

Chan, S. T. H., and Yeung, W. S. B. (1983). Sex control and sex reversal in fish under natural conditions. *In* "Fish Physiology" (W. S. Hoar, D. J. Randall, and E. M. Donaldson, Eds.) Vol IX, Part B, pp. 171–222. Academic Press, New York.

Corley-Smith, G. E., Lim, C. J., and Brandhorst, B. P. (1995). Delayed *in vitro* fertilization using coho salmon ovarian fluid. *In* "The Zebrafish Book," 3rd Ed. University of Oregon Press, Eugene, OR.

Driever, W., Solnica-Krezel, L., Schier, A. F., Neuhauss, S. C. F., Malicki, J., Stemple, D. L., Stainier, D. Y. R., Zwartkruis, F., Abdelilah, S., Rangini, Z., Belak, J., and Boggs, C. (1996). A genetic screen for mutations affecting embryogenesis in zebrafish. *Development* **123,** 37–46.

Eaton, R. C., and Farley, R. D. (1974). Spawning cycle and egg production of zebrafish, *Brachydanio rerio,* in the laboratory. *Copeia* **1,** 195–204.

Freeman, G., and Lundelius, J. W. (1982). The developmental genetics of dextrability and sinistrality in the gastropod *Lymnaea peregra. Wilhelm Roux's Arch. Dev. Biol.* **191,** 69–83.

Gerhart, J., Danilchick, M., Doniach, T., Roberts, S., Rowning, B., and Steward, R. (1989). Cortical rotation of the *Xenopus* egg: Consequences for the anteroposterior pattern of embryonic dorsal development. *Development* (1989 Suppl.) 37–51.

Gerhart, J. C. (1980). Mechanisms regulating pattern formation in the amphibian egg and early embryo. *In* "Biological Regulation and Development" (R. F. Goldberg, Ed.), pp. 133–150. Plenum Press, NY.

Gestl, E. E., Kauffman, E. J., Moore, J. L., and Cheng, K. C. (1997). New conditions for generation of gynogenetic half-tetrad embryos in the zebrafish (*Danio rerio*). *J. Heredity* **88,** 76–79.

Haffter, P., Granato, M., Brand, M., Mullins, M. C., Hamerschmidt, M., Kane, D. A., Odenthal, J., van Eeden, F. J. M., Jiang, Y.-J., Heisenberg, C.-P., Kelsh, R. N., Furutani-Seiki, M., Vogelsang, E., Beuchle, D., Schach, U., Fabian, C., and Nüsslein-Volhard, C. (1996). The identification of genes with unique and essential functions in the development of the zebrafish, *Danio rerio. Development* **123,** 1–36.

Hammerschmidt, M., Blader, P., and Strähle, U. (1998). Strategies to perturb zebrafish development. *In* "Biology of the Zebrafish" (H. W. Detrich, M. Westerfield, and L. I. Zon, Eds.) Methods in Cell Biology, Vol. 59, pp. 87–115. Academic Press, San Diego.

Hart, N. H., Becker, K. A., and Wolenski, J. S. (1992). The sperm site during fertilization of the zebrafish egg: Localization of actin. *Mol. Reprod. Dev.* **32,** 217–228.

Hart, N. H., and Collins, G. C. (1991). An electron microscope and freeze-fracture study of the egg cortex of *Brachydanio rerio. Cell Tiss. Res.* **265,** 317–328.

Heasman, J. (1997). Patterning the *Xenopus blastula. Development* **124,** 4179–4191.

Hyatt, B. A., Lohr, J. L., and Yost, H. J. (1996). Initiation of vertebrate left-right axis formation by maternal Vg1. *Nature* **384,** 62–65.

Jesuthasan, S., and Strähle, U. (1996). Dynamic microtubules and specification of the zebrafish embryonic axis. *Curr. Biol.* **7,** 31–42.

Jones, E. A., and Woodland, H. R. (1987). The development of animal cap cells in *Xenopus:* A measure of the start of animal cap competence to form mesoderm. *Development* **101,** 557–563.

Kageura, H. (1997). Activation of dorsal development by contact between the cortical dorsal determinant and the equatorial core cytoplasm in eggs of *Xenopus laevis. Development* **124,** 1543–1551.

Kane, D. A., and, Kimmel, C. B. (1993). The zebrafish midblastula transition. *Development* **119,** 447–456.

Kemphues, K. J., Kusch, M., and Wolf, N. (1988). Maternal-effect lethal mutations on linkage group II of *Caenorhabditis elegans. Genetics* **120,** 977–986.

Kemphues, K. J., and Strome, S. (1997). Fertilization and establishment of polarity in the embryo. *In* "*C. elegans* II" (D. L. Riddle, T. Blumenthal, B. J. Meyer and J. R. Priess, Eds.), pp. 335–359. Cold Spring Harbor Laboratory Press, Cold Spring Harbor, NY.

Kessel, R. G., Roberts, R. L., and Tung, H. N. (1988). Intercellular junctions in the follicular envelope of the teleost, *Brachydanio rerio. J. Submicrosc. Cytol. Pathol.* **20,** 415–424.

Kessel, R. G., Tung, H. N., Roberts, R., and Beams, H. W. (1985). The presence and distribution of gap junctions in the oocyte-follicle cell complex of the zebrafish, *Brachydanio rerio. J. Submicrosc. Cytol. Pathol.* **17,** 239–253.

Larabell, C. A., Torres, M., Rowning, B. A., Yost, C., Miller, J. R., Wu, M., Kimelman, D. T., and Moon, R. T. (1997). Establishment of the dorsoventral axis in *Xenopus* embryos is presaged by early asymmetries in β-catenin that are modulated by the Wnt signalling pathway. *J. Cell Biol.* **136,** 1123–1134.

Lehmann, R., and Nüsslein-Volhard, C. (1986). Abdominal segmentation, pole cell formation, and embryonic polarity require the localized activity of *oskar,* a maternal gene in *Drosophila. Cell* **47,** 141–152.

Long, W. (1983). The role of the yolk syncytial layer in determination of the plane of bilateral symmetry in the rainbow trout, *Salmo gairdneri* Richardson. *J. Exp. Zool.* **228,** 91–97.

Mizuno, T., Yamaha, E., Wakahara, M., Kuroiwa, A., and Takeda, H. (1996). Mesoderm induction in zebrafish. *Nature* **383,** 131–132.

Moon, R. T., Brown, J. D., and Torres, M. (1997). WNTs modulate cell fate and behavior during vertebrate development. *TIGS* **13,** 157–162.

Mullins, M. C., Hammerschmidt, M., Haffter, P., and Nüsslein-Volhard, C. (1994). Large-scale mutagenesis in the zebrafish: In search of genes controlling development in a vertebrate. *Curr. Biol.* **4,** 189–202.

Mullins, M. C., Hammerschmidt, M., Kane, D. A., Odenthal, J., Brand, M., van Eeden, F. J. M., Furutani-Seiki, M., Granato, M., Haffter, P., Heisenberg, C.-P., Jiang, Y.-J., Kelsh, R. N., and Nüsslein-Volhard, C. (1996). Genes establishing dorsoventral pattern formation in the zebrafish embryo: The ventral specifying genes. *Development* **123,** 81–93.

Neuhauss, S. (1996). Craniofacial development in zebrafish (*Danio rerio*): Mutational analysis, genetic characterization, and genomic mapping. Ph.D. Thesis. Fakultät für Biologie, Eberhard-Karl-Universität Tübingen.

Nieuwkoop, P. D. (1969). The formation of the mesoderm in urodelean amphibians, II: The origin of the dorso-ventral polarity of the mesoderm. *Wilhelm Roux's Arch. Dev. Biol.* **163,** 298–315.

Odenthal, J., Rossnagel, K., Haffter, P., Kelsh, R. N., Vogelsang, E., Brand, M., van Eeden, F. J. M., Furutani-Seiki, M., Granato, M., Hammerschmidt, M., Heisenberg, C.-P., Jiang, Y.-J., Kane, D. A., Mullins, A. C., and Nüsslein-Volhard, C. (1996). Mutations affecting xanthophore pigmentation in the zebrafish, *Danio rerio. Development* **123,** 391–398.

Oppenheimer, J. M. (1936). The development of isolated blastoderms of *Fundulus heteroclitus. J. Exp. Zool.* **72,** 247–269.

Sakai, M. (1996). The vegetal determinants required for the Spemann organizer move equatorially during the first cell cycle. *Development* **122,** 2207–2214.

Sakai, N., Burgess, S., and Hopkins, N. (1997). Delayed *in vitro* fertilization of zebrafish eggs in Hank's saline containing bovine serum albumin. *Mol. Mar. Biotechnol.* **6,** 84–87.

Schnabel, R., and Priess, J. R. (1997). Specification of cell fates in the early embryo. *In* "*C. elegans* II" (D. L. Riddle, T. Blumenthal, B. J. Meyer, and J. R. Priess, Eds), pp. 361–382. Cold Spring Harbor Laboratory Press, Cold Spring Harbor, NY.

Schneider, S., Steinbeisser, H., Warga, R. M., and Hausen, P. (1996). β-catenin translocation into nuclei demarcates the dorsalizing centers in frog and fish embryos. *Mech. Dev.* **57,** 191–198.

Schulte-Merker, S., Smith, J. C., and Dale, L. (1994). Effects of truncated activin and FGF receptors and of follistatin on the inducing activities of BVg1 and activin: Does activin play a role in mesoderm induction? *EMBO J.* **13,** 3533–3541.

Selman, K., Petrino, T. R., and Wallace, R. A. (1994). Experimental conditions for oocyte maturation in the zebrafish, *Brachydanio rerio. J. Exp. Zool.* **269,** 538–550.

Selman, K., and Wallace, R. A. (1989). Cellular aspects of oocyte growth in teleosts. *Zool. Sci.* **6,** 211–231.

Selman, K., Wallace, R. A., Sarka, A., and Qi, X. (1993). Stages of oocyte development in the zebrafish, *Brachydanio rerio. J. Morphol.* **218,** 203–224.

Signoret, J., and Lefresne, J. (1971). Contribution a l'etude de la segmentation de l'oeuf d'axolotl: I. Definition de la transition blastuleenne. *Ann. Embryol. Morphog.* **4,** 113.

Slack, J. M. W. (1994). Inducing factors in *Xenopus* early embryos. *Curr. Biol.* **4,** 116–126.

Solnica-Krezel, L., Schier, A. F., and Driever, W. (1994). Efficient recovery of ENU-induced mutations from the zebrafish germline. *Genetics* **136,** 1401–1420.

St. Johnston, D., and Nüsslein-Volhard, C. (1992). The origin of pattern and polarity in the Drosophila embryo. *Cell* **68,** 201–219.

Streisinger, G., Walker, C., Dower, N., Knauber, D., and Singer, F. (1981). Production of clones of homozygous diploid zebra fish (*Brachydanio rerio* I). *Nature* **291,** 293–296.

Streisinger, G., Singer, F., Walker, C., Knauber, D., and Dower, N. (1986). Segregation analyses and gene-centromere distances in zebrafish. *Genetics* **112,** 311–319.

Thorgard, G. H. (1983). Chromosome set manipulation and sex control in fish. *In* "Fish Physiology" (W. S. Hoar, D. J. Randall, and E. M. Donaldson, Eds.) Vol IX, Part B, pp. 405–434. Academic Press, New York.

Westerfield, M. (1993). "The Zebrafish Book." University of Oregon Press, Eugene, OR.

Whittington, P. M., and Dixon, K. E. (1975). Quantitative studies of germ plasm and germ cells during early embryogenesis of *Xenopus laevis*. *J. Embryol. Exp. Morph.* **33,** 57–74.

Wittbrodt, J., and Rosa, F. M. (1994). Disruption of mesoderm and axis formation in fish by ectopic expression of activin variants: The role of maternal activin. *Genes Dev.* **8,** 1448–1462.

Wylie, C., Kofron, M., Payne, C., Anderson, R., Hosobuchi, M., Joseph, E., and Heasman, J. (1996). Maternal β-catenin establishes a 'dorsal signal' in early *Xenopus* embryos. *Development* **122,** 2987–2996.

Yoon, C., Kawakami, K., and Hopkins, N. (1997). Zebrafish *vasa* homologue RNA is localized to the cleavage planes of 2- and 4-cell-stage embryos and is expressed in the primordial germ cells. *Development* **124,** 3157–3166.

CHAPTER 2

Developmental Mutant Screens in the Zebrafish

Fredericus J. M. van Eeden,[*] Michael Granato,[†] Jörg Odenthal,[‡] and Pascal Haffter[‖]

[*] Wellcome/CRC Institute
Cambridge CB2 1QR, United Kingdom

[†] University of Pennsylvania
Philadelphia, Pennsylvania 19104-6058

[‡] University of Sheffield
Sheffield S10 2TN, United Kingdom

[‖] Max-Planck Institut für Entwicklungsbiologie
72076 Tübingen, Germany

METHODS IN CELL BIOLOGY, VOL. 60

0091-679X/99 $30.00

I. Introduction

The goal of large-scale saturation screens is to identify genes which define developmental pathways. Such screens provide multiple entry points to molecular analysis of developmental mechanisms via cloning of the mutated genes. With this goal in mind, it seems very desirable to be able to carry out mutational screens at the largest scale possible, which requires the optimization of all steps in the screening procedure to highest efficiency. Systematic mutation screens have been carried out in a number of invertebrate and plant model organisms such as *Drosophila melanogaster* (Nüsslein-Volhard and Wieschaus, 1980; Jürgens *et al.*, 1984; Nüsslein-Volhard *et al.*, 1984; Wieschaus *et al.*, 1984), *Caenorhabditis elegans* (Brenner, 1974; Kemphues *et al.*, 1988) and *Arabidopsis thaliana* (Mayer *et al.*, 1991). In these screens, large numbers of lines derived from mutagenized individuals were inbred for several generations to obtain homozygous mutant progeny that could be scored for defective morphological traits.

Several characteristics of the zebrafish make it a suitable organism to carry out large-scale mutational screens. These include a short generation time, high fecundity, and rapid development of externally fertilized embryos that are translucent (reviewed by Driever *et al.*, 1994; Haffter and Nüsslein-Volhard, 1996). The translucent nature of the zebrafish embryo allows for easy visual and noninvasive detection of morphological defects *in vivo* while development proceeds. Synchronous development allows efficient recognition of morphological defects by simple comparison of siblings. The large number of progeny that are usually obtained in a single clutch of embryos makes it possible to distinguish Mendelian traits from nongenetic defects.

Large-scale, systematic mutation screens in the zebrafish are now possible due to the development of efficient ways to breed and maintain large numbers of individual lines (Driever *et al.*, 1996; Haffter *et al.*, 1996). The aquarium systems and feeding regimes used in the large scale screen carried out in Tübingen have been described previously by Mullins *et al.* (1994). The aquarium systems and mating boxes for setting up large numbers of crosses have since become commercially available (Schwarz Aquarienbau, Göttingen, Germany) and will not be described here. In this review, we will focus on the chemical mutagenesis procedure and screening strategies for efficient induction and recovery of developmental mutations in the zebrafish.

II. Mutagenesis

The choice of the zebrafish line to be mutagenized depends on the assay and the breeding scheme to be used for the screen. Tü and AB are the two lines that have mainly been used for mutational screens in the past (Driever *et al.*, 1996; Haffter *et al.*, 1996). For specific screens (e.g., allele screens), genetically marked lines such as TL (Haffter *et al.*, 1996) may be more suitable for an efficient screening strategy (see following discussion). For screens requiring the squeezing of females to obtain unfertilized oocytes such as haploid screens or early pressure screens (see following discussion), the AB line seems more suitable because it more readily gives oocytes that can be fertilized *in vitro.* Whatever line is used, it should be free of lethal and other mutations. For more elaborate screening assays, it is advisable to perform the desired assay on inbred progeny of nonmutagenized fish before using that line for the mutagenesis. To screen for existing mutations in the background of wild-type lines to be used, families are raised from single-pair matings. The progeny of crosses carried out among siblings of these families are then screened for mutant phenotypes as described in the breeding scheme that follows. Depending on the number of families screened and the fraction of the genomes screened in each family, one can calculate the chance that the line to be used is free of mutations (see the calculations that follow).

Efficient mutagenesis methods and high mutation rates are prerequisites for performing large-scale mutational screens. Several mutagenic reagents, including gamma rays (Chakrabarti *et al.*, 1983; Walker and Streisinger, 1983), UV light (Grunwald and Streisinger, 1992a), and ethylnitrosourea (ENU) (Grunwald and Streisinger, 1992b) have been used to mutagenize zebrafish sperm. In these procedures, sperm are isolated from males, treated with the mutagen, and then used to fertilize eggs *in vitro.* Alternatively, males have been mutagenized with ENU and immediately mated to females following the mutagenesis (Riley and Grunwald, 1995). F_1 fish derived from ENU mutagenized sperm are mosaic in their germline for the induced mutations because the mutation carried by one DNA strand is not resolved prior to fertilization. For an efficient diploid screen, it is important that F_1 fish are not mosaic for mutations in the germline because this will result in fewer F_2 fish carrying the mutation which results in a less-efficient identification of the mutation in the screening procedure.

For the induction of mutations affecting single genes, chemical mutagens are generally preferred over gamma rays or X rays (Singer and Grunberger, 1983). X rays and gamma rays result in large deletions or chromosomal rearrangements and induce recoverable mutations at an approximately tenfold lower rate than ENU (Mullins *et al.*, 1994). Using chemical mutagenesis to induce point mutations also offers the possibility that besides complete loss-of-function mutations, partial loss-of-function and gain-of-function alleles can be isolated which may be very useful for elucidating the function of the mutated gene. Ethylmethanesulfonate (EMS), although the mutagen of choice in *Drosophila* and *C. elegans,* does not produce high rates of recoverable mutations in the zebrafish (Mullins *et al.*, 1994; Solnica-Krezel *et al.*, 1994). ENU was found to be the most efficient chemical mutagen for inducing point mutations that can be recovered in an F_2 breeding scheme (Mullins *et al.*, 1994;

Solnica-Krezel *et al.,* 1994). To avoid mosaicism of ENU-induced mutations in the F_1 generation, it is important that germ cells are mutagenized at premeiotic stages during spermatogenesis (Jenkins, 1967; Mullins *et al.,* 1994; Solnica-Krezel *et al.,* 1994). A point mutation induced in one DNA strand of a premeiotic germ cell is fixed in both strands during DNA replication. For most types of screens, ENU mutagenesis of premeiotic germ cells is the method of choice, and we describe this procedure in greater detail later.

A. ENU Mutagenesis of Premeiotic Germ Cells

The ENU mutagenesis protocol described here is a slightly modified version of previously published ENU mutagenesis protocols (Mullins *et al.*, 1994; Solnica-Krezel *et al.*, 1994). ENU is a harmful substance, and the mutagenesis is therefore carried out in a fume hood. ENU (Sigma) is dissolved in a final concentration of 10m*M* acetic acid and stored in aliquots at −20°C at a concentration of 100–125m*M*. The precise concentration is determined by measuring the optical density at 238 nm at pH 6.0 (extinction coefficient = $5830 M^{-1}$ cm^{-1}). For the mutagenesis, ENU is diluted to a final concentration of 3.0m*M* in 10m*M* sodium phosphate buffer, pH 6.6 (Be aware that the pH of phosphate buffers varies with concentration). The mutagenesis is carried out in the dark at 21–22.5°C in 1-l cylinders, within which a second cylinder is placed. The bottom of the inner cylinder has been removed and replaced by a fine-mesh stainless-steel screen in order to allow transfer of fish for exchanging the water.

Prior to the first mutagenesis, males are mated to females, and only males producing fertilized offspring are selected for the mutagenesis. Up to 5 males are placed in 300 ml of aqueous ENU solution for 1 hour. Following the mutagenesis, the inner cylinder containing the mutagenized fish is transferred twice to aquarium water for 1 hour each and then returned into their aquaria at 27°C for the night. All ENU solutions, wash water and items in contact with ENU are treated with 1*M* NaOH for the inactivation of ENU as described (Ashburner, 1989). This mutagenesis treatment is repeated at weekly intervals up to six times.

A strong correlation has been reported between the number of mutagenic treatments and the frequency of induced mutations (Solnica-Krezel *et al.,* 1994; Driever *et al.,* 1996). The frequency of ENU-induced mutations is also dosage-dependent; increasing the ENU concentration from 2.0 to 3.0m*M* results in a significant increase in the mutation frequency (Solnica-Krezel *et al.,* 1994), but ENU doses that are significantly higher than 3.0m*M* result in too high lethality. The fish are very nervous during the treatment, and it therefore seems to be critical to carry out the mutagenesis in the dark with as little disturbance to the fish as possible. The frequency of lethality can be variable between different experiments. Other than ENU concentration and number of treatments, only the density of fish during the treatment strictly correlates with lethality in our hands. Increasing pH or temperature can cause an increase in lethality (Mullins *et al.,* 1994; Solnica-Krezel *et al.,* 1994), but the effect of these factors on mutation rate and lethality has not been assessed systematically. To obtain 20–30 fertile males after six treatments with 3.0m*M* ENU we usually start with 100 healthy fertile males.

Following the last mutagenesis, the mutagenized males are mated at weekly intervals. F_1 progeny obtained later than 3 weeks after the last mutagenic treatment are nonmosaic for ENU-induced mutations and are raised (Mullins *et al.,* 1994; Solnica-Krezel *et al.,* 1994). The overall mutation rate of ENU-induced mutations does not change significantly in F_1 progeny obtained several weeks after the last mutagenesis, indicating the induction of mutations in spermatogonial stem cells (Mullins *et al.,* 1994; Solnica-Krezel *et al.,* 1994). Assuming that the number of spermatogonial stem cells is 500–1000 (Mullins *et al.,* 1994), we usually limit the number of F_1 fish derived from each male to 500 to reduce the risk of raising progeny from spermatogonial clones.

B. Assessment of Mutation Frequency

The efficiency of mutagenesis is determined by the frequency of induced mutations at one of several pigmentation loci. This is done in a noncomplementation assay in which the mutagenized males are mated to females that are homozygous or heterozygous mutant carriers for one or several of these pigmentation loci. The frequency of abnormally pigmented individuals in the progeny of these mating allows an estimate of the mutation frequency. In the past, mutations in *golden* (*gol*), *brass* (*brs*), *albino* (*alb*) and *sparse* (*spa*) have been used for this test (Mullins *et al.,* 1994; Solnica-Krezel *et al.,* 1994). The ENU mutagenesis treatment described above usually yielded mutation rates of $0.9–3.3 \times 10^{-3}$ for these loci.

C. Keeping Track of Progeny

It is important that mutagenized founder fish are kept and that F_1 progeny and subsequent F_2 families can be traced back to their respective founder fish. We have in the past used a nomenclature system in which each mutagenized founder male is given a different letter designation that is also used for the F_1 fish. For the F_2 families, a number is added, and for each mutant found in a particular F_2 family, another letter is added. The *sonic-you* allele syu^{tq252a}, for example, derives from the mutagenized founder male *tq,* which gave rise to an F_2 family termed *tq252* in which the mutation *tq252a* was identified (Haffter *et al.,* 1996; van Eeden *et al.,* 1996). One reason for keeping track of F_1 progeny and subsequent F_2 families is to know whether multiple mutant alleles of the same gene have been induced independently of each other. A second reason is that if a mutant allele is sequenced, one can go back and sequence the corresponding region in the genome of the founder fish to demonstrate that this mutation was not present before the mutagenesis and was probably induced by the treatment with ENU. For that reason, it is important that mutagenized founder fish are frozen away once enough F_1 progeny has been obtained.

III. Breeding Scheme

Several genetic methods have been developed to screen for mutations in progeny directly obtained from F_1 females. The advantage of these methods is that only limited breeding is required following the mutagenesis which is an attractive option if large breeding capacity is not available. Haploid embryos are obtained by activating eggs from such females *in vitro* using UV-irradiated sperm (Streisinger *et al.,* 1981; Walker, this volume). Of the haploid embryos, 50% derived from a female heterozygous for a specific mutation will reveal the phenotype of this mutation. However, the drawback of this method is that haploid embryos may develop abnormally 2 days after fertilization, which makes it difficult or impossible to screen for mutations causing subtle defects or mutations affecting later stages of development. Methods have also been developed that allow the generation of diploid parthenogenetic progeny from females by treatment of activated eggs with heat shock or early pressure (Streisinger *et al.,* 1981, 1986; Beattie *et al.,* 1998). However, both procedures decrease the survival of the embryos and cause abnormal development in a substantial fraction of the embryos to be screened. In addition, early pressure also poses the problem that genetic recombination between the centromere and loci that are located distal to the centromere can occur. This results in a decreased efficiency of screening for mutations affecting distal loci.

To circumvent the problems associated with the screening of haploid or parthenogenetic diploids, a classical three-generation crossing scheme is used to generate homozygous diploid embryos (Fig. 1). Such a standard crossing scheme was employed successfully in large-scale mutational screens (Driever *et al.,* 1996; Haffter *et al.,* 1996). After the mutagenesis, the mutagenized males are bred at weekly intervals with wild-type females, and F_1 fish are raised starting 3 weeks after the last mutagenesis. F_2 families are derived from single-pair matings of F_1 fish and 60–80 larvae per F_2 family are raised to adulthood. Half of the fish in such an F_2 family are heterozygous for any particular mutation carried by either of the two F_1 parents. Sibling crosses among F_2 fish will match two carriers of such a mutation in a quarter of the matings, and a recessive mutant phenotype will be displayed by 25% of the F_3 embryos from such an egglay.

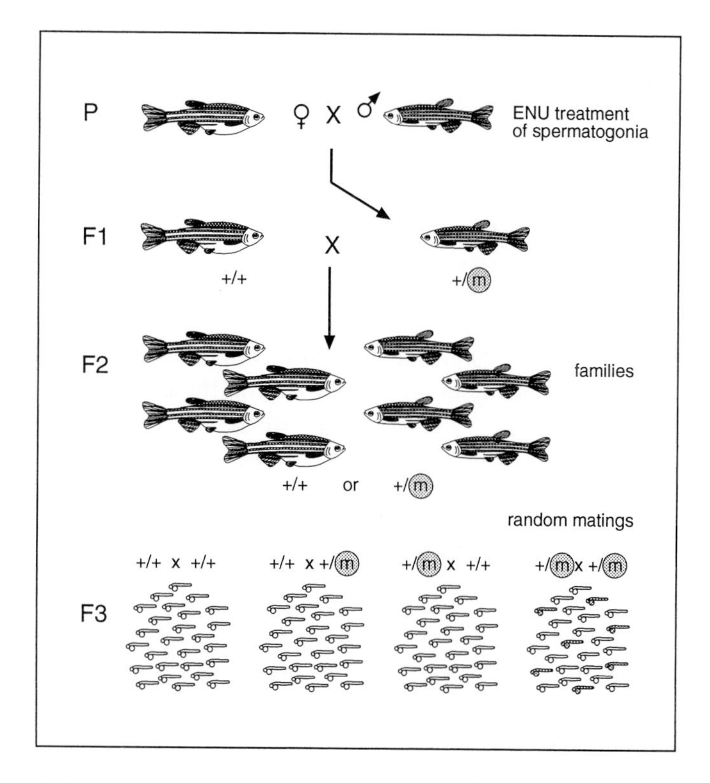

Fig. 1 Classical three-generation crossing scheme for the generation of homozygous diploid embryos (from Haffter *et al.*, 1996). Males are mutagenized with ENU and mated to wild-type females. The F_1 progeny raised from matings 3 or more weeks after the last mutagenesis are heterozygous for one mutagenized genome. F_2 families are raised from single pair matings between F_1 fish. A mutation **m** present in one of the F_1 parents is shared by 50% of the fish in the F_2 family. In 1/4 of random matings between F_2 siblings, both F_2 parents will be heterozygous carriers for **m** and 25% of the resulting F_3 embryos from such a cross will be homozygous for **m** and display the mutant phenotype. Because both F_1 fish carry one mutagenized genome, the F_2 family will contain two mutagenized genomes. The fraction of the two mutagenized genomes screened depends on the number of successful F_2 sibling crosses. In six such crosses, for example, 82% are screened resulting in 1.64 mutagenized genomes screened.

IV. Screening Procedure

Each F_2 family derived from two F_1 fish represents two mutagenized genomes. However, because only 25% of these are made homozygous in each F_2 sibling cross, the actual number of genomes screened in an F_2 family is calculated using the formula $2 \times (1 - 0.75^n)$, where n is the number of successful F_2 sibling crosses. Ideally, one should aim for six successful sibling crosses (1.64 genomes) per F_2 family, and an F_2 family giving less during the first set-up is set up a second time. However, if a family has given six or more successful crosses during the first setup, it is inefficient to set it up again because fewer, than 0.1 genomes are being screened in every cross obtained in addition to the first six.

The F_2 parents and the progeny of each successful cross are given identical labels indicating the F_2 family and the cross number. The F_2 parents can be kept as pairs in single boxes containing approximately 1 l of aquarium water for up to 10 days without feeding. If screens take longer than 10 days they should be separated and attached to running water.

At least 20 and up to 100 fertilized F_3 embryos from each egglay are sorted into E3 medium (see appendix) on the day they are laid. A convenient setup is to sort 2–3 × 40 fertilized embryos into 60 mm Petri dishes that are grouped together in 120 mm × 120-mm squared dishes. If a mutant phenotype is found, the mutant embryos can be resorted among the dishes and inspected again at various times during development.

A. Visual Screens

The translucent nature of the zebrafish embryo allows us to screen for many different traits in live embryos under the dissecting microscope. The screening procedure outlined later is a description of a large-scale visual screen that was carried out for mutations causing defects during the first 5 days of zebrafish development (Haffter *et al.,* 1996). In this screen, 12 embryos or larvae per egglay were inspected under the dissecting microscope at three different time points during development. If a mutant phenotype is present in 25% of the embryos of a particular egglay, it should be found with a chance of more than 96% when screening 12 individuals. If a mutant phenotype was found, the remaining embryos of that clutch were inspected, and mutant embryos were counted to check for Mendelian segregation of that mutation. At 24–36 hpf, embryos were scored for defects of the eyes, brain, notochord, spinal chord and somites. At 48–60 hpf, embryos were first checked for response to touch and anesthetesized with MESAB (see Appendix) before further scoring for fins and cardiovascular defects. After this screen, the medium was replaced to remove the anesthetic and the chorions left behind by the hatched larvae. Around 96 hpf, larvae were again checked for touch response and motility before being anesthetized again. This time, embryos were scored for defects of pigmentation, pectoral fins, jaw and branchial arches, blood, ear, liver, and gut. Depending on the particular focus of the screen, it may seem appropriate in some cases, to choose different timepoints. For mutations affecting gastrulation and early axis formation, it may be worth screening at 10–12 because some of the mutations affecting these early processes die before 24 hpf (Kane *et al.,* 1996; Mullins *et al.,* 1996; Solnica-Krezel *et al.,* 1996). It may also seem more appropriate to screen for cardiovascular mutants before 48 hpf because pericardial and general edema frequently obscure specific cardiovascular defects at later stages (Chen *et al.,* 1996; Stainier *et al.,* 1996).

During the screening procedure outlined previously, the score sheet shown in Fig. 2 was used. One of the most important features of this score sheet is that it not only allows for entries and descriptions of particular phenotypes but also requires that decisions are made immediately on the score sheet itself. On the top left, the name of the F_2 family is entered, and the remainder of the top row is used to decide whether a particular F_2 family should be set up a second time, kept, or discarded. This score sheet allows for scoring up to eight crosses in a family (for more crosses, more score sheets are added). This score sheet also allows for the simultaneous description of several phenotypes that may be found in different crosses of the same family. Besides boxes for specific morphological traits to be scored, there are also special boxes for scoring pleiotropic phenotypes that are found at rather high frequencies such as necrosis (ncr), edema (oed), bent body axis (bnt), small eyes (bde), retardation (rtd), and death (ded). Retardation, edema, general necrosis, and brain necrosis represented about two-thirds of all mutants and were not kept because they were considered to be rather general and nonspecific (Mullins *et al.,* 1994; Haffter *et al.,* 1996).

B. Other Screens

Besides a simple visual screen under the dissecting microscope, more specific screening methods may be useful for discovering specific mutant phenotypes. Examples of screening procedures that have actually been performed include a simple touch response assay for mutations affecting locomotion behavior (Granato *et al.,* 1996), an optokinetic response test for mutations affecting the visual system (Brockerhoff *et al.,* 1995), and injection of fluorescent tracer dyes into the eyes of fixed larvae to detect mutations affecting the retinotectal axon projections (Baier *et al.,* 1996). Whole mount staining techniques on zebrafish embryos using antibodies (Henion *et al.,* 1996; Eisen *et al.,* this issue) or RNA probes (Moens *et al.,* 1996) have also been applied successfully to identify mutations causing more subtle defects in particular developmental processes. A number of whole mount staining procedures such as skeletal stainings on fixed embryos using Alcian Blue (Piotrowski *et al.,* 1996) or 2-4-dimethyl-aminostyryl-N-ethyl pyridinium iodide (DASPEI) live staining of lateral line hair cells (Whitfield *et al.,* 1996) may be interesting assays for more specific defects in future screens. In any event, it is a good idea to include a visual screen in addition to the specific screening assay and to include embryos displaying visible phenotypes in the assay. In the screen for mutations affecting the retinotectal projections, for example, of the 37 identified complementation groups in the retinotectal screen, 34 (92%) had an associated visible phenotype. Noticing an associated phenotype can aid in (pre)selecting further interesting candidate mutants by similarity in visual phenotype (Karlstrom *et al.,* 1996; Trowe *et al.,* 1996).

tq252 crosses set up **4** crosses successful **4** crosses 2nd round **0** keep family **X** discard family ☐

		lost	OK	bhv	hat	pgm	ird	jaw	eye	arc	ear	brn	hrt	cir	liv	gut	swb	not	flp	ntu	som	fin	ncr	oed	bnt	bde	rtd	ded	A	B	C	D	E	remarks	Init
keep___	cross **1**	1d	**X**																																YJ
discard	total **80**	2d	**X**																																MM
return	date **19.1.**	5d	**X**																																DK
keep___	cross **2**	1d	**X**																																YJ
discard	total **26**	2d	**X**																																MM
return	date **19.1.**	5d																						**B**	**B**					6 /26					DK
keep_ **A**_	cross **3**	1d																			**A**								19/82						YJ
discard	total **80**	2d		**A**	**A**	**A**								**A**							**A**														MM
return	date **19.1.**	5d																										**A**							DK
keep___	cross **4**	1d	**X**																																DK
discard	total **80**	2d	**X**																																MG
return	date **20.1.**	5d	**X**																																DK
keep___	cross	1d																																	
discard	total	2d																																	
return	date	5d																																	
keep___	cross	1d																																	
discard	total	2d																																	
return	date	5d																																	
keep___	cross	1d																																	
discard	total	2d																																	
return	date	5d																																	
keep___	cross	1d																																	
discard	total	2d																																	
return	date	5d																																	

A d1: u-shaped somites, no myoseptum, d2: reduced motility, no lateral melanophore stripe, not hatched, reduced circ.

var uni

B d5 tail curled down, edema

var uni

C

var uni

D

var uni

E

The more laborious screening procedures may make it difficult or even impossible to screen large numbers of F_3 individuals. Mendelian inheritance is one of the important criteria for keeping a particular mutant and it therefore seems useful to collect (and fix if necessary) a larger number of individuals that can be subsequently assayed to verify a particular mutant phenotype.

V. Allele Screens

Sequencing mutant alleles frequently provides good evidence that a mutant phenotype is the result of a mutation in a particular gene. The majority of complementation groups identified in mutational screens in the zebrafish are represented by single alleles only (Driever *et al.*, 1996; Haffter *et al.*, 1996). This poses the problem that the single allele may not carry a mutation in the open reading frame. In many cases, it will therefore be necessary to identify additional alleles of a particular gene. Identifying additional alleles may also be necessary to obtain null alleles causing a complete loss of function phenotype. One such example is *syu*, for which only a single weak allele was identified in the large-scale mutational screen (van Eeden *et al.*, 1996). We identified new strong alleles of *syu* in an allele screen by noncomplementation of syu^{tq252} (Schauerte *et al.*, 1997).

The strategy of this allele screen is outlined in Fig. 3. A tester fish population of 1000 fish was prepared by crossing together mutations in four genes, *syu*, *you-too* (*yot*), *fused somites* (*fss*), and *unplugged* (*unp*) (Granato *et al.*, 1996; van Eeden *et al.*, 1996). The idea behind this strategy was to screen simultaneously for new alleles in four genes. Each of the four mutations should, in theory, be represented by 50% of the tester fish population. This was tested by setting up 172 crosses among the tester fish. We found that the effective carrier frequency of *yot* was only 25%, which is probably the result of a dominant effect of the yot^{ty119} allele which reduces viability of heterozygous carriers for that allele (van

Fig. 2 Example of a scoresheet used in the large-scale Tübingen screen. The name of the F_2 family (*tq252*) is indicated on the top left. Four crosses were set up and all four laid; no second round of crosses was necessary. From cross 2, 26 embryos were obtained; the other crosses gave 80 embryos each. Phenotype A (now termed syu^{tq252a}) was found after 1 day of development, redescribed on subsequent days, and kept. Phenotype B was found after 4–5 days of development and discarded. Letters A and B were entered into boxes corresponding to the affected trait or tissue. Shaded boxes were checked only if a mutant phenotype was found; black boxes are traits that cannot be screened at that stage of development. The ratios of mutant to sibling embryos (19/82 and 6/26) are indicated in the panels below the letters corresponding to the respective phenotypes A and B. A was circled because it was considered a mutant phenotype; B was not circled because it could be the result of a syndrome associated with the small size of the clutch. Further descriptions of phenotypes A and B are given in the lower part of the scoresheet. uni was circled since both phenotypes appeared to be uniform. This scoresheet is available in Microsoft Word5 format upon request (pascal.haffter@tuebingen.mpg.de).

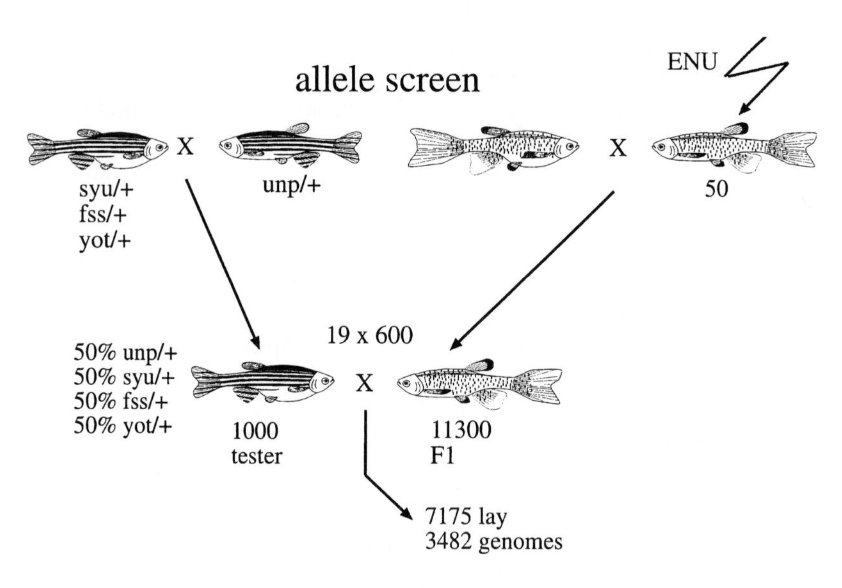

Fig. 3 Crossing scheme for identifying newly induced alleles of *syu, fss, yot,* and *unp.* By mating a *syu, fss, yot* triple heterozygous carrier to an *unp* heterozygous carrier, 1000 tester fish were generated. TL males that were homozygous for the leo^{t1} mutation causing a spotted pigmentation in adult fish and the dominant mutation $loft^{t2}$ causing long fins were mutagenized with ENU and mated to TL females to generate 11,300 F_1 fish. Among these F_1 fish, carriers for newly induced alleles of *syu, fss, yot,* or *unp* were identified by noncomplementation in single-pair matings with the striped tester fish. Approximately 600 crosses were set up in weekly intervals for 19 weeks; 7175 gave eggs resulting in an estimated 3482 mutagenized genomes screened. The estimated number of mutagenized genomes screened is fewer than half of the number of egglays because some of the F_1 fish were tested twice.

Eeden *et al.*, 1996). The effective carrier frequency of *unp* was 39%, whereas the effective carrier frequency of *syu* and *fss* was 50 and 49%, respectively (Table I).

The tester fish populations are set up at weekly intervals in single-pair matings with F_1 fish that were derived from ENU mutagenized males. These F_1 fish were generated in a TL background, and again a nomenclature was used that would

Table I
Allele Screen Calculations

Mutation	Effective carrier frequency in tester fish	Corrected number of genomes screened	Number of new alleles	Calculated allele frequency
sonic you	0.50	3481.5	2	5.7×10^{-4}
you-too	0.25	1741	2	1.1×10^{-3}
fused somites	0.49	3412	2	5.9×10^{-4}
unplugged	0.39	2759	13	4.8×10^{-3}

allow to trace back newly identified alleles to the respective mutagenized founder fish. The TL strain is homozygous for the recessive mutation *leo*^{t1} mutation causing a spotted pigmentation in adult fish and also carries the dominant mutation *lof*^{t2} causing long fins (Haffter *et al.*, 1996). The tester fish population has normal fins and normal striped pigmentation; this allows reliable and efficient separation of the fish after mating. This is crucial because contamination of the mutagenized population with tester fish will falsely lead to identification of "new alleles" in the next round of testing.

A total of 11,300 crosses were set up of which 7175 were successful. The calculated number of mutagenized genomes screened by noncomplementation was corrected for by the effective carrier frequency of each mutation in the tester population. Some of the F_1 fish were tested twice, and the calculated number of mutagenized genomes was also corrected assuming that the second setup tested only half as many mutagenized genomes (Table I). This allele screen yielded 2 new alleles each for *syu, yot,* and *fss,* whereas for *unp* 13 new alleles were identified.

The calculated allele frequencies of *syu* (5.7×10^{-4}) and *fss* (5.9×10^{-4}) are significantly lower, whereas that of *unp* (4.8×10^{-3}) is significantly higher than the average mutation rate of 2.3×10^{-3} that was previously determined for four different pigmentation loci using ENU as a mutagen under similar conditions. It is tempting to speculate that the low number of alleles found for *syu* and *fss* in the original large-scale screen was the result of the low mutability of these genes since the phenotypes of these mutants is quite obvious (van Eeden *et al.*, 1996). The reduced survival rate of *yot* heterozygotes could explain the low allele frequency for this gene in the large-scale screen. Mutants for *unp,* however, display a subtle and transient phenotype (Granato *et al.*, 1996), which could easily be missed in a large-scale screen for many different phenotypes. This could explain the high number of alleles found in an allele screen aimed at specifically detecting this mutant phenotype versus the low number of alleles found in the large-scale screen, in which many different traits were screened. This shows that allele frequency is coupled not only to the mutation rate but also to the detectability of a particular phenotype in the screening procedure.

The high frequency of *unp* alleles identified in the allele screen could in theory be the result of spermatogonial clones. In three cases, two *unp* alleles were isolated from the same mutagenized founder fish. We estimated the number of spermatogonial clones as a function of [SP − (TP × TGPF/TG)]/[1 − (TGPF/TG)], where SP is the number of second positives (3), TP is the total number of positives (13), TG is the total number of genomes screened (2759), and TGPF is the total genomes of screened that were derived from founder fish giving rise to a positive (578). This calculation yields a theoretical value of 0.3 for the number of spermatogonial clones; therefore, we consider it to be rather unlikely that, in any of the three cases, the two *unp* alleles actually derive from spermatogonial clones.

An alternative strategy for identifying ENU-induced point mutations at specific loci in the zebrafish was described by Riley and Grunwald (1995). In this procedure, males are mutagenized once with a lower dose of ENU and immediately mated to wild-type females following the mutagenesis. The resulting F_1 offspring are mosaic for ENU mutations induced at postmitotic stages of spermatogenesis. The mosaic F_1 fish are screened for mutations by noncomplementation of mutant carriers to identify mosaic carriers for newly induced alleles. It has been reported that, under optimal conditions, this method yielded mutation rates that were 10- to 15-fold higher than previously described. Our attempts with this method have not yielded equally high mutation rates suggesting that optimization of several parameters (e.g., ENU dose) may be more critical.

VI. Dominant Screens

A number of mutations that were identified in screens for recessive mutations cause dominant phenotypes. These are either mutations causing transient phenotypes or dominant mutations that do not cause lethality. Among these are, for example, mutations affecting genes involved in dorsoventral axis formation (Mullins *et al.*, 1996), mutations affecting the cardiovascular system (Chen *et al.*, 1996), and also mutations affecting pigmentation and shape of the adult (Haffter *et al.*, 1996). Individual heterozygous carriers for such dominant phenotypes can be recognized among F_1 embryos and raised separately to adulthood. Dominant mutations are, however, more reliably identified in allele screens or while generating F_2 families from single pairs of F_1 fish, in which 50% of the offspring display the dominant phenotype.

In a screen involving over 18,000 F_1 derived from ENU mutagenized males, 414 individual F_1 embryos displaying a phenotype were sorted and raised separately. Of these, 150 (36%) reached fertility, and upon crossing to wild type, 19 (5%) rescreened positively and displayed the expected dominant phenotype in their respective offspring. The screening of progeny from 7175 F_1 fish within the allele screen described previously yielded 10 additional dominant candidate mutations, of which 9 were found to be real in the next generation. Inspection of over 11,000 adult F_1 fish yielded another 25 putative dominant mutations causing adult phenotypes, of which 5 (20%) rescreened positively in the subsequent generation. This simple visual screen therefore yielded a total of 33 dominant mutations.

Several of these dominant mutations were found to be semilethal, and although many of them display much stronger and much more obvious recessive phenotypes, they are likely to be underrepresented in a screen involving a classical breeding scheme resulting from their reduced survival rate if raised together with other F_1 fish. The majority of the work involved in a dominant screen is not the screening procedure itself but rather the separate raising of hundreds of individuals from early developmental stages to fertility.

VII. Managing the Outcome of Screens

A. Criteria for Keeping Mutants

The identification of mutant phenotypes during the screening procedure itself is not necessarily a scientifically challenging task. It merely involves good eyes that are able to compare siblings and recognize a difference occurring in a significant fraction, which is similar to finding ten differences between two almost identical drawings. The much more challenging task is to establish solid scientific criteria to decide on which mutants to keep. Because the work involved in maintaining and characterizing mutants is intensive, it is very important to spend considerable time on this issue.

Among the most important criteria are uniformity of a phenotype and Mendelian segregation. Only a small fraction of mutations causing variable phenotypes were reidentified in subsequent generations, and mutations with variable penetrance such as *momo* (Odenthal *et al.,* 1996), for example, are still among the most difficult to study. A significant number of egglays usually produces a variable number of abnormal embryos; therefore, it is critical that enough F_3 individuals are screened to establish Mendelian segregation. The most solid criterion for Mendelian segregation comes about if a mutation is identified in more than one cross from the same F_2 family.

Two-thirds of all mutations identified in large-scale mutational screens were discarded because their phenotypes were pleiotropic and did not appear to be involved in any defined aspect of early development (Mullins *et al.,* 1994; Driever *et al.,* 1996; Haffter *et al.,* 1996). These were mutations causing retardation, edema, general necrosis, or brain necrosis. Criteria for keeping mutants based on their phenotype may also be modified during the screen, and several mutants that are considered worth keeping during initial phases of the screen may not be kept later on. Visual screens carried out in addition to screening assays focusing on particular aspects of development should help in establishing criteria on which mutants to keep. It is unlikely that a mutation affecting overall morphology of the embryo is specifically required for development of a particular cell type that is screened by whole mount antibody staining. Mutations causing subtle defects in particular cell types or tissues are much more likely to be involved in specific processes of development.

All mutations that are considered worth keeping during the screening procedure should be rescreened immediately. During the rescreen, embryos should be scored by at least two people to establish consistent criteria for keeping mutations. During the large-scale screen, 25% of all mutants that were kept in the initial screen were discarded after the rescreen, either because the mutant phenotype was not confirmed or because it did not pass the standard quality for a mutation to be kept (Haffter *et al.,* 1996).

Mutations to be kept are outbred to wild-type strains and raised. To avoid problems associated with biased sex ratios in the following generation, it is advisable to raise two independent outcrosses of separate F_2 parents. It is not recommended to raise wild-type siblings of the F_3 generation at this point because of the mutagenic load carried in the background of these fish. The number of late lethals was estimated to be similar to the sum of embryonic and larval lethals identified in an ENU screen (Fredericus J. M. van Eeden and Pascal Haffter, unpublished), and with the risk of additional mutations reducing viability or fertility in subsequent generations, it seems best to dilute the mutagenic load as soon as possible.

B. Complementation Analysis, Mapping

After the identification of mutant carriers in the next generation, complementation groups can be established by crosses between independent mutants with similar phenotypes. This may become a difficult task if phenotypic groups are very large. Among mutations causing late notochord defects (Odenthal *et al.*, 1996), for example, complementation analysis still has not been completed yet. Complementation analysis involving mutant strains from other laboratories poses additional problems.

Mapping of mutations immediately following their identification should enormously facilitate such complementation analysis in the future. Genetic maps of molecular markers have been established (Postlethwait *et al.*, 1994; Knapik *et al.*, 1996), and over 100 mutations identified in developmental screens have been placed on the map (Robert Geisler, Jörg Rauch, and Pascal Haffter unpublished). One interesting possibility would be to use a polymorphic reference line for breeding F_1s from the mutagenized founder fish. This would allow immediate mapping of the mutations using mutant and sibling embryos found in the F_3 generation of the screen. This is also an interesting option for screens involving laborious assays because molecular markers that are closely linked to the mutation would simplify the identification of mutant carriers in subsequent generations.

C. Databases

We have successfully used FileMaker Pro databases to assist in coordinating screens and managing the outcome. Different data for each mutant are entered in separate fields that can be presented in different layouts depending on specific needs. In addition, separate databases can be interconnected using relational fields, which is a very useful feature for future projects involving the mutants identified in the screen.

D. Calculating Saturation

Ideally, the aim of a mutational screen is to identify mutations in all mutable genes participating in a particular developmental pathway. The aim is, therefore, to carry out the screen to a reasonably high saturation level. None of the past two large-scale screens has approached saturation, however, which is obvious from the large number of complementation groups that are represented by single alleles only (Driever *et al.,* 1996; Haffter *et al.,* 1996). The saturation level of a mutational screen not only depends on the mutation frequency and the number of genomes screened but it also depends on the frequency with which a particular mutant phenotype is detected and kept. Saturation levels are therefore usually better for mutations with obvious phenotypes that can easily be detected. Most of the complementation groups containing many alleles fall into this category.

It is nevertheless possible to at least calculate a theoretical saturation level using the formula $1 - [1 - (mr \times dc]^{sg}$, where *mr* is the mutation rate, *dc* is the detectability (the frequency, with which a mutant phenotype is actually recognized if it is present), and *sg* is the number of mutagenized genomes screened. Assuming an average mutation rate of 2.3×10^{-3} as determined for ENU for four pigmentation loci (Mullins *et al.,* 1994) and 100% detectability, the screening of 1000 mutagenized genomes would reach a saturation level of 90% which means that 90% of all mutants with such properties should be identified in that screen. However, if we assume a mutation rate of 5.8×10^{-4} as determined for *syu* and *fss* in the allele screen and assuming a detectability of 80% for these mutants, 83% saturation would be reached after screening 3857 mutagenized genomes (the size of the Tübingen screen; Haffter *et al.,* 1996), 1.8 alleles on average would be expected for each gene, and 17% of the genes with such an allele frequency still remain to be identified.

VIII. Potential of Future Screens

Positional cloning of genes identified by mutations in the zebrafish has become feasible and will become a standard procedure in the near future (Talbot and Schier, this volume; Zhang *et al.,* 1998). Although still several-fold below the mutation frequency obtained by ENU, insertional mutagenesis using viral insertion in the zebrafish has become feasible and already yielded a number of promising mutations (Gaiano *et al.,* 1996; Hopkins, this volume). Recent advancements in developing transposons for insertional mutagenesis may also make this a feasible option in the future (Ivics *et al.,* 1997, Ivics *et al.,* this volume). Insertional screens offer the advantage over chemical mutagenesis that the affected gene can rapidly be cloned since the insertion serves as a molecular tag.

Using mutational screens in the zebrafish for the discovery of novel genes will be open to any imaginative assay that is feasible. In *Drosophila,* enhancer and suppressor screens have been very successful for identifying additional components in a particular developmental pathway. Numerous mutations in the zebrafish are homozygous viable and should be amenable to such screens. However, because the number of mutagenized genomes that can be screened in the zebrafish is much lower than those that can be reached with screens in *Drosophila,* these screens are likely to be limited to searching for dominant enhancers or suppressors. In the allele screen outlined earlier, we have attempted to identify dominant enhancers of the weak dominant effect of *yot*^{ty119} and identified several candidate mutations (Fredericus van Eeden, Pascal Haffter, Heike Schauerte, and Cornelia Fricke, unpublished). However, because of the variability of the dominant phenotype of *yot*^{ty119}, which is probably caused by the genetic heterogeneity of the zebrafish lines used in that screen, it seems difficult to finally demonstrate a specific genetic interaction. Isogenic lines of zebrafish will be a prerequisite to minimize the effects of variability in genetic backgrounds in such screens.

Recently, it has become possible to generate transgenic zebrafish lines showing tissue- and cell-specific expression of green fluorescent protein (Meng *et al.*, this volume). Such lines offer the exciting opportunity to screen for mutations affecting particular tissues in live embryos. The advantages over screens employing whole mount *in situ* hybridization or antibody staining as assays are that the work associated with fixing and staining procedures is reduced, screening on live embryos can be done continuously throughout development, and dominant screens can be performed on F_1s followed by raising of candidate individuals. The generation of new lines expressing tissue- and cell-specific reporter genes for screening purposes is likely to boost the field.

Appendix—Recipes and Media

E3 medium

1× concentration:
5m*M* NaCl
0.17m*M* KCl
0.33m*M* $CaCl_2$
0.33m*M* $MgSO_4$

Make up as a 60× stock solution. Add 10^{-5}% Methylene Blue before use.

MESAB

10 × stock solution:
0.2% 3-aminobenzoic acid ethyl ester, methanesulfonate salt
(Sigma A-5040)
Adjust to pH 7.5 with 1*M* Tris pH 9.0.

Acknowledgments

We thank Christiane Nüsslein-Volhard and Angeles Ribera for critical comments on the manuscript.

References

Ashburner, M. (1989). Mutation and mutagenesis. *In* "*Drosophila,* a Laboratory Handbook" (M. Ashburner, Ed.), pp 299–418. Cold Spring Harbor Press, Cold Spring Harbor, NY.

Baier, H., Klostermann, S., Trowe, T., Karlstrom, R. O., Nüsslein-Volhard, C., and Bonhoeffer, F. (1996). Genetic dissection of the retinotectal projection. *Development* **123,** 415–425.

Brenner, S. (1974). "The genetics of *Caenorhabditis elegans.*" *Genetics* **77,** 71–94.

Brockerhoff, S. E., Hurley, J. B., Janssen-Bienhold, U., Neuhauss, S. C., Driever, W., and Dowling, J. E. (1995). "A behavioral screen for isolating zebrafish mutants with visual system defects." *Proc. Natl. Acad. Sci. USA* **92,** 10545–10549.

Chakrabarti, S., Streisinger, G., Singer, F., and Walker, C. (1983). Frequency of gamma-ray induced specific locus and recessive lethal mutations in mature germ cells of the zebrafish, *Brachydanio rerio. Genetics* **103,** 109–124.

Chen, J.N., Haffter, P., Odenthal, J., Vogelsang, E., Brand, M., van Eeden, F. J. M., Furutani-Seiki, M., Granato, M., Hammerschmidt, M., Heisenberg, C.-P., Jiang, Y.-J., Kane, D. A., Kelsh, R. N., Mullins, M. C., and Nüsslein-Volhard, C. (1996). Mutations affecting the cardiovascular system and other internal organs in zebrafish. *Development* **123,** 293–302.

Driever, W., Solnica-Krezel, L., Schier, A. F., Neuhauss, S. C. F., Malicki, J., Stemple, D. L., Stainier, D. Y. R., Zwartkruis, F., Abdelilah, S., Rangini, Z., Belak, J., and Boggs, C. (1996). A genetic screen for mutations affecting embryogenesis in zebrafish. *Development* **123,** 37–46.

Driever, W., Stemple, D., Schier, A., and Solnica-Krezel, L. (1994). Zebrafish: Genetic tools for studying vertebrate development. *Trends Genet.* **5,** 152–159.

Gaiano, N., Amsterdam, A., Kawakami, K., Allende, M., Becker, T., and Hopkins, N. (1996). Insertional mutagenesis and rapid cloning of essential genes in zebrafish. *Nature* **383,** 829–832.

Granato, M., van Eeden, F. J. M., Schach, U., Trowe, T., Brand, M., Furutani-Seiki, M., Haffter, P., Hammerschmidt, M., Heisenberg, C.-P., Jiang, Y.-J., Kane, D. A., Kelsh, R. N., Mullins, M. C., Odenthal, J., and Nüsslein-Volhard, C. (1996). Genes controlling and mediating locomotion behaviour of the zebrafish embryo and larva. *Development* **123,** 399–413.

Grunwald, D. J., and Streisinger, G. (1992a). Induction of mutations in the zebrafish with ultraviolet light. *Genet. Res.* **59,** 93–101.

Grunwald, D. J., and Streisinger, G. (1992b). Induction of recessive lethal and specific locus mutations in the zebrafish with ethyl nitrosourea. *Genet. Res.* **59,** 103–116.

Haffter, P., Granato, M., Brand, M., Mullins, M. C., Hammerschmidt, M., Kane, D. A., Odenthal, J., van Eeden, F. J. M., Jiang, Y.-J., Heisenberg, C.-P., Kelsh, R. N., Furutani-Seiki, M., Vogelsang, E., Beuchle, D., Schach, U., Fabian, C., and Nüsslein-Volhard, C. (1996). The identification of genes with unique and essential functions in the development of the zebrafish, *Danio rerio. Development* **123,** 1–36.

Haffter, P., and Nüsslein-Volhard, C. (1996). Large scale genetics in a small vertebrate, the zebrafish. *Int. J. Dev. Biol.* **40,** 221–227.

Haffter, P., Odenthal, J., Mullins, M. C., Lin, S., Farrell, M. J., Vogelsang, E., Haas, F., Brand, M., van Eeden, F. J. M., Furutani-Seiki, M., Granato, M., Hammerschmidt, M., Heisenberg, C.-P., Jiang, Y.-J., Kane, D. A., Kelsh, R. N., Hopkins, N., and Nüsslein-Volhard, C. (1996). Mutations affecting pigmentation and shape of the adult zebrafish. *Dev. Genes Evolut.* **206,** 260–276.

Henion, P. D., Raible, D. W., Beattie, C. E., Stoesser, K. L., Weston, J. A., and Eisen, J. S. (1996). "Screen for mutations affecting development of zebrafish neural crest." *Dev. Genet.* **18,** 11–17.

Ivics, Z., Hackett, P. B., Plasterk, R. H., and Izsvák, Z. (1997). Molecular reconstruction of *Sleeping Beauty,* a *Tcl*-like transposon from fish, and Its transposition in human cells. *Cell* **91,** 501–510.

Jenkins, J. B. (1967). The induction of mosaic and complete *dumpy* mutants in *Drosophila melanogaster* with ethylmethanesulfonate. *Mutat. Res.* **4,** 90–92.

Jürgens, G., Wieschaus, E., and Nüsslein-Volhard, C. (1984). Mutations affecting the pattern of the larval cuticle in *Drosophila melanogaster* II. Zygotic loci on the third chromosome." *Roux's Arch. Dev. Biol.* **193,** 283–295.

Kane, D. A., Hammerschmidt, M., Mullins, M. C., Maischein, H.-M., Brand, M., van Eeden, F. J. M., Furutani-Seiki, M., Granato, M., Haffter, P., Heisenberg, C.-P., Jiang, Y.-J., Kelsh, R. N., Odenthal, J., Warga, R. M., and Nüsslein-Volhard, C. (1996). The zebrafish epiboly mutants. *Development* **123,** 47–55.

Karlstrom, R. O., Trowe, T., Klostermann, S., Baier, H., Brand, M., Crawford, A. D., Grunewald, B., Haffter, P., Hoffman, H., Meyer, S. U., Müller, B. K., Richter, S., van Eeden, F. J. M., Nüsslein-Volhard, C., and Bonhoeffer, F. (1996). Zebrafish mutations affecting retinotectal axon pathfinding. *Development* **123,** 427–438.

Kemphues, K. J., Priess, J. R., Morton, D. G., and Cheng, N. S. (1988). "Identification of genes required for cytoplasmic localization in early *C. elegans* embryos. *Cell* **52,** 311–320.

Knapik, E. W., Goodman, A., Atkinson, O. S., Roberts, C. T., Shiozawa, M., Sim, C. U., Weksler-Zangen, S., Trolliet, M. R., Futrell, C., Innes, B. A., Koike, G., McLaughlin, M. G., Pierre, L., Simon, J. S., Vilallonga, E., Roy, M., Chiang, P.-W., Fishman, M., Driever, W., and Jacob, H. J. (1996). A reference cross for zebrafish (*Danio rerio*) anchored with simple sequence length polymorphisms. *Development* **123,** 451–460.

Mayer, U., Ruiz, R. A. T., Berleth, T., Miséra, S., and Jürgens, G. (1991). "Mutations affecting body organization in the *Arabidopsis* embryo" *Nature* **353,** 402–407.

Moens, C. B., Yan, Y. L., Appel, B., Force, A. G., and Kimmel, C. B. (1996). *valentino*: A zebrafish gene required for normal hindbrain segmentation. *Development* **122,** 3981–3990.

Mullins, M. C., Hammerschmidt, M., Haffter, P., and Nüsslein-Volhard, C. (1994). Large-scale mutagenesis in the zebrafish: In search of genes controlling development in a vertebrate. *Curr. Biol.* **4,** 189–202.

Mullins, M. C., Hammerschmidt, M., Kane, D. A., Odenthal, J., Brand, M., Furutani-Seiki, M., Granato, M., Haffter, P., Heisenberg, C.-P., Jiang, Y.-J., Kelsh, R. N., van Eeden, F. J. M., and Nüsslein-Volhard, C. (1996). "Genes establishing dorsoventral pattern formation in the zebrafish embryo: the ventral determinants." *Development* **123,** 81–93.

Nüsslein-Volhard, C., and Wieschaus, E. (1980). Mutations affecting segment number and polarity in *Drosophila. Nature* **287,** 795–801.

Nüsslein-Volhard, C., Wieschaus, E., and Kluding, H. (1984). Mutations affecting the pattern of the larval cuticle in *Drosophila melanogaster* I. zygotic loci on the second chromosome. *Roux's Arch. Dev. Biol.* **193,** 267–282.

Odenthal, J., Haffter, P., Vogelsang, E., Brand, M., van Eeden, F. J. M., Furutani-Seiki, M., Granato, M., Hammerschmidt, M., Heisenberg, C.-P., Jiang, Y.-J., Kane, D. A., Kelsh, R. N., Mullins, M. C., Warga, R. M., Allende, M. L., Weinberg, E. S., and Nüsslein-Volhard, C. (1996). Mutations affecting the formation of the notochord in the zebrafish, *Danio rerio. Development* **123,** 103–115.

Piotrowski, T., Schilling, T. F., Brand, M., Jiang, Y.-J., Heisenberg, C.-P., Beuchle, D., Grandel, H., van Eeden, F. J. M., Furutani-Seiki, M., Granato, M., Haffter, P., Hammerschmidt, M., Kane, D. A., Kelsh, R. N., Mullins, M. C., Odenthal, J., Warga, R. M., and Nüsslein-Volhard, C. (1996). Jaw and branchial arch mutants in zebrafish II: Anterior arches and cartilage differentiation. *Development* **123,** 345–356.

Postlethwait, J. H., Johnson, S. L., Midson, C. N., Talbot, W. S., Gates, M., Ballinger, E. W., Africa, D., Andrews, R., Carl, T., Eisen, J. S., Horne, S., Kimmel, C. B., Hutchinson, M., Johnson, M., and Rodriguez, A. (1994). A genetic linkage map for the zebrafish. *Science* **264,** 699–703.

Riley, B. B., and Grunwald, D. J. (1995). "Efficient induction of point mutations allowing recovery of specific locus mutations in zebrafish." *Proc. Natl. Acad. Sci. U.S.A.* **92,** 5997–6001.

Schauerte, H. E., van Eeden, F. J. M., Fricke, C., Odenthal, J., Strähle, U., and Haffter, P. (1997). *Sonic hedgehog* is not required for the induction of medical floor plate cells in the zebrafish. *Development* **125,** 2983–2993.

Singer, B., and Grunberger, D. (1983). Molecular Biology of Mutagens and Carcinogens. Plenum Press, New York.

Solnica-Krezel, L., Schier, A. F., and Driever, W. (1994). Efficient recovery of ENU-induced mutations from the zebrafish germline. *Genetics* **136,** 1401–1420.

Solnica-Krezel, L., Stemple, D. L., Mountcastle-Shah, E., Rangini, Z., Neuhauss, S. C. F., Malicki, J., Schier, A. F., Stainier, D. Y. R., Zwartkruis, F., Abdelilah, S., and Driever, W. (1996). Mutations affecting cell fates and cellular rearrangements during gastrulation in zebrafish. *Development* **123,** 67–80.

Stainier, D. Y. R., Fouquet, B., Chen, J.-N., Warren, K. S., Weinstein, B. M., Meiler, S. E., Mohideen, M.-A. P. K., Neuhauss, S. C. F., Solnica-Krezel, L., Schier, A. F., Zwartkruis, F., Stemple, D. L., Malicki, J., Driever, W., and Fishman, M. C. (1996). "Mutations affecting the formation and function of the cardiovascular system in the zebrafish embryo." *Development* **123,** 285–292.

Streisinger, G., Singer, F., Walker, G., Knauber, D., and Dower, N. (1986). Segregation analysis and gene-centromere distances in zebrafish. *Genetics* **112,** 311–319.

Streisinger, G., Walker, C., Dower, N., Knauber, D., and Singer, F. (1981). Production of clones of homozygous diploid zebra fish (*Brachydanio rerio*). *Nature* **291,** 293–296.

Trowe, T., Klostermann, S., Baier, H., Granato, M., Crawford, A. D., Grunewald, B., Hoffmann, H., Karlstrom, R. O., Meyer, S. U., Richter, S., Nüsslein-Volhard, C., and Bonhoeffer, F. (1996). "Mutations disrupting the ordering and topographic mapping of axons in the retinotectal projection of the zebrafish, *Danio rerio.*" *Development* **123,** 439–450.

van Eeden, F. J. M., Granato, M., Schach, U., Brand, M., Furutani-Seiki, M., Haffter, P., Hammerschmidt, M., Heisenberg, C.-P., Jiang, Y.-J., Kane, D. A., Kelsh, R. N., Mullins, M. C., Odenthal, J., Warga, R. M., Allende, M. L., Weinberg, E. S., and Nüsslein-Volhard, C. (1996). Mutations affecting somite formation and patterning in the zebrafish *Danio rerio. Development* **123,** 153–164.

Walker, C., and Streisinger, G. (1983). "Induction of mutations by gamma-rays in pregonial germ cells of zebrafish embryos." *Genetics* **103,** 125–136.

Whitfield, T. T., Granato, M., van Eeden, F. J. M., Schach, U., Brand, M., Furutani-Seiki, M., Haffter, P., Hammerschmidt, M., Heisenberg, C.-P., Jiang, Y.-J., Kane, D. A., Kelsh, R. N., Mullins, M. C., Odenthal, J., and Nüsslein-Volhard, C. (1996). Mutations affecting development of the zebrafish inner ear and lateral line. *Development* **123,** 241–254.

Wieschaus, E., Nüsslein-Volhard, C., and Jürgens, G. (1984). Mutations affecting the pattern of the larval cuticle in *Drosophila melanogaster* III. Zygotic loci on the X-chromosome and fourth chromosome. *Roux's Arch. Dev. Biol.* **193,** 296–307.

Zhang, J., Talbot, W. S., and Schier, A. F. (1998). Positional cloning identifies zebrafish *one-eyed pinhead* as a permissive EGF-related ligand required during gastrulation. *Cell* **92,** 241–251.

CHAPTER 3

Haploid Screens and Gamma-Ray Mutagenesis

Charline Walker

Institute of Neuroscience
University of Oregon
Eugene, Oregon 97403-1254

METHODS IN CELL BIOLOGY, VOL. 60

0091-679X/99 $30.00

I. Introduction

Mutational analysis has been a productive way to dissect, at the cellular and molecular levels, pathways and key events of development. A number of large-scale mutagenesis screens have been conducted to identify developmental mutations in zebrafish (Haffter *et al.*, 1996; Driever *et al.*, 1996). In addition, smaller screens are ongoing in many other laboratories (Kimmel, 1989; Riley and Grunwald, 1995). Regardless of which type of mutagen is used, it is important to screen as efficiently as possible in terms of time, space, and the number of fish used. In this chapter, I describe the use of haploids to screen for early developmental mutations in mutagenized zebrafish, with an emphasis on the use of gamma-rays as a mutagen. Screening haploids allow rapid identification of mutation-bearing females in a parental P or an F_1 screen, avoiding the necessity of raising several generations of fish stocks prior to screening (Fig. 1).

Gynogenetic haploid embryos are produced when eggs are fertilized by UV-irradiated sperm; the resulting haploid embryos develop solely from maternal genetic information. Although the UV-irradiated sperm provide no male genetic contribution to the embryo, they are necessary to activate embryonic development. Because recessive mutations are no longer masked by diploidy, the use of haploids is a very powerful tool to identify newly induced mutations in a mutagenesis screen. A relatively lethal-free genetic background is particularly important for a haploid screen.

Haploid embryos live about 4 days postfertilization, a stage at which their diploid counterparts are free swimming and feeding larvae; this allows screening for mutations that affect a large number of developmental processes. However, because haploid development is not completely normal, there are limitations to their use in certain types of screens.

Developmental mutations uncovered in a haploid screen are then characterized in diploids. It is reassuring that, in the majority of cases, the homozygous diploid phenotype is very similar to the original haploid phenotype.

Gamma rays are an effective mutagen in producing early developmental mutations (e.g., Grunwald *et al.*, 1988; Halpern *et al.*, 1993; Fritz *et al.*, 1996; Talbot *et al.*, 1998). They generate a whole range of lesions from just a few base pair changes to chromosomal rearrangements and large deletions (Russell and Russell, 1959; Chakrabarti *et al.*, 1983; Kubota *et al.*, 1992; Fritz *et al.*, 1996; Talbot *et al.*, 1998). Here I present results from mutagenesis studies comparing gamma irradiation at different stages: the 1000 cell blastula, mature sperm, and spermatogonia, with the aim of producing recoverable mutations in the germline of zebrafish.

II. Production of Haploid Embryos

Zebrafish eggs can be fertilized *in vitro,* which makes it possible to manipulate the ploidy of the resulting embryos. One can produce gynogenetic haploid,

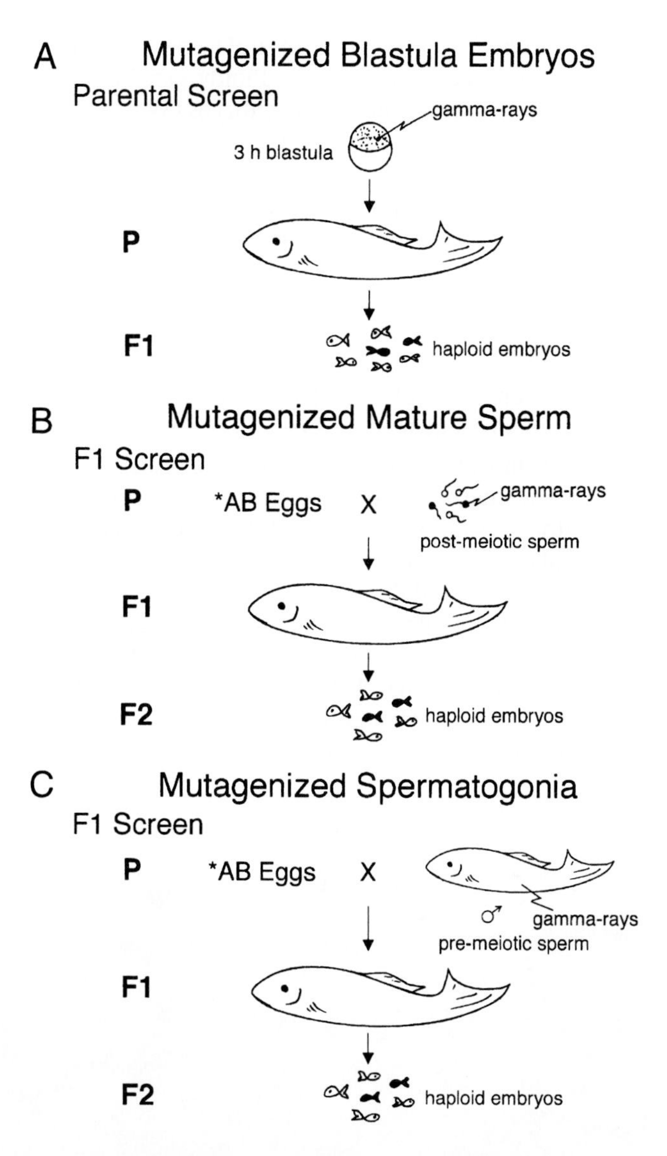

Fig. 1 Haploid screens and gamma-ray mutagenesis strategies. In (A), the primordial germ cells are targeted in the blastula stage embryo, mutagenized fish are raised to adulthood, and the haploid progeny of mutagenized females are screened for mutations. This parental screen is contrasted to the F_1 screens shown in (B) and (C). In (B), mature sperm is collected in a small vial, mutagenized, and used to fertilize normal eggs. Haploid progeny of the resulting F_1 females are screened for mutations. In (C), adult males are mutagenized, sperm from mutagenized spermatogonia are used to produce an F_1 generation, and haploid progeny from female F_1 adults are screened for mutations. The mutagen used in each case is gamma-ray irradiation, but any mutagen can be used in these screening strategies.

gynogenetic diploid, triploid, and tetraploid embryos using techniques developed by Streisinger *et al.* (1981). Ploidy can be verified by preparing chromosome spreads from 24-hour whole embryos (Westerfield, 1995) as shown in Fig. 2, which compares a haploid to a diploid chromosome spread. Androgenetic haploids can also be produced by using normal sperm to fertilize eggs irradiated with large doses of gamma rays, X rays, or UV light. Gynogenetic and androgenetic haploids look virtually the same (Corley-Smith *et al.,* 1996; Corley-Smith *et al.,* Ch. 5, Volume 59).

A. *In Vitro* Fertilization

1. Females are separated from males late in the afternoon before the day of the experiment so that they will not lay their eggs early.

2. On the morning of the experiment, sperm is collected from males. Males are anesthetized in tricaine (0.2 mg/ml 3-aminobenzoic acid ethyl ester, methanesulfonate salt, brought to pH 7 with 0.4 mg/ml Na_2HPO_4) until rapid gill movements slow down, rinsed, and placed upside down on a damp sponge. The genital region is blotted dry with a tissue. After parting the anal fins, sperm is extruded by gentle pressure and stroking on the sides of the abdomen with smooth forceps and collected in a 20 μl microcapillary. The amount of sperm collected per male varies (on average 1–5 μl); sperm is collected and pooled in Hank's saline (Westerfield, 1995) until the solution is cloudy. Sperm held in Hank's on ice (4°C) maintains the capacity to fertilize eggs efficiently for at least 90 minutes.

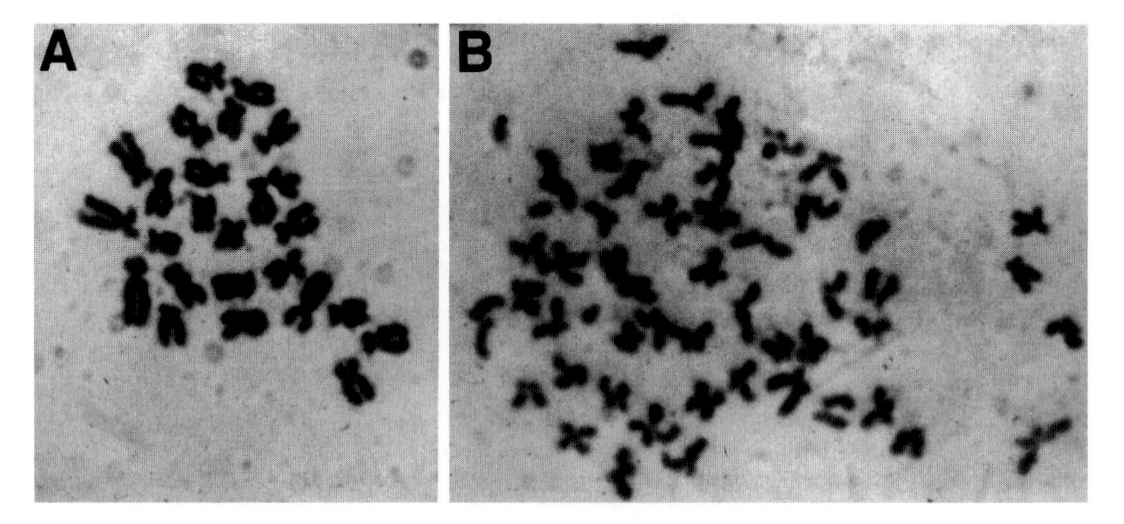

Fig. 2 Chromosome spreads from 24-hr whole embryos. (A) haploid spread, 1n = 25 chromosomes; (B) diploid spread, 2n = 50 chromosomes.

Usually there is a great excess of sperm. At this point, sperm can be inactivated by UV-irradiation as described later.

3. Starting at dawn in the fish light cycle, each female is similarly anesthetized, rinsed, blotted on a paper towel, and laid in a 35-mm plastic dish. Eggs are manually extruded or "squeezed" from the female by using one finger to brace the back, and gently pressing the belly with a finger of the other hand. Eggs should come out easily, and after using a spatula to move the eggs away from the female, she is placed into a recovery container. Care should be taken to avoid excessive wetness, because when water contacts an egg, the chorion swells and the egg can no longer be fertilized.

4. The eggs are immediately fertilized by adding about 25 μl sperm in Hank's and 0.75 ml water. After about 1 minute, more water is added to fill the dish three-quarters full. Sperm in Hank's saline remain quiescent, but as soon as the saline is diluted with water, the sperm are activated and able to fertilize eggs, remaining active for only about 1 minute. Thus, all the eggs in a clutch are fertilized essentially synchronously. Sperm in Hank's can be gently mixed with eggs before adding water. In this case, the time of fertilization is when water is added. Premature contact of sperm or eggs with water should be avoided.

Besides the successful females which give good eggs, other females may not be prepared to release eggs, and they should not be pressed too hard. Still other females may give eggs which have already broken down (bad eggs), and these eggs are not worth fertilizing.

Both male and female fish are significantly stressed when handled in an *in vitro* fertilization experiment. This can lead to illness or poor production of gametes. We find that by allowing males a minimum of 3 weeks and females a minimum of 4 weeks to recover before using them again, the fish are more likely to continue producing eggs and sperm. They can be set up for natural matings during this waiting period.

B. UV-Irradiation to Make Haploid Embryos

UV-irradiation cross-links DNA. When sperm is UV-irradiated, the DNA is destroyed although the sperm retains the ability to activate the egg. The haploid embryo, resulting from an egg fertilized by UV-irradiated sperm, has no genetic contribution from the male. A 0.5-ml aliquot of freshly collected sperm in Hank's saline is spread in a watchglass that sits on top of ice (ice should never touch the sperm) and is exposed for 2 minutes to UV light from an 18 in. (43 cm) Sylvania germicidal lamp (254 nm) at a distance of 15 in. (38 cm) from the watchglass. After UV-irradiation, the sperm is put into a clean vial and kept on ice. UV-sperm is used just like normal sperm to fertilize freshly squeezed eggs.

Care must be taken to shield oneself from UV by wearing safety glasses and latex gloves. Sperm should not be too concentrated, and solid material such as feces should be removed, to prevent shielding the sperm from the UV source.

III. Development of Haploid Embryos

Haploid embryos live for about 4 days post fertilization. The schedule of early development is approximately the same for haploids and diploids. Interestingly, whereas early cleavage cycles are the same, haploids enter midblastula transition (when cells slow their division rates and zygotic transcription begins) one cleavage cell cycle later than diploids (Kane and Kimmel, 1993). Thereafter, haploids contain more numerous and smaller cells than diploids (especially evident for pigment cells).

The smaller cells, in addition to gene haplo-insufficiency, probably contribute significantly to the inviable haploid syndrome (Fig. 3) which is fairly consistent from embryo to embryo: the haploid body is short and stocky, yet has the appropriate number of muscle segments. Morphogenesis of the brain is abnormal, the brain looks less clearly defined and has more cell death than in diploids. From dorsal view, the brain folds or kinks from side to side; possibly this defect is related to the general body shortening. The eye is incompletely formed at the choroid fissure; the otic vesicles are always present but may be duplicated on one or both sides of the body. Blood cells seem too large for the blood vessels, and most haploid embryos have circulation problems. As development proceeds, certain parts of the body such as the pericardial cavity start swelling. Eventually, there is generalized edema, and the embryos die. However, as you can see in Fig. 3, all the major body parts can be identified in haploids, including features such as the notochord, floorplate, heart, and blood.

Another feature of normal haploid clutches is that in each batch there are a variable number of defective embryos, easily distinguished from the well-developed haploids (Fig. 4). Because deleterious genes are unmasked in haploids, the genetic background of the fish influences the frequency of defective embryos. Egg quality has an effect as well, and if the mother is unhealthy, the haploids she produces do not develop well. Such defective embryos are easily recognizable, and when screening a batch of haploids, it is useful to sort embryos into groups based on how well formed they are. We sort embryos into three categories, although the severity of defects occurs along a continum. Table I shows, in

Fig. 3 Comparison of day 1 and day 2 haploid and diploid embryos, lateral views. Anterior is to the left, and dorsal is to the top. (A) day 1 haploid; (B) day 1 diploid. Details of trunk region in (C) day 1 haploid and (D) day 1 diploid. (E) day 2 haploid; (F) day 2 diploid. Head detail of (G) day 2 haploid and (H) day 2 diploid. cc, central canal; fp, floorplate; n, notochord; y, yolk extension; e, ear; m, melanocyte; hc, heart cavity; hgc, hatching gland cells.

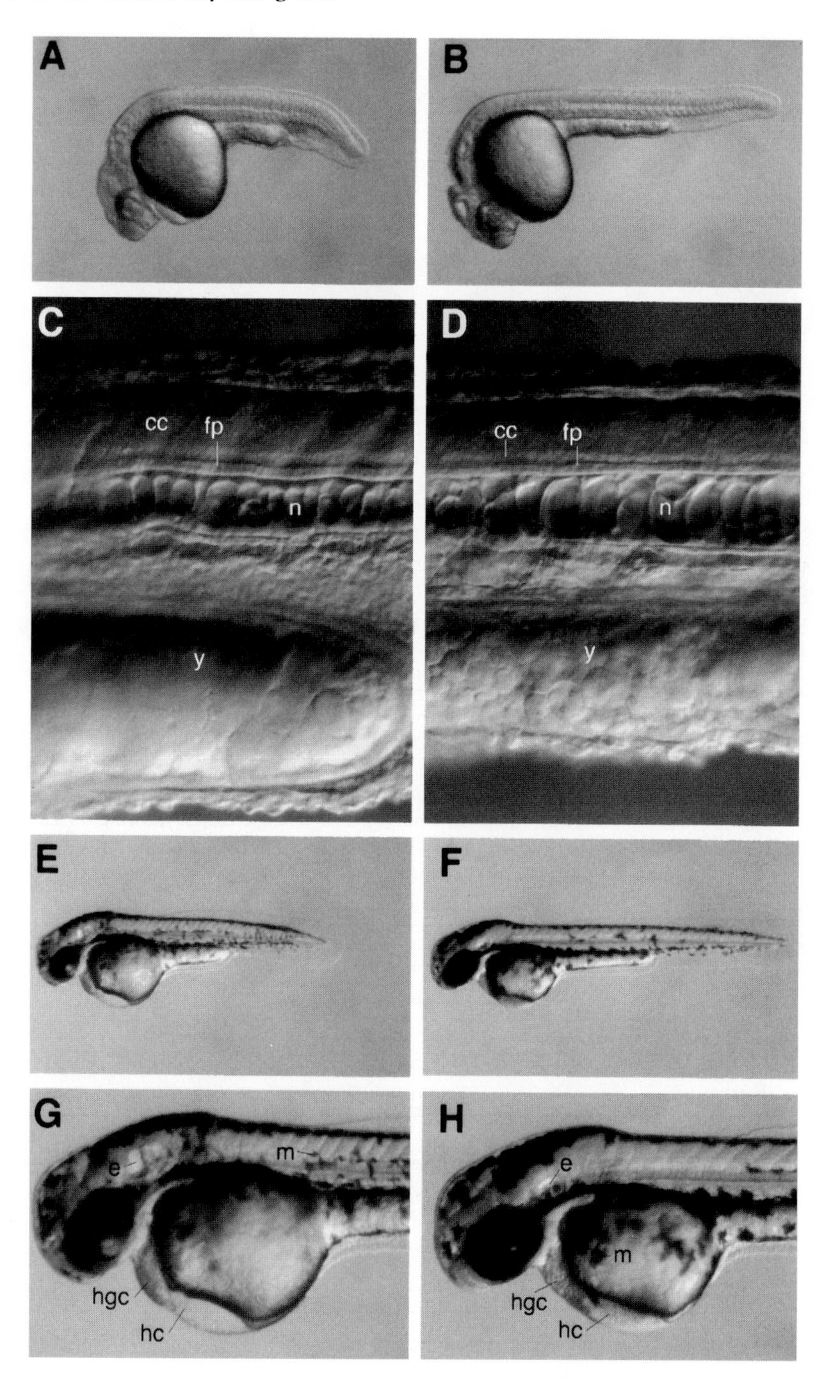
A
B
C
cc
fp
n
y
D
cc
fp
n
y
E
F
G
e
m
hgc
hc
H
e
m
hgc
hc

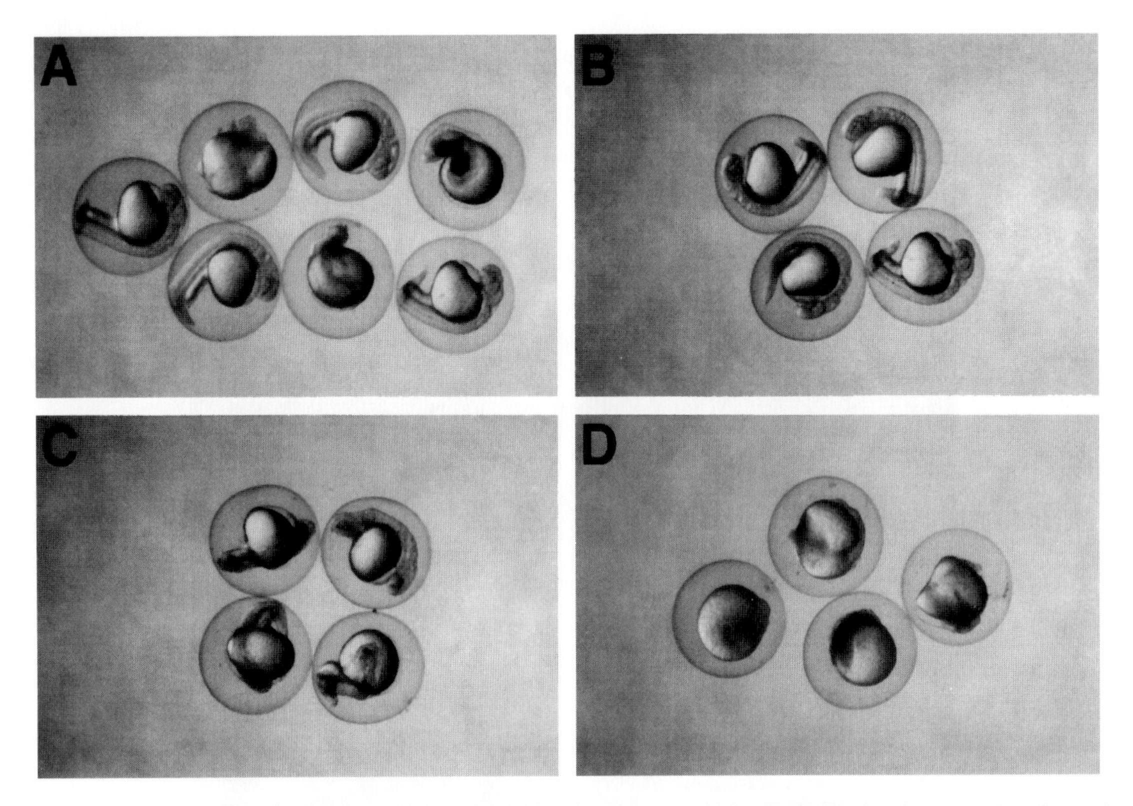

Fig. 4 Examples of day 1 haploid embryos from a single clutch showing the categories of "good-looking" and "defective" embryos typical of haploids. The proportion of **A** embryos in the entire clutch is higher than shown. (A) unsorted haploids; (B) sorted **A** embryos; (C) sorted **B** embryos; (D) sorted **C** embryos.

Table I
Clonal and Selected Wild-Type Strains Have More Than 50% A Haploid Embryos

Strain	Number of females	Average clutch size	Haploid			
			% A embryos	% B embryos	% C embryos	% dead
Clonal[a]	14	56	57	6	7	30
AB[b]	99	212	53	12	13	22
*AB Founder[c]	19	219	69	6	12	13

[a] The clonal strains are homozygous and embryonic lethal-free (see Section IV).
[b] The AB strain is the semiselected Oregon wild-type line (see text).
[c] The *AB strain is the gynogenetically derived wild-type line currently being used in the Oregon mutagenesis screens.

various backgrounds, the proportion of embryos belonging in each category: **A** or normal looking embryos; **B** embryos, which have a twisted, short axis but still have a recognizable head and tail; and **C** embryos, which are bits of tissue on a live yolk. Figure 4 shows examples of each category.

During a morphological screen, it is sometimes difficult to decide whether a few similarly defective embryos are mutant or not, but the variability of the haploid defective embryos helps to distinguish them from real mutant embryos, which usually have a characteristic phenotype.

Within their limitations, haploids are consistent enough to detect mutations and have been very useful in finding mutations which affect early development. Haploids may not be appropriate for some screens (see Section VI); for example, a screen for phenotypes later than day 3 would have to be done in diploids. However, haploid analysis has the advantage that mutant progeny are represented in the same proportion as is the mutation in the gametes of the mother.

IV. Genetic Background

A consistent background free of lethal mutations is advantageous for genetic screens; there is less chance of interactions with pre-existing mutations, and one can be confident that a new mutation has been induced by the mutagen. A good genetic background is particularly important for a haploid mutagenesis screen because it is beneficial to have a high proportion of **A** embryos to screen.

A. Clonal Lines

In the 1970s, George Streisinger produced lethal-free clonal lines of fish with the idea that the ideal genetic background would be one in which all the fish were identical to each other prior to mutagenesis. These lines were gynogenetically derived first by making fish homozygous diploid with heat shock (HS) which inhibits the first mitotic division, then by making identical clonal progeny from a single homozygous female with early pressure (EP) which inhibits the second meiotic division of the eggs (Streisinger *et al.,* 1981).

To produce homozygous fish, eggs are fertilized with UV-irradiated sperm to eliminate the male genetic contribution. During the first mitosis, the maternal chromosomes replicate. Heat shock prevents the separation of the replicated chromosomes and prevents the first cell division, establishing diploidy to the embryo. Subsequent replications and cell divisions are allowed to proceed (method described in Corley-Smith *et al.,* Ch. 5, Volume 59).

Fertilization triggers the start of the second meiotic division of the egg. A diploid number of condensed chromosomes align on a spindle in preparation for the meiotic reduction division which normally results in a haploid pronucleus and a polar body which is extruded. High hydrostatic pressure applied at this time breaks down the spindle and keeps the full diploid complement of chromosomes together. With no further intrusion, a high percentage of embryos (60–90%) go through normal development. Specifically, eggs are fertilized with UV-irradiated sperm then immediately put in a small glass vial capped with a rubber sheet. At 1.4 minutes postfertilization, 8000 lb/in.^{2} is applied by a French press. At 6 minutes postfertilization, pressure is relieved, and the eggs are allowed to continue developing without disturbance for 3–4 hours.

The first batches of homozygous fish produced from the wild-type strain were of very poor quality, but by crossing the best homozygous males and females and performing several successive generations of homozygosing, clonal lines were produced which were nearly as good as wild-type strains in terms of viability and fecundity.

Gamma-ray mutagenesis screens were conducted with clonal lines for a number of years, but as the scope of the screen increased in size, the lack of vigor of these fish proved to be a problem; the clutch sizes were too small (less than 100 eggs), and the quality of the eggs was often poor, resulting in too many defective embryos.

B. AB Wild Type

Turning to the nonhomozygous Oregon wild-type strain, AB, from which the clones were derived, we noticed that some females produced beautiful haploid embryos. This line had already gone through a preselection process to enrich the stock with females that produced more than 50% **A** haploid embryos by discarding females that produced poor-quality haploids. Males were unselected. Once established, the AB strain was kept closed for over 50 generations; however, at each generation, gametes from many individuals (20–30 females and 60 males) were mixed in an overt attempt to maintain heterozygosity within the closed line. Clutch sizes from AB females are over double that of clonal females (Table I).

C. *AB-Derived Wild Type

Unlike the homozygotes, the AB strain was not lethal-free, and the haploid screen revealed a number of background mutations with visible embryonic phenotypes which interfered with the screen. In 1991–1992, by a single round of stringent selection we produced a new line, called *AB, from gynogenetic diploid progeny of 21 AB founder females. Haploid embryos from 106 AB females were put through the entire 3-day morphological mutagenesis screen (see Section V). Out of those, the selected group of 21 females produced haploid embryos which were normal at all screening points and which had 50% or more **A** embryos at 24 hours. Gynogenetic diploid progeny were made for each of these females by the EP technique described earlier. The EP progeny of the 21 selected females were male and female. Crossing all possible combinations of males and females together established the *AB line. Although this line has a limited gene pool, care is taken to keep it as diverse as possible, so like the AB line, 20–30 females are crossed to about 60 males each generation to conserve the hybrid vigor within the population. Haploids from the *AB line have an improved genetic background as seen by the higher percentage of **A** embryos in Table I.

The *AB strain is not homozygous nor is it necessarily lethal-free; it has been selected to produce good haploids for the first 3 days of development.

V. The Haploid Screen

Our haploid screen covers the first 3 days of development at 28.5°C. We are interested in mutations affecting early development, especially the patterning of embryonic mesoderm and nervous system.

Eggs are squeezed from P or F_1 potentially mutation-bearing females and fertilized *in vitro* with UV-irradiated sperm to make haploids (Fig. 1). Each female is kept in a 1-l individual container until the screen is completed. After a few hours, fertile eggs are sorted into embryo medium (Westerfield, 1995) containing 100,000 units penicillin and 100 mg streptomycin per liter, at about 100 embryos per 100-mm plastic Petri dish. (We do not use antibiotics to raise stocks of fish, but they help the haploid embryos survive the entire screening procedure). At 24 hours postfertilization, dead embryos are counted and removed. The remaining embryos are taken through a screening regimen which is described in detail later. Females carrying potential mutations of interest are then outcrossed with wild-type *AB males for recovery of the mutation.

Our approach is to combine several screening methods to increase the probability of finding an interesting mutation. The morphological screen and the early qualitative behavioral screen pick up *de novo* mutations; the RNA *in situ* hybridization screen and the PCR-based screen find mutations affecting known genes.

A. Morphology Screen

The morphological screen is conducted using a dissecting microscope. Haploid embryos at four developmental stages of the morphological screen, tailbud, day 1, day 2, and day 3, are shown in Fig. 5.

1. At the tailbud stage (10–12 hours), the cells in the embryo are just completing gastrulation movements (Kimmel *et al.,* 1995). Mutations affecting epiboly, axis, and cell cycle can be distinguished at this time.
2. On day 1 (24–30 hours, or the pharyngula stage), the embryo has formed many of its primary organs. At this time, we screen for morphological features including: body shape, eyes, ears, hatching gland, notochord, floorplate, somite shape and approximate number, and cell death.

Screening is made easier by sorting the **A, B,** and **C** embryos into separate groups so that your eye can pick out patterns. Mutants usually have a characteris-

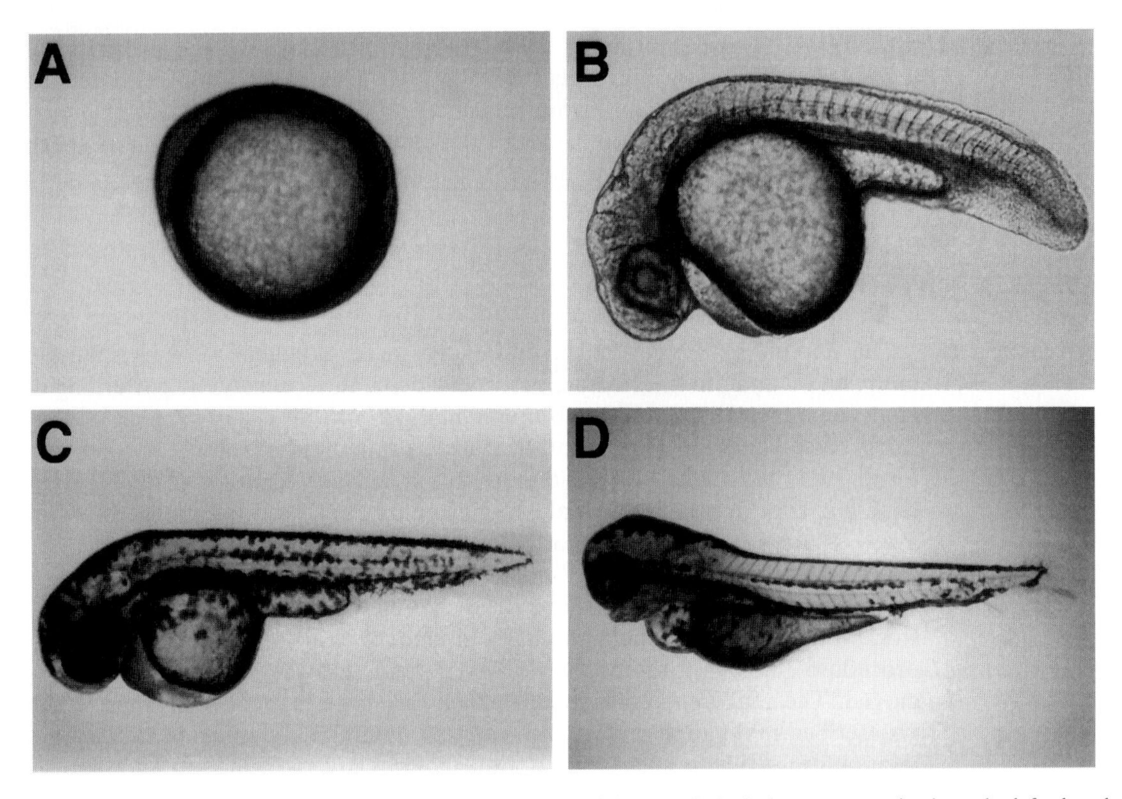

Fig. 5 Haploid embryos at key stages of the morphological screen; anterior is to the left, dorsal is to the top. (A) tailbud stage; (B) day 1; (C) day 2; (D) day 3.

tic phenotype, and this distinct "look" helps to distinguish mutants from variably defective haploids (Fig. 4).

3. At day 2 (48–54 hours), the embryos are screened for heartbeat, blood, pigment cells (melanophores and xanthophores), pectoral fins, new cell death, and body shape.

4. By day 3 (72–78 hours), some of the haploid embryos have begun to swell, but we can screen the good-looking haploids for jaw defects, the presence of Mauthner cell axons, and the presence of neuromasts of the lateral line; the two latter screens are conducted with a compound microscope.

B. Behavioral Screen

Although haploids never perform normal swimming movements, we can screen at earlier times for motility and sensory defects. The following early behavioral screens test motility in 24-hour embryos and sensory reflexes in 30-hour embryos, the tests are noninvasive and are conducted using a dissecting microscope.

Haploid embryos, like diploids, produce spontaneous flexing movements at 24 hours of development, which later cease (Grunwald *et al.,* 1988). It is easy to look systematically through a batch of 24-hour **A** embryos for ones that do not flex. This lack of behavior could be caused by problems with neurons, muscles, or neuromuscular junctions. An example of a mutant which failed this test is the recessive lethal mutation *fibrils unbundled* (*fub1*^{b45}) (Felsenfeld *et al.,* 1990). Embryos homozygous for the *fub1* mutation have unstriated skeletal muscles. The myofibrils in somitic myotubes are severely disorganized; the major contractile proteins are present, but not properly aligned. These fish barely move during the "wiggle" test.

A test of the early sensory system can be performed at about 30 hours. Haploids, like diploids, respond reflexively with a sharp flex of the body to light touches on the head or body with a tiny probe of nylon monofilament. This screen detects mutants that block or alter development of the early functional sensory-motor pathway. We did not identify any mutants using this screen on several thousand embryos. Nevertheless, such a screen holds promise because the hyperactive mutants *techno trousers* (*tnt*) and *roller coaster* (*roc*) were identified in another touch test in the Germany diploid mutagenesis screen (Granato *et al.,* 1996).

C. *In Situ* Screen

RNA *in situ* hybridization can be used efficiently to screen for mutations that affect the normal expression pattern of known genes; these mutations may be in the gene itself or in an upstream gene. 20 embryos are fixed in 4% paraformaldehyde at each of two developmental stages, 10 and 22 hours. The embryos are processed essentially following the *in situ* hybridization protocol of Oxtoby and Jowett (1993).

For the tailbud screen, embryos are fixed at late gastrula, about 10 hours, and are incubated with a mix containing riboprobes for a number of genes whose expression patterns do not overlap or interfere with each other. At present we examine the expression pattern of the following genes which mark the following structures: *hatching gland1* (*hgg1*) in the prechordal plate mesoderm, *floating head* (*flh*) in the epiphysis and notochord, *valentino* (*val*) in hindbrain rhombomeres 5 and 6, *forkhead6* (*fkh6*) in presumptive neural crest, *hairy-enhancer split-related1* (*her1*) in trunk and tailbud, and *paxb* in the presumptive midbrain–hindbrain junction (Fig. 6, *hggl* and *paxb* expression not shown).

A second set of embryos, fixed at 22 hours, are screened with the following probes: *sonic hedgehog* (*shh*) in floorplate and forebrain, *myoD* in the somites, *distalless2* (*dlx2*) in developing pharyngeal arches 1, 2, and 3, *krox20* in hindbrain rhombomeres 3 and 5, *engrailed3* (*eng3*) at the midbrain–hindbrain border, and *lim5* in the diencephalon (Fig. 6).

The mutation *valentino*^{b361} was identified in the 22-hour screen because it alters the *krox20* expression pattern. Normally *krox20* is expressed in two very

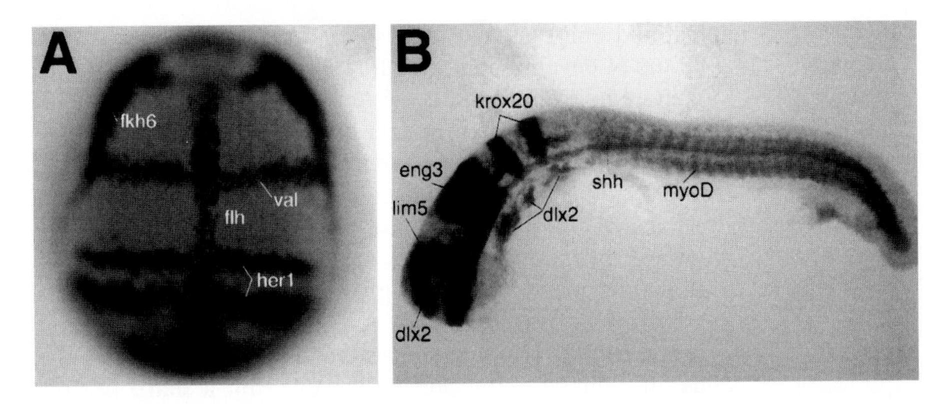

Fig. 6 In the haploid *in situ* hybridization screen, a cocktail of riboprobes reveals patterns of gene expression at late gastrula stage (A), and at 22 hours (B) (photos courtesy of Sharon Amacher).

sharp bands of cells in the position of rhombomeres 3 and 5 in the hindbrain. In *val,*^{b361} the *krox20* expression in r5 is reduced whereas expression in r3 is normal (Moens *et al.,* 1996). In this case, a mutation in an upstream gene, *valentino,* modified the expression of the known gene, *krox20.* The *valentino* gene was cloned molecularly by homology to mouse *kreisler* (Moens *et al.,* 1998), and a probe for that gene is now used in the early screen to mark the r5 and 6 territory.

D. PCR Screen

A polymerase chain reaction-based (multiplex PCR) assay is used to screen individual haploid embryos for deletions of known gene sequences (Fritz *et al.,* 1996).

DNA is isolated from approximately ten haploid embryos per clutch at 24 hours. Combinations of the following PCR primers are used to amplify specific DNA sequences from each embryo: *axial, dlx2, dlx3, elrC, eng1, eng2, goosecoid, hoxb4, hoxb5, hoxb6, msxA, msxB, msxD, noggin, tenascinC, shh, hoxc3.* An F_1 female which produces a clutch of embryos where one or more embryos are missing a specific PCR amplification product is likely to bear a mutation in that gene. The mutation *T*(*msxB*)^{b220} was identified because two of ten haploid progeny from the original F_1 female were missing the *msxB* PCR product (Fritz *et al.,* 1996). Further analysis showed that the mutation was the result of a translocation and that other genes in addition to *msxB* were missing. Figure 7 shows another example where three of nine haploid progeny were missing the *goosecoid* PCR product.

E. Haploid versus Diploid Phenotype

Usually the phenotype of a mutant in a haploid is strikingly similar to the homozygous diploid phenotype. The tail is always shorter, and the phenotype is more severe in a haploid; two such mutants are shown in Figure 8 C–F.

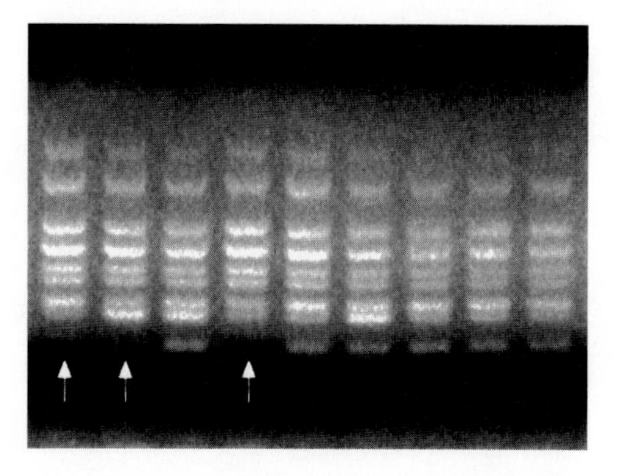

Fig. 7 The PCR-based screen uses a specific mixture of compatible primers to amplify gDNA sequences of randomly selected individual haploid embryos of potential mutation-bearing females. Here, 3 out of 9 haploid embryos from a single female are missing the bottom *goosecoid* band. PCR products used in this reaction, from the top band down, are: *krox20, msxD, hoxb4, msxA, eng2, hoxc3, dlx3, dlx2, and goosecoid (gsc)* (photo courtesy of Andreas Fritz).

Exceptions to this rule are the mutations producing neural degeneration (ned) haploid phenotypes. Sometimes these mutants can look very different as diploids, usually being less severe, and they may not be necrotic at all. An example of a *ned* mutation as a haploid and a diploid is *b225* phenotype 1 (Fig. 8 A,B).

VI. Limitations of Haploids

Because of irregularities caused by the haploid syndrome discussed earlier, it is not possible to screen for all features in haploids. Because, for example, haploids may have more than one otic vesicles on each side, and otoliths may be absent, they can be screened for mutations that block ear development completely, but not for mutations that perturb subtleties of how it develops.

In diploids, a single pair of Mauthner neurons develop in the hindbrain, extending their axons into the spinal cord where they can be recognized uniquely in live embryos by their large diameters (Kimmel *et al.,* 1978). With a background of less than 1%, well-formed haploids always have Mauthner neurons that project into the spinal cord, as we determined from Nomarski screens involving 4600 embryos. However, the axon diameters are smaller than normal and axons run along an incorrect spinal pathway in more than 10% of well-formed haploids, and in about the same fraction there are one or more extra Mauthner cells. Hence, one can use haploids to identify a mutation that deletes the Mauthner neuron or that changes the course of the Mauthner axon so that it does not enter the spinal cord, but not for mutations that duplicate the cells or change the axonal pathway within the spinal cord.

In haploids, morphogenesis of the brain is abnormal as previously described. Identifying details of brain structure are problematic in the morphology screen;

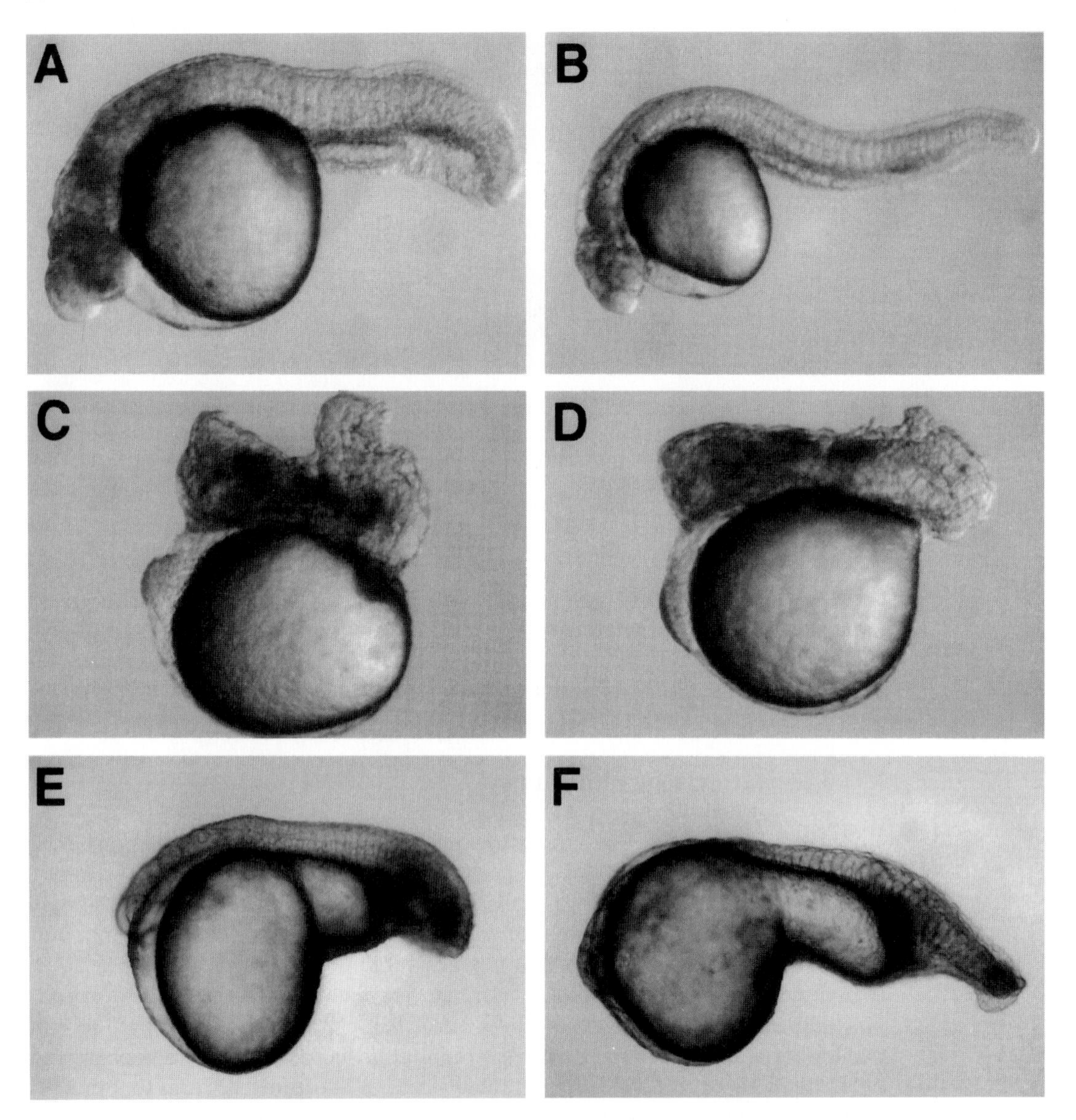

Fig. 8 Three mutants recovered from gamma-ray mutagenesis screens, as a haploid on the left and a diploid on the right. The mutation *b225* is likely to be a translocation, and there are two phenotypes in a clutch of embryos from a mutation-bearing female. (A,B) *b225* phenotype 1 would be considered a neural degeneration (ned) phenotype (also see Table II). (C,D) *b225* phenotype 2 would be classified in the "tail" category although there is also considerable cell death. (E,F) *cerebum*^{b305} would also be put in the "tail" category (*cerebum*^{b305} photos courtesy of Marnie Halpern).

however, the expression domains of markers such as *krox20* or *engrailed3* in the *in situ* screen are quite consistent in haploids and allow for screening developing regions of the haploid brain. In fact, the *valentino* mutation, which alters *krox20* expression in rhombomere 5, also has a morphological defect which a trained eye can pick out in diploid embryos, but it is unlikely that this mutation would have been found in the haploid morphology screen. This is an example of how a subtle phenotype can be found in a haploid, using the appropriate screen.

Swim bladders do not inflate in haploids, but many organ systems such as the heart, liver, and gut can be screened as being present or absent before the embryo dies; again, subtle changes will be missed.

In summary, haploid screens do have limitations; ingenious screens can circumvent this to some degree, and haploid screens are successfully producing interesting developmental mutations.

VII. Mosaicism in F_1 Screens

Given the previously stated limitations, the haploid approach has the advantage that mutations are represented in a larger proportion of a clutch of embryos than diploids. If a female, heterozygous for a Mendelian recessive mutation, is crossed to a male heterozygous for the same mutation, 25% of the progeny are homozygous mutant. In contrast, haploid progeny of the same female would be 50% mutant.

The use of haploids is particularly beneficial in an F_1 screen, compared to F_2 and F_3 family screens where one or two further generations are required (Mullins *et al.,* 1994; Solnica-Krezel *et al.,* 1994). Often the germline of the mutant-bearing female in an F_1 screen is mosaic for the mutation, whereas in the F_2 and F_3 family screen, mutation-bearing fish are fully heterozygous. This mosaicism can arise from several sources, for example, the target of mutagenesis or the nature of the mutation. If the target of mutagenesis is to make a mutation in a pregonial cell (see Section VIII.A), depending on how many cells contribute to the germline, that mutation will be represented in the haploid progeny of the F_1 female by a clone of mutant embryos, usually less than 50%. Also, depending on the mutagen, it may take one or two rounds of DNA replication before a mutation is fully fixed, as in the case of chemical mutagenesis of mature sperm. The F_1 embryo and its germline, produced from that sperm, will be mosaic.

Each screening strategy has its advantages and disadvantages. In F_2 and F_3 screens, the germline of mutation-bearing fish is nonmosaic. But the F_1 screen is faster, saving one or two generations of time and considerable space. EP treatment of progeny from F_1 mutagenized females allows one to screen F_2 parthenogenetic diploids. (Henion *et al.,* 1996; Beattie *et al.,* Ch. 4, this volume). This strategy, as in F_1 screens in general, is rapid, can be conducted in a small-scale mutagenesis screen, and has the advantages of screening diploids. The disadvantage of this method is that the proportion of mutants in a batch of embryos depends on the gene–centromere distance of the locus of the mutation (Streisinger *et al.,* 1986). Chromosomes undergo recombination during meiosis I, and EP keeps meiotic half tetrads (sister chromatids) together, producing heterozygous F_2 diploids. High chiasmic interference makes double crossovers rare (Postlethwait and Talbot, 1997). The chance of a gene undergoing recombination is inversely proportional to its gene–centromere distance, so that mutations close to the centromere are represented by homozygous mutant EP diploids up to 50%; however this proportion goes down for more distal loci, and for mutations near the ends of chromosomes, there are few homozygous mutants. Thus, the mutant clone size varies depending on the gene-centromere distance. Mosaicism in the F_1 germline further reduces the probability of finding the most distal genes.

Because haploids provide a direct phenotypic expression of the germline of the mutagenized female, haploid analysis is a great advantage in an F_1 screen for genes regulating early development because haploids produce the largest possible mutant clone size.

VIII. Haploid Screens and Gamma-Ray Mutagenesis

Here I discuss results of a small-scale haploid screen of gamma-ray mutagenized fish, conducted from 1986 to 1995. The strategies we used were designed to produce germline mutations where a mutation in a single germ line cell is represented by a clone of mutant progeny.

A. Gamma-Irradiated Blastula Stage Embryos

For the sake of efficiency, one wants as large a target size (number of chromosome sets) as possible while maintaining a workable mutant clone size. In the blastula, the target of mutagenesis are the early progenitor cells of the germline, and mutations can be induced independently in one or more of those cells. In the early blastula, there appears to be relatively few cells that make up the germline lineage. A mutation in one of those cells will be amplified to the extent of the contribution to the germ cells that the progenitor makes. The 3-hour blastula (1000 cells) was chosen for mutagenesis because at this time the cells are especially sensitive to gamma rays. After this period, the embryos become more refractory to mutagenesis (Walker-Durchanek, 1980).

Gamma-ray mutagenesis was conducted by exposing 3-hour blastula embryos in a 5-ml glass vial to irradiation from a cesium-137 source in a model M Gammator (Radiation International) at a dose of 250–350 rads. Batches of embryos fertilized within about 5 min. of each other were pooled for irradiation. The embryos which survived irradiation were raised to adulthood (P) and their progeny screened as haploids (Fig. 1). At this dose, the mutation rate of 0.35 mutations per female (Table II) was observed.

Table II
Frequency of Mutation and Clone Size: Gamma Irradiation at Different Stages

Strain Stage irradiated	# females screened	# putative mutants	Mutations per female	Average clone size, % ± s.d.	Of mutant batches, % with multiple phenotypes[a]
Clonal[b] blastula stage	232	75	0.32	28.6 ± 18.0	10.4
AB[c] blastula	166	57	0.34	24.2 ± 15.3	24.4
AB+ *AB[d] mature sperm	109	66	0.61	29.5 ± 14.1	38.6
*AB spermatogonia	86	23	0.27	30.4 ± 14.0	22.2

[a] Mutant clutches with more than one discernible phenotype.
[b] The clonal strains are homozygous and embryonic lethal-free (see Section IV).
[c] The AB strain is the semiselected Oregon wild-type line (see text).
[d] The *AB strain is the gynogenetically derived wild-type line currently being used in the Oregon mutagenesis screens.

Two different studies suggest that only a few cells in the 1000-cell embryo produce the germline of the adult fish. First, Walker and Streisinger (1983) showed that gamma-ray induction of mutations at the *golden* locus in P fish irradiated as 3-hour blastulae, produced clones of *golden* mutant progeny that made up 5–50% of F_1 clutches of embryos. A mutation at the *golden* locus can be viable, produces a pigment defect in which melanocytes are incompletely filled with melanin, and is easily scored at day 2. Mutagenized fish can be tested for newly induced alleles of the *golden* gene by crossing P parents to gol^{b1}/gol^{b1} diploids. Because the mutagenized pregonial cells are diploid, the gametes are haploid, and the gamma rays are likely to induce just one lesion in a single cell (in one chromosome set), an F_1 clutch with 50% *golden* embryos would mean that all the germ cells came from the daughters of a single cell of the blastula stage embryo! Calculating backward from many clones of *golden* embryos, we determined that approximately five cells, present in the 3-hour blastula, contribute to the adult germline. The range of clone sizes (5–50%) in fish mutagenized at the blastula stage suggests that perhaps not all pregonial cells contribute equally to the gametes.

In the second study, Yoon *et al.* (1997) cloned *vas,* the zebrafish homologue to the *Drosophila vasa* gene, which is specifically expressed in the germ cell lineage. Using *in situ* hybridization, they found *vas* RNA in a very limited number of cells in the early zebrafish embryo. Their data suggest that cells which inherit *vas* RNA may be the primordial germ cells. At the 1000 cell stage (3 hours), *vas* RNA was localized to just four cells. This number of cells agrees remarkably well with the estimate from the earlier study. Thus, in the 3-hour blastula, approximately four primordial germ cells (eight genomes) are the target of mutagenesis, and a lesion in one of these cells will generate a clone of mutation-bearing germ cells.

Males are generated in a dose-dependent fashion by gamma-irradiation of blastula stage embryos. This was especially apparent when we irradiated clonal lines where most offspring of clonal intercrosses normally were female (Table III). Most of these induced males were fertile, although sterility also increased with dose. Production of males became even more significant when we started using the AB strain for mutagenesis where only 50% of the stock was female before irradiation. The extra males are useful in screening for new alleles of recovered mutations by natural crosses, and they could be used in a haploid androgenetic screen (Corley-Smith *et al.,* Ch. 5, Volume 59).

B. Gamma-Irradiated Sperm

Because excess males, produced by irradiating blastulae, are not useful in our gynogenetic haploid screen, we turned to irradiating mature sperm. Sperm is collected from *AB males as usual and pooled into Hank's saline in a small glass vial. The sperm, sitting in a beaker of ice, is exposed to 350–450 rads from a cesium-137 source. The mutagenized sperm is then used to fertilize normal *AB eggs (Fig. 1). Approximately 25–50% of the embryos survive to adulthood establishing the F_1 generation. Irradiation of mature sperm has less of an impact in conversion of fish to males (Table III); therefore, we can use a higher dose for mutagenesis.

Interestingly, when we compared gamma-irradiated blastulae with gamma-irradiated sperm, the frequency of mutation at the *golden* locus (Table IV) is slightly lower in sperm, even though the dose is higher. Because these numbers are small, it is difficult to know if the difference is significant. This potential difference in mutation frequency may be because in gamma-irradiated sperm the target size is only one chromosome set instead of approximately eight chromosome sets in gamma-irradiated blastula.

Table III
Gamma Rays Convert Fish to Males; Blastula Stage Embryos Are Particularly Sensitive

Strain Stage irradiated	Dose (rads)	Number of fish	% Female
Clonal[a]	0	81	97
Blastula	90	198	76
	180	210	62
	250	932	47
	370	67	31
AB[b]	0	111	46
Blastula	240	610	26
AB	0	155	42
Mature sperm	380	382	32
	575	116	29

[a] The clonal strains are homozygous and embryonic lethal-free (see Section IV).
[b] The AB strain is the semiselected Oregon wild-type line (see text).

Table IV
Frequency of Gamma-Ray-Induced Mutation at the *golden* Locus as an Indicator of Mutagenesis Efficiency

Strain Stage irradiated	Dose (rads)	Number golden[a]	Total	% golden
Clonal[e]	215	5	161	3.1
Blastula[b]	425	2	43	4.6
AB[f]	380	8	306	2.6
Sperm[c]	575	5	237	2.1
*AB[g] Spermatogonia[d]	?	2	361	0.5

[a] Nonmosaic embryos.
[b] Mutagenized females were crossed to homozygous *golden* males.
[c] Irradiated mature sperm fertilized *golden* eggs.
[d] Sperm from irradiated males fertilized *golden* eggs.
[e] The clonal strains are homozygous and embryonic lethal-free.
[f] The AB strain is the semiselected Oregon wild-type line (see text).
[g] The *AB strain is the gynogenetically derived wild-type line currently being used in the Oregon mutagenesis screens.

The mutation frequency in the gamma-irradiated sperm mutagenesis screen is greater than in the gamma-irradiated blastula screen (Table II), however, the frequency of multiple phenotypes in a single mutagenized female also is higher. The mutation frequency might be influenced by translocations which produce two phenotypes for a single event.

The average size of mutant clones in irradiated mature sperm is 30% (Table II). One would expect a clone size of 50% for gamma-ray-induced mutations because gamma-rays make double strand breaks, but this number is reduced by addition of the smaller translocation clones (see Section IX). Moreover, some gamma-ray-induced lesions may require an additional round of DNA replication before they are stable, thus a mosaic germline may result as discussed before.

C. Gamma-Irradiated Spermatogonia

In an attempt to produce more deletion mutations relative to translocations, we mutagenized young adult males to target spermatogonia, a stem cell population whose DNA is undergoing replication and meiosis.

The males are very sensitive to irradiation. In one experiment, we produced a P stock by irradiating 40 males five times for 2 minutes at an estimated dose of 700–750 rad/minute, with a rest period of 9–10 days between exposures. After waiting for mutagenized mature sperm to be flushed out, 2 months after the final irradiation, 20 (50%) of the fish were still alive, and of these, 13 mutagenized males gave sperm. An aliquot of this pooled sperm was used in a control experiment to measure site-specific mutation frequency at the *golden* locus (Table IV), and the remainder were used to fertilize normal eggs to produce an F_1 stock. There is no apparent bias toward inducing males using spermatogonia irradiation (data not shown).

Table II shows that the rate of mutagenesis using gamma-irradiated spermatogonia is lower than gamma-irradiated sperm. The frequency of site-specific mutation at the *golden* locus is also lower (Table IV).

Many gamma-ray-induced mutations were too pleiotropic to be useful, but from 140 mutation-bearing females of interest, 87 mutations were successfully recovered in the F_2 outcross generation. Because there was no attempt to recover mutations from 23, the percentage of mutations successfully recovered is 87/117 or 74%.

IX. Nature of Gamma-Ray-Induced Mutations

The nature of the majority of gamma-ray-induced mutations is not known, although a few have been characterized by genetic mapping or cloning. One expects a range of disruptions from a few base pair changes to large deletions and translocations.

Table V is a compilation of approximately 200 mutations, gathered in the morphology screen. The majority of mutations were found in day 1 embryos (65%), but a significant number were identified in the day 2 screen (24%) (Table VA); thus we find the day 2 screen to be very worthwhile.

We have observed a great variety of mutant phenotypes in our haploid gamma-ray mutagenesis screens. Table VB summarizes mutations broadly classified by the tissue affected. The most common phenotype is cell death (necrosis) in the nervous system or the body. However, patterns of cell death can be very different, helping to distinguish between mutations.

The most interesting mutations affect specific tissue types or patterning of the embryo. Despite the short tail, characteristic of haploids, mutations involving the tail were the most frequent of nondegenerative phenotypes. Pigment pattern mutants were also frequent because they are easily identified. Mutations that affect specific body parts (the jaw, notochord, etc.) were less frequent. Studies using ENU (*N*-ethyl-*N*-nitrosourea) as a mutagen have shown similar proportions of mutations in these broad phenotypic categories (Mullins *et al.,* 1994). There may be a higher proportion of necrotic or neural degeneration (ned) mutations generated by gamma-ray mutagenesis. This is expected because deletions and translocations typical of gamma-ray mutations often remove multiple genes. Also, as mentioned earlier, a mutant may have a ned phenotype as a haploid, but not as a diploid.

A. Characterized Mutations

Only a few gamma-ray-induced mutations have been characterized at the molecular level to date. The nature of the screen biases what kind of mutations

Table V
A. Gamma-Ray-Induced Mutations Disrupt Development at Various Times[a]

Developmental stage	# mutants	% mutants
12 h	15	7.5
day 1	131	65.5
day 2	47	23.5
day 3	7	3.5
Total	200	

B. Classification of Gamma-Ray-Induced Mutations by Phenotypic Categories[b]

Phenotype	# of mutants
Cell death	70
Neural degeneration	55
Tail	53
Pigment	19
Eye	9
Jaw	7
Notochord	4
Floorplate	3
Total	220

[a] This data set, collected from the morphology screen, shows when the mutant phenotype was observed.
[b] This data set was collected from the morphology screen.

are saved. Mutants kept from the morphology, behavioral, and *in situ* screens tend to be more specific rather than pleiotropic. For example, the *no tail*^{b160} mutation (*no tail* gene is the homologue of mouse *brachyury*), found in the morphology screen, alters only two nucleotides in the open reading frame which were replaced by six nucleotides producing a frameshift and a truncated protein product (Halpern, 1993; Schulte-Merker, 1994). The haploid *no tail*^{b160} embryo belongs in the **B** category because it lacks a notochord and tail, but it is otherwise quite good-looking.

The nicotinic acetylcholine receptor mutation *nic1*^{b107} was discovered in the early behavioral (wiggle) screen. It results from a deletion of about 1.5 kb in the gene encoding the acetylcholine receptor (Sepich *et al.,* 1998). The haploid embryo is paralyzed, but is otherwise a beautiful **A** embryo.

A well-characterized gene identified in the *in situ* screen is *valentino* (Moens *et al.,* 1998); gamma mutations *val*^{b361} and *val*^{b475} were both detected by their altered *krox20* rhombomere 5 stripe. Initially in both alleles, the segregation ratio of mutants to wild types was distorted, suggesting that the mutations were translocations. Both mutations resolved to Mendelian inheritance in subsequent generations. *val*^{b361} is a deletion at the distal end of linkage group 23 in which a close proximal gene, *silent heart,* is still present. *val*^{b475} is a larger deletion which removes markers on both sides of *valentino* including *silent heart.*

In the polymerase chain reaction (PCR)-based screen, the probability of finding missing sequences is greater for a translocation or large deficiency than for a small lesion. It is not surprising, then, that Fritz *et al.* (1996) find translocations and complex rearrangements for the majority of mutations recovered in the PCR screen. T(*msxb*)^{b220} is a translocation identified in the PCR screen that subsequently resolved into a large deficiency of 15–40 cM. On the other hand, *tenascin C Df(ten)*^{b345} is a deletion of less than 10 cM. Thus the PCR-based screen identifies a large range of molecular defects.

B. Translocations

Many gamma-ray-induced mutations show a distortion of the expected segregation pattern in the original screened clutch and in subsequent generations. Although clones of mutations in progeny of F_1 mutagenized fish may vary in size, mutations recovered in the F_2 generation should be 50% mutant in haploid offspring from a heterozygous female, or 25% mutant in a diploid intercross. Often, however, gamma-ray induced mutations show distorted segregation patterns with a second phenotype in the clutch. This suggests that the mutation may be a reciprocal translocation, each translocation partner displaying a distinct phenotype. Multiple mutations from a single mutation-bearing female are common, and some of these mutations act independently; for example, both single and double mutants will be represented in a clutch of haploids, and the segregation will be 50% for each mutation. In translocations, double mutants do not occur in the haploid clutch, and often the frequency of the two phenotypes added together will account for 50% of the clutch in a ratio of 2:1:1 (wild type, phenotype 1, phenotype 2) (Table VI).

Sometimes, in the first outcross generation or in subsequent generations, a female will produce haploid progeny exhibiting just one of the phenotypes in 50% of the clutch. The translocation has become unbalanced, and one of the partners has segregated away from the other. At this point, the mutation acts like a deletion and has the proper Mendelian segregation ratios. An observation, which we cannot explain, is that sometimes when a translocation unbalances, the phenotype of the resulting mutant is more severe than was the same phenotype in clutches with its partner, such as the case of *b173* and *b344* (Table VI).

Table VI gives three examples of translocation mutations recovered from gamma-ray mutagenesis haploid screens.

oep^{b173}, is a gamma-ray induced translocation allele of the gene *one eyed pinhead* (Schier *et al.,* 1997). Two distinct mutant phenotypes, *b173* and *b188,* are present in a clutch with no double-mutant embryos. Mutant gametes are

Table VI
Examples of Translocation Mutations in Haploid Clutches

Genotype of Mom	% wild-type WT/Total	% phenotype 1 Ø1/Total	% phenotype 2 Ø2/Total	Unbalanced phenotype[a]
1. *oep*^{b173}; *b188*	64	18	18	
	483/756	139/756	134/756	
oep^{b173}	50	50		Severe
unbalanced	329/655	326/655		
b188	50		50	Same
unbalanced	220/440		220/440	
2. *b344*	52	24	24	
	381/729	176/729	172/729	
b344	50	50		Severe
unbalanced	51/102	51/102		
3. *cyc*^{b213b}	49	25	26	
	307/624	156/624	161/624	

[a] The translocation has unbalanced when one altered chromosome separates from its reciprocal partner. The resulting haploid phenotype is often more severe.
[b] From Talbot *et al.* (1998).

underrepresented, 18% rather than 25% for both phenotypes, but the segregation ratio is close to 2 : 1 : 1 as expected for a translocation. When either of the partners segregate away from the other, the mutation, present in 50% of the gametes, acts in a Mendelian fashion (Table VI). When *b173* is unbalanced, the *oep* mutant phenotype becomes much more severe, whereas when the translocation partner, *b188,* is unbalanced, there is little or no change in phenotype.

b344 is an example of a classic translocation according to the foregoing criteria. Both partners have neural degeneration phenotypes but are easily distinguished. The segregation ratio in haploid embryos is 2:1:1, and when one phenotype segregated away from the other, it was observed at the expected 1:1 ratio. The phenotype of the unbalanced mutation is more severe.

The most well characterized of the three translocation mutations is *cyclops*^{b213}. Talbot *et al.* (1998) has shown by mapping that this mutation is the result of a reciprocal translocation of a distal piece of linkage group 12, that measures about 33 cM and contains the *cyclops* gene, with a distal piece of linkage group 2, that measures about 125 cM. Diploid embryos which inherit both rearranged chromosomes appear wild type.

X. Conclusion

I have discussed the use of haploids for a gamma-ray mutagenesis scheme; however, haploids can be used to screen for mutations induced by any mutagen.

Also, the efficiency of using haploids makes it possible to conduct screens on a small scale and, thus, suitable for many laboratories.

Although large-scale diploid screens have identified hundreds of genes regulating early zebrafish development, it is clear that there are many more genes yet to be identified (Driever *et al.,* 1996; Haffter *et al.,* 1996). Particularly powerful are directed screens focusing on a specific aspect of development. Subtle mutations, such as the mutant *valentino* (Moens *et al.,* 1996) found in the *in situ* screen, would have been missed in a screen relying solely on morphological criteria. Thus, a multidirectional approach, such as combining morphological, behavioral, *in situ,* and PCR screens, enhances the chances of finding interesting developmental mutations.

In the near future, the zebrafish genome project will identify many potential developmental regulatory genes, characterized by their sequence, their gene expression pattern, and their position on the genetic map. We will want to elucidate the *in vivo* role of such genes by finding mutations which disrupt their function. Reverse genetic strategies, such as the gamma-ray based screen of Fritz *et al.* (1996), identify deficiencies which completely eliminate function of a specific targeted gene. Then, new alleles can be identified quickly by noncomplementation screens to ENU-mutagenized or gamma-ray mutagenized stocks (Riley and Grunwald, 1995). Mapping a gamma-ray-induced deletion is rapid because there is a high probability that closely linked markers will be entirely absent (Postlethwait and Talbot, 1997); thus, a whole set of deficiencies can be assembled which, when mapped, define a gene and its neighborhood (see, for example, Fisher *et al.,* 1997).

By designing new imaginative screens, we will learn more about gene interactions involved in early developmental pathways. Eventually, we will come to understand the early development of a simple vertebrate, the zebrafish, and learn a great deal about our own development in the process.

Acknowledgments

I especially thank Charles Kimmel for continuing and developing the gamma-ray mutagenesis screens after the death of George Streisinger. I am grateful to Sharon Amacher, Monte Westerfield, Nathalia Glickman, Patricia Edwards, and Graham Corley-Smith for helpful comments on the manuscript, and acknowledge my companion-in-screening, Marnie Halpern.

References

Chakrabarti, S., Streisinger, G., Singer, F., and Walker, C. (1983). Frequency of gamma-ray induced specific locus and recessive lethal mutations in mature germ cells of the zebrafish, *Brachydanio rerio. Genetics* **103,** 109–124.

Corley-Smith, G. E., Lim, C. T. J., and Brandhorst, B. P. (1996). Production of androgenetic zebrafish (*Danio rerio*). *Genetics* **142,** 1265–1276.

Driever, W., Solnica-Krezel, L., Schier, A. F., Neuhauss, S. C. F., Malicki, J., Stemple, D. L., Stainier, D. Y. R., Zwartkruis, F., Abdelilah, S., Rangini, Z., Belak, J., and Boggs, C. (1996). A genetic screen for mutations affecting embryogenesis in zebrafish. *Development* **123,** 37–46.

Felsenfeld, A. L., Walker, C., Westerfield, M., Kimmel, C., and Streisinger, G. (1990). Mutations affecting skeletal muscle myofibril structure in the zebrafish. *Development* **108,** 443–459.

Fisher, S., Amacher, S. L., and Halpern, M. E. (1997). Loss of *cerebum* function ventralizes the zebrafish embryo. *Development* **124,** 1301–1311.

Fritz, A., Rozowski, M., Walker, C., and Westerfield, M. (1996). Identification of selected gamma-ray induced deficiencies in zebrafish using multiplex polymerase chain reaction. *Genetics* **144,** 1735–1745.

Granato, M., van Eeden, F. J. M., Schach, U., Trowe, T., Brand, M., Furutani-Seiki, M., Haffter, P., Hammerschmidt, M., Heisenberg, C. P., Jiang, Y. J., Kane, D. A., Kelsh, R. N., Mullins, M. C., Odenthal, J., and Nüsslein-Volhard, C. (1996). Genes controlling and mediating locomotion behavior of the zebrafish embryo and larva. *Development* **123,** 399–413.

Grunwald, D. J., Kimmel, C. B., Westerfield, M., Walker, C., and Streisinger, G. (1988). A neural degeneration mutation that spares primary neurons in the zebrafish. *Dev. Biol.* **126,** 115–128.

Haffter, P., Granato, M., Brand, M., Mullins, M. C., Hammerschmidt, M., Kane, D. A., Odenthal, J., van Eeden, F. J. M., Jiang, Y. J., Heisenberg, C. P., Kelsh, R. N., Furutani-Seiki, M., Vogelsang, E., Beuchle, D., Schach, U., Fabian, C., and Nüsslein-Volhard, C. (1996). The identification of genes with unique and essential functions in the development of the zebrafish, *Danio rerio. Development* **123,** 1–36.

Halpern, M. E., Ho, R. K., Walker, C., and Kimmel, C. B. (1993). Induction of muscle pioneers and floor plate is distinguished by the zebrafish *no tail* mutation. *Cell* **75,** 99–111.

Henion, P. D., Raible, D. W., Beattie, C. E., Stoesser, K. L., Weston, J. A., and Eisen, J. S. (1996). Screen for mutations affecting development of zebrafish neural crest. *Dev. Genet.* **18,** 11–17.

Kane, D. A., and Kimmel, C. B. (1993). The zebrafish midblastula transition. *Development* **119,** 447–456.

Kimmel, C. B., Sessions, S. K., and Kimmel, R. J. (1978). Radiosensitivity and time of origin of Mauthner neuron in the zebra fish. *Dev. Biol.* **62,** 526–529.

Kimmel, C. B. (1989). Genetics and early development of zebrafish. *Trends Genet.* **5,** 283–288.

Kimmel, C. B., Ballard, W. W., Kimmel, S. R., Ullmann, B., and Schilling, T. F. (1995). Stages of embryonic development of the zebrafish. *Dev. Dyn.* **203,** 253–310.

Kubota, Y., Shimoda, A., and Shima, A. (1992). Detection of gamma-ray-induced DNA damages in malformed dominant lethal embryos of the Japanese medaka (*Oryzios latipes*) using AP-PCR fingerprinting. *Mutat. Res.* **283,** 263–270.

Moens, C. B., Yan, Y.-L., Appel, B., Force, A. G., and Kimmel, C. B. (1996). *valentino:* A zebrafish gene required for normal hindbrain segmentation. *Development* **122,** 3981–3990.

Moens, C. B., Cordes, S. P., Giorgianni, M. W., Barsh, G. S., and Kimmel, C. B. (1998). Equivalence in the genetic control of hindbrain segmentation in fish and mouse. *Development* **125,** 381–391.

Mullins, M. C., Hammerschmidt, M., Haffer, P., and Nüsslein-Volhard, C. (1994). Large-scale mutagenesis in the zebrafish: In search of genes controlling development in a vertebrate. *Curr. Biol.* **4,** 189–202.

Oxtoby, E., and Jowett, T. (1993). Cloning of the zebrafish *krox-20* gene (*krx-20*) and its expression during hindbrain development. *Nucl. Acids Res.* **21,** 1087–1095.

Postlethwait, J. H., and Talbot, W. S. (1997). Zebrafish genomics: From mutants to genes. *Trends Genet.* **13,** 183–190.

Riley, B. B., and Grunwald, D. J. (1995). Efficient induction of point mutations allowing recovery of specific locus mutations in zebrafish. *Proc. Natl. Acad. Sci. USA* **92,** 5997–6001.

Russell, W. L., and Russel, L. B. (1959). The genetic and phenotypic characteristics of radiation-induced mutations in mice. *Radiat. Res.* (Suppl.) **1,** 296–305.

Schier, A. F., Neuhauss, S. C. F., Helde, K. A., Talbot, W. S., and Driever, W. (1997). The *one-eyed pinhead* gene functions in mesoderm and endoderm formation in zebrafish and interacts with *no tail. Development* **124,** 327–342.

Schulte-Merker, S., van Eeden, F. J. M., Halpern, M. E., Kimmel, C. B., and Nüsslein-Volhard, C. (1994). *no tail* (*ntl*) is the zebrafish homologue of the mouse T (Brachyury) gene. *Development* **120,** 1009–1015.

Sepich, D. S., Wegner, J., O'Shea, S., and Westerfield, M. (1998). An altered intron inhibits synthesis of the acetylcholine receptor alpha-subunit in the paralyzed zebrafish mutant *nic1*. *Genetics* **148,** 361–372.

Solnica-Krezel, L., Schier, A. F., and Driever, W. (1994). Efficient recovery of ENU-induced mutations from the zebrafish germline. *Genetics* **136,** 1401–1420.

Streisinger, G., Walker, C., Dower, N., Knauber, D., and Singer, F. (1981). Production of clones of homozygous diploid zebrafish (*Brachydanio rerio*). *Nature* **291,** 293–296.

Streisinger, G., Singer, F., Walker, C., Knauber, D., and Dower, N. (1986). Segregation analysis and gene-centromere distances in zebrafish. *Genetics* **112,** 311–319.

Talbot, W. S., Egan, E. S., Gates, M. A., Walker, C., Ullmann, B., Neuhauss, S. C. F., Kimmel, C. B., and Postlethwait, J. H. (1998). Genetic analysis of chromosomal rearrangements in the cyclops region of the zebrafish genome. *Genetics* **148,** 373–380.

Walker, C., and Streisinger, G. (1983). Induction of mutations by gamma-rays in pregonial germ cells of zebrafish embryos. *Genetics* **103,** 125–136.

Walker-Durchanek, R. C. (1980). Induction of germ line mutations by γ-irradiation of zebrafish embryos. Thesis. Department of Biology, University of Oregon, Eugene, OR.

Westerfield, M. (1995). "The Zebrafish Book: A Guide for the Laboratory Use of Zebrafish (*Danio rerio*)." University of Oregon Press, Eugene, OR, p. 335.

Yoon, C., Kawakami, K., and Hopkins, N. (1997). Zebrafish *vasa* homologue RNA is localized to the cleavage planes of 2- and 4-cell-stage embryos and is expressed in the primordial germ cells. *Development* **124,** 3157–3165.

CHAPTER 4

Early Pressure Screens

Christine E. Beattie,[*] David W. Raible,[†] Paul D. Henion,[‡] and Judith S. Eisen[‖,1]

[*] Neurobiotechnology Center and Department of Pharmacology
Ohio State University
Columbus, Ohio 43210

[†] Department of Biological Structure
University of Washington
Seattle, Washington 98195

[‡] Neurobiotechnology Center and Department of Cell Biology and Neuroanatomy
Ohio State University
Columbus, Ohio 43210

[‖] Institute of Neuroscience
University of Oregon
Eugene, Oregon 97403

I. Introduction

Genetic screens for mutations affecting early embryonic phenotypes have provided a powerful approach for unraveling developmental mechanisms in or-

[1] To whom correspondence should be addressed.

0091-679X/99 $30.00

ganisms as such the fruit fly, *Drosophila melanogaster,* and the nematode worm, *Caenorhabditis elegans.* However, a systematic genetic approach has been more problematic in vertebrates for a number of reasons. First, it is difficult to drive a mutation to homozygosity in animals that have relatively long generation times and often yield few offspring. Second, vertebrates carry ancient gene duplications that may cause redundancy and, in some cases, functional polyploidy (J. H. Postlethwait, personal communication). Finally, it is difficult to screen for early-acting mutations in animals, such as placental mammals, in which all of early development is hidden within the mother's uterus. Despite these obstacles, the first paper describing a systematic, saturation screen for developmental mutations in flies (Nüsslein-Volhard and Weischaus, 1980) and a paper demonstrating the potential for similar screens in a vertebrate, the zebrafish, *Danio (Brachydanio) rerio* (Streisinger *et al.*, 1981) were published less than a year apart. The zebrafish has recently begun to fulfill this early promise by becoming a model for genetic studies of vertebrate development (Eisen, 1996). In this chapter, we review our studies using gynogenetic diploid zebrafish embryos produced by the early pressure (EP) method to screen for mutations affecting early embryogenesis, and we compare these screens with those of haploid embryos and of diparental diploids.

II. Rationale for Use of Early Pressure to Produce Diploid Embryos for Genetic Screens

Because recessive alleles are hidden in heterozygous diploids, genetic studies present a significant challenge in many eukaryotes. In a classical two-generation breeding scheme, mutations induced in premeiotic germ cells result in heterozygous F_1 founders which are then bred to wild-type individuals to create F_2 families, half of the members of which are heterozygous for the mutation. Mutations are revealed by crossing F_2 siblings and screening their F_3 progeny, one-quarter of which should be homozygous for the mutation. Undertaking such an F_3 screen in zebrafish requires many months to raise the F_2 families, as well as large numbers of aquaria to accommodate enough fish for screening all 25 chromosomes simultaneously. Although balancer chromosomes would allow screening chromosome by chromosome, these genetic tools have not yet been generated in zebrafish. Despite these difficulties, numerous, exciting mutations have been isolated from F_3 screens recently undertaken in zebrafish in the Nüsslein-Volhard and Driever labs (Haffter *et al.,* 1996; Driever *et al.,* 1996).

Because "heterozygosity in diploid eukaryotes often makes genetic studies cumbersome" (Streisinger *et al.*, 1981), Streisinger was prompted to devise simple methods to produce homozygous zebrafish parthogenetically (Fig. 1, see color plate; Streisinger *et al.,* 1981). Using these methods, the F_2 progeny of mutagenized fish can be screened directly for recessive lethal mutations, obviating the necessity for backcrosses or intercrosses to drive a mutation to homozygosity prior to screening. This ability significantly reduces the numbers of fish and aquaria required for screening relative to the numbers needed for the more standard F_3 screens described earlier. Two types of embryos that immediately display recessive phenotypes can be produced using Streisinger's methods, haploids, and parthenogenetic diploids. Haploid embryos are produced by using genetically impotent, UV-irradiated sperm (Streisinger *et al.,* 1981) to fertilize eggs extruded from females heterozygous for mutagenized chromosomes. Screens of haploid embryos have identified mutations affecting patterning and morphogenesis of the early embryo (Kimmel, 1989; Halpern *et al.,* 1993; Moens *et al.,* 1996; Schilling *et al.,* 1996), as well as aspects of later development such as motility (Westerfield *et al.,* 1990; Felsenfeld *et al.,* 1990).

Even though screening haploid embryos is very convenient, mutant phenotypes of homozygous diploid embryos from the same mother are not always identical to those seen in the original haploid screen. In addition, haploid embryos have phenotypes that make them unsuitable for certain types of screens. For example, we find that haploid embryos have a variable "neurogenic" phenotype in which there is a slight overproduction of some primary neurons (C. B. Kimmel, B. Appel, R. Cornell, Z. Varga, and J. S. Eisen, unpublished); this phenotype could easily lead to isolation of "false-positives" or mask mutations in genes that alter patterning of primary neurons. Furthermore, haploid embryos are not viable and show significant degeneration at 3 days, precisely the time at which some of the cell types we are interested in can first be recognized.

Screening homozygous diploid embryos circumvents these problems. As with haploids, gynogenetic diploid embryos are produced by fertilizing eggs with UV-irradiated sperm; the resulting eggs are then made diploid by preventing one of the next two cell divisions (Fig. 1). Normally, eggs complete the first meiotic division about the time of ovulation and finish the second meiotic division after fertilization, extruding the second polar body before sperm and egg nuclei fuse to form a diploid zygotic nucleus that begins the mitotic divisions of cleavage. Gynogenetic diploid embryos are generated by inhibiting the second meiotic division with early hydrostatic pressure (EP) or the first mitotic division by application of heat shock (HS) (Streisinger *et al.,* 1981). Individuals produced by HS at the fist cleavage division are homozygous for maternal alleles at all loci. Individuals produced by EP are not homozygous for all loci, however, for the following reason. During meiosis, recombination occurs between homologous chromosomes while they are aligned as tetrads, creating individual chromatids that are partly maternal and partly paternal in origin. The first meiotic division separates homologous chromosomes, and the second meiotic division separates sister chromatids. When the second meiotic division is blocked, the resulting embryo is heterozygous for loci distal to the crossover point closest to the centromere and homozygous for loci proximal to this crossover (see Postlethwait and Talbot, 1997); thus, these individuals are diploid gynogenetic half-tetrads. Because of chiasma interference, there tends to be a single crossover event per chromosome arm (Streisinger *et al.,* 1986; Johnson *et al.,* 1995b). Thus, EP embryos will be homozygous only for loci proximal to the first crossover event on each chromosome arm. Previous screens of gynogenetic diploid embryos have isolated mutations affecting early embryonic patterning and morphogenesis (Kimmel *et al.*, 1989; Hatta *et al.*, 1991).

Several considerations prompted our choice to screen embryos produced by the EP method. Space limitations in our fish facility necessitated screening embryos from F_1 parents rather than raising these embryos to adulthood and screening their progeny. Screening haploid embryos requires less work than producing gynogenetic diploids. However, the morphology of motoneurons and arrangement of dorsal root ganglia (DRG) is often irregular in haploid embryos, whereas it is entirely normal in EP diploid embryos (Fig. 2). HS embryos are homozygous at all loci and, thus, would be preferable to EP embryos, except that only 10–20% of HS embryos survive (Streisinger *et al.*, 1981).

EP has advantages, but it also has potential limitations. For example, some EP-generated embryos exhibit a specific syndrome of epigenetic defects (Streisinger *et al.*, 1981), such as cyclopia, loss of anterior structures, or notochord defects (Hatta and Kimmel, 1993) which resemble those of ventralized embryos

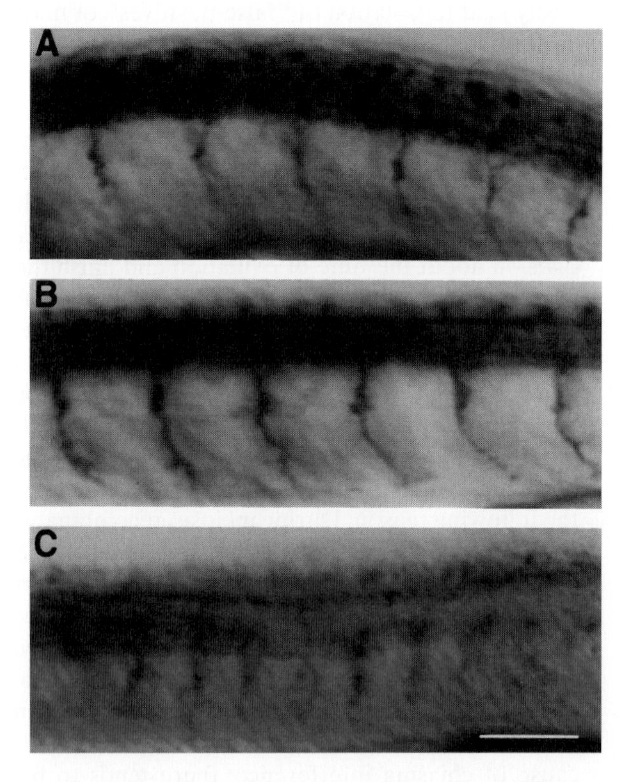

Fig. 2 Motor axons are defective in haploid embryos. Motor axons labeled with the zn-1 and znp-1 monoclonal antibodies are normal in wild-type (A) and EP diploid (B) embryos, but aberrant in haploid (C) embryos. Side views of 24-h embryos, anterior to the left and dorsal to the top. Scale bar 50 μm.

(Solnica-Krezel *et al.,* 1996; Hammerschmidt *et al.,* 1996). In practice, we can distinguish this syndrome of epigenetic defects from many types of mutations, and, in particular, from those mutations in which we are specifically interested. Within the large clutches of embryos we obtained from each female in our studies (Table I), embryos with epigenetic defects represented only a small fraction, about 20%. Further, the phenotypes of these embryos were highly variable, ranging from embryos entirely lacking heads and structures derived from dorsal mesoderm to embryos with only mild cyclopia. However, in contrast to mutant embryos which typically have a similar phenotype, the entire range of phenotypes was characteristically found within a clutch. Thus, unless a mutation has a variable phenotype resembling these epigenetic defects, it can be distinguished easily from EP effects on development. However, EP embryos are probably unsuitable for screening for mutations in dorsoventral patterning, since it would be difficult to distinguish embryos carrying bona fide mutations from embryos with EP-induced defects.

Another problem in screening EP embryos is that mutations far from the centromere may be obscured because of mitotic recombination and high chiasma interference (Streisinger *et al.,* 1986; Johnson *et al.,* 1995b). In practice, this means that the fraction of embryos homozygous for a mutation in a gene will be reduced in relationship to the distance of that gene from the centromere. For example, *golden-1 (gol1)* is at the distal tip of linkage group 18 (LG18; S. Street and J. H. Postlethwait, personal communication) and females heterozygous for *gol1* produce homozygous *gol1*$^{-}$ EP diploid embryos in 5% of their offspring (Streisinger *et al.,* 1986). In contrast, *floating head (flh)* is located near the centromere of LG10, and homozygous *flh*$^{-}$ EP diploids appear in 50% of the offspring of a heterozygous female (Talbot *et al.,* 1995). This problem can be addressed by screening sufficient numbers of embryos per clutch. In F_3 screens, the probability of any embryo being homozygous for a mutation is one in four, so it is necessary to screen only 11 embryos per clutch to be 95% confident that a mutant will be observed if its parents are in fact both heterozygous [calculated from the cumulative binomial distribution function (Weast, 1970)]. Seeing only a single mutant seems unreliable for scoring the phenotype, but screening 17 embryos allows one to see two homozygous mutant embryos with the same level of confidence. In an EP screen, we can calculate the gene–centromere distance from the fraction

Table I
Production of Gynogenetic Diploid Embryos by the EP Method[a]

	Clutch size	Normal (2 h)	Live (24 h)	Normal (24 h)
n (Ave ± SD)	156.9 ± 75.4	124.8 ± 69.4	103.4 ± 57.1	81.9 ± 48.9
%		79	83	79

[a] Numbers calculated for 279 clutches of embryos from F_1 females heterozygous for mutagenized chromosomes.

of homozygous mutants in a clutch (Streisinger *et al.,* 1986). If we consider a mutation distal on the chromosome for which EP yields only 10% homozygotes, one would need to examine 28 embryos to be 95% confident of seeing one mutant, and 46 embryos to be 95% confident of seeing two mutants. By inspecting at least this many embryos, we have been able to identify reliably mutations in genes that we know are very distal from their centromeres. For example, we isolated a new allele of *spadetail (spt)* (S. Amacher, unpublished). Gene–centromere distance analysis from EP *spt* embryos yielded 12% homozygous mutants (D. A. Kane, C. Walker, and C. B. Kimmel, unpublished), consistent with mapping data (S. Amacher, unpublished) showing that the *spt* gene is located distally on its linkage group.

III. Rationale for Use of Screens Focused on Specific Phenotypes

Most of the screens described thus far in zebrafish have been designed to find mutations affecting many aspects of early embryogenesis. Although such screens may yield a wealth of mutations [see Haffter *et al.* (1996) and Driever *et al.* (1996) for overviews of recent large-scale screens], they may not reveal subtle phenotypes in the specific processes one desires to study. Work in flies (Seeger *et al.,* 1993; Van Vactor *et al.,* 1993) has clearly demonstrated the utility of antibody markers to screen for very specific defects, for example inappropriate axonal pathfinding. We devised similar screens using antibodies to recognize defects in motoneuron pathfinding (Beattie *et al.,* 1995; Beattie and Eisen, 1996) and in neural crest-derived neurons (Raible *et al.,* 1994; Henion *et al.,* 1996). RNA probes have also been used to screen for mutations affecting the development of specific CNS regions (Moens *et al.,* 1996).

IV. Design of Our Screen and Mutant Phenotypes Isolated

A. Methods of Mutagenesis

Wild-type adult males were mutagenized with *N*-ethyl-*N*-nitrosourea (ENU), to induce point mutations (Solnica-Krezel *et al.,* 1994) and bred to wild-type adult females to produce heterozygous F_1 fish. Eggs were extruded from F_1 females, activated with UV-irradiated sperm, and subjected to EP by treating them with high pressure to stop the second meiotic division, which occurs after fertilization. Resulting embryos were observed using a stereomicroscope. During cleavage stages [2 hours postfertilization (h)] obviously abnormal embryos were removed. After completion of somitogenesis (24 h), each clutch of embryos was enzymatically dechorionated (Westerfield, 1993) and screened for visible phenotypes; embryos with putative epigenetic defects were then removed. Each clutch of embryos was then split in half. Half the embryos were fixed immediately and processed for immunohistochemistry to screen for defects in primary motoneurons. The remaining embryos were left to develop until 48 h, when they were screened for defects in the appearance and patterning of neural crest-derived pigment cells and at 72 h for defects in neural crest-derived pigment cells and cranial cartilages. These embryos were then processed for immunohistochemistry to screen for defects in DRG neurons. Females whose progeny had abnormal motoneurons or neural crest derivatives were outcrossed to wild-type males to produce an outcross stock. Mutations were recovered from this stock by pairwise intercross matings. Half the individuals in such a stock are expected to carry the mutation in heterozygous form. Hence, a quarter of random intercrosses are expected to yield clutches in which a quarter of the embryos are homozygous for the mutation, consistent with what we observed.

We screened clutches from 341 females heterozygous for mutagenized chromosomes, from which we identified 34 mutations affecting development of motoneurons or neural crest (see Table II). In addition, we isolated several new alleles of previously identified mutations, including *spt, cyclops (cyc),* and *no tail (ntl).* Other mutations were discarded. All mutations originally identified were subsequently recovered from the first outcross. We reidentified the phenotypes of all mutations in the F_2 generation to check that they were identical to those originally observed in the EP screen, and all mutations have been carried for two to four generations with no change in phenotype.

B. Screen for Mutations Affecting Primary Motoneurons

1. The Motoneuron Phenotype in Wild-Type Embryos

As in other vertebrates (Tanabe and Jessell, 1996), zebrafish motoneurons appear to be specified by signaling from notochord or floorplate (Beattie *et al.,* 1997). Like other fish and amphibians, zebrafish have two classes of motoneurons, primary motoneurons, a small population of large cells that form early, near the end of gastrulation, and secondary motoneurons, a larger population of small cells that form later (Myers, 1985; Myers *et al.,* 1986; Kimmel and Westerfield,

Table II
Summary of Phenotypes of Recovered Neural Crest and Motoneuron Mutations from 341 Clutches Analyzed

Neural crest mutations[a]						Motoneuron mutations		
Total	DRG, M, CC	DRG, CC	DRG	M	CC	Total	Short CaP axon	Disorganized axons
32	3	8	4	3	14	5	4	1

[a] Cell types affected: DRG, dorsal root ganglion neurons; M, melanophores; CC, cranial cartilages.

1990; Kimmel *et al.,* 1994). Primary motoneurons can be individually identified in each segment of the zebrafish trunk and tail by their cell body positions within the spinal cord, axonal trajectories along cell-specific pathways on the medial surface of the overlying myotome (Eisen *et al.,* 1986; Myers *et al.,* 1986), and patterns of gene expression (Appel *et al.,* 1995; Tokumoto *et al.,* 1995). Both primary and secondary motoneurons appear to be part of a spinal cord domain in which the Delta/Notch signaling pathway regulates the number of motoneurons and interneurons (Appel and Eisen, 1998) and the identities of individual primary motoneurons appear to be specified by local positional cues (Eisen, 1991; Appel *et al.,* 1995). After these identities are established, each specific primary motoneuron extends an axon along a cell-specific pathway based on recognition between its growth cone and pathway-specific guidance cues (Gatchalian and Eisen, 1992; Beattie and Eisen, 1997).

2. Screen Design

Although the cell bodies of primary motoneurons can be recognized under the compound microscope using Nomarski differential interference contrast (DIC) optics with a 40× or 50× objective (Eisen *et al.,* 1986), axons cannot be visualized in this way, and neither cell bodies nor axons can be seen easily at lower magnification. Screening embryos using high-power objectives and DIC optics is extremely labor-intensive. Instead, we needed a dependable method to visualize motoneurons reliably and quickly in wholemount embryos under the stereomicroscope. We chose to screen for mutations affecting primary motoneurons using a combination of antibodies that recognize both their cell bodies (zn-1; Trevarrow *et al.,* 1990) and their axonal trajectories (zn-1 and znp-1; Melançon, 1994), making cellular morphology visible under the stereomicroscope. We reasoned that this approach would reveal mutations affecting both patterning in the spinal cord (e.g., changes in the number or position of a specific primary motoneuron cell body) and axonal pathfinding (e.g., inappropriate pathway selection by a specific primary motoneuron). After each half clutch of embryos was processed for immunohistochemistry, the embryos were kept together in PBS-glycerol, and all embryos within a half clutch were examined together under the stereomicroscope. If necessary, embryos were manipulated so that motoneuron cell bodies and axons could be visualized along the entire length of the trunk. This was especially important in the case of the axon of the identified primary motoneuron, MiP, which is often difficult to see.

3. Examples of Mutations Affecting Primary Motoneurons

All five of the motoneuron mutations we isolated alter primary motoneuron axonal trajectories (Beattie *et al.,* 1995; Beattie and Eisen, 1996). Four of the mutations have similar phenotypes in which the axon of a single primary motoneuron, CaP, fails to extend normally along its cell-specific pathway (Table II;

Fig. 3, see color plate). Three of these are recessive, embryonic lethal mutations which complementation testing showed to be in different genes. Of these, *stumpy (sty)*^{b398} has the most severe phenotype. *sty*^{b398} fails to complement a partially dominant, viable mutation with a similar motoneuron phenotype, *sty*$^{b393;}$ thus these mutations are probably allelic. The fifth mutation, *loose ends (loe)*^{b420}, is also an embryonic lethal but has a quite different phenotype in which the axons of all motoneurons appear disorganized.

Surprisingly, despite the severely altered axonal trajectories, none of the mutants we isolated have obvious motility defects, although we cannot rule out subtle defects that might only be visible using high-speed video microscopy. This lack of motility defects may arise because, although primary motor axons form an exquisitely stereotyped projection pattern in the axial musculature in which each primary motoneuron arborizes in a discrete territory that does not overlap with the arborization of any other primary motoneuron, muscle fibers may be electrically coupled and thus able to respond to activation by motoneurons that do not innervate them directly. By labeling individual muscle cells with low-molecular-weight fluorescent dyes, we have found that, during the first several days of development, muscles fibers innervated by distinct primary motoneurons are dye-coupled, both within single myotomes and between adjacent myotomes on the same side of the embryo (J. S. Eisen, S. H. Pike, and M. Westerfield, unpublished; C. E. Beattie, E. Melançon, and J. S. Eisen, in preparation). Because dye-coupling between cells often signifies that they are electrically connected by low-resistance junctions (Stewart, 1978), activation of a single muscle fiber by a motoneuron may also activate muscle fibers innervated by other motoneurons. If this is the case, normal movements might be achieved, even in mutant embryos, as long as the axons of some primary motoneurons make functional connections with muscle fibers.

4. Comparison with Motoneuron Mutations Isolated in Other Screens

How do the phenotypes of the motoneuron mutants isolated in our EP screen compare with those isolated in other screens? A number of mutations isolated in an F_3 screen by virtue of altered motility also affect primary motoneuron morphology (Granato *et al.,* 1996). For example, in *unplugged (unp)* mutants CaP motoneurons have altered axonal morphology and in *diwanka (diw)* mutants both CaP and MiP axons are shorter than normal. Both of these mutants have motility defects: *unp* mutants are initially immobile, although they show some recovery later and in *diw* mutants both sides of the embryo contract simultaneously, rather than showing the alternation characteristic of wild types. These movement defects seem unlikely to result directly from changes in motor axon projections because the mutations isolated in our screen had similar defects in motor axon projections, but showed no movement defects.

All the mutants we identified appear to have the correct number of primary motoneurons. Two types of mutations affecting primary motoneuron number have been isolated in other screens: First, mutants deficient in midline structures have fewer primary motoneurons (Brand *et al.,* 1996; Beattie *et al.,* 1997). Other mutants, such as *mindbomb* (*mib* aka *white tail;* Jiang *et al.,* 1996; Schier *et al.,* 1996) have a so-called neurogenic phenotype in which primary motoneuron number is augmented and the number of other neurons, such as secondary motoneurons, is decreased. Interestingly, although *mib* mutants show a brain phenotype, the most conspicuous phenotype is lack of trunk and tail pigmentation.

An approach similar to ours has also been taken to isolate mutations affecting motoneuron pathfinding in *D. melanogaster* (Van Vactor *et al.,* 1993). In this case, immunohistochemistry was used to visualize motor axons and growth cones in embryos derived from a mutagenized parental stock. As in our screen, the robustness of the antibody-labeling made the labor-intensive process of mutant identification feasible, because whole mount embryos could be screened directly under the stereomicroscope.

C. Screen for Mutations Affecting Neural Crest Cells

1. Normal Neural Crest Development

The vertebrate neural crest consists of a small set of easily recognized cells that segregates from the neural tube, migrates on specific pathways, and gives rise to distinct derivatives. Zebrafish neural crest cell behavior is similar to that of other vertebrates (Raible *et al.,* 1992; Eisen and Weston, 1993; Schilling, 1993; Schilling and Kimmel, 1994; Raible and Eisen, 1994). After neural crest formation, the cells divide as they migrate to widely dispersed locations in the embryo and differentiate into a number of different cell types, including neurons and glia of peripheral ganglia, pigment cells, and in the head, elements of the cranial skeleton. The issue of how the neural crest generates such a diverse set of derivatives is unresolved and currently the focus of considerable investigation in several vertebrate species.

2. Screen Design

We designed our screen to recognize the various neural crest derivatives, including neurons of the dorsal root ganglia, pigment cells, and elements of the cranial skeleton. Pigment cells (Streisinger *et al.,* 1981, 1989; Walker and Streisinger, 1983; Chakrabarti *et al.,* 1983; Johnson *et al.,* 1995a; Kelsh *et al.,* 1996; Odenthal *et al.,* 1996) and cranial cartilages (Schilling and Kimmel, 1997; Schilling, 1993), can be recognized directly under the stereomicroscope; however, other derivatives, such as neurons and glia, are much more difficult to visualize directly. For example, sensory neurons of the DRG cannot be seen by direct microscopic examination of living embryos because of their relatively small numbers, small size, and particular locations. Thus, we used the anti-Hu monoclonal antibody

(Marusich *et al.,* 1994) to reveal DRG neurons in whole mount embryos under the steromicroscope.

3. Examples of Mutations Affecting Neural Crest Development

Some of the neural crest mutations we isolated were pleiotropic, affecting many neural crest derivatives, whereas others affected only specific subsets of derivatives (Table II; Henion *et al.,* 1994, 1996; Raible *et al.,* 1994). Some mutations affected the development of both trunk and cranial neural crest derivatives. For example, *colgate (col)*^{b382} mutants have defects in pigment development and patterning, DRG neuron development and crest-derived head cartilages, although none of these derivatives is completely absent. Other mutations appear to affect subsets of neural crest derivatives. For example, *end zone (enz)*^{b403} mutants have severely reduced pigmentation along the body axis where neural crest-derived melanophores are normally located, whereas crest-derived cranial cartilages and DRG neurons appear normal.

Mutations affecting DRG neurons have a variety of phenotypes (Fig. 4, see color plate). In some cases, for example *laughing man (lam)*^{b300} and *col* mutants, altered DRG neuron development is associated with altered development of other crest derivatives. Other mutations, such as *nosedive (ndv)*^{b302} affect development of DRG neurons but spare other crest derivatives. However, this mutation also causes cell death in the central nervous system (CNS), which becomes apparent on day 2. We have identified many other mutations which have CNS cell death, but in which differentiated DRG neurons are present and appear normal, showing that central and peripheral nervous system cell death are not necessarily linked.

Some mutations that disrupt cranial cartilages appear to be rather general. In *lam* mutants, cranial cartilages appear to be missing or delayed in development. In *gobbler (gob)*^{b374} mutants, distinct cartilage elements form, but are abnormally shaped (Fig. 5, see color plate). Other mutations affecting cranial cartilages have more specific defects which identify genes that probably regulate development of different cranial cartilages differentially. For example, *flat Top (ftp)*^{b369} mutants appear to lack visceral cartilages, whereas neurocranial cartilages are present. In contrast, *helix (hlx)*^{b392} mutants lack many neurocranial cartilages but have most visceral cartilaginous elements. In both of these mutations, pigmentation and DRG neurons appear to develop normally.

4. Comparison with Neural Crest Mutations Isolated in Other Screens

How do the neural crest phenotypes isolated in our EP screen compare with neural crest mutations isolated in other screens? The first mutation described in zebrafish, *goll,* affects crest-derived pigment cells; it was initially isolated as a spontaneous recessive mutation (Streisinger *et al.,* 1981) and later used to establish the efficiency of gamma rays for inducing mutations at specific loci (Walker and Streisinger, 1983). Mutations affecting cranial cartilages have also been isolated in screens of the haploid progeny of gamma-ray mutagenized zebrafish (Schilling *et al.,* 1996). More recently, numerous mutations affecting zebrafish cranial cartilages (Schilling *et al.,* 1996; Piotrowski *et al.,* 1996; Neuhauss *et al.,* 1996) and crest-derived pigment cells (Kelsh *et al.,* 1996; Odenthal *et al.,* 1996; Driever *et al.,* 1996) were isolated in screens of the F_3 progeny of ENU-mutagenized males. Undoubtedly, some of the mutations found in these other screens will be allelic with mutations isolated in our screen, as will be revealed by future complementation analysis. Thus far, no other mutations affecting DRG neurons have been described. Although numerous mutations affecting murine neural crest derivatives have been described (Jackson, 1994), isolation of these mutations occurred over many years and they were derived from many sources, rather than from systematic saturation screens. Most of these mutations arose spontaneously and were identified because heterozygotes or homozygous mutants had alterations in easily visible phenotypes such as pigment pattern.

D. Screen for Mutations with Visible Phenotypes

During the course of screening for mutations affecting motoneurons and neural crest, we observed living embryos at 24 h to separate out those with severe abnormalities from those without visible phenotypes. Because our screen was focused on very specific types of mutations, and because we looked at essentially every embryo processed for immunohistochemistry, we wanted to remove clutches of embryos that had clear developmental abnormalities that affected other body regions, to save the effort of processing and examining them. This part of the screen also afforded us the opportunity to pick out embryos with visible phenotypes which we did not plan to pursue but which were of interest to others. Even with our fairly cursory examination of embryos at this stage, we isolated alleles of *spt, ntl,* and *cyc,* three mutations affecting early development previous isolated in the Kimmel lab (Kimmel *et al.,* 1989; Hatta *et al.,* 1991; Halpern *et al.,* 1993). Thus, EP can readily be used to screen for visible phenotypes, as well as in a screen focused on subtle, specific phenotypes.

V. Closing Remarks

We have successfully isolated mutations affecting motoneuronal pathfinding and neural crest development using a screen in which mutations are identified in gynogenetic diploid embryos by whole mount immunohistochemistry as well as visible phenotypes. Comparison of the types of mutations we have isolated relative to those isolated in other screens reveals that screening F_2 diploid embryos produced by EP compares favorably with screening F_3 diploid embryos produced by normal matings, particularly when space is a consideration, although EP embryos are unsuitable for screening for some very specific features of embryogenesis, such as dorsovental patterning. The aspect of our screen that provided the most important difference from other screens was our ability to focus on particular phenotypes revealed by molecular markers for specific cell types. This approach is likely to have widespread application because it focuses on a single aspect of development, a strategy which has been very important for isolation of specific mutations in other eukaryotes, and it reveals subtle mutations that may be cryptic and, thus, invisible without markers. Together with screens for visible phenotypes in haploid (Kimmel, 1989) and diparental diploid (Haffter *et al.,* 1996; Driever *et al.,* 1996) embryos, this approach should provide new insights into the mechanisms underlying vertebrate development.

Acknowledgments

We thank Sharon Amacher, Bruce Appel, Allan Force, Charles Kimmel, Cecilia Moens, Peter O'Day, and John Postlethwait for critical comments on the manuscript; Sharon Amacher, Bruce Appel, Robert Cornell, Don Kane, Charles Kimmel, Ellie Melançon, Susan Pike, John Postlethwait, Sunny Street, Zoltan Varga, Charline Walker and Monte Westerfield for sharing unpublished data; Kirsten Stoesser for participation in the mutagenesis screen; the staff of the University of Oregon Zebrafish Facility for animal care, and Pat Edwards for typing references. Supported by Dysautonomia Foundation, American Cancer Society, American Heart Association and NIH grants HD22486 and NS23915.

References

Appel, B., and Eisen, J. S. (1998). Regulation of neuronal specification in the zebrafish spinal cord by Delta function. *Development* **125,** 371–380.

Appel, B., Korzh, V., Glasgow, E., Thor, S., Edlund, T., Dawid, I. B., and Eisen, J. S. (1995). Motoneuron fate specification revealed by patterned LIM homeobox gene expression in embryonic zebrafish. *Development* **121,** 4114–4125.

Beattie, C. E., Stoesser, K. L., and Eisen, J. S. (1995). Motoneuronal mutants in embryonic zebrafish. *Soc. Neurosci. Abstr.* **21,** 1511.

Beattie, C. E., and Eisen, J. S. (1996). Mutations affecting motoneuronal pathfinding in embryonic zebrafish. *Soc. Neurosci. Abstr.* **22,** 1716.

Beattie, C. E., and Eisen, J. S. (1997). Notochord alters the permissiveness of myotome for pathfinding by an identified motoneuron in embryonic zebrafish. *Development* **124,** 713–720.

Beattie, C. E., Hatta, K., Halpern, M. E., Liu, H., Eisen, J. S., and Kimmel, C. B. (1997). Temporal separation in specification of primary and secondary motoneurons in zebrafish. *Dev. Biol.* **187,** 171–182.

Brand, M., Heisenberg, C.-P., Warga, R., Pelegri, F., Karlstrom, R. O., Beuchle, D., Picker, A., Jiang, Y.-J., Furutani-Seiki, M., van Eeden, F. J. M., Granato, M., Haffter, P., Hammerschmidt, M., Kane, D., Kelsh, R., Mullins, M., Odenthal, J., and Nüsslein-Volhard, C. (1996). Mutations affecting development of the midline and general body shape during zebrafish embryogenesis. *Development* **123,** 129–142.

Chakrabarti, S., Streisinger, G., Singer, F., and Walker, C. (1983). Frequency of gamma-ray induced specific locus and recessive lethal mutations in mature germ cells of the zebrafish, *Brachydanio rerio. Genetics* **103,** 109–124.

Driever, W., Solnica-Krezel, L., Schier, A. F., Neuhauss, S. C. F., Malicki, J., Stemple, D. L., Stainier, D. Y. R., Zwartkruis, F., Abdelilah, S., Rangini, Z., Belak, J., and Boggs, C. (1996). A genetic screen for mutations affecting embryogenesis in zebrafish. *Development* **123,** 37–46.

Eisen, J. S. (1991). Determination of primary motoneuron identity in developing zebrafish embryos. *Science* **252,** 569–572.

Eisen, J. S. (1996). Zebrafish make a big splash. *Cell* **87,** 969–977.

Eisen, J. S., Myers, P. Z., and Westerfield, M. (1986). Pathway selection by growth cones of identified motoneurons in live zebrafish embryos. *Nature* **320,** 269–271.

Eisen, J. S., and Weston, J. A. (1993). Development of the neural crest in the zebrafish. *Dev. Biol.* **159,** 50–59.

Felsenfeld, A. L., Walker, C., Westerfield, M., Kimmel, C., and Streisinger, G. (1990). Mutations affecting skeletal muscle myofibril structure in the zebrafish. *Development* **108,** 443–459.

Gatchalian, C. L., and Eisen, J. S. (1992). Pathway selection by growth cones of ectopic motoneurons in embryonic zebrafish. *Neuron* **9,** 105–112.

Granato, M., van Eeden, F. J. M., Schach, U., Trowe, T., Brand, M., Furutani-Seiki, M., Haffter, P., Hammerschmidt, M., Heisenberg, C.-P., Jiang, Y.-J., Kane, D. A., Kelsh, R. N., Mullins, M. C., Odenthal, J., and Nüsslein-Volhard, C. (1996). Genes controlling and mediating locomotion behavior of the zebrafish embryo and larva. *Development* **123,** 399–413.

Haffter, P., Granato, M., Brand, M., Mullins, M. C., Hammerschmidt, M., Kane, D. A., Odenthal, J., van Eeden, F. J. M., Jiang, Y.-J., Heisenberg, C.-P., Kelsh, R. N., Furutani-Seiki, M., Vogelsang, E., Beuchle, D., Schach, U., Fabian, C., and Nüsslein-Volhard, C. (1996). The identification of genes with unique and essential functions in the development of the zebrafish, *Danio rerio. Development* **123,** 1–36.

Halpern, M. E., Ho, R. K., Walker, C., and Kimmel, C. B. (1993). Induction of muscle pioneers and floor plate is distinguished by the zebrafish *no tail* mutation. *Cell* **75,** 99–111.

Hammerschmidt, M., Pelegri, F., Mullins, M. C., Kane, D. A., van Eeden, F. J. M., Granato, M., Brand, M., Furutani-Seiki, M., Haffter, P., Heisenberg, C.-P., Jiang, Y.-J., Kelsh, R. N., Odenthal, J., Warga, R. M., and Nüsslein-Volhard, C. (1996). *dino* and *mercedes,* two genes regulating dorsal development in the zebrafish embryo. *Development* **123,** 95–102.

Hatta, K., and Kimmel, C. B. (1993). Midline structures and central nervous system coordinates in zebrafish. *Pers. Dev. Neurobiol.* **1,** 257–268.

Hatta, K., Kimmel, C. B., Ho, R. K., and Walker, C. (1991). The cyclops mutation blocks specification of the floor plate of the zebrafish CNS. *Nature* **350,** 339–341.

Henion, P. D., Raible, D. W., Beattie, C. E., Stoesser, K. L., Weston, J. A., and Eisen, J. S. (1996). Screen for mutations affecting development of zebrafish neural crest. *Dev. Genet.* **18,** 11–17.

Henion, P. D., Raible, D. W., Eisen, J. S., and Weston, J. A. (1994). Generation and identification of zebrafish neural crest mutants. *Dev. Biol.* **163,** 548.

Horstadius, S. (1950). "The Neural Crest: Its Properties and Derivatives in the Light of Experimental Research." Oxford University Press, New York.

Jackson, I. J. (1994). Molecular and developmental genetics of mouse coat color. *Ann. Rev. Genet.* **28,** 189–217.

Jiang, Y.-J., Brand, M., Heisenberg, C.-P., Beuchle, D., Furutani-Seiki, M., Kelsh, R. N., Warga, R. M., Granato, M., Haffter, P., Hammerschmidt, M., Kane, D. A., Mullins, M. C., Odenthal, J., van Eeden, F. J. M., and Nüsslein-Volhard, C. (1996). Mutations affecting neurogenesis and brain morphology in the zebrafish, *Danio rerio. Development* **123,** 205–216.

Johnson, S. L., Africa, D., Walker, C., and Weston, J. A. (1995a). Genetic control of adult pigment stripe development in zebrafish. *Dev. Biol.* **167,** 27–33.

Johnson, S. L., Africa, D., Horne, S., and Postlethwait, J. H. (1995b). Half-tetrad analysis in zebrafish: Mapping the *ros* mutation and the centromere of linkage group I. *Genetics* **139,** 1727–1737.

Kelsh, R. N., Brand, M., Jiang, Y.-J., Heisenberg, C.-P., Lin, S., Haffter, P., Odenthal, J., Mullins, M. C., van Eeden, F. J. M., Furutani-Seiki, M., Granato, M., Hammerschmidt, M., Kane, D. A., Warga, R. M., Beuchle, D., Vogelsang, L., and Nüsslein-Volhard, C. (1996). Zebrafish pigmentation mutations and the processes of neural crest development. *Development* **123,** 369–389.

Kimmel, C. B. (1989). Genetics and early development of zebrafish. *Trends Genet.* **5,** 283–288.

Kimmel, C. B., Kane, D. A., Walker, C., Warga, R. M., and Rothman, M. B. (1989). A mutation that changes cell movement and cell fate in the zebrafish embryo. *Nature* **337,** 358–362.

Kimmel, C. B., Warga, R. M., and Kane, D. A. (1994). Cell cycles, clonal strings, and the origin of the zebrafish central nervous system. *Development* **120,** 265–276.

Kimmel, C. B., and Westerfield, M. (1990). Primary neurons of the zebrafish. *In* "Signals and Sense: Local and Global Order in Perceptual Maps." (G. M. Edelman, W. E. Gall, and W. M. Cowan, Eds.), pp 561–588. Wiley, New York.

Marusich, M. F., Furneaux, H. M., Henion, P. D., and Weston, J. A. (1994). Hu neuronal proteins are expressed in proliferating neurogenic cells. *J. Neurobiol.* **25,** 143–155.

Melançon, E. (1994). The Role of Muscle Pioneers in Pathfinding of Primary Motoneurons in the Embryonic Zebrafish. M.S. Thesis, University of Oregon.

Moens, C. B., Yan, Y.-L., Appel, B., Force, A., and Kimmel, C. B. (1996). *valentino:* A zebrafish gene required for normal hindbrain segmentation. *Development* **122,** 3981–3990.

Myers, P. Z. (1985). Spinal motoneurons of the larval zebrafish. *J. Comp. Neurol.* **263,** 555–561.

Myers, P. Z., Eisen, J. S., and Westerfield, M. (1986). Development and axonal outgrowth of identified motoneurons in the zebrafish. *J. Neurosci.* **6,** 2278–2289.

Neuhauss, S. C. F., Solnica-Krezel, L., Schier, A. F., Zwartkruis, F., Stemple, D. L., Malicki, J., Abdelilah, S., Stainier, D. Y. R., and Driever, W. (1996). Mutations affecting craniofacial development in zebrafish. *Development* **123,** 357–367.

Nüsslein-Volhard, C., and Wieschaus, E. (1980). Mutations affecting segment number and polarity in *Drosophila. Nature* **287,** 795–801.

Odenthal, J., Rossnagel, K., Haffter, P., Kelsh, R. N., Vogelsang, E., Brand, M., van Eeden, F. J. M., Furutani-Seiki, M., Granato, M., Hammerschmidt, M., Heisenberg, C.-P., Jiang, Y.-J., Kane, D. A., Mullins, M. C., and Nüsslein-Volhard, C. (1996). Mutations affecting xanthophore pigmentation in the zebrafish, *Danio rerio. Development* **123,** 391–398.

Piotrowski, T., Schilling, T. F., Brand, M., Jiang, Y.-J., Heisenberg, C.-P., Beuchle, D., Grandel, H., van Eeden, F. J. M., Furutani-Seiki, M., Granato, M., Haffter, P., Hammerschmidt, M., Kane, D. A., Kelsh, R. N., Mullins, M. C., Odenthal, J., Warga, R. M., and Nüsslein-Volhard, C. (1996). Jaw and branchial arch mutants in zebrafish II: Anterior arches and cartilage differentiation. *Development* **123,** 345–356.

Postlethwait, J. H., and Talbot, W. S. (1997). Zebrafish genomics: From mutants to genes. *Trends Genet.* **13,** 183–190.

Raible, D. W., and Eisen, J. S. (1994). Restriction of trunk neural crest cell fate in the embryonic zebrafish. *Development* **120,** 495–503.

Raible, D. W., Henion, P. D., Stoesser, K. S., Weston, J. A., and Eisen, J. S. (1994). Genetic analysis of zebrafish neural crest. *Soc. Neurosci. Abstr.* **20,** 653.

Raible, D. W., Wood, A., Hodsdon, W., Henion, P., Weston, J. A., and Eisen, J. S. (1992). Segregation and early dispersal of neural crest cells in the embryonic zebrafish. *Dev. Dyn.* **195,** 29–42.

Schier, A. F., Neuhauss, S. C. F., Harvey, M., Malicki, J., Solnica-Krezel, L., Stainier, D. Y. R., Zwartkruis, F., Abdelilah, S., Stemple, D. L., Rangini, Z., Yang, H., and Driever, W. (1996). Mutations affecting the development of the embryonic zebrafish brain. *Development* **123,** 165–178.

Schilling, T. F. (1993). Cell Lineage and Mutational Studies of Cranial Neural Crest Development in the Zebrafish Embryo. Ph.D. Thesis, University of Oregon.

Schilling, T. F., and Kimmel, C. B. (1994). Segment and cell type lineage restrictions during pharyngeal arch development in the zebrafish embryo. *Development* **120,** 483–494.

Schilling T. F., and Kimmel, C. B. (1997). Musculoskeletal patterning in the pharyngeal segments of the zebrafish embryo. *Development* **124,** 2945–2960.

Schilling, T. F., Piotrowski, T., Grandel, H., Brand, M., Heisenberg, C.-P., Jiang, Y.-J., Beuchle, D., Hammerschmidt, M., Kane, D. A., Mullins, M. C., van Eeden, F. J. M., Kelsh, R. N., Furutani-Seiki, M., Granato, M., Haffter, P., Odenthal, J., Warga, R. M., Trowe, T., and Nüsslein-Volhard, C. (1996). Jaw and branchial arch mutants in zebrafish I: branchial arches. *Development* **123,** 329–344.

Schilling, T. F., Walker, C., and Kimmel, C. B. (1996). The *chinless* mutation and neural crest cell interactions in zebrafish jaw development. *Development* **122,** 1417–1426.

Seeger, M., Tear, G., Ferres-Marco, D., and Goodman, C. S. (1993). Mutations affecting growth cone guidance in *Drosophila:* Genes necessary for guidance toward and away from the midline. *Neuron* **10,** 409–426.

Solnica-Krezel, L., Schier, A. F., and Driever, W. (1994). Efficient recovery of ENU-induced mutations from the zebrafish germline. *Genetics* **136,** 1401–1420.

Solnica-Krezel, L., Stemple, D. L., Mountcastle-Shah, E., Rangini, Z., Neuhauss, S. C. F., Malicki, J., Schier, A. F., Stainier, D. Y. R., Zwartkruis, F., Abdelilah, S. and Driever, W. (1996). Mutations affecting cell fates and cellular rearrangements during gastrulation in zebrafish. *Development* **123,** 67–80.

Stewart, W. W. (1978). Functional connections between cells as revealed by dye-coupling with a highly fluorescent naphthalimide tracer. *Cell* **14,** 741–759.

Streisinger, G., Coale, F., Taggart, C., Walker, C., and Grunwald, D. J. (1989). Clonal origin of cells in the pigmented retina of the zebrafish eye. *Dev. Biol.* **131,** 60–69.

Streisinger, G., Singer, F., Walker, C., Knauber, D., and Dower, N. (1986). Segregation analysis and gene-centromere distances in zebrafish. *Genetics* **12,** 311–319.

Streisinger, G., Walker, C., Dower, N., Knauber, D., and Singer, F. (1981). Production of clones of homozygous diploid zebra fish *(Brachydanio rerio). Nature* **291,** 293–296.

Talbot, W. S., Trevarrow, B., Halpern, M. E., Melby, A. E., Far, G., Postlethwait, J. H., Jowett, T., Kimmel, C. B., and Kimelman, D. (1995). A homeobox gene essential for zebrafish notochord development. *Nature* **378,** 150–157.

Tanabe, Y., and Jessell, T. M. (1996). Diversity and pattern in the developing spinal cord. *Science* **274,** 1115–1123.

Tokumoto, M., Gong, Z., Tsubokawa, T., Hew, C. L., Uyemura, K., Hotta Y., and Okamoto, H. (1995) Molecular heterogeneity among primary motoneurons and within myotomes revealed by the differential mRNA expression of novel islet-1 homologs in embryonic zebrafish. *Dev. Biol.* **171,** 578–589.

Trevarrow, B., Marks, D. L., and Kimmel, C. B. (1990). Organization of hindbrain segments in the zebrafish embryos. *Neuron* **4,** 669–679.

Van Vactor, D., Sink, H., Fambrough, D., Tsoo, R., and Goodman, C. S. (1993). Genes that control neuromuscular specificity in *Drosophila. Cell* **73,** 1137–1153.

Walker, C., and Streisinger, G. (1983). Induction of mutations by gamma-rays in pregonial germ cells of zebrafish embryos. *Genetics* **103,** 125–136.

Weast, R. C., Ed. (1970). "Handbook of Tables for Mathematics." 4th ed. Chemical Rubber Co., Cleveland, OH.

Westerfield, M. (1993). "The Zebrafish Book: A Guide for the Laboratory Use of Zebrafish (*Brachydanio rerio*)," 2nd ed. University of Oregon Press, Eugene, OR.

Westerfield, M., Liu, D. W., Kimmel, C. B., and Walker, C. (1990). Pathfinding and synapse formation in a zebrafish mutant lacking functional acetylcholine receptors. *Neuron* **4,** 867–874.

CHAPTER 5

Retrovirus-Mediated Insertional Mutagenesis in Zebrafish

Adam Amsterdam and Nancy Hopkins

Center for Cancer Research and Department of Biology
Massachusetts Institute of Technology
Cambridge, Massachusetts 02139

I. Introduction

Large-scale chemical mutagenesis screens have resulted in the isolation of thousands of mutations which affect zebrafish embryonic development (Haffter *et al.,* 1996; Driever *et al.,* 1996). These screens have used an alkylating agent, ethyl nitrosourea (ENU), to induce mutations, primarily by causing base-pair substitutions. Some of the genes disrupted by these mutations have been isolated, primarily via a candidate gene approach (e.g., Schulte-Merker *et al.,* 1994; Talbot *et al.,* 1995; Brand *et al.,* 1996; Kishimoto *et al.,* 1997), and at least one has been cloned by pure positional cloning (Zhang *et al.,* 1998). These examples demonstrate that the cloning of genes disrupted by chemical mutagenesis is possible, and many other chapters in this volume are devoted to this task. However, positional cloning remains an arduous task.

Insertional mutagenesis is an alternative to chemical mutagenesis in which exogenous DNA is used as the mutagen. Even though this inserted DNA is a less efficient mutagen than ENU, it serves as a molecular tag to aid in the isolation of the mutated gene. Several methods might be employed to insert DNA into

METHODS IN CELL BIOLOGY, VOL. 60

0091-679X/99 $30.00

the zebrafish genome, including DNA microinjection (Stuart *et al.,* 1988; Culp *et al.,* 1991), or microinjection aided by retroviral integrases (Ivics *et al.,* 1993) or a transposable element's transposase (Raz *et al.,* 1997). However, to date, by far the most efficient way to make a large number of independent insertions in the zebrafish genome is to use a pseudotyped retrovirus.

Retroviruses have an RNA genome and, upon infection of a cell, will reverse transcribe their genome to a DNA molecule, the provirus. The provirus integrates into a host cell chromosome where it remains stably and thus is inherited by all the descendants of that cell because infection is not lethal for the cell. Replication-defective retroviral vectors, unlike nondefective retroviruses, are infectious agents which can integrate like retroviruses, but whose genetic material lacks the coding sequences for the proteins required to make progeny virions. Retroviral vectors are made in "split genome" packaging cells in which the genome of the retroviral vector is expressed from one integrated set of viral sequences, whereas the retroviral genes required for packaging, for infection, for reverse-transcription, and for integration are expressed from another locus. The most widely used retroviral vectors have been derived from a murine retrovirus, Moloney Murine Leukemia Virus (MoMLV), resulting in replication-defective viruses that can be produced at very high titers; however, these retroviruses were initially only capable of infecting mammalian cells.

Retroviruses have a host range, or *tropism,* which is frequently determined by their envelope protein, which recognizes and binds to some specific component, usually a protein, on the surface of the cell to be infected. Cell types which have an appropriate receptor can be infected by the retrovirus, those that do not are refractory to infection. The tropism of a virus can be changed by pseudotyping, a process in which virions acquire the genome and core proteins of one virus but the envelope protein of another. One way to enable this situation in split genome packaging cells is to simply substitute the gene encoding the alternative envelope protein for the usual one. Although there is some specificity as to which envelope proteins can encapsidate specific viral genomes (any given genome can only be pseudotyped with a limited number of envelope proteins), one such combination that is particularly useful is the encapsidation of MoMLV with the envelope glycoprotein (G-protein) of vesicular stomatitis virus (VSV) (Weiss *et al.,* 1974). VSV is a rhabdovirus which is apparently pantropic; it can infect cells of species as diverse as insects and mammals (Wagner, 1972). MoMLV vectors pseudotyped with vesicular stomatitis virus G protein (VSV-G) possess two qualities essential for their use in high-frequency germline transgenesis in zebrafish: the extended host range allows for the infection of fish cells, and the VSV-G pseudotyped virions are unusually stable, allowing viruses to be concentrated 1000-fold by centrifugation.

When pseudotyped retroviral vectors are injected into zebrafish blastulae, many of the cells become independently infected, producing a mosaic organism in which different cells harbor insertions at different chromosomal sites. When cells destined to give rise to the germline are infected, some proportion of the progeny of the injected fish will contain one or more insertions (Lin *et al.,* 1994). When sufficiently high titer virus is used, one can infect a very high proportion of the germline of injected fish. On average, about a dozen independent insertions can be inherited from a single founder, although any given insertion will be present only in about 2–5% of the offspring (Gaiano *et al.,* 1996a). However, the progeny are nonmosaic for the insertions and transmit them in a Mendelian fashion to 50% of their progeny. Furthermore, because more than one virus can infect a single cell, some germ cells contain multiple insertions, and thus offspring can be born with two, three, or even more independently segregating insertions (Gaiano *et al.,* 1996a). This remarkable transgenesis rate made it possible to conduct a pilot insertional mutagenesis screen (Gaiano *et al.,* 1996b; Allende *et al.,* 1996; Amsterdam *et al.,* 1997) and should allow for large-scale screens capable of isolating hundreds of insertional mutants.

II. Insertional Mutagenesis: A Pilot Screen

In order to establish the efficacy of retroviral vectors as mutagens in the fish, we inbred about 600 proviral insertions and screened for recessive phenotypes which could be visually scored in the first 5 days of embryonic development. We found six recessive embryonic lethal mutations, a frequency of about one mutation per 100 insertions (Gaiano *et al.,* 1996b; Allende *et al.,* 1996; Amsterdam *et al.,* 1997). We also found one viable dominant insertional mutation. Here we explain how this pilot screen was conducted, briefly describe the phenotypes of the mutants, and describe how the affected genes were cloned and what criteria were used to establish that the correct genes had been cloned.

A. Finding Insertional Mutants

This screen was conducted using a "single insertion" breeding scheme (see Fig. 1). Injected fish (founders) were outcrossed to nontransgenic fish to make F_1 pools. Because the germlines of the founders are mosaic, any given insertion would be inherited by a small proportion of their progeny, generally from 1 to 15%. Transgenic F_1 offspring were identified by polymerase chain reaction (PCR) analysis of their DNA from fin clips, and Southern analysis was then used to classify fish by the different insertions that they harbored by diagnostic junction fragments. For insertions inherited by several fish, all the fish with that insertion were kept as a family for inbreeding. For insertions harbored by few F_1 fish, a single fish was outcrossed to generate an F_2 family, half of which inherited the insertion; transgenic F_2 fish were subsequently identified by PCR.

In either case, there were small families of fish for each insertion. The progeny of inbred pairs were then screened for phenotypic abnormalities in 25% of the clutch. However, because the starting population of fish had background mutations, when a mutation was found, it was essential to confirm that it was

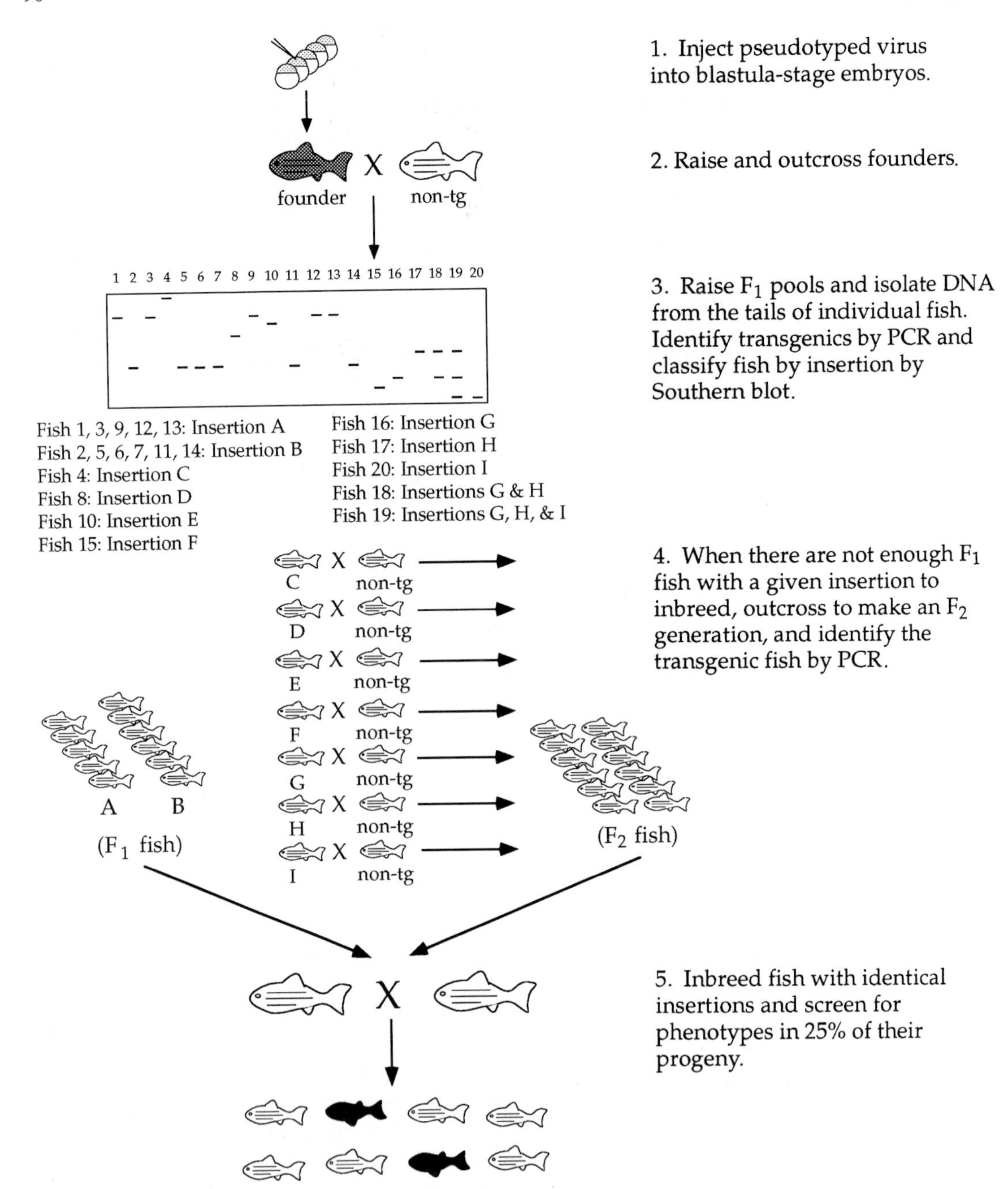

Fig. 1 Insertional mutagenesis: pilot screen (single insertion breeding scheme).

caused by the insertion. Because a background mutation would very rarely be linked by chance to an insertion, most could be easily distinguished from true insertional mutants by one of two means. First, only half of the fish in the family would carry a mutation if it was not linked to the insertion, thus only one-quarter of crosses amongst fish in the family should be expected to produce the phenotype if it is a background mutation. Thus when it was possible to get several mating pairs from the family, background mutations could be easily identified as those which were not found in the progeny of all pairs. Second, if the mutation and the insertion were linked, then all the phenotypic embryos would be transgenic; if the mutation and the insertion were unlinked, then only 75% of the phenotypic embryos would be transgenic. This was determined by performing PCR upon DNA prepared from individual mutant and wild-type embryos. Of course, these criteria demonstrate only whether or not the insertion and the mutation are completely unlinked, which would be expected for at least 98% of the background mutations. However, it is possible for an insertion to be linked to a background mutation by as much as 20 cM and still appear linked by these criteria. Thus while these crude linkage tests were useful for sifting out nearly all the background mutations, they were not considered proof that a mutation was caused by an insertion; this required much finer linkage analysis, as described next.

After a mutation which was likely to be insertionally induced was found, the next step was to use inverse PCR to clone a fragment of genomic DNA flanking the insertion. This DNA served two purposes. First, it was used as a probe on Southern blots to distinguish transgenic from nontransgenic chromosomes. This allows for finer linkage analysis to demonstrate that the insertion is tightly linked to the mutation and thus likely to be its cause. Second, it was used as the starting point for identifying the mutated gene.

Tight linkage was established by genotyping individual wild-type and mutant progeny from a cross between two heterozygotes. Since transgenic and nontransgenic chromosomes could be distinguished in Southern blots probed with the appropriate junction fragment, individuals could be genotyped as nontransgenic, transgene–heterozygous, or transgene–homozygous. If the mutation and insertion were tightly linked, each mutant would be homozygous for the insertion, and all of the wild types would be either heterozygous or nontransgenic. Every mutant analyzed is the equivalent to observing two meiotic events (one from each parent); every wild type analyzed is the equivalent of observing two-thirds of a meiosis (only one in three recombination events between a mutation and a marker in a dihybrid cross will lead to a wild-type embryo which is homozygous for the marker; thus scoring for wild types which are homozygous for a marker detects only one-third of the recombination events between these loci). If any recombinants are observed between the mutation and the insertion, the insertion could not be the cause of the mutation. For each mutation, several hundred meioses were analyzed in this fashion without finding any recombinants, demonstrating in each case that the mutation and the insertion were very closely linked (less than 1 cM) and thus likely to be identical.

The phenotypes of the seven insertional mutants are shown in Table I. Interestingly, similar to the distribution of phenotypes found in the chemical screens, 3/7 had nonspecific pleiotropic defects, 1/7 exhibited extensive cell death in the central nervous system as its primary phenotype, and 3/7 had specific defects.

The mutations with pleiotropic phenotypes for which the genes have been cloned are *no arches* (*nar*) and *pescadillo* (*pes*). Embryos homozygous for the *nar* mutation have reduced head structures as well as smaller pectoral fins and internal organs such as the liver and gut (Gaiano *et al.*, 1996b). The most dramatic feature of the phenotype is the nearly complete absence of neural-crest-derived cartilage in the head: both the pharyngeal arches and the anterior portion of the skull are absent or severely reduced and deformed. However, not all head structures are equally affected: the otic vesicles in the mutant are comparable in size to those in the wild type, as is the mesoderm-derived portion of the skull. Embryos mutant for *pes* exhibit a reduction in size or absence of many structures; they have a small head, eyes, and fin buds, underdeveloped liver and gut, reduced jaw arches and an absence of gill arches (Allende *et al.*, 1996). The common feature of all of these defects appears to be that each of these structures appears to be initially established but then fails to grow.

The *dead eye* (*dye*) mutation shares the phenotype of nearly 20% of the mutants found in the chemical screens: extensive apoptosis in the central nervous system (Allende *et al.*, 1996). Apoptosis can be seen initially in the tectum and hindbrain by the second day of development, and degeneration spreads thereafter.

The three specific phenotypes include *not really finished* (*nrf*), *891,* and *D1* (Amsterdam *et al.*, 1997). Embryos homozygous for the *nrf* mutation fail to develop normal photoreceptor cells in the retina. Even though the rest of the retina appears normal, the photoreceptor layer is largely absent; the few photoreceptors that are there appear misshapen (T. Becker, S. Burgess, and N. Hopkins, unpublished). The *891* mutation affects the yolk; fish homozygous for this insertion die apparently as a result of rotting yolk. By 36 hours postfertilization, the yolk begins to darken, followed by degeneration of the yolk extension. Eventually the yolk becomes even darker, and the embryo dies. *D1* is the viable dominant mutation. Fish heterozygous or homozygous for the *D1* mutation have an abnormal adult pigment pattern; the black melanophore stripes are interrupted and branched in the trunk region of adult fish (K. Kawakami and N. Hopkins, unpublished).

B. Cloning the Affected Genes

The DNA flanking the mutagenic insertions that we obtained by inverse PCR was used as a starting point to clone the mutated genes. So far we have cloned the genes associated with six of the seven insertional mutations (see Table I). For four of these, all that was required to find the genes was to sequence fewer than 2 kb of genomic sequences adjacent to the provirus, an amount easily cloned by inverse PCR. When these sequences were used in BLAST searches of the

Table I
Retrovirally Induced Insertional Mutations in Zebrafish

	Mutant	Phenotype	Disrupted gene
Nonspecific phenotypes	*no arches* (*nar*)[a]	Reduction of most head structures; absence of pharyngeal arches. Onset at day 3, embryonic lethal.	Homologue of bovine CPSF (cleavage and polyadenylation specificity factor);[b] also homologus to *Drosophila clipper,* an endoribonuclease.[c] Homologus expressed sequence tags found in mammals. Transcript maternally supplied and expressed zygotically in flies and fish.
	pescadillo (*pes*)[d]	Impaired growth of many structures, including the eyes, ears, arches, fins, liver, and gut. Onset at day 3, embryonic lethal.	Homologous to genes of unknown function in human, mouse, and yeast; contains BRCT superfamily motif.[e] Transcript supplied maternally and expressed zygotically in in embryonic sites which fail to grow in mutants.
	80A[a]	Reduction of head structures. Onset at day 2, embryonic lethal.	Unidentified.
Apoptosis phenotype	*dead eye* (*dye*)[d]	Excessive apoptosis in the eyes, brain, and neural tube. Onset at day 2, embryonic lethal.	Homologue of genes in human[f] and *Xenopus,*[g] similarity to yeast nuclear pore component NIC96.[h] Transcript supplied maternally and expressed in developing central nervous system in both frogs and fish.
Specific phenotypes	*not really finished* (*nrf*)[i]	Retinal photoreceptors absent or abnormal. Postembryonic lethal.	Homologue of transcription factors NRF-1[j] (human) and IBF-1[k] (chicken).
	D1[l]	Interrupted stripe pattern in adult pigmentation; dominant phenotype. Onset at 3 weeks. Homozygous viable.	Homologous to genes of unknown function in mammals. WD repeat domain containing protein.
	891[m]	Yolk darkens, yolk extension withers away, yolk rots. Onset at day 1, embryonic lethal.	Homologous to genes of unknown function in *C. elegans, Drosophila,* and mammals. Weak homology to bacterial transporters. Expressed in the cells surrounding the yolk and yolk extension.

[a] Gaiano *et al.,* 1996b
[b] Barabino *et al.,* 1997
[c] Bai and Tolias, 1996
[d] Allende *et al.,* 1996
[e] Bork *et al.,* 1997
[f] Nagase *et al.,* 1995
[g] Hudson *et al.,* 1996
[h] Grandi *et al.,* 1993
[i] T. Becker, S. Burgess, and N. Hopkins, unpublished
[j] Evans and Scarpulla, 1990
[k] Gomez-Cuadrado *et al.,* 1995
[l] K. Kawakami and N. Hopkins, unpublished
[m] M. Allende, H. Wang, S. Marty, S. Lin, and N. Hopkins, unpublished

public database, homologies were found to parts of coding regions of genes from other organisms, both genes of known function and the enormous number of expressed sequences tags and other random cDNAs coming out of assorted genome projects. These regions of homology were then used as a putative exon to fish out the rest of the gene, either by using 3′ and 5′ RACE (rapid amplification of cDNA ends) or by screening a cDNA library.

In the two cases in which a putative exon could not be found by using a few kilobases of flanking sequences to BLAST search, larger fragments of DNA had to be isolated. The flanking sequence was used as a probe to isolate either lambda clones or bacterial artificial chromosomes (BACs). Further sequencing and continued BLAST searches eventually found a putative exon for one; both RACE and the isolation of a cDNA lambda clone confirmed that this exon was genuine and allowed the mapping of most of the other exons of this gene, which showed that the insertion was present in the middle of a large intron. The use of an exon-trapping kit was also useful in finding exons in both of these cases. Once an exon was found with the trap, PCR was used with one primer in the putative exon and one in the provirus to map the distance from the exon to the insertion. The exon was then used to probe a cDNA library to find the entire gene.

Cloning a gene that is proximal to a mutagenic insertion is not, of course, absolute proof that the correct gene has been identified. The ultimate proof would be rescue: if the gene could be reintroduced in *trans* into mutant embryos and rescue the phenotype. Rescue is not easily accomplished in stable transgenics, but it can be done transiently for some mutations. Demonstration that the expression of the gene is abrogated in mutants makes a compelling case, short of the rescue experiment. This was done for all of the recessive mutations using either reverse transcriptase-PCR (RT-PCR) or northern analysis to analyze RNA prepared from pools of wild type and mutant embryos.

In several cases, elimination of gene expression in insertional mutants was also demonstrated by whole mount *in situ* hybridization. If a clutch of embryos from a heterozygote cross is hybridized with an antisense probe, one quarter of the embryos should reflect any loss of gene expression which results from the mutation. This has two advantages. One is that the loss of gene expression may be visible before the onset of the phenotype. This could be beneficial if the gene is normally expressed predominantly in tissues which are absent or reduced in the mutant. A reduction of RNA (as analyzed by northern) in mutants relative to wild type could be posited to be the result of the reduction of these tissues in mutant embryos, not reduction of gene expression per se. Thus the ability to see a reduction or absence of gene expression prior to this would be stronger evidence that the loss of gene expression is the cause, not consequence, of the mutation. Second, if the mechanism of the mutation is that expression in only part of the embryo is affected, *in situ* hybridization would be more likely to indicate this than analysis of whole embryo RNA. In several cases, *in situ* hybridization also revealed that the expression pattern of the gene in wild-type embryos was very suggestive that the correct gene had been identified. In three cases, *dye, pes,* and *891,* the gene is expressed in exactly the tissues which are affected in the mutants prior to the onset of the phenotype (Allende *et al.,* 1996; Amsterdam, 1998; H. Wang, S. Marty, and S. Lin, personal communication).

The only case in which expression of the cloned gene did not appear affected in the mutant is the dominant mutation *D1;* northern analysis indicates that wild-type, heterozygous, and homozygous fish all contain the same amount of transcript for the cloned gene. Because the mutation is dominant, it would not be surprising if it is a neomorph, rather than a null allele. The provirus resides in a very large downstream intron of the clone gene. It is possible that expression of the gene is affected in a subset of tissues in a way that we have not yet been able to detect, as has been demonstrated in the case of a retrovirally induced mouse mutation in which the provirus had integrated in a downstream intron (Seperack *et al.,* 1995). However, it is also possible that another, as of yet unidentified, gene is actually responsible for this mutation.

Of the six insertional mutants for which we have cloned the affected genes, only two were in "known" genes: the *nar* gene encodes the smallest subunit of the splicing and polyadenylation specificity factor (Barabino *et al.,* 1997), and the *nrf* gene encodes a transcription factor known as NRF-1 or IBF-1 in humans and chickens, respectively (Evans and Scarpulla, 1990; Gomez-Cuadrado *et al.,* 1995). Additionally, the *dye* gene has some homology to the *S. cerevisiae* NIC96 gene, which encodes a component of the nuclear pore, but the homology is not strong enough to ensure that it fulfills the same function (Grandi *et al.,* 1993). The other three genes are novel; even though they have homologues in other organisms, ranging from *C. elegans* to human, these are only sequences with no known function.

III. Considerations for a Large-Scale Screen

The results from the pilot screen suggest that insertional mutagenesis could be a very powerful tool for mutating and cloning genes in zebrafish. However, the question remains whether it would be possible to conduct a screen on a sufficient scale to approach the number of mutations isolated by chemical mutagenesis. At a mutagenic frequency of one mutation per 100 insertions, insertional mutagenesis is 100 times less efficient than chemical mutagenesis. However, if every family (and thus every fish tank) contained many more insertions, insertional mutagenesis could be much more efficient. One way of achieving such a "multiple insertion" screen is shown in Fig. 2. This scheme takes advantage of the fact that founders often experience multiple proviral integration events in the same primordial germ cell. Here, one generates F_1 fish with multiple insertions (at least three) which are then used to make F_2 families with six or more segregating insertions.

In order to enrich for fish with multiple insertions, founders are crossed to each other so that insertions can be inherited from each parent. At least 20%

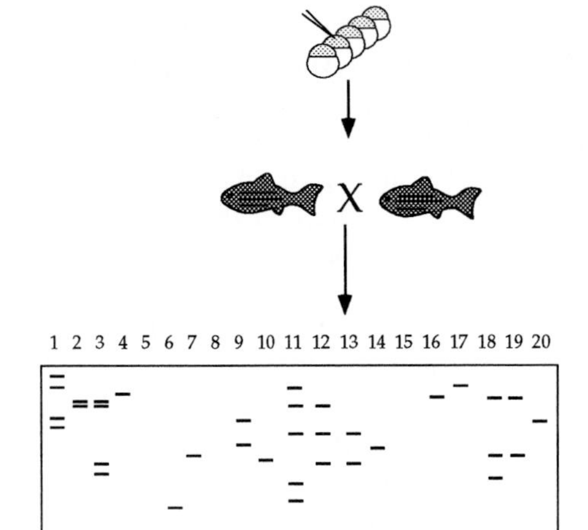

1. Inject pseudotyped virus into blastula-stage embryos.

2. Raise and inbreed founders.

3. Raise F_1 pools and isolate tail DNA from individual fish. Use Southern analysis to identify fish with multiple nonoverlapping insertions.

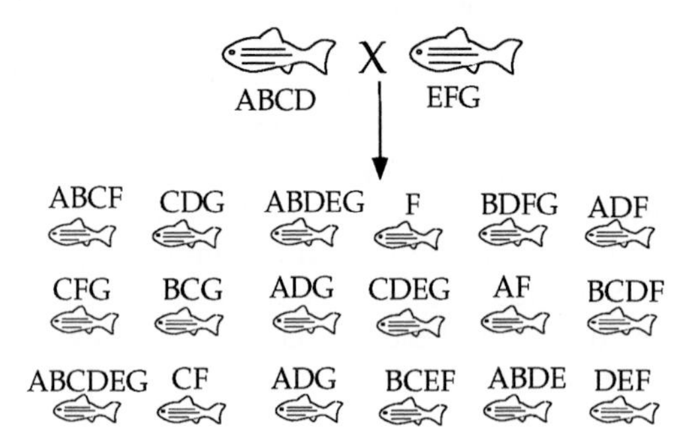

4. Cross multi-insert F_1 fish to each other to generate F_2 pools with 6-8 insertions, in which half of the fish have any given insertion.

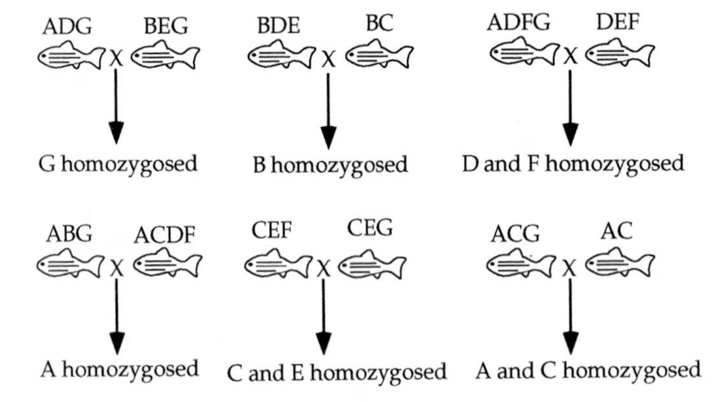

5. Screen at least six crosses within each F_2 pool. In this manner each insertion will be homozygosed in at least one of the crosses.

Fig. 2 Insertional mutagenesis: multiple insertion breeding scheme.

of the offspring of two crossed founders should harbor three or more insertions. Southern analysis can be used to identify these multiinsertion F_1 fish and to identify those with mostly nonoverlapping insertions. These multiinsert F_1 can then be crossed to each other to generate F_2 families in which there are six or more insertions. Multiple crosses are then set up within each F_2 family; any given insertion will be in half the fish in an F_2 family; thus one-quarter of the crosses will homozygose any given insertion. If six crosses are screened per F_2 family, then each insertion will have an 82% probability of being homozygosed. While in the chemical screens, at least one mutation was found for each F_2 family screened, in a multiple insertion screen one mutation should be found in every 10 to 15 families.

We estimate that a lab with 20 people and 4000 tanks could screen 60,000 to 90,000 insertions in 2 years. Based on the frequency of insertions found in the pilot screen, a 75,000 insertion screen should generate 750 to 1,000 mutations. If all of the estimated 2400 embryonic essential genes are mutable by proviral insertion, this should represent insertional alleles of 650 to 820 genes (based upon a random Poisson distribution). Because the chemical screens were thought to reach about half-saturation, at least half of these can be expected to be easily clonable alleles of previously identified loci, whereas the rest should be novel mutants.

References

Allende, M. L., Amsterdam, A., Becker, T., Kawakami, K., Gaiano, N., and Hopkins, N. (1996). Insertional mutagenesis in zebrafish identifies two novel genes, *pescadillo* and *dead eye,* essential for embryonic development. *Genes Dev.* **10,** 3141–3155.

Amsterdam, A., Yoon, C., Allende, M. Becker, T., Kawakami, K., Burgess, S., Gaiano, N., and Hopkins, N. (1997). Retrovirus-mediated insertional mutagenesis in zebrafish and identification of a molecular marker for embryonic germ cells. *Cold Spring Harbor Symp. Quant. Biol.* **62,** 437–450.

Amsterdam, A. (1998). Use of a pseudotyped retroviral vector to accomplish insertional mutagenesis in zebrafish. PhD Thesis. Massachusetts Institute of Technology.

Bai, C., and Tolias, P. P. (1996). Cleavage of RNA hairpins mediated by a developmentally regulated CCCH zinc-finger protein. *Mol. Cell Biol.* **16,** 6661–6667.

Barabino, S. M. L., Hubner, W., Jenny, A., Minvielle-Sebastia, L., and Keller, W. (1997). The 30-kD subunit of mammalian cleavage and polyadenylation specificity factor and its yeast homolog are RNA-binding zinc finger proteins. *Genes Dev.* **11,** 1703–1716.

Bork, P., Hofmann, K., Bucher, P., Neuwald, A. F., Altschul, S. F., and Koonin, E. V. (1997). A superfamily of conserved domains in DNA damage-responsive cell cycle checkpoint proteins. *FASEB* **11,** 68–76.

Brand, M., Heisenberg, C.-P., Jiang, Y.-J., Beuchle, D., Lun, K., Furutani-Seiki, M., Granato, M., Haffter, P., Hammerschmidt, M., Kane, D., Kelsh, R., Mullins, M., Odenthal, J., van Eeden, F. J. M., and Nusslein-Volhard, C. (1996). Mutations in zebrafish genes affecting the formation of the boundary between midbrain and hindbrain. *Development* **123,** 179–190.

Culp, P., Nusslein-Volhard, C., and Hopkins, N. (1991). High-frequency germ-like transmission of plasmid DNA sequences injected into fertilized zebrafish eggs. *Proc. Natl. Sci. USA* **88,** 7953–7957.

Driever, W., Solnica-Krezel, L., Schier, A. F., Neuhauss, S. C. F., Malicki, J., Stemple, D. L., Stainier, D. Y. R., Zwartkruis, F., Abdelilah, S., Rangini, Z., Belak, J., and Boggs, C. (1996). A genetic screen for mutations affecting embryogenesis in zebrafish. *Development* **123,** 37–46.

Evans, M. J., and Scarpulla, R. C. (1990). NRF-1: A trans-activator of nuclear-encoded respiratory genes in animal cells. *Genes Dev.* **4,** 1023–1034.

Gaiano, N., Allende, M., Amsterdam, A., Kawakami, K., and Hopkins, N. (1996a). Highly efficient germ-line transmission of proviral insertions in zebrafish. *Proc. Natl. Acad. Sci. USA* **93,** 7777–7782.

Gaiano, N., Amsterdam, A., Kawakami, K., Allende, M., Becker, T., and Hopkins, N. (1996b). Insertional mutagenesis and rapid cloning of essential genes in zebrafish. *Nature* **383,** 829–832.

Gomez-Cuadrado, A., Martin, M., Noel, M., and Ruiz-Carrillo, A. (1995). Initiation binding receptor, a factor that binds to the transcription initiation site of the histone *h5* gene, is a glycosylated member of a family of cell growth regulators. *Mol. Cell. Biol.* **15,** 6670–6685.

Grandi, P., Doye, V., and Hurt, E. C. (1993). Purification of NSP1 reveals complex formation with 'GLFG' nucleoporins and a novel nuclear pore protein NIC96. *EMBO J.* **12,** 3061–3071.

Haffter, P., Granato, M., Brand, M., Mullins, M. C., Hammerschmidt, M., Kane, D. A., Odenthal, J., van Eeden, F. J. M., Jiang, Y-J., Heisenberg, C-P., Kelsh, R. N., Furutani-Seiki, M., Vogelsang, E., Beuchle, D., Schach, U., Fabian, C., and Nusslein-Volhard, C. (1996). The identification of genes with unique and essential functions in the development of the zebrafish. *Danio rerio. Development* **123,** 1–36.

Hudson, J. W., Alarcon, V. B., and Elinson, R. P. (1996). Identification of new localized RNAs in the Xenopus oocyte by differential display PCR. *Dev. Genet.* **19,** 190–198.

Ivics, Z., Izsvak, Z., and Hackett, P. B. (1993). Enhanced incorporation of transgenic DNA in zebrafish chromosomes by a retroviral integration protein. *Mol. Mar. Biol. Biotech.* **2,** 162–173.

Kishimoto, Y., Lee, K. H., Zon, L., Hammerschmidt, M., and Schulte-Merker, S. (1997). The molecular nature of zebrafish *swirl:* BMP2 function is essential during early dorsoventral patterning. *Development* **124,** 4457–4466.

Lin, S., Gaiano, N., Culp, P., Burns, J. C., Friedmann, T., Yee, J-K., and Hopkins, N. (1994). Integration and germ-line transmission of a pseudotyped retroviral vector in zebrafish. *Science* **265,** 666–669.

Nagase, T., Miyajima, N., Tanaka, A., Sazuka, T., Seki, N., Sato, S., Tabata, S., Ishikawa, K., Kawarabayasi, Y., Kotani, H., and Nomura, N. (1995). Prediction of the coding sequences of unidentified human genes III: The coding sequences of 40 new genes (KIAAA0081-0120) deduced by analysis of cDNA clones from human cell line KG-1. *DNA Res.* **2,** 37–43.

Raz, E., van Luenen, H. G. A. M., Schaerringer, B., Plasterk, R. H. A., and Driever, W. (1997). Transposition of the nematode *Caenorhabditis elegans Tc3* element in the zebrafish *Danio rerio. Curr. Biol.* **8,** 82–88.

Schulte-Merker, S., van Eeden, F. J. M., Halpern, M. E., Kimmel, C. B., and Nusslein-Volhard, C. (1994). *no tail* (*ntl*) is the zebrafish homologue of the mouse T (*Brachyury*) gene. *Development* **120,** 1009–1015.

Seperack, P. K., Mercer, J. A., Strobel, M., Copeland, N. G., and Jenkins, N. A. (1995). Retroviral sequences located within an intron of the *Dilute* gene after *dilute* expression in a tissue-specific manner. *EMBO* **14,** 2326–2332.

Stuart, G. W., McMurray, J. V., and Westerfield, M. (1988). Replication, integration and stable germ-like transmission of foreign sequences injected into early zebrafish embryos. *Development* **103,** 403–412.

Talbot, W. S., Trevarrow, B., Halpern, M. E., Melby, A. E., Farr, G., Postlethwait, J. H., Jowett, T., Kimmel, C. B., and Kimmelman, D. (1995). A homeobox gene essential for zebrafish notochord development. *Nature* **378,** 150–157.

Wagner, R. R. (1972). Rhabdoviridae and their replication. *In* "Fundamental Virology" (B. N. Fields and D. M. Knipe, eds.), pp. 489–503. New York: Raven Press.

Weiss, R. A., Boettiger, D., and Love, D. N. (1974). Phenotypic mixing between vesicular stomatitis virus and avian RNA tumor viruses. *Cold Spring Harbor Symp. Quant. Biol.* **39,** 913–918.

Zhang, J., Talbot, W. S., and Schier, A. F. (1998). Positional cloning identifies zebrafish *one-eyed pinhead* as a permissive EGF-related ligand required during gastrulation. *Cell* **92,** 241–251.

CHAPTER 6

Genetic Applications of Transposons and Other Repetitive Elements in Zebrafish

Zoltán Ivics,* Zsuzsanna Izsvák,* and Perry B. Hackett[†,1]

* Division of Molecular Biology
The Netherlands Cancer Institute
Amsterdam 1066CX, The Netherlands

† Institute of Human Genetics and
Department of Genetics and Cell Biology
University of Minnesota
St. Paul, Minnesota 55108

I. Introduction

Vertebrate genomes are estimated to encode approximately 50,000 to 100,000 genes that are either unique or belong to small families of fewer than about ten members. These genes require about 2×10^8 base pairs of information which, in the case of the zebrafish, reside in a haploid genome that is estimated to be

[1] Corresponding author: Department of Genetics and Cell Biology, University of Minnesota, 1445 Gortner Avenue, St. Paul, Minnesota 55108-1095

METHODS IN CELL BIOLOGY, VOL. 60

0091-679X/99 $30.00

about 1.7×10^9 base pairs (Chapter 1; Johnson *et al.*, 1996). Much of the remaining 90% of the zebrafish genome is composed of repetitive sequences that come in a wide range of types, sizes, and stabilities. Surprisingly, for an organism that has become a primary model system for studying the molecular basis of early vertebrate development, several fundamental features of the zebrafish genome are not known, including the contributions of families of repetitive elements. Classical examinations of the rates at which fragmented denatured DNA segments of vertebrate genomes reassociate, called C_0t analysis (Fig. 1), showed that vertebrate genomes are largely composed of repetitive classes of DNA called foldback DNA, highly repetitive DNA, middle-repetitive DNA, and single-copy DNA (Britten and Kohne, 1968). Several highly repetitive DNA sequences occur

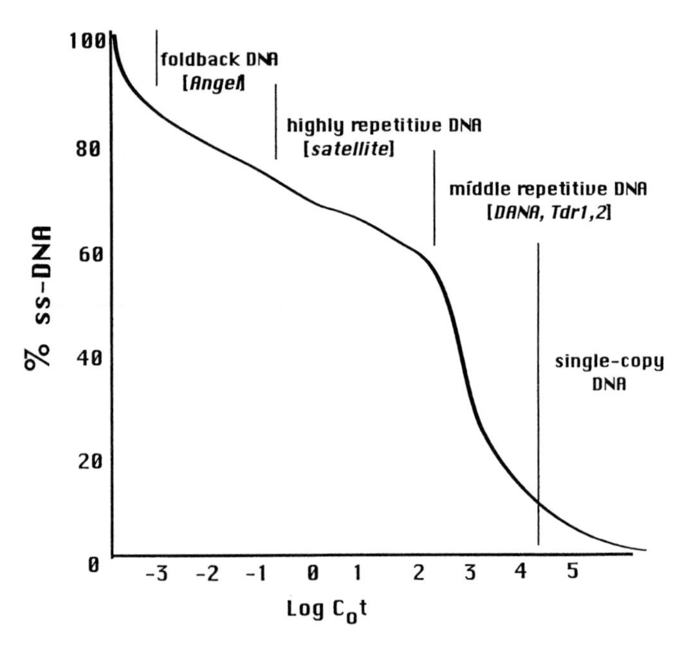

Fig. 1 A representative reassociation profile of vertebrate DNA. Purified DNA is sheared, boiled to separate the double helix into single strands, and then allowed to reanneal through gradual cooling. The percentage of single-stranded DNA declines as complementary sequences reassociate. The reannealing of the DNA strands is dependent on their original concentration (C_0) and the time over which they can pair with their complementary sequences (t), hence the name C_0t curve. The most rapidly reassociated DNA fragments are palindromic DNAs whose single strands can, for practical experimental purposes, instantly form *intra*-molecular stem loops called foldback DNA (e.g., Angel sequences in Fig. 2). Highly repetitive, satellite DNAs generally are composed of very short, reiterated sequences (e.g., CA repeats) where the alignment is not critical for reassociation because slippage between strands will still provide stable duplexes and tandemly repeated structural sequences such as those that comprise centromeres present at 10^4 to 10^5 copies or more per genome. Middle repetitive sequences are present in genomes at 10^2 to 10^4 copies whereas the single-copy sequences vary from about 1–10 copies per genome. The sequences discussed in this review are in brackets.

in arrays of hundreds to hundreds of thousands of tandem, short monomeric units. When chromatin from teleost fish is sheared and centrifuged in equilibrium-density gradients, these highly reiterated tandem repeats often form bands of unique density leading to the name "satellite" DNAs (Franck *et al.,* 1991).

Repetitive DNA sequences are not passive, simple sequences that stuff genomes. Several types such as those that compose centromeres and telomeres play essential roles in chromosomal dynamics during cellular replication. Many other repetitive elements are mobile. Some move or amplify by known mechanisms, whereas others behave in unpredictable and unknown ways. Although repetitive elements can interfere with experiments when their presence is not known ahead of time (e.g., during chromosome walks), there are circumstances when they can be used for experimental purposes. Here we review the types of repetitive elements that have been found in the zebrafish genome and how these sequences can be used as genetic tools for molecular and developmental studies in zebrafish. Consequently, we do not cover certain families of repetitive elements that encode well-studied gene products such as rRNAs, tRNAs, snRNAs, histones, and other gene families; these repetitive DNA sequences are classical genes which are conventionally reviewed in the contexts of their encoded products. Likewise, simple sequence repetitive elements such as CA repeats are reviewed in other chapters of this volume in terms of their specific uses in zebrafish.

The types, numbers, conservation, and chromosomal locations of repetitive elements provide important clues about chromosome dynamics, evolutionary forces, and mechanisms for exchange of genetic information between organisms. In particular, one class of repetitive elements, DNA-type transposable elements, have many potential applications in genetic research. They are routinely used for insertional mutagenesis, gene mapping, gene tagging, and gene transfer in simpler model systems such as bacteria, yeast, nematodes, and flies. These methods are being developed for zebrafish to facilitate the identification, mapping, and isolation of genes involved in vertebrate growth and development as well as the investigation of the evolutionary processes that have been shaping vertebrate genomes. Accordingly, we and others have embarked on an intensive search to identify, characterize, and explore the genetic uses of repetitive elements in zebrafish.

Of special importance are transgenic vectors based on transposons. The synthetic transposons described in this chapter should find wide use in genetic studies of the functions of particular genes during vertebrate development.

II. Classification of Repetitive Elements

There are several ways to classify repetitive elements. Repetitive elements are characterized in terms of their lengths, copy numbers, arrangements, functions, modes of amplification, and mobilities within a single genome and between genomes. Figure 2 shows our classification of the repetitive sequences reviewed in this chapter. This classification scheme is based on both function and mode of mobility of the repetitive elements.

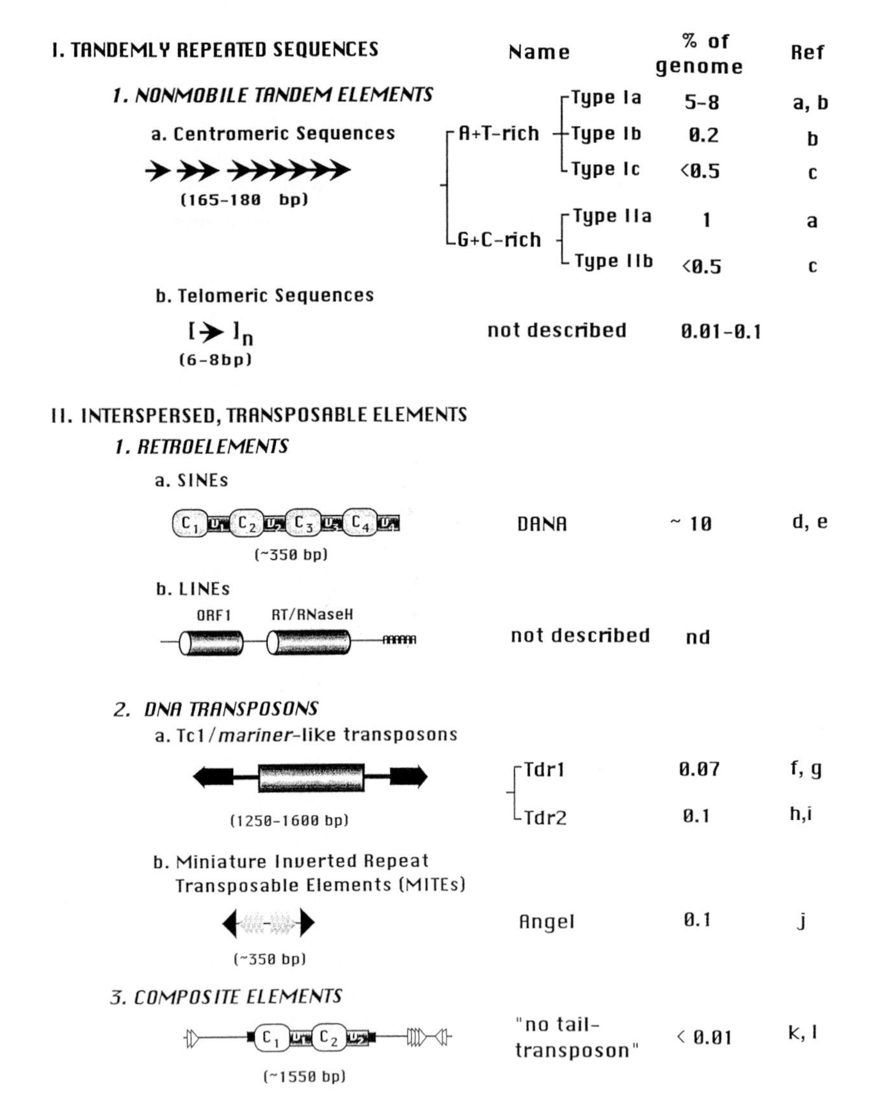

Fig. 2 Repetitive elements in zebrafish. Characteristic motifs are schematized. Elements designated by arrows indicate different lengths and/or variations in sequence. The references: a, He *et al.*, 1992; b, Ekker *et al.*, 1992; c, Martin *et al.*, personal communication; d, Izsvák *et al.*, 1996; e, Shimoda *et al.*, 1996a; f, Radice *et al.*, 1994; g, Izsvák *et al.*, 1995; h, Ivics *et al.*, 1996; i, Lam *et al.*, 1996b; j, Izsvák *et al.*, 1998; k, Schulte-Merker *et al.*, 1994; l, A. Fritz, personal communication [modified from Izsvák *et al.* (1997)].

A. Nonmobile, Tandemly Repeated, Structural Sequences

Two structures present in every zebrafish chromosome are the centrally located centromeres and the telomeres at both ends of every chromosome. These structures are made up of tandem repeats of relatively simple DNA sequences. Because of their simple, highly repetitive nature, these sequences are referred to as *satellite* sequences.

1. Satellite Elements

The first identified repetitive DNA sequences in zebrafish were evident in agarose gels in which genomic DNA, cleaved with the restriction endonucleases *Alu*I, *Mbo*I, and *Sma*I, produced small, intensely stained ethidium bromide-bands (He *et al.,* 1992). The tandem orientation of these repetitive sequences was evident from the ladders of bands that could be generated by incomplete cleavage of the DNA with the enzymes. Because they were produced by the restriction enzyme *Alu*I, these bands were called *Alu* sequences based on precedent that similar, *Alu*I-generated fragments found in human DNA were called *Alu* elements (Schmid and Jelinek, 1982). This was most unfortunate because the elements in these two vertebrates are not related in any way other than they can be seen on gels following cleavage of the genomes with certain enzymes. The human *Alu* elements are SINE-type sequences (discussed in Section II.C.1.a) that are mobile and dispersed throughout the genome rather than being tandemly reiterated.

The features of the zebrafish "Alu elements" are similar to those of animal DNA sequences identified several decades ago by density gradient analysis. In these experiments, a fraction of the DNA from some animals produced satellite bands in CsCl gradients. Satellite DNA sequences are tandemly reiterated from about 10^5 to more than 10^6 times. The lengths of the repeating units vary from a few to several hundred base pairs. Satellite DNA is located primarily in the constitutively transcriptionally inactive, heterochromatic regions of chromosomes usually near centromeres and telomeres. These DNA elements are generally stable in animal genomes (Britten, 1994). The following two types of tandemly repeated DNA elements have been identified in zebrafish genomes.

a. [A + T]-Rich Tandem Elements

A family of highly repetitive elements comprise approximately 5–8% of the zebrafish genome (Ekker *et al.,* 1992; He *et al.,* 1992). These sequences were first called "Alu" elements but now are known by their subtypes, named Type Ia and Type Ib (He *et al.,* 1992). The zebrafish Type I elements have been localized to centromeric regions of zebrafish chromosomes by fluorescence *in situ* hybridization (Ekker *et al.,* 1996; R. Phillips, unpublished). The monomeric units are about 180 bp, about 65% [A + T], and approximately 90% conserved. A third sequence called ZTR8 with 72.5% [A + T], and a monomer length of 191 bp has been found; it has a low abundance of less than 0.5% in the zebrafish genome (Martin *et al.,* personal communication). In Fig. 2 it is named Type Ic to clarify its relationship with other similar elements. Though similar in overall properties, the specific sequences of Types Ia, Ib, and Ic DNAs are distinct.

b. [G + C]-Rich Tandem Elements

Alu/Type II elements have a [G + C] content of about 65% (He *et al.,* 1992) and monomer lengths of around 180 bp and are generally found in a tandem arrangement, although they are also found dispersed in the genome. They localize to the centromeres of most if not all chromosomes (Martin *et al.,* personal communication). Type IIa sequences make up about 1% of the zebrafish genome. A second Type II sequence called ZTR22 with about 50% [G + C] and a monomer length of 183 bp has a lower abundance of less than about 0.5% in the zebrafish genome (Martin *et al.,* personal communication). In Fig. 2 we refer to this sequence as Type IIb on the basis of its relationship with the Type IIa sequences. Together, the Type I and Type II DNAs make up about 10% of the zebrafish genome.

2. Telomeric Elements

There are no reported sequences for zebrafish telomeres. However, they must be present. And, on the basis of the conservation of the telomeric sequence $(TTAGGG)_n$ in many vertebrates including representative orders of bony fish (Cypriniformes and Salmoniformes), reptiles, amphibians, birds, and mammals (Meyne *et al.,* 1989), it is highly likely that zebrafish will have a similar repeated sequence. Generally, these sequences are reiterated several thousand fold at each end of a chromosome. Moreover, the motif can be found distributed at nontelomeric sites in vertebrate chromosomes, although the tandem arrays are not as extensive as at the ends of chromosomes (Meyne *et al.,* 1990). Accordingly, we have listed telomeric repeats in Fig. 2 because they are certain to exist. If their abundance in zebrafish is similar to that in other animals, then we expect 2×10^5 to 10^6 copies of the monomeric sequence per genome, amounting to about 0.1–0.5% of the chromosomal DNA.

B. Nonstructural, Tandemly Repeated Sequences

In contrast to the stable, long tandemly repeated DNA sequences that make up essential components of chromosomes, microsatellite (simple) sequences are short (2–6 bp), whereas minisatellites (variable number of tandem repeat, VNTR) are longer (>15 bp) nucleotide repeats (Charlesworth *et al.,* 1994). Microsatellites and VNTRs are highly variable and are found in the euchromatin. Unlike centromeric and telomeric elements, these sequences are often unstable.

1. Microsatellite Repeat Sequences

Microsatellite (MS) sequence repeats have been identified from sequencing of genomic DNAs. Microsatellite sequences are thought to arise by "slippage" of DNA polymerase during replication (Levinson and Gutman, 1987; Schlotterer and Tautz, 1992). A-Rich elements and certain dinucleotide MS families such as the CA-repeats are located preferentially in intercistronic sequences, introns, and untranslated regions of genes. The MS sequences are being detected at approximately the same frequencies in zebrafish chromosomes as in other vertebrate genomes. A preliminary survey to identify expressed sequence tags (ESTs) associated with differential gene expression in zebrafish, using developmental stage-specific and tissue-specific libraries, uncovered 18 MS sequences in 1189 partial cDNAs, 6 of which were CA-repeats (Gong *et al.,* 1997). The MS elements have also been found as integral parts of other, more elaborate repetitive elements in zebrafish that are discussed later. Thus, as in other vertebrates, MS sequences are present in the zebrafish genome and are being exploited to greater extents as their identities and map positions are determined.

2. Unclassified Tandem Repeat Elements

The myocyte enhancer factor 2C mRNA contains a tandemly repeated 130-bp motif starting 17 bp ahead of its translational termination codon. Consequently, this stop codon-containing repeat continues into the 2.3 kb 3′-UTR. The 130-bp repeated unit has 20 full or partial copies in a head-to-tail orientation (Breitbart, personal communication). The repeated region is followed by another type of repetitive element called DANA, which is described in the section on SINE sequences (Section II.C.1.a). The enhancer factor 2C element is not similar to any other identified repetitive element in zebrafish and might represent a single amplification event. If this motif is present elsewhere in the genome, its association with the DNA repetitive element could place it in the category of Mobile Composite Elements (Fig. 2).

C. Interspersed Transposable Elements

The most interesting class of repetitive elements are transposons because of their variability, their dynamic activities that can affect gene expression as well as chromatin structure, and their potential use as instruments for genetic and developmental studies. Transposable elements can be divided into two major groups, DNA transposons that move directly as DNA sequences and retrotransposons that require an RNA intermediate plus RNA-dependent DNA polymerase (reverse transcriptase) for their mobilization and amplification. Retroelements are generally more numerous, up to 5×10^5/genome, compared to DNA transposons that may exist from a few up to 10^4/genome. In zebrafish, a screen of about 4000 3′ cDNA sequences revealed 112 homologies to recognized human repetitive retroelements (Genebank, 1998), a frequency of about 2%. The mechanism of mobility of one class of abundant transposable elements, the miniature inverted-repeat transposable elements (MITEs, Section II.C.3), has not been experimentally verified. Some transposons do not move, but they are prone to being amplified. Consequently, because of their relative permanence and dispersion, these elements are extremely useful for evolutionary studies as well as gene mapping (discussed further in Section III.A.1.b). Mass transposition of the elements, perhaps as a result of "genomic stress" (Arnault and Dufournel, 1994), may lead to gross chromosomal rearrangements and speciation (Fondevila, 1993; Sheen *et al.,* 1993; Kim and Simmons, 1994; Kloeckener-Gruissem and Freeling, 1995; Kidwell and Lisch, 1997). In the following sections, we describe the characteristics of the transposable elements identified so far in zebrafish and how they can be used as tools for investigating gene expression during development in zebrafish.

1. Retroelements

Retroelements/retrotransposons transpose via an RNA intermediate. Two groups have been categorized, those with long terminal repeats (LTR-retrotransposons) and those without (non-LTR retrotransposons). LTR-retrotransposons can be subdivided into the Ty1/*copia* and Ty3/*gypsy*-type elements, named after their resemblance to retrotransposons in yeast and *Drosophila* (Voytas and Boeke, 1993). Ty1/*copia* and Ty3/*gypsy*-type elements have been found in some fish (Flavell and Smith, 1992; Britten *et al.,* 1995). In zebrafish, remnants of about eight human LTR-related elements have been discovered by homology screening of about 3972 zebrafish 3′ cDNA sequences from a variety of sources (GeneBank, 1998). A partial reverse transcriptase, *pol,* like cDNA sequence has been uncovered, suggesting that retrotransposition may be active.

There are three categories of non-LTR retrotransposons: SINEs (short interspersed nuclear elements) or retroposons. LINEs (long interspersed nuclear elements), and processed pseudogenes (Weiner *et al.,* 1986). These elements are characterized by the absence of repeated termini, the existence of an A-rich 3′-tail, and target-site duplications flanking their insertion points. Retroelements are dispersed nonrandomly in genomes; their integration sites are often close to or inside another retrotransposon or a Pol III-transcriptional unit (SanMiguel *et al.,* 1996; Voytas, 1996). Both SINEs and LINEs are thought to evolve rapidly as a sequential series of "master" elements that reproduce themselves via RNA intermediates (Deininger *et al.,* 1992; Takasaki *et al.,* 1994; Preston, 1996). Consequently, each species acquires its own subfamilies of related elements. This property of retroelements is useful for evolutionary studies and chromosome marking.

a. SINEs

SINEs are characterized by their relatively small size, up to about 500 bp, and high repetition, from 10^3 to 5×10^5 copies/genome. Most SINEs are thought to be derivatives of tRNAs (Daniels and Deininger, 1985). The first SINE element found in zebrafish is called DANA; it has a most novel structure in contrast to SINEs found in other vertebrates (Izsvák *et al.,* 1996). More recently, screens of ESTs have turned up a substantial number of partial sequences with homologies to the MER (medium reiteration-frequency repetitive sequences, 200 to 10^4

copies/genome) class of repetitive elements found in human genomes (Kaplan *et al.,* 1991). Of nearly 4000 partial zebrafish cDNAs, 47 had MER-related sequences in the 3′ ends (GeneBank, 1998), a frequency of about 1%. These reiterated sequences can interfere with the identification of ESTs. There may be many more sequences than this survey implies because many SINE elements prefer to integrate in nested clusters (Voytas, 1996).

b. DANA: A Composite Retroelement in Zebrafish

The DANA (also called *mermaid,* Shimoda *et al.,* 1996a) retroposon exhibits all the hallmarks of a tRNA-derived SINE, but it has an unusual substructure of distinct cassettes. In contrast to generic SINEs, DANA appears to have been assembled by insertions of pairs of short sequences, one of which is a "constant" or C-region and the other a "variable" or v-region sequence (C_v blocks in Fig. 3A), into a progenitor, tRNA-derived element (C_1v_1 block). This unit apparently was amplified further as a transposable unit. Three of the v-regions are composed

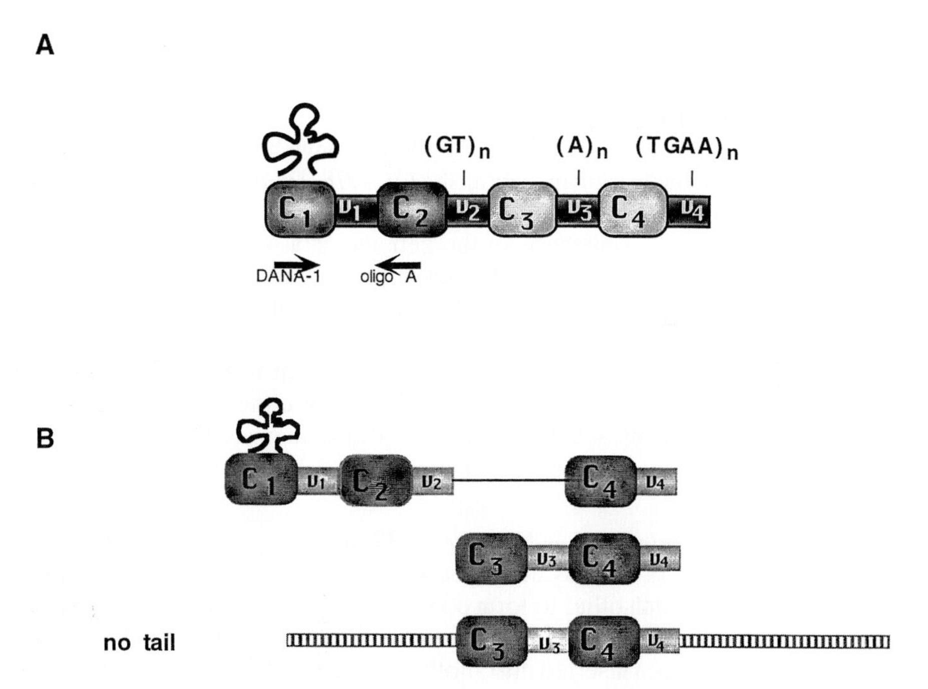

Fig. 3 DANA: a zebrafish retro-(SINE) element. (A) The four conserved regions (C_1–C_4) are indicated by the large shaded boxes, whereas the narrower boxes represent the variable regions (v_1v_4). The tRNA-related sequence is shown as a cloverleaf structure above C_1 and the simple sequence repeats found in variable regions 2, 3 and 4 are displayed. Recommended PCR primers for the IRS-PCR described in Figures 8 and 11 and Table I. (B) Partial and composite DANA elements that contain partial DANA sequences found in zebrafish. The $C_3v_3C_4v_4$ cassette that inserted in the composite *ntl*-transposon is shown at the bottom. [Modified from Izsvák *et al.* (1997).]

Table I
Primers for IRS-PCR of Repetitive Elements Found in the Zebrafish Genome

Repetitive element	Oligonucleotide primer	Reference
Angel (Angel-2)	5′-TTTCAGTTTTGGGTGAACTATCC	Izsvák *et al.* (1998)
DANA (DANA-1)	5′-GGCGACRCAGTGGCGCAGTRGG	Izsvák *et al.* (1996)
(DANA-A)	5′-AGAAYRTGCAAACTCCACACAGA	Izsvák *et al.* (1996)
Tdr1 (T1)	5′-TCCATCAGACCACAGGACAT	Izsvák *et al.* (1996)
(T2)	5′-TGTCAGGAGGAATGGGCCAAAATTC	Izsvák *et al.* (1996)

of mono-(v_3), di- (v_2), or tetranucleotide (v_4) microsatellite repeats. However, the composition of many DANA elements varies from the consensus structure, with different combinations of C_nv_n blocks (Fig. 3B). In one very interesting case, a block of DANA has apparently merged with another zebrafish repetitive element to form a new composite transposon, one of which caused an insertional mutation in the *ntl* gene (described in more detail in Section II.C.1.d). The possible evolution of the DANA elements has been reviewed (Izsvák *et al.,* 1997).

c. LINEs

Partial L1 LINE-like sequences have been found in EST screens but not fully characterized in zebrafish. Of 3978 *Danio rerio* cDNA sequences, 21 (0.5%) contained human LINE-1-like sequences (GeneBank, 1998). Extrapolating this rate of occurrence to the genome as a whole, there could be as many as 2000 expressed RNAs with such sequences and five- tenfold more for the genome as a whole. LINE elements are present at 10^4 to 10^5 copies in other vertebrate genomes. Active, full-length LINEs have two open reading frames, ORF-1 and ORF-2; ORF-2 appears to be equivalent to the *pol* (reverse transcriptase) gene of retroviruses (Fig. 2). LINEs may either be remnants of or precursors to retroviruses. Long terminal repeat sequences, the hallmark of retroviruses, have also been found in EST screens of zebrafish mRNAs.

d. Unclassified and Composite Elements

As previously noted for SINEs, transposable elements are capable of inserting into each other to form composite elements that can be amplified as new transposons. One example has been found in zebrafish in which the C_3C_4 block of DANA has been inserted into another sequence (Fig. 3B) that subsequently was amplified and inserted into the *ntl* gene (Schulte-Merker *et al.,* 1994; Izsvák *et al.,* 1996). The particular composite element that integrated into the *ntl* gene is amplified about 80- to 100-fold in the zebrafish genome (Fig. 2). This "ntl-transposon" has not been categorized because its mechanism of transposition is not known. Other, unidentified, composite elements surely exist in the zebrafish genome; their structures and modes of amplification remain to be determined.

2. DNA Transposons

Two types of DNA transposable elements that do not involve an RNA intermediate have been found in zebrafish. The first type, first found by Radice *et al.* (1994), is similar to the Tc1 element first identified in *Caenorhabditis elegans,* whereas the second type is a family of miniature, inverted-repeat transposable elements (Izsvák *et al.,* 1998). These two types of transposon have very distinctive features and transpose by different mechanisms. The Tc1-like elements are of special interest because they have provided the template for construction of powerful transposon vectors that are functional in fish as well as other vertebrates (Ivics *et al.,* 1997).

a. Tc1-like Transposable Elements (TcEs)

The Tc1/*mariner* superfamily of transposons has been found in many eukaryotic organisms, from protozoa to vertebrates—including humans (Doak *et al.,* 1994; Oosumi *et al.,* 1995; Avancini *et al.,* 1996). The best-studied members are the Tc1 and Tc3 transposons from *C. elegans* (Plasterk and van Luenen, 1997). Full-length TcEs are about 1.6 kb although lengths vary from 1.1 to 2.3 kb. These transposons, when complete, contain a single gene encoding a transposase. The gene is flanked by terminal inverted-repeats (IRs, Fig. 4). These transposons spread in ether of two ways (Fig. 5). First, the transposase can catalyze the complete excision of the transposon from its original location and promote its reintegration elsewhere in the genome by a "cut-and-paste" mechanism. Second, the transposon can be amplified as a result of the normal cellular process of double-stranded DNA break repair, during which the excision osite is repaired using the transposon-containing homologous chromosome as template (Plasterk, 1993). TcEs always integrate into a TA dinucleotide target site, which is duplicated upon insertion. However, not all TA sites are used with the same efficiency; there are insertion "hot spots" (van Luenen and Plasterk, 1994; Ketting *et al.,* 1997).

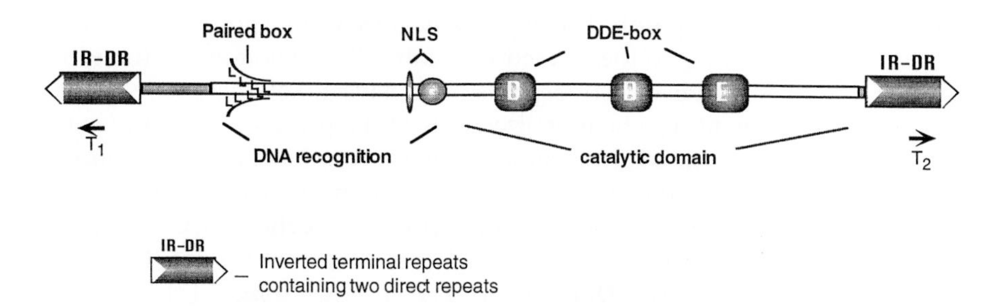

Fig. 4 Structural motifs of Tc1/*mariner*-like transposons from fish, with the conserved paired box motif (L–L–L), the bipartite NLS, the catalytic DDE domain, and the IR/DR inverted terminal-repeat sequences. T_1 and T_2 represent primers, discussed in Table I and Fig. 11.

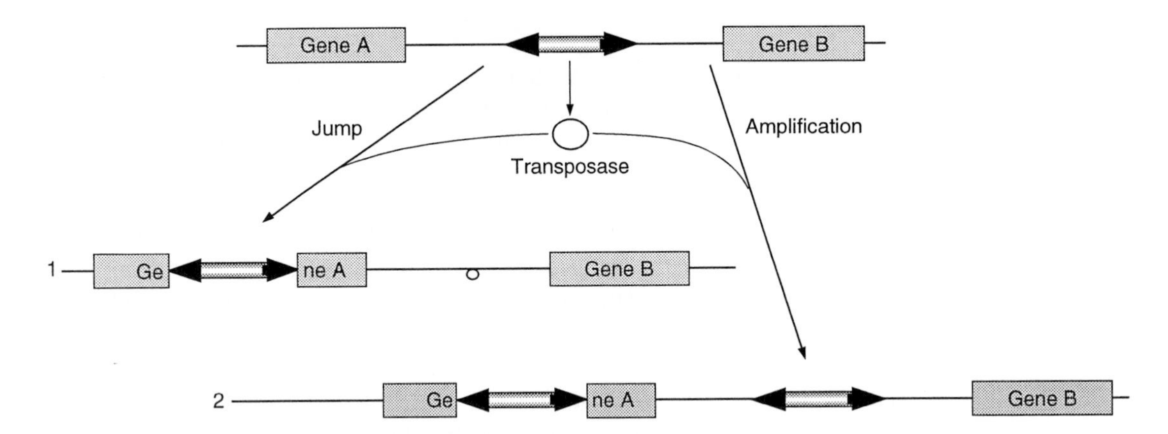

Fig. 5 Transposition and amplification of Tc1/*mariner*-like transposons from fish, with the conserved paired box motif (L–L–L), the bipartite NLS, the catalytic DDE domain, and the IR/DR inverted terminal-repeat sequences. (1) The jump to the left illustrates excision of the transposon from one chromosomal locus to another (in this case into gene A), leaving a "hole" or footprint in its former position. (2) Alternatively, during DNA replication, single-strand nicks by transposase can allow duplication of the transposon at another site (in this case, also in gene A). The transposase shown coming from the transposon could come as well from another transposase gene in another transposon.

Because of their extreme simplicity, TcEs are able to spread horizontally, invading genomes of many species, while leaving many others untouched. The evolutionary relationships of host genomes and their resident TcEs is often divergent (Izsvák *et al.,* 1996; 1997). Indeed, this spotty appearance of TcEs in some but not all related species is a cardinal signal of their abilities (and power) as molecular parasites. Nevertheless, most transposons do not jump or move except in rare circumstances. This is because many, if not most transposons, are inactive in their genomes for either of two reasons. First, there may not be a functional transposase gene in the genome because of genetic drift. Second, mutations can accumulate in the transposase-binding sites within the IRs of the transposon. Usually mobile DNA transposons are divided into two types according to their functional status. The *autonomous* transposable elements encode their own transposase and have functional IRs, whereas *nonautonomous* elements can be mobilized only in the presence of an active transposase, if their IRs are intact, from another transposon. The inactivity of most TcEs is mandated by their parasitic nature; if their jumping were not controlled, they would destabilize their host genomes. As discussed in Section III, consideration of these aspects is important in designing TcE-based vectors for genetic analyses.

Functional Domains of TcE Transposons. TcEs have three regions, the transposase gene that encodes the enzyme for mobilization and two flanking inverted repeats. Each of these regions has subdomains that are important for mobilization.

Domains of TcE Transposases. The transposase enzymes encoded by TcEs consist of several major motifs that are critical for transposition. Figure 4 shows the domains of the transposase gene of the major TcE in zebrafish, Tdr1. At the N-terminus of many (but not all) TcEs, there is a bipartite structure reminiscent of the *paired* domain found in the pax family of transcription factors (Vos and Plasterk, 1994) containing a leucine-zipper-like motif (Franz *et al.,* 1994; Ivics *et al.,* 1996). This motif may allow transposase molecules at either end of the transposon to interact with one another. Within the first 120 amino-terminal residues of the transposase is the DNA-binding domain which is specific for a particular transposon (Colloms *et al.,* 1994; Vos and Plasterk, 1994). Overlapping with the DNA-recognition domain, a nuclear localization signal (NLS) is conserved throughout the entire Tcl family, indicating that transposons can take advantage of the transport machinery of host cells for nuclear uptake of their transposases. The NLS is flanked by casein kinase II phosphorylation sites, suggesting that the activity of these transposons may be regulated through the signaling system of the host cell (Ivics *et al.,* 1996). The C-terminal halves contain a DDE domain, so named because of three conserved acidic amino acids in the domain, that catalyzes transposition; although mariner-like elements have a DDD domain (Hart *et al.,* 1997). The DDE domain is found in integrase and transposase proteins encoded by retroviruses, retrotransposons, and certain bacterial IS elements, which suggests a common mechanism of DNA cleavage and joining (Craig, 1995).

Domains of the Inverted Repeats, the IR/DR Motif. The prototype Tc1 transposon from *C. elegans* has short, 54-bp IRs flanking its transposase gene (Emmons *et al.,* 1983; Rosenweig *et al.,* 1983) and the IRs of other TcEs are often about 20–80 bp. In contrast, most TcEs from fish have longer, 210- to 250-bp, IRs at their termini and directly repeated DNA sequence (DR) motifs at the ends of each IR (hence IR/DRs in Fig. 4). In this respect, fish TcEs are similar to the *Minos* (Franz and Savakis, 1991), *Paris* (Petrov *et al.,* 1995), and *S* elements (Merriman *et al.,* 1995) from drosophilid fly species, and the TXz and TXr families in *Xenopus laevis* (Lam *et al.,* 1996b). These IR/DR elements form a group of TcEs on the basis of their organization (Izsvák *et al.,* 1995). Both direct repeats in the IR/DR flanks are the cores of the binding sites for transposase (Ivics *et al.,* 1997). Although similar, the IR sequences are not exact repeats. For instance, the right-hand IR contains a poly(A) cleavage/addition site that is missing in the left-hand IR.

As mentioned previously, suppression of transposition is critical for the survival of the host cell and the invasive transposon. Consequently, there must be regulation of transposition that is probably autogenously mediated by the transposase and its binding to the IRs. Multiple binding sites for transposase in the IR/DRs suggest a regulatory mechanism of transposition activity. In this context, there may be a correlation between the leucine-zipper/paired domain and the IR/DR structure (Ivics *et al.,* 1996). The N-terminus of transposase could allow transposase molecules to oligomerize before or during DNA-binding and thereby control of transposase expression and the mobilization of the TcE (Izsvák *et al.*, 1997).

Two TcE Subfamilies Exist in the Zebrafish Genome. There are two distinct subfamilies of TcEs in the zebrafish genome (Fig. 2), Tdr1 (Radice *et al.*, 1994) and Tdr2 (Ivics *et al.*, 1996), which has also been named Tzf (Lam *et al.*, 1996a). Relatives of both of these TcEs exist in other fish as well (Ivics *et al.*, 1996), including Atlantic salmon (*Salmo salar*), Tss1 (Radice *et al.*, 1994), and Tss2 (also called SALT1, Goodier and Davidson, 1994). Both Tdr subfamilies have been amplified to approximately 500–1000 copies that are dispersed throughout the zebrafish genome (Izsvák *et al.*, 1995; Lam *et al.*, 1996a). Elements from the same subfamily can be more highly conserved between different species (e.g., 70% nucleic acid identity between Tss1 and Tdr1) than the conservation between members of two different subfamilies of elements in the same species. Tdr1 and Tdr2/Tzf are characteristically different in their encoded transposases and their inverted repeat sequences, and they share only about 30% nucleic acid identity (Izsvák *et al.*, 1995; Lam *et al.*, 1996a). The evolutionary consequences of horizontal hopping of TcEs between zebrafish and other fish genomes has been reviewed (Izsvák *et al.*, 1997). There may be other Tdr-like sequences yet to be identified.

Although we estimate that there are about $1–2 \times 10^3$ Tdr transposons per zebrafish genome, we and others have yet to isolate an active element from zebrafish or any other fish (Radice *et al.*, 1994; Goodier and Davidson, 1994; Izsvák *et al.*, 1995; Ivics *et al.*, 1996; Lam *et al.*, 1996a). We noted earlier that TcEs do not often jump. However, Lam *et al.* (1996a) reported that Tdr2/Tzf RFLPs can be detected in the progeny from pairwise zebrafish matings, which they suggest is evidence that Tdr2/Tzf elements actively transpose in zebrafish. Their calculated frequency of mobilization of about 8 transpositions per offspring is not insignificant; it is comparable to the average background mutation rate of 0.01% found in screens for induced mutations in zebrafish (Mullins *et al.*, 1994). Besides suggesting that there are at least several active transposons, the claim of high-level transposition also suggests that the sites from or into which the transposons jumped should exist; but, none have been identified. At this time, we do not even know whether transposase from Tdr1 could mobilize a Tdr2 element or vice versa. The zebrafish genome contains several, high copy number, potentially transposable, repetitive elements that frequently associate with each other. These elements also can affiliate with TcEs (Izsvák *et al.*, 1996; Shimoda *et al.*, 1996a). Hence, other repetitive elements could very well cause the observed RFLPs observed by Lam *et al.* (1996a,b) as well as contribute to the background mutation rate. The existence of active transposons in the zebrafish genome remains to be demonstrated.

3. Miniature Inverted-Repeat Transposable Elements

MITEs are palindromic repetitive DNA elements, ranging in size from approximately 80–500 bp, that are interspersed in eukaryotic genomes. They were first described in plants (Bureau and Wessler, 1992) but have since been found in as wide a range of organisms as transposons (Yeadon and Catcheside, 1995; Unsal and Morgan, 1995; Smit and Riggs, 1996). In spite of their prominence in eukaryotic genomes, the literature describing MITEs is relatively sparse because their mechanism of spread has not been established. This leaves open their classification in terms of mechanism of transposition. Although MITEs have terminal inverted-repeats, which partially resemble the IRs of other repetitive element families such as DNA-transposons, they do not encode a recombinase and are therefore nonautonomous. Nevertheless, many genomes accumulate on the order of 10^4 copies of MITEs, suggesting that they use a common cellular mechanism for their amplification. MITEs may not be innocuous sequences. On the basis of their high numbers, distribution, and proximity to genes, MITEs may affect gene expression as well as genome organization. In these respects, MITEs share some features with both DNA-transposons and retroposons. The basis for this hypothesis comes from our analysis of MITEs in zebrafish. The first such sequence reported in the zebrafish genome, *Oops,* which was referred to as a foldback element Ivics *et al.,* 1995), subsequently has been renamed *Angel* (Izsvák *et al.,* 1998).

a. Angel: An Abundant MITE in Zebrafish

We estimate that there are about 10^3–10^4 *Angel* elements that have been amplified and scattered throughout the zebrafish genome over the course of evolution. As in other organisms, *Angel* elements are found in the vicinity of fish genes. *Angel* elements and amphibian MITEs (Unsal and Morgan, 1995) share common motifs in their terminal inverted-repeats, suggesting a common evolutionary origin and related amplification and/or transpositional mechanisms of these elements in fish and frogs. *Angel* elements have the potential to form partially paired stem-loop structures *in vitro* by intramolecular base-pairing of the individual single strands of duplex DNA (Fig. 6). Theoretically, the stem structures were perfect palindromes when *Angels* first formed and then they later accumulated base changes as a consequence of evolutionary drift. It also appears that duplication of degenerate *Angel* elements, in an inverted orientation, may form new and energetically more stable MITEs. These conclusions lead us to propose that MITEs transpose via a DNA intermediate, using the cellular DNA replication machinery for their amplification (Izsvák *et al.,* 1998).

Reconstruction of orthologous loci containing MITEs in closely related species has not revealed a single instance of somatic or germinal excision of any MITE (Wessler *et al.,* 1995). For this reason, MITEs can be useful as chromosomal and evolutionary markers, as we discuss later.

III. Genetic Applications of Repetitive Elements

A. Genetic Mapping and Phylogenetic Analyses

Repetitive elements can be used as genetic tools for several important purposes such as to (1) define or distinguish zebrafish chromosomal fragments in somatic

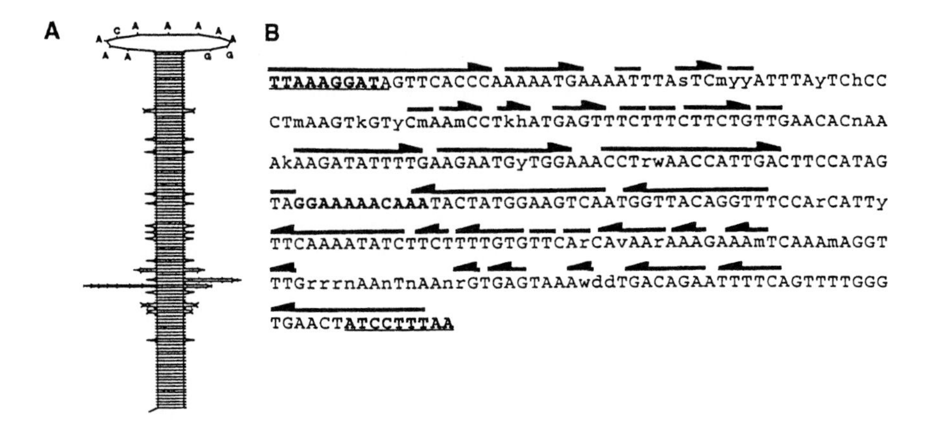

Fig. 6 Structure and sequence of the zebrafish MITE *Angel,* an imperfect palindromic sequence predicted to form an intramolecular stem-loop structure. (A) Predicted stem-loop structure of one strand of the consensus zebrafish *Angel* sequence shown in B. (B) Consensus *Angel* sequence. Black arrows indicate regions of complementarity between the left and the right halves of the element, and the underlined sequences indicate terminal indirect repeat sequences. [Modified from Izsvák, Z., Ivics, Z., Mohn, D., Okamoto, H., and Hackett, P. B. (1998). Short inverted repeat, transposable elements in teleost fish and implications for their mechanism of amplification. J. Mol. Evol. **47**.]

cell hybrids, (2) define evolutionary lineages, (3) provide sites for PCR amplification of dispersed regions of zebrafish chromosomes and thereby identify polymorphic sites in different strains of fish, and (4) provide dispersed sequence tagged sites. The Tdr transposons have served a more useful purpose; they have been used to construct transgenic vectors that can deliver new genetic constructs into zebrafish chromosomes for gain-of-function activities, inactivate and tag genes for identification of function, and function as gene/enhancer-trap constructs for gene identification. Microsatellite sequences and VNTRs are also often used for linkage analysis (Hearne *et al.,* 1992) and in population genetics (Zhivotovsky and Feldman, 1995). In the following sections we describe how repetitive elements are being used in zebrafish developmental genetics.

1. Repetitive Elements as Evolutionary and Population Markers

Zebrafish, as a model system for developmental vertebrate genetics, has a need for high-resolution genetic linkage maps. The first such map for zebrafish was constructed using 652 random amplified polymorphic DNA (RAPD) markers and a handful of identified mutant genes (Postlethwait *et al.,* 1994) and has been followed by more detailed maps (Johnson *et al.,* 1996). The average spacing of the markers on the map is 4.3 cM. An alternative approach has been to position VNTRs, especially CA microsatellite repeats. A CA-repeat map with 102 sites spaced an average of 30 cM has been constructed, and the primers have been available for general use (Kauffman *et al.,* 1995; Knapik *et al.,* 1996). Assuming that the genetic map is about 3200 cM (S. Johnson, personal communication) and that the zebrafish haploid genome is about 1.7×10^9 bp (Hinegardner and Rosen, 1972), there will be almost 600 kbp/cM. Thus, if a gene has been mapped between two markers separated by 2 cM, then the gene would reside on average within a region of more than 1000 kbp, a very long distance for chromosome walking to find a mutant locus. With intensive crossing of marker genomes, loci can be mapped to about 0.1–0.2 cM or about 60–120 kbp. Clearly, a richer, denser map with additional markers is needed.

a. Chromosome Marking Using Repetitive DNA Sequences

The satellite DNAs and the highly repetitive DNAs such as DANA should be useful for identifying zebrafish chromosomes. Somatic cell hybrids and radiation hybrids are two particularly powerful methods for linking genes onto large fragments of zebrafish DNA. The satellite DNA sequences have been used to identify zebrafish chromosomal fragments in hybrids with mouse cells (Ekker *et al.,* 1996; Shimoda *et al.,* 1996b; Martin *et al.,* personal communication). In these experiments, biotinylated probes and fluorescenated (avidin-FITC) probes were used in fluorescence *in situ* hybridization (FISH) to define hybrid chromosomes containing zebrafish sequences. Tdr transposons numbering about 10^3 interspersed copies/genome are also useful markers that can be used alone or in combinations with other repetitive elements.

b. Detection of Repetitive Element Polymorphisms in Genes

Repetitive elements, especially SINEs and MITEs have been successfully used as genetic markers for population studies and evolutionary analyses (Deininger *et al.,* 1992; Turner *et al.,* 1992). Unlike DNA-transposons, SINEs (Murata *et al.,* 1993) and MITES (Izsvák *et al.,* 1998) appear to remain fixed in the genome after they have inserted and, therefore, provide excellent evolutionary and phylogenetic markers. The zebrafish DANA and MITE elements can be used as genetic markers because they have high copy-numbers (about 10^5 and 10^3 per genome, respectively), are dispersed throughout the zebrafish genome, and have Mendelian segregation patterns. Figure 7 shows one example of how both MITEs and DANA elements have been used to define the phylogenetic relationships between three closely related *Danio* species. Because MITEs and SINEs do not relocate in genomes, the sequential addition of these repeated elements into the fourth intron of one gene for translational initiation factor 4E (eIF4E) suggests that *Danio rerio* is descended from the pearl danio (*Danio albolineatus*) which in turn was derived from the giant danio (*Danio aequipinnatus*) lineage (Fig. 7A). The MITE (*Angel*) was the first to insert, followed later by its being split by a DANA element. Another example is shown in Fig. 7B where an *Angel* sequence has interrupted the second intron of a carp growth hormone gene. The first clue is a variation in the sizes of the introns as determined by restriction enzyme polymorphisms and then clarification of the identities of the intruding elements. In these two instances, the integrity of the genes was not damaged as a result

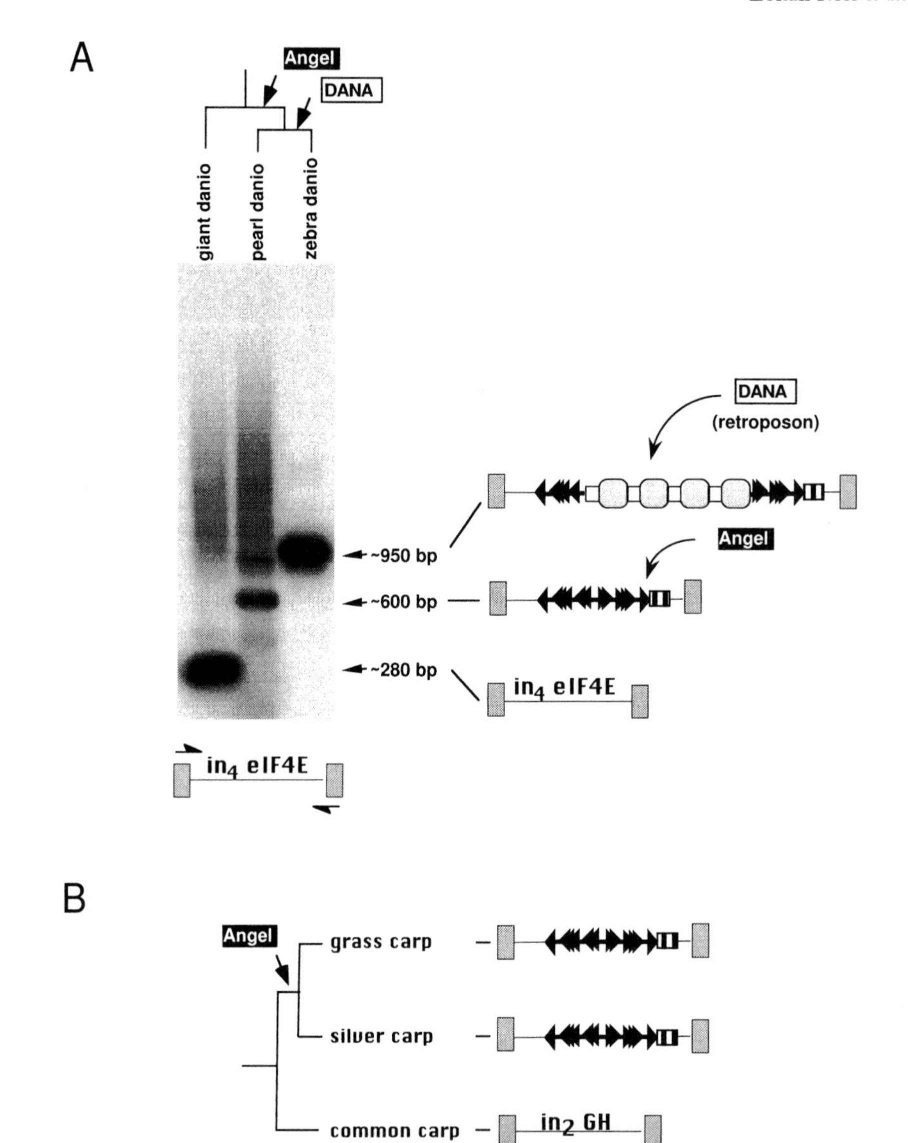

Fig. 7 Phylogenetic positioning of closely related species by repetitive element mapping. (A) Southern blot of PCR products obtained from genomic DNA samples of zebrafish, pearl danio and giant danio, using primers bracketing the fourth intron of the zebrafish eIF4E gene. The blot was hybridized with an approximately 950-bp fragment encompassing this intron in zebrafish. At the top, the phylogenetic lineage of the three *Danio* species (Meyer *et al.,* 1993) is shown. The proposed sequential insertion of *Angel* and DANA elements into the intron is depicted on the right. (B) The second introns of the growth hormone genes of three carp species are depicted to illustrate the presence of *Angel*-related elements in both the silver and grass carp introns, and the absence of such element in the common carp intron.

of insertion into an intron. However, this is not always the case. In the *ntl*^{b195} mutant strain of zebrafish, a composite repetitive element that contains a partial DANA sequence (Figs. 2 and 3B) has inactivated the gene (Schulte-Merker *et al.,* 1994).

c. Detection of Repetitive Element Polymorphisms in Genomes

Polymorphic DNA fragments can be generated by DANA or *Angel*-specific primers using the technique of interspersed repetitive sequence-PCR (IRS-PCR) (Sinnett *et al.,* 1990; Zietkiewicz *et al.,* 1992; Kass and Batzer, 1995). IRS-PCR amplifies genomic DNA flanked by repetitive elements using repeat-specific primers (Table I) to produce polymorphic fragments (Fig. 8). In this illustration,

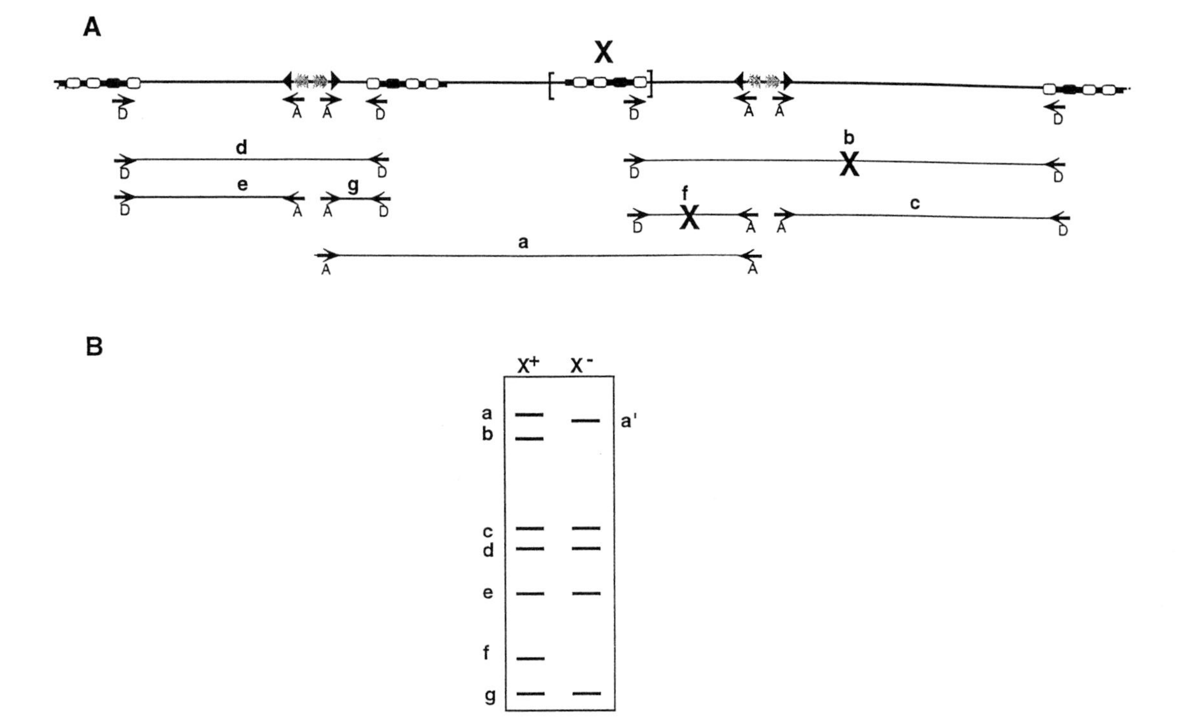

Fig. 8 IRS-PCR fingerprinting and identification of zebrafish lines. (A) Illustration of a chromosomal region in the zebrafish genome containing DANA (D) and *Angel* (A) sequences. The arrows above the elements represent specific PCR primers. The X superimposed on the central DANA element is meant to represent a variable (or missing) element or a mutated primer binding site in the genome of another zebrafish strain. The various amplified sequences are identified by lowercase letters, beginning with the longest detectable PCR product. The products branded with an X will not be produced in the X$^-$ strain. Elements separated by more than 5000 bp will not be efficiently amplified nor will elements that have the wrong orientations relative to each other. (B) Schematic of the two sets of DNA amplification products from both genomes with (X$^+$) and without (X$^-$) the DANA-X element [modified from Izsvák *et al.* (1997)].

the DANA element X exists or not in two different strains of zebrafish. Note that in this example not only will there be an increase in DANA-primed bands (X-related sequences *b* and *f*), but fragments generated by Angel primers may change in size as well (fragment *a* is larger than *a′*). In general, efficient generation of IRS-PCR products is limited to primers in the correct orientations separated by 5 kb or less. The number of detectable polymorphic bands may be significantly increased by the combination of various primers to repetitive sequences in the zebrafish genome, including Tdr1 and Tdr2. For instance, Hong Kong, Indonesia, Singapore, *gold,* C-32, and leopard/C-32 zebrafish strains, as well as two wild-type strains used in our lab, all have distinct IRS-PCR fingerprints with multiple banding differences (Izsvák *et al.,* 1995, 1998). Polymorphic fragments can be recovered from gels and cloned to provide sequence-tagged sites (STSs) for mapping mutations. Figure 8B illustrates the general principles and constraints for using IRS-PCR to generate STSs. We have estimated that approximately 0.1% of the zebrafish genome can be directly analyzed by IRS-PCR, using only the four primers given in Table I (Izsvák *et al.,* 1997). Different PCR primers to either the C_1 or C_2 boxes have been used for scanning zebrafish genomes for polymorphisms (Fig. 3A; Izsvák *et al.,* 1996; Shimoda *et al.,* 1996b) and for developing STSs (Burgtorf and Lehrach, personal communication). Shimoda *et al.,* (1996b) have proposed cloning IRS-PCR products derived from somatic cell hybrids to construct chromosome-specific libraries.

B. Transposons for Gene Identification and Gene Transfer in Zebrafish

Repetitive elements have considerable influence on the genomes they invade and occupy through amplification. A liberated transposable element can integrate into either of two types of chromatin, functional DNA sequences where it may have a deleterious effect as a result of insertional mutagenesis or nonfunctional chromatin where it may not have much of a consequence (Fig. 9). This power has been exploited in simpler model systems for nearly two decades (Bingham *et al.,* 1981; Bellen *et al.,* 1989). Transposon tagging is an old technique in which transgenic DNA is delivered to cells so that it will integrate into genes, thereby inactivating them by insertional mutagenesis. In the process, the inactivated genes are tagged by the transposable element which then can be used to recover the mutated allele. A related technique is to use a tagging element in "gene or enhancer-trap" screens. A marker gene, such as the green fluorescence protein (GFP), can act as a reporter for genomic transcriptional enhancer-like elements located sufficiently close to the inserted transposon. This technique can be used to find and identify tissue- and development-specific transcriptional regulatory signals as well as genes. Insertion of a transposable element may disrupt the function of a gene which can lead to a characteristic phenotype. This is how new, active transposons are often discovered in various organisms.

As illustrated in Fig. 9, because insertion is approximately random, the same procedures that generate insertional, loss-of-function mutants can often be used

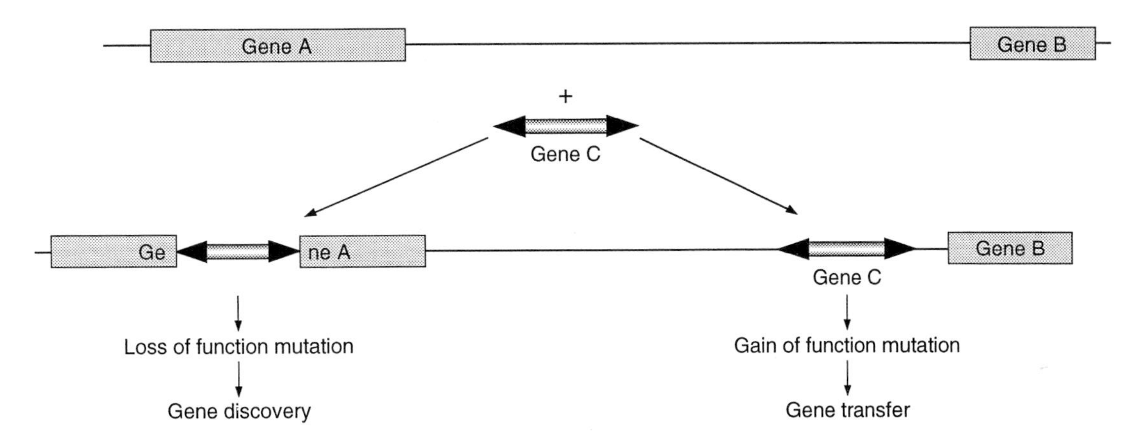

Fig. 9 Parallel uses of transposable elements. Depending on the integration site of a mobile element, here a DNA transposon, the effect can be either a loss-of-function or gain-of-function mutation. A loss-of-function mutation can be accompanied, as shown here, by a gain-of-function (gene C). Both types of activity can be exploited as described in the text.

to deliver genes that will confer new phenotypes to cells. Gain-of-function mutants can be used to understand the roles that gene products play in growth and development as well as the importance of their regulation. A decade ago it seemed like it would be pretty simple to deliver exogenous DNA to zebrafish chromosomes; after all, the single fertilized cell was unusually conspicuous, large, and encased in an optically clear chorion that would permit rapid microinjection of a DNA solution. Alas, transgenic technology in zebrafish has been deceptively difficult. The delivery of DNA into the cell was about as trivial an exercise as predicted, but the DNA generally fails to move into chromosomes (Westerfield *et al.,* 1992; Hackett, 1993; Iyengar *et al.,* 1996). This is probably the result of two factors. First, the nuclear membrane may not permit entry of most of the exogenous DNA, except when added by some type of transporter, such as peptides with nuclear localization signals (Collas *et al.,* 1996; Collas and Alestrom, 1997). Second, the extraordinarily rapid rate of DNA synthesis (about 50- to 100-fold faster than in mammalian cells) during the first ten replication cycles may render the chromatin impervious to outside attack by any recombinase molecule. Thus, although about 10^6–10^7 DNA molecules are often delivered to zebrafish one- or two-cell stage embryos, generally integration of only a very few DNAs in a few cells occurs over a prolonged period during development. This low-level penetrance into just a few cells results in mosaic founder animals with low frequencies (often only 1–5%) of germline transmission of the transgene to offspring. Consequently, there were attempts to find mechanisms to deliver DNA more efficiently into zebrafish chromosomes.

In the meantime, chemical mutagenesis screens using the random mutagen ethylnitrosourea (ENU) were set up to determine the essential genes required for growth and development. The outcomes of two major screens were largely successful; more than 6000 mutants and about 600 loci in the zebrafish genome were identified (Haffter *et al.,* 1996; Driever *et al.,* 1996). There was a problem: the ENU-induced point mutations made identification of the responsible genes practically impossible without chromosomal walks of more than 1 Mbp. Hence, there was a strong incentive to find procedures for introducing equivalent mutations that could be quickly identified and cloned. Transgenic technologies offered a solution, using some type of mobile DNA to tag genes for insertional mutagenesis.

1. Retroviruses as Mobile Vectors for Gene Transfer in Zebrafish

Retroviruses are effective vectors for transgenic constructs for some species (Jaenisch, 1988). However, no natural retrovirus has been identified that infects zebrafish. Nevertheless, because the principles governing integration of proviral DNA genomes into cellular chromosomes are fairly conserved (Craig, 1995), we reasoned that we could harness some features of mammalian retroviruses for transgenesis in zebrafish. It was clear that because retroviruses recognize specific receptors on the surfaces of specific cells, we could not use recombinant viruses for infection. But with the availability of microinjection, penetration of the embryonic cell membranes was not an issue. Whole viruses were not necessary, just their integration machinery. Accordingly, we introduced both transgenic DNA and retroviral integrase protein from the Moloney murine leukemia virus (MoMLV) into zebrafish embryos and found that integrase protein did indeed enhance expression and integration of transgenes (Ivics *et al.,* 1993). However, the integration activity of the protein preparation was much lower in microinjected zebrafish embryos than in virus-infected cells, possibly because of the necessity for some host factors that were not adequate in zebrafish cells. This spurred an interest in finding a recombinase that did not require host factors so that a more efficient *in vitro* system could be developed. While this was being done, an alternative approach was tested, using complete but modified retroviruses.

In order to obtain a retrovirus that could infect zebrafish cells, the Hopkins lab developed a pseudotyped retroviral vector that contained an MoMLV-based genome packaged in the envelope protein of the vesicular stomatitis virus (VSV), which is able to fuse with nearly all cellular membranes including those of zebrafish cells (Burns *et al.,* 1993; Lin *et al.,* 1994). In their protocol, $1–2 \times 10^4$ virus pseudotype particles are injected into blastula-stage embryos (512–2000 cells) because the viruses are not able to penetrate embryonic cell membranes. Proviruses are able to integrate into the germlines of zebrafish at less than a 0.1% efficiency (Allende *et al.,* 1996; Gaiano *et al.,* 1996; Gaiano and Hopkins, 1996). Because of the delay in the delivery of the pseudotype virus, all the embryos obtained by this protocol will be highly mosaic. There was a range of from about 5 to 22 different proviral insertions per F_1 fish. Nevertheless, for screening of mutations that affect growth and development, this is a useful strategy because it allows a number of mutations to be screened simultaneously per fish which subsequently can be individually segregated by outcrossing (Schier *et al.,* 1996). However, there are some drawbacks to these viruses. First, it is necessary to acquire very high titers of virus because so little volume can be injected; obtaining the required titers is difficult. Second, retroviral vectors can accommodate only a limited length of DNA. Third, retroviral vectors have not been shown to be able to deliver genes that can be stably expressed. Fourth, the virus must be contained and handled with extreme care because it can infect the experimenter as easily as the zebrafish cells.

Accordingly, an alternative means of inserting transgenic DNA into zebrafish genomes was needed that would (1) have an early and narrow window of activity to reduce mosaicism, (2) have a high efficiency of integration, (3) be able to accommodate variable lengths of transgenic DNA, and (4) be safe and easy to use by any lab. Transposons in other systems could meet these challenges.

2. Zebrafish Transposons as Vectors

Except for the restriction fragment length polymorphism (RFLP) screens of Lam *et al.* (1996a) described in Section II.C.2.a, there has not been a single case of a spontaneous mutation caused by a TcE insertion reported in fish. Because of a variety of mutations in their transposase genes, none of the fish TcEs found so far, and the search has been intensive, encode an active transposase. However, there are other ways of obtaining an active transposon; one can be made (Ivics *et al.,* 1997). But on what should it be based? The two DNA transposons isolated so far from zebrafish, Tdr1 and Tdr2/Tzf, have relatively high copy numbers in zebrafish which would certainly interfere with most applications such as insertional mutagenesis and transgenesis. Even though most, if not all, of the Tdr elements have lost their capacity to produce transposase, many of them presumably still have intact IR sequences and thus the ability to transpose in the presence of an transposase. This could cause massive transposition, genetic instability, and interference with experimental objectives.

This problem could be alleviated by using a heterologous transposon system. However, in animals, a major obstacle to transferring an active transposon system from one species to another is the requirement for factors produced by the natural host for mobilization of the element. Thus, attempts to use the P-element transposon of *Drosophila melanogaster* for genetic transformation of non-Drosophilid insects, zebrafish, and mammalian cells have been unsuccessful (Gibbs *et al.,* 1994; Rio *et al.,* 1988). However, TcEs presumably do not require species-specific factors for transposition because transposase made in *E. coli* is the only protein required for *in vitro* transposition (Vos *et al.,* 1996; Lampe *et al.,* 1996). Moreover, their history for horizontal invasion of genomes (e.g., Robertson and Lampe, 1995; Kidwell, 1992; Kordis and Gubensek, 1995; Lohe *et al.,* 1995) plus the finding that gene vectors based on the *Minos* element, a TcE endogenous to *Drosophila hydei,* could be successfully used for germline transformation of the agricultural pest *Ceratitis capitata* (Loukeris *et al.*, 1995), suggested that Tc1/*mariner* type transposons would function in zebrafish. However, early tests showed that the transposons were not completely independent of their hosts; we found negligible activity of the Tc1 transposon in zebrafish using assays described in the following section. There were other candidate transposons. The Tdr elements found in zebrafish, as well as related transposons in other fish were members of the Tc1/*mariner* family (Izsvák *et al.*, 1996) that did not appear to require host factors. Thus, because the Tdr elements found in zebrafish are members of the Tc1/*mariner* family, the genomes of other fish seemed to be the best hunting grounds to find a transposon that could be developed into a wide host range transformation vector.

3. Sleeping Beauty, a Transposon System Active in Zebrafish

We looked for a heterologous fish transposon that could be revived and used in zebrafish, but we could not find a single teleost TcE with an uninterrupted transposase open reading frame. So, we used phylogenetic data to reconstruct a transposase gene, named *Sleeping Beauty* (SB) from bits and pieces of inactive elements from salmonid fish (Ivics *et al.*, 1997). Two basic types of assays were used to demonstrate activity of the system (Fig. 10). The system requires two parts, a substrate transposon that has a gene sandwiched between functional IR sequences and a source of transposase. Three sources of transposase can be delivered to cells—the protein itself, its mRNA, or the transposase gene. The assay system shown in Fig. 10A is designed for tissue-cultured cells. Two plasmids are co-delivered (e.g., by lipofection) to recipient cells. When transposase is expressed from the plasmid on the right, it enzymatically excises the transposon from the plasmid on the left and directs its reintegration into chromosomal DNA at some frequency. The rate of transfer of a neomycin-resistant gene into chromosomal DNA was about 20-fold enhanced in the presence of the plasmid encoding the transposase. What about zebrafish, our main concern? For this, an inter-plasmid transposition assay was used (Fig. 10B). Here the ability of the SB transposase to excise a marker TcE from one plasmid and insert it into another was tested. The experiment was conducted by co-microinjecting transposase mRNA and the two plasmids into zebrafish embryos at the one-cell stage. Low-molecular-weight DNA containing the extrachromosomal plasmids was prepared from the embryos at about 50% epiboly, transformed into *E. coli* cells, and amp/kan-resistant colonies were selected. The number of recombinant plasmids was about 20-fold higher than control transformations lacking transposase mRNA. The physical translocation of the marked TcE was confirmed by Southern hybridization.

For transgenesis, stability of expression is important. Both the SB and Tc3 transposon systems have been shown to deliver a GFP expression in both F0 and F1 offspring (Raz *et al.*, 1997; Mohn *et al.*, unpublished). Moreover, other Tc1/*mariner* transposons have been developed for transgenesis. Tc3 and Tc1

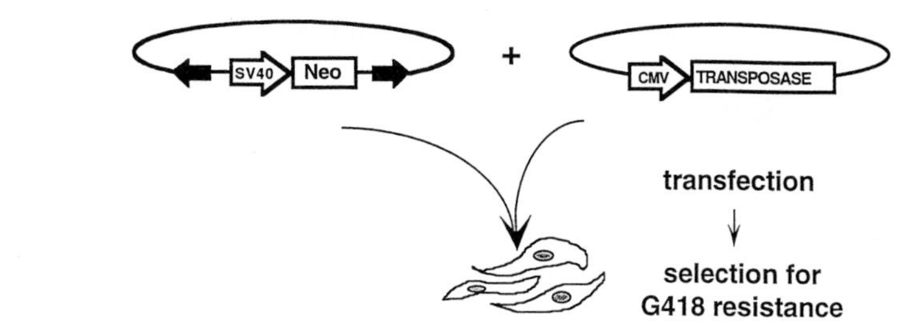

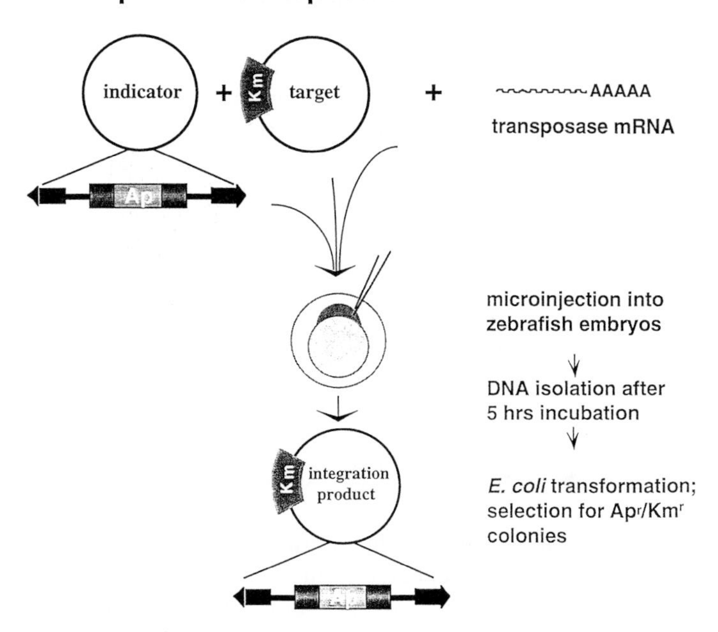

Fig. 10 Assays for transposon mobilization in cultured cells and in zebrafish embryos. (A) Genetic assay for *Sleeping Beauty*-mediated transgene integration in cultured cells. (B) Inter-plasmid assays for excision and integration. The two plasmids plus an mRNA encoding the transposase protein are coinjected into the one-cell zebrafish embryo. One of the plasmids has an ampicillin-resistance gene (A_p) flanked by IR/DR sequences which can be recognized by the transposase. At 5 hours after fertilization and injection, the low-molecular-weight DNA is isolated from the embryos and used to transform *E. coli.* The bacteria are grown on selective media containing ampicillin and kanamycin to select for bacteria harboring single plasmids containing both the Km and Ap antibiotic-resistance markers. The plasmids from doubly resistant cells are examined. Because the amount of DNA in injected plasmid is almost equal to that of the genome, the number of integrations of Ap-transposons into target plasmids should approximate the number of integrations into the genome. [Modified from Ivics *et al.* (1997), © Cell Press, and Izsvák *et al.* (1997).]

have been able to serve as vectors for gene delivery into zebrafish and mammalian cells (Raz *et al.,* 1997; van Luenen *et al.,* personal communication). In many cases, especially insertional mutagenesis, identifying the locus of integration can be quite important. Figure 11 schematizes one way that IRS-PCR and probing with suitable sequences can allow identification of the mutated locus. Similar to Fig. 8 (the same region of a chromosome is shown), depending on the primers used, there will be new bands, and there will be shifts of other bands to reflect the increase in length caused by the integration of the vector. In particular, PCR products *e* and *g* will be good STSs for the locus of insertion, whereas sequences *a'* and *d'* will contain the region but will be less informative about the actual site of entry.

These experiments demonstrate that the synthetic SB system of vector plus transposase can deliver transgenes for transgenesis as well as insertional mutagenesis and gene/promoter trapping and gene mapping in vertebrates. But, there are several parameters of the system that need defining and/or improving. The efficiency of transfer of transposons of different sizes is not known. There still is a problem with early integration. The relative efficiencies of the natural Tc1/3 and *mariner* transposons compared to the SB transposon need examination. For insertional mutagenesis, transposons may be either less efficient or less random in inducing mutations compared to chemical mutagens so that not all genes may be identified (Schier *et al.,* 1996; Spradling *et al.,* 1995). Nevertheless, insertional mutagenesis is such a powerful technique for the generation of recoverable mutations that transposons should become a useful genetic tool for molecular and developmental biologists working with zebrafish.

IV. Summary and Perspectives

About 15–20% of the zebrafish genome has been characterized in terms of several types of repeated elements which are amplified and mobilized by different mechanisms. There are probably several more families of elements yet to be found, including endogenous retroviral sequences, LTR-retrotransposons and LINE elements. Many of the early elements were found accidentally through sequence comparisons that showed they were present in and around other genes.

Fig. 11 IRS-PCR identification of sites of insertion of transposable element T into the zebrafish genome. (A) Top, illustration of a chromosomal region in the zebrafish genome containing DANA (D) and *Angel* (A) sequences as in Fig. 8. The arrows above the elements represent specific PCR primers. After integration of transposon T, the IRS-PCR products will change, as shown in B. The various amplified sequences are identified by lowercase letters. Specific primers for the IRs in transposon T will allow identification of the locus into which T inserted; which will be contained in fragments *g* and *e*. (B) Schematic of the two sets of DNA amplification products from both genomes with (lane 1) and without (lane 2) the transposon insertion.

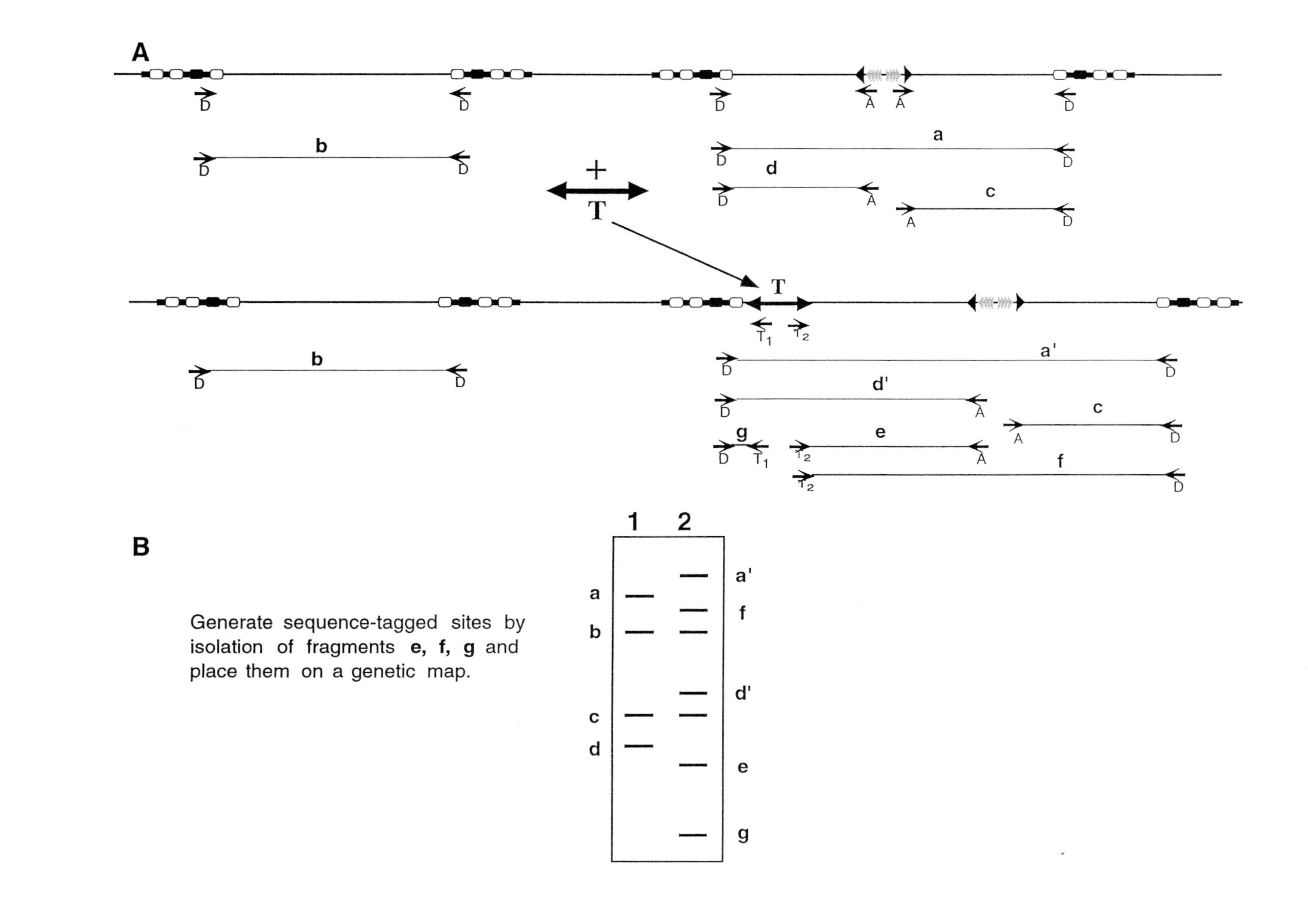
A
D
D
D
A
A
D
b
a
d
c
+
T
T
T1
T2
b
a'
d'
c
g
e
f
B
1
2
a
b
c
d
a'
f
d'
e
g
Generate sequence-tagged sites by isolation of fragments e, f, g and place them on a genetic map.

In this respect they were a nuisance. But, now they have become genetic tools for the molecular cell biologist. For the first time, transposons can be used in vertebrates as vehicles for bringing new phenotypes into genomes (gain-of-function mutations) as well as destroying endogenous genes (loss-of-function) in order to determine function and importance in developmental pathways. The endogenous elements will be used as landmarks and reference points for gene mapping.

Acknowledgments

We appreciate preliminary unpublished results from Drs. A. Fritz, M. Halpern, M. Ekker, R. Phillips, Z. Gong, R. Bartbreit, and C. Bergtorf. The work in the authors' laboratory was supported by grants from NIH, USDA and SeaGrant.

References

Allende, M. L., Amsterdam, A., Becker, T., Kawakami, K., Gaiano, N., and Hopkins, N. (1996). Insertional mutagenesis in zebrafish identifies two novel genes, pescadillo and dead eye, essential for embryonic development. *Genes Dev.* **10,** 3141–3155.

Arnault, C., and Dufournel, I. (1994). Genome and stresses: Reactions against aggressions, behavior of transposable elements. *Genetica* **93,** 149–160.

Avancini, R. M., Walden, K. K., and Robertson, H. M. (1996). The genomes of most animals have multiple members of the Tc1 family. *Genetica* **98,** 131–140.

Bellen, H. J., O'Kane, C. J., Wilson, C., Grossniklaus, U., Pearson, R. K., and Gehring, W. J. (1989). P-element-mediated enhancer detection: a versatile method to study development in *Drosophila. Genes Dev.* **3,** 1288–1300.

Bingham, P. M., Levis, R., and Rubin, G. M. (1981). Cloning of DNA sequences from the white locus of *D. melanogaster* by a novel and general method. *Cell* **25,** 693–704.

Britten, R. J. and Kohne, D. E. (1968). Repeated sequences in DNA. *Science* **161,** 529–540.

Britten, R. J. (1994). Evolutionary selection against change in many *Alu* repeat sequences interspersed through primate genomes. *Proc. Natl. Acad. Sci. USA* **91,** 5992–5996.

Britten, R. J., McCormick, T. J., Mears, T. L., and Davidson, E. H. (1995). *Gypsy/Ty3*-class retrotransposons integrated in the DNA of herring, tunicate, and echinoderms. *J. Mol. Evol.* **40,** 13–24.

Bureau, T. E., and Wessler, S. R. (1992). *Tourist:* A large family of small inverted repeat elements frequently associated with maize genes. *Plant Cell* **4,** 1283–1294.

Burns, J. C., Friedmann, T., Driever, W., Burrascano, M., and Yee, J. K. (1993). Vesicular stomatitis virus G glycoprotein pseudotyped retroviral vectors: concentration to very high titer and efficient gene transfer into mammalian and nonmammalian cells. *Proc. Natl. Acad. Sci. USA* **90,** 8033–8037.

Charlesworth, B., Sniegowski, P., and Stephan, W. (1994). The evolutionary dynamics of repetitive DNA in eukaryotes. *Nature* **371,** 215–220.

Collas, P., and Alestrom, P. (1997). Rapid targeting of plasmid DNA to zebrafish embryo nuclei by the nuclear localization signal of SV40 T antigen. *Mol. Mar. Biol. Biotechnol.* **6,** 48–58.

Collas, P., Husebye, H., and Alestrom, P. (1996). The nuclear localization sequence of the SV40 T antigen promotes transgene uptake and expression in zebrafish embryo nuclei. *Trans. Res.* **5,** 451–458.

Colloms, S. D., van Luenen, H. G., and Plasterk, R. H. (1994). DNA binding activities of the *Caenorhabditis elegans* Tc3 transposase. *Nucl Acids. Res.* **22,** 5548–5554.

Craig, N. L. (1995). Unity in transposition reactions. *Science* **270,** 253–254.

Daniels, G. R., and Deininger, P. L. (1985). Repeat sequence families derived from mammalian tRNA genes. *Nature* **317,** 819–822.

Deininger, P. L., Batzer, M. A., Hutchison, C. A. R., and Edgell, M. H. (1992). Master genes in mammalian repetitive DNA amplification. *Trends Genet* **8,** 307–311.

Doak, T. G., Doerder, F. P., Jahn, C. L., and Herrick, G. (1994). A proposed superfamily of transposase genes: Transposon-like elements in ciliated protozoa and a common "D35E" motif. *Proc. Natl. Acad. Sci. USA* **91,** 942–946.

Driever, W., Solnica-Krezel, L., Schier, A. F., Neuhaus, S. C. F., Malicki, J., Stemple, D. L., Stanier, D. Y. R., Zwartkruis, F., Abdelilah, S., Rangini, Z., Belak, J., and Boggs, C. (1996). A genetic screen for mutation affecting embryogenesis in zebrafish. *Development* **123,** 37–46.

Ekker, M., Fritz, A., and Westerfield, M. (1992). Identification of two families of satellite-like repetitive DNA sequences from the zebrafish (*Brachydanio rerio*). *Genomics* **13,** 1169–1173.

Ekker, M., Speevak, M. D., Martin, C. C., Joly, L., Giroux, G., and Chevrette, M. (1996). Stable transfer of zebrafish chromosome segments into mouse cells. *Genomics* **33,** 57–64.

Emmons, S. W., Yesner, L., Ruan, K. S., and Katzenberg, K. (1983). Evidence for a transposon in *Caenorhabditis elegans. Cell* **32,** 55–65.

Flavell, A. J., and Smith, D. B. (1992). A *Ty1-copia* group retrotransposon sequence in a vertebrate. *Mol. Gen. Genet.* **233,** 322–326.

Fondevila, A. (1993). Genetic instability and rapid speciation: Are they coupled? *In* "Transposable Elements and Evolution" (J. F. McDonald, ed.) pp. 242–254.

Franck, J. P. C., Harris, A. S., Bentzen, P., Denovan-Wright, E. and Wright, M. (1991). Organization and evolution of satellite, minisatellite, and microsatellite DNAs in telost fishes. *In* "Oxford Surveys on Eukaryotic Genes" (N. Maclean, ed.), Vol. 7, pp. 51–82.

Franz, G., Loukeris, T. G., Dialektaki, G., Thompson, C. R., and Savakis, C. (1994). Mobile Minos elements from *Drosophila hydei* encode a two-exon transposase with similarity to the paired DNA-binding domain. *Proc. Natl. Acad. Sci. USA* **91,** 4746–4650.

Franz, G., and Savakis, C. (1991). Minos, a new transposable element from *Drosophila hydei,* is a member of the Tc1-like family of transposons. *Nucl. Acids. Res.* **19,** 6646.

Gaiano, N., Allende, M., Amsterdam, A., Kawakami, K., and Hopkins, N. (1996). Highly efficient germ-line transmission of proviral insertions in zebrafish. *Proc. Natl. Acad. Sci. USA* **93,** 7777–7782.

Gaiano, N., and Hopkins, N. (1996). Insertional mutagenesis and rapid cloning of essential genes in zebrafish. *Nature* **383,** 829–832.

GeneBank (1998). Zebrafish EST sequences available at: http://www.ncbi.nlm.nih.gov/htbinpost/Taxonomy/wgetorg?name=Danio+rerio&lvl=0.

Gibbs, P. D., Gray, A., and Thorgaard, G. (1994). Inheritance of P element and reporter gene sequences in zebrafish. *Mol. Mar. Biol. Biotechnol.* **3,** 317–326.

Gong, Z., Yan, T., Liao, J., Lee, S. E., and Hew, C. L. (1997). Rapid identification and isolation of zebrafish genes. submitted.

Goodier, J. L., and Davidson, W. S. (1994). Tc1 transposon-like sequences are widely distributed in salmonids. *J. Mol. Biol.* **241,** 26–34.

Hackett, P. B. (1993). The molecular biology of transgenic fish. *In* "Biochemistry and Molecular Biology of Fishes" (T. P. Hochachka and P. Mommsen, eds.), Volume 2, pp. 207–240. Elsevier Science Publishers B. V. Amsterdam.

Haffter, P., Granato, M., Brand, M., Mullins, M. C., Hammerschmidt, M., Kane, D. A., Odenthal, J., van Eeden, F. J. M., Jiang, Y.-J., Heisenberg, C.-P., Kelsh, R. N., Furutani-Seiki, M., Vogelsang, E., Beuchle, D., Schach, U., Fabian, C., and Nüsslein-Volhard, C. (1996). The identification of genes with unique and essential functions in the development of the zebrafish. *Danio rerio. Development* **123,** 1–36.

Hartl, D. L., Lozovskaya, E. R., Nurminsky, D. I., and Lohe, A. R. (1997). What restricts the activity of mariner-like transposable elements? *Trends Genet.* **13,** 197–201.

He, L., Zhu, Z., Faras, A. J., Guise, K. S., Hackett, P. B., and Kapuscinski, A. R. (1992). Characterization of AluI repeats of zebrafish (*Brachydanio rerio*). *Mol. Mar. Biol. Biotechnol.* **1,** 125–135.

Hearne, C. M., Ghosh, S., and Todd, J. A. (1992). Microsatellites for linkage analysis of genetic traits. *Trends Genet.* **8,** 288–294.

Hinegardner, R., and Rosen, D. E. (1972). Cellular DNA content and the evolution of teleostean fishes. *Am. Nat.* **166,** 621–644.
Ivics, Z., Izsvák, Z., and Hackett, P. B. (1993). Enhanced incorporation of transgenic DNA into zebrafish chromosomes by a retroviral integration protein. *Mol. Mar. Biol. Biotechnol.* **2,** 162–173.
Ivics, Z., Izsvák, Z., and Hackett, P. B. (1995). Repeated sequence elements in zebrafish and their use in molecular genetic studies. *Zebrafish Monitor* **3,** 1–4.
Ivics, Z., Izsvák, Z., and Hackett, P. B. (1997). Molecular reconstruction of *Sleeping Beauty,* a Tc1-like transposon system from fish and its transposition in human cells. *Cell* **91,** 501–510.
Ivics, Z., Izsvák, Z., Minter, A., and Hackett, P. B. (1996). Identification of functional domains and evolution of Tc1-like transposable elements. *Proc. Natl. Acad. Sci. USA* **93,** 5008–5013.
Iyengar, A., Müller, F., and Maclean, N. (1996). Regulation and expression of transgenes in fish—A review. *Trans. Res.* **5,** 147–166.
Izsvák, Z., Ivics, Z., Garcia-Estefania, D., Fahrenkrug, S. C., and Hackett, P. B. (1996). DANA elements: A family of composite, tRNA-derived short interspersed DNA elements associated with mutational activities in zebrafish (*Danio rerio*). *Proc. Natl. Acad. Sci. USA* **93,** 1077–1081.
Izsvák, Z., Ivics, Z., and Hackett, P. B. (1995). Characterization of a Tc1-like transposable element in zebrafish (*Danio rerio*). *Mol. Gen. Genet.* **247,** 312–322.
Izsvák, Z., Ivics, Z., and Hackett, P. B. (1997). Repetitive elements and their genetic applications in zebrafish. *Biochem. Cell Biol. J.* **75,** 507–523.
Izsvák, Z., Ivics, Z., Shimoda, N., Mohn, D., Okamoto, H., and Hackett, P. B. (1998). Short inverted repeat transposable elements in teleost fish and implications for their mechanism of amplification. *J. Mol. Evol.* **47**.
Jaenisch, R. (1988). Transgenic animals. *Science* **240,** 1468–1474.
Johnson, S. L., Gates, M. A., Johnson, M., Talbot, W. S., Horne, S., Bai, K., Rude, S., Wong, J. R., and Postlethwait, J. H. (1996). Centromere-linkage analysis and consolidation of the zebrafish genetic map. *Genetics* **142,** 1277–1288.
Kaplan, D. J., Jurka, J., Solus, J. F., and Duncan, C. H. (1991). Medium reiteration frequency repetitive sequences in the human genome. *Nucl. Acids Res.* **19,** 4731–4738.
Kass, D. H., and Batzer, M. A. (1995). Inter-Alu polymerase chain reaction: Advancements and applications. *Anal. Biochem.* **228,** 185–193.
Kauffman, E. J., Gestl, E. E., Kim, D. J., Walker, C., Hite, J. M., Yan, G., Rogan, P. K., Johnson, S. L., and Cheng, K. C. (1995). Microsatellite-centromere mapping in the zebrafish (*Danio rerio*). *Genomics* **30,** 337–341.
Ketting, R. F., Fischer, S. E. J., and Plasterk, R. H. A. (1997). Target choice determinants of the Tc1 transposon of *Caenorhabditis elegans. Nucl. Acids Res.* **25,** 4041–4047.
Kidwell, M. G. (1992). Horizontal transfer. *Curr. Opin. Genet. Dev.* **2,** 868–873.
Kidwell, M. G., and Lisch, D. (1997). Transposable elements as sources of variation in animals and plants. *Proc. Natl. Acad. Sci. USA* **94,** 7704–7711.
Kim, J. K., and Simmons, M. J. (1994). Gross chromosomal rearrangements mediated by transposable elements in *Drosophila melanogaster. BioEssays* **16,** 269–275.
Kloeckener-Gruissem, B., and Freeling, M. (1995). Transposon-induced promoter scrambling: A mechanism for the evolution of new alleles. *Proc. Natl. Acad. Sci. USA* **92,** 1836–1840.
Knapik, E. W., Goodman, A., Atkinson, O. S., Roberts, C. T., Shiozawa, M., Sim, C. U., Weksler-Zangen, S., Trolliet, M. R., Futrell, C., Innes, B. A., Koike, G., McLaughlin, M. G., Pierre, L., Simon, J. S., Villalonga, E., Roy, M., Chiang, P. W., Fishman, M. C., Driever, W., and Jacob, H. J. (1996). A reference cross DNA panel for zebrafish (*Danio rerio*) anchored with simple sequence length polymorphisms. *Development* **123,** 451–460.
Kordis, D., and Gubensek, F. (1995). Horizontal SINE transfer between vertebrate classes. *Nat. Genet.* **10,** 131–132.
Lam, W. L., Lee, T. S., and Gilbert, W. (1996a). Active transposition in zebrafish. *Proc. Natl. Acad. Sci. USA* **93,** 10870–10875.
Lam, W. L., Seo, P., Robison, K., Virk, S., and Gilbert, W. (1996b). Discovery of amphibian Tc1-like transposon families. *J. Mol. Biol.* **257,** 359–366.

Lampe, D. J., Churchill, M. E., and Robertson, H. M. (1996). A purified mariner transposase is sufficient to mediate transposition in vitro. *EMBO J.* **15,** 5470–5479.

Levinson, G., and Gutman, G. A. (1987). Slipped-strand mispairing: A major mechanism for DNA sequence evolution. *Mol. Biol. Evol.* **4,** 203–221.

Lin, S., Gaiano, N., Culp, P., Burns, J. C., Friedmann, T., Yee, J. K., and Hopkins, N. (1994). Integration and germ-line transmission of a pseudotyped retroviral vector in zebrafish. *Science* **265,** 666–669.

Lohe, A. R., Moriyama, E. N., Lidholm, D. A., and Hartl, D. L. (1995). Horizontal transmission, vertical inactivation, and stochastic loss of *mariner*-like transposable elements. *Mol. Biol. Evol.* **12,** 62–72.

Loukeris, T. G., Livadaras, I., Arca, B., Zabalou, S., and Savakis, C. (1995). Gene transfer into the medfly, *Ceratitis capitata,* with a *Drosophila hydei* transposable element. *Science* **270,** 2002–2005.

Martin, C. C., Ye, F., Tellis, P., Ghrioux, H., Joly, L., Chevrette, M., and Ekker, M. (personal communication). Analysis of repetitive DNA sequences from the zebrafish *Danio rerio* using zebrafish mouse somatic cell hybrids.

Merriman, P. J., Grimes, C. D., Ambroziak, J., Hackett, D. A., Skinner, P., and Simmons, M. J. (1995). S elements: A family of Tc1-like transposons in the genome of *Drosophila melanogaster. Genetics* **141,** 1425–1438.

Meyne, J., Ratliff, R. L., and Moyzis, R. K. (1989). Conservation of the huyman telomere sequence (TTAGGG)n among vertebrates. *Proc. Natl. Acad. Sci. USA* **86,** 7049–7053.

Meyne, J., Baker, R. J., Hobart, H. H., Hsu, T. C., Ryder, O. A., Ward, O. G., Wiley, J. E., Wurster-Hill, D. H., Yates, T. L., and Moyzis, R. K. (1990). Distribution of non-telomeric sites of the (TTAGGG)n telomeric sequence in vertebrate chromosomes. *Chromosoma* **99,** 3–10.

Mullins, M. C., Hammerschmidt, M., Haffter, P., and Nüsslein-Volhard, C. (1994). Large-scale mutagenesis in the zebrafish: In search of genes controlling development in a vertebrate. *Curr. Biol.* **4,** 189–202.

Murata, S., Takasaki, N., Saitoh, M., and Okada, N. (1993). Determination of the phylogenetic relationships among Pacific salmonids by using short interspersed elements (SINEs) as temporal landmarks of evolution. *Proc. Natl. Acad. Sci. USA* **90,** 6995–6999.

Oosumi, T., Belknap, W. R., and Garlick, B. (1995). *Mariner* transposons in humans. *Nature* **378,** 873.

Petrov, D. A., Schutzman, J. L., Hartl, D. L., and Lozovskaya, E. R. (1995). Diverse transposable elements are mobilized in hybrid dysgenesis in *Drosophila virilis. Proc. Natl. Acad. Sci. USA* **92,** 8050–8054.

Plasterk, R. H. (1993). Molecular mechanisms of transposition and its control. *Cell* **74,** 781–786.

Plasterk, R. H. A. (1995). Reverse genetics: from gene sequence to mutant worm. *In* "Methods in Cell Biology," Vol. 48, pp. 59–80. Academic Press, San Diego.

Plasterk, R. H. A. and van Luenen, H. G. A. M. (1997). Transposons. *In "C. elegans"* (D. C. Riddle, T. Blumenthal, B. J. Meyer, and J. R. Pries, eds). Cold Spring Harbor Laboratory, Cold Spring Harbor, NY, pp. 97–116.

Postlethwait, J., Johnson, S., Midson, C. N., Talbot, W. S., Gates, M., Ballinger, D., Africa, E. W., Andrews, R., Carl, T., Eisen, J. S., Horne, S., Kimmel, C. B., Hutchinson, M., Johnson, M., and Rodriguez, A. (1994). A genetic map for zebrafish. *Science* **264,** 699–703.

Preston, B. D. (1996). Error-prone retrotransposition: Rime of the ancient mutators. *Proc. Natl. Acad. Sci. USA* **93,** 7427–7431.

Radice, A. D., Bugaj, B., Fitch, D. H., and Emmons, S. W. (1994). Widespread occurrence of the Tc1 transposon family: Tc1-like transposons from teleost fish. *Mol. Gen. Genet.* **244,** 606–612.

Raz, E., van Luenen, H. G. A. M., Schaerringer, B., Plasterk, R. H. A., and Driever, W. (1997). Transposition of the mematode *Caenorhabditis elegans Tc3* element in the zebrafish *Danio rerio. Curr. Biol.* **8,** 82–88.

Rio, D. C., Barnes, G., Laski, F. A., Rine, J., and Rubin, G. M. (1988). Evidence for *Drosophila* P element transposase activity in mammalian cells and yeast. *J. Mol. Biol.* **200,** 411–415.

Robertson, H. M., and Lampe, D. J. (1995). Recent horizontal transfer of a *mariner* transposable element among and between Diptera and Neuroptera. *Mol. Biol. Evol.* **12,** 850–862.

Rosenweig, B., Liao, L. W., and Hirsh, D. (1983). Sequence of the *C. elegans* transposable element Tc1. *Nucl. Acids Res.* **11,** 4201–4209.

SanMiguel, P., Tikhonov, A., Jin, Y.-K., Motchoulskaia, N., Zakharov, D., Melake-Berhan, A., Springer, P. S., Edwards, K. J., Lee, M., Avramova, Z., and Bennetzen, J. L. (1996). Nested retrotransposons in the intergenic regions of the maize genome. *Science* **274,** 765–768.

Schier, A. F., Joyner, A. L., Lehmann, R., and Talbot, W. S. (1996). From screens to genes: Prospects for insertional mutagenesis in zebrafish. *Genes Dev.* **10,** 3077–3080.

Schlotterer, C., and Tautz, D. (1992). Slippage synthesis of simple sequence DNA. *Nucl. Acids Res.* **20,** 211–215.

Schmid, C. W., and Jelinek, W. R. (1982). The *Alu* family of dispersed repetitive sequences. *Science* **216,** 1065–1070.

Schulte-Merker, S., van Eeden, F. J., Halpern, M. E., Kimmel, C. B., and Nüsslein-Volhard, C. (1994). *no tail (ntl)* is the zebrafish homologue of the mouse *T (Brachyury)* gene. *Development* **120,** 1009–1015.

Sheen, F.-M., Lim, J. K., and Simmons, M. J. (1993). Genetic instability in *Drosophila melanogaster* mediated by *hobo* transposable elements. *Genetics* **133,** 315–334.

Shimoda, N., Chevrette, M., Ekker, M., Kikuchi, Y., Hotta, Y., and Okamoto, H. (1996a). *Mermaid:* A family of short interspersed repetitive elements widespread in vertebrates. *Biochem. Biophys. Res. Commun.* **220,** 226–232.

Shimoda, N., Chevrette, M., Ekker, M., Kikuchi, Y., Hotta, Y., and Okamoto, H. (1996b). *Mermaid,* a family of short interspersed repetitive elements, is useful for zebrafish genome mapping. *Biochem. Biophys. Res. Commun.* **220,** 233–237.

Sinnett, D., Deragon, J. M., Simard, L. R., and Labuda, D. (1990). Alumorphs—Human DNA polymorphisms detected by polymerase chain reaction using Alu-specific primers. *Genomics* **7,** 331–334.

Smit, A. F., and Riggs, A. D. (1996). Tiggers and DNA transposon fossils in the human genome. *Proc. Natl. Acad. Sci. USA* **93,** 1443–1448.

Spradling, A. C., Stern, D. M., Kiss, I., Roote, J., Laverty, T., and Rubin, G. M. (1995). Gene disruptions using P transposable elements: an integral component of the *Drosophila* genome project. *Proc. Natl. Acad. Sci. USA* **92,** 10824–10830.

Takasaki, N., Murata, S., Saitoh, M., Kobayashi, T., Park, L., and Okada, N. (1994). Species-specific amplification of tRNA-derived short interspersed repetitive elements (SINEs) by retroposition: A process of parasitization of entire genomes during the evolution of salmonids. *Proc. Natl. Acad. Sci. USA* **91,** 10153–10157.

Turner, B. J., Elder, J. F., Jr., Laughlin, T. F., Davis, W. P., and Taylor, D. S. (1992). Extreme clonal diversity and divergence in populations of a selfing hermaphroditic fish. *Proc. Natl. Acad. Sci. USA* **89,** 10643–10647.

Unsal, K., and Morgan, G. T. (1995). A novel group of families of short interspersed repetitive elements (SINEs) in *Xenopus:* Evidence of a specific target site for DNA-mediated transposition of inverted-repeat SINEs. *J. Mol. Biol.* **248,** 812–823.

van Luenen, H. G., and Plasterk, R. H. (1994). Target site choice of the related transposable elements Tc1 and Tc3 of *Caenorhabditis elegans. Nucl. Acids. Res.* **22,** 262–269.

Vos, J. C., and Plasterk, R. H. (1994). Tc1 transposase of *Caenorhabditis elegans* is an endonuclease with a bipartite DNA binding domain. *EMBO J.* **13,** 6125–6132.

Vos, J. C., De Baere, I., and Plasterk, R. H. (1996). Transposase is the only nematode protein required for in vitro transposition of Tc1. *Genes Dev.* **10,** 755–761.

Voytas, D. F. (1996). Retroelements in genome organization. *Science* **274,** 737–738.

Voytas, D. F., and Boeke, J. D. (1993). Yeast retrotransposons and tRNAs. *Trends Genet.* **9,** 421–427.

Weiner, A. M., Deininger, P. L., and Efstratiadis, A. (1986). The reverse flow of genetic information. *Annu. Rev. Biochem.* **55,** 631–666.

Wessler, S. R., Bureau, T. E., and White, S. E. (1995). LTR-retrotransposons and MITEs: Important players in the evolution of plant genomes. *Curr. Opin. Genet. Dev.* **5,** 814–821.

Westerfield, M., Wegner, J., Jegalian, B. G., deRobertis, E. M., and Puschel, A. W. (1992). Specific activation of mammalian *Hox* promoters in mosaic transgenic zebrafish. *Genetics* **103,** 125–136.

Yeadon, P. J., and Catcheside, D. E. (1995). *Guest:* A 98 bp inverted repeat transposable element in Neurospora crassa. *Mol. Gen. Genet.* **247,** 105–109.

Zhivotovsky, L. A., and Feldman, M. W. (1995). Microsatellite variability and genetic distances. *Proc. Natl. Acad. Sci. USA* **92,** 11549–11552.

Zietkiewicz, E., Labuda, M., Sinnett, D., Glorieux, F. H., and Labuda, D. (1992). Linkage mapping by simultaneous screening of multiple polymorphic loci using Alu oligonucleotide-directed PCR. *Proc. Natl. Acad. Sci. USA* **89,** 8448–8451.

CHAPTER 7

Transgenesis

Anming Meng, Jason R. Jessen, and Shuo Lin
Institute of Molecular Medicine and Genetics
Medical College of Georgia
Augusta, Georgia 30912

I. Introduction

Transgenic technology is useful for many biological studies (Hanahan, 1989; Jaenisch, 1988). The ectopic expression of transgenes in whole animals allows one to study gain-of-function phenotypes. Alternatively, disruption of endogenous genes by random transgene insertion or through targeted homologous recombination allows the study of loss-of-function phenotypes, an approach that can elucidate the biological role of a gene. Transgenic technology is often used as a tool for identifying mutant genes after they have been mapped to specific chromosomal loci (Antoch *et al.*, 1997). By employing reporter genes under the control of specific regulatory sequences, transgenic techniques make possible the functional dissection of the *cis*-acting elements responsible for spatial and temporal gene expression patterns. In addition, tissues or cells expressing a reporter transgene can be used in cell lineage analysis and transplantation experiments.

METHODS IN CELL BIOLOGY, VOL. 60

0091-679X/99 $30.00

Stuart *et al.* (1988) generated the first transgenic zebrafish by microinjecting plasmid DNA into the cytoplasm of one- to four-cell staged embryos. Since this seminal achievement, numerous papers describing transgenic zebrafish have been published (Amsterdam *et al.*, 1995; Culp *et al.*, 1991; Gibbs *et al.*, 1994; Higashijima *et al.*, 1997; Lin *et al.*, 1994b; Long *et al.*, 1997; Stuart *et al.*, 1990; Westerfield *et al.*, 1992). In addition to plasmid DNA microinjection methods, approaches using retroviral infection (Lin *et al.*, 1994a), microprojectiles (Zelenin *et al.*, 1991) and electric field-mediated gene transfer (Muller *et al.*, 1993; Powers *et al.*, 1992) have been developed to introduce DNA into zebrafish. In this chapter, we describe the procedures we use to generate transgenic zebrafish by microinjection of plasmid DNA. The efficiency for germline transmission of plasmid DNA is usually below 10% (Amsterdam *et al.*, 1995; Bayer and Campos-Ortega, 1992; Lin *et al.*, 1994b; Stuart *et al.*, 1988) but can be higher (Culp *et al.*, 1991). However, this method has proven to be reliable for producing transgenic zebrafish that express transgenes in multiple generations (Amsterdam *et al.*, 1995; Bayer and Campos-Ortega, 1992; Higashijima *et al.*, 1997; Lin *et al.*, 1994b; Long *et al.*, 1997; Stuart *et al.*, 1990).

II. Materials

A. Equipment

1. Flaming/Brown Micropipette Puller: Model P-91, Sutter Instrument Company.
2. Microinjection capillaries (1 mm), Cat. No. GC100-10, Clark Electromedical Instruments.
3. Micromanipulator: model MN-151, Narishige Scientific Instrument Lab, Japan [telephone: (03) 3308-8233]; Magnetic stand: model GJ-1; CT-1 teflon tubing; HI-6-1 stainless steel injection holder (for 1-mm glass capillary).
4. MICRO-MATE Interchangeable Syringe (10 cc): Popper and Sons, Inc.
5. Plastic two-way stopper for 1-mm Teflon tubing.
6. Microscopes: SV6 or SV11 Stereomicroscope; Axioplan 2 microscope with FITC filter set 09 (excitation BP 450-490, beamsplitter FT510, emission LP520); AttoArc microscope illuminator with HBO100w; ACHROPLAN 40x/0,75W, 63x/0,90W, plan-NEOFLUOR 10x/0,30, 20x/0,50; Color video camera: ZVS-3C75DE, Carl Zeiss, Inc.
7. Kodak digital camera, DCS 420.
8. Breeding traps and dividers: Cat. No. CD-203, Aquaculture Supply, Florida [telephone: (904) 567–8540].
9. Polycarbonate mouse cage (24 × 14 × 13 cm): Cat. No. 6601177, Nalgene.
10. Microinjection plates (see Method).

11. Glass Pasteur pipettes with rubber bulbs.
12. Glass microscope slides (5 × 7.5 cm).
13. Plastic cover slips.
14. 100 × 20 mm and 60 × 15 mm Petri dishes, 24-well tissue culture dishes

B. Reagents, Buffers, and Kits

1. Holtfreter's solution: 7.0 g NaCl, 0.4 g sodium bicarbonate, 0.2 g $CaCl_2$ (anhydrous) or 0.24 g $CaCl_2 \cdot 2H_2O$, 0.1 g KCl. Dissolve in 2-L ddH_2O, add 30 μl concentrated HCl, pH should be 6.5–7.0. Make fresh Holtfreter's solution daily.
2. Pronase: Cat. No. P 5147, Sigma. Make 30 mg/ml stock solution in ddH_2O, pH 7.0. Self-digest the stock solution at 37°C for 1 hour and store 10-ml aliquots at −20°C.
3. GENECLEAN II Kit: Cat. No. 1001-400, Bio101, Inc.
4. KCl: 1*M* in ddH_2O, autoclave and store as 1 ml aliquots at −20°C.
5. Tetramethyl-rhodamine dextran: Cat. No. D-1817, Molecular Probes. Stock solution is 12.5% in ddH_2O, store at −20°C.
6. Qiagen brand or other plasmid DNA preparation kits.
7. DNA constructs containing the Green Fluorescent Protein (GFP) or *lacZ* reporter gene.
8. Expand Long Template PCR System: Cat. No. 1 681 842, Boehringer Mannheim.
9. Tricaine (3-aminobenzoic acid ethyl ester): Cat. No. A 5040, Sigma. Stock solution is 0.16% in ddH_2O (adjust to pH 7.0 using 1.0*M* Tris base). Store at 4°C.
10. DNA extraction buffer: 10m*M* Tris pH 8.2, 10m*M* EDTA, 200m*M* NaCl, 0.5% SDS, and 200 μg/ml pronase.
11. Light mineral oil: Cat. No. 0121-1, Fisher Scientific.
12. X-gal staining solutions
    a. Fix Solution (store at room temperature): 12.5% glutaraldehyde in PBS.
    b. Wash Solution (store at room temperature): 0.1*M* K_2HPO_4/KH_2PO_4 buffer, pH 7.4, 1m*M* $MgCl_2$, 0.3% Triton-X 100.
    c. 20× KCN (store at 4°C in the dark): Dissolve 0.33 g $K_3Fe(CN)_6$ and 0.42 g $K_4Fe(CN)_6 \cdot H_2O$ in 10 ml 0.1 M K_2HPO_4/KH_2PO_4 buffer, pH 7.4.
    d. Staining Solution (make fresh each time): Mix 1 ml 0.1*M* K_2HPO_4/KH_2PO_4 buffer (pH 7.4), 50 μl 20 × KCN, 2 μl 1*M* $MgCl_2$, 20 μl 50× X-gal (40 mg/ml in DMSO, store at −20°C).

14. Zebrafish Wnt5A and GFP primers

Wnt5A (sense primer)	5′ CAGTTCTCACGTCTGCTACTTGCA 3′
Wnt5A (antisense primer)	5′ ACTTCCGGCGTGTTGGAGAATTC 3′
GFP (sense primer)	5′ AATGTATCAATCATGGCAGAC 3′
GFP (antisense primer)	5′ TGTATAGTTCATCCATGCCATGTG 3′

III. Methods

A. General Protocols for Generating Transgenic Zebrafish

1. Preparation of Plasmid DNA for Microinjection

Plasmid DNA is prepared using a Qiagen kit or other plasmid DNA preparation procedure following the manufacturer's instruction. DNA for microinjection is linearized with the appropriate restriction enzyme(s), but supercoiled DNA can be used for transient gene expression assays. If possible, insert DNA fragments should be purified free of vector sequences. Digested plasmid DNA is run on an agarose gel containing 0.5 μg/ml ethidium bromide and the desired DNA fragment is excised and purified using a GENECLEAN II Kit. At the final step of purification, DNA is dissolved in nuclease-free 1× TE (pH 8.0) at a concentration five to ten times higher than the final concentration to be used for microinjection. The DNA solution is spun for 30 seconds using a microcentrifuge, and two-thirds of the volume is transferred to a new 1.5-ml Eppendorf microcentrifuge tube. Effort should be made to avoid taking any precipitate which can later clog the microinjection needle during DNA loading. To make sure that the DNA is pure and structurally intact, 2 μl of the DNA should be checked by electrophoresis and ethidium bromide staining. The purity of the DNA to be microinjected is a critical factor in determining the number of embryos that survive the microinjection procedure.

For microinjection, the DNA should be diluted to 50–100 μg/ml in 0.1*M* KCl containing tetramethyl-rhodamine dextran (0.125%). This dye serves as an internal control by allowing the amount of microinjected DNA to be monitored so that embryos are consistently microinjected with a similar amount of DNA. Our experiences indicate that higher concentrations of DNA can increase gene expression levels to a certain degree but often result in lower postinjection embryo survival rates. The prepared DNA solution can be stored at 4°C for up to 2 weeks or at −20°C for a longer time.

2. Maintenance of Fish for Microinjection

The postinjection embryo survival rate is also influenced by the quality of zebrafish eggs obtained. To ensure that there is a sufficient number of healthy eggs for microinjection, dedicated stocks of healthy and highly productive breeding fish should be established. When young fish reach sexual maturity, a preselection can be made based on a pairwise mating test to determine which fish will provide the highest quality eggs. The selected males and females are maintained separately and fed four to five smaller meals daily as opposed to three larger meals. We have observed that fish fed smaller and more frequent meals tend to produce more eggs. The fish are fed a diet consisting of both brine shrimp and various flake foods. We keep 14 tanks of stock fish (7 male tanks and 7 female tanks), each containing approximately 20 fish, solely for the purpose of microinjection. The microinjection fish tanks are labeled with the days of the week and the fish for microinjection are used for mating only once a week. Fish often produce no eggs or bad eggs after resting for several weeks without mating. Therefore, periodical mating of the microinjection fish is necessary to sustain the productivity of breeding stocks. This should be done even during weeks when no eggs are needed for microinjection.

3. Preparation of Microinjection Plates

Both the lid and bottom dish of a 60-mm Petri dish can be used to make microinjection plates. A plate is formed by pouring 12.5 ml of warm 1.2% agarose (prepared in ddH_2O) into a lid or dish and placing a glass microscope slide (5×7.5 cm) onto the agarose at a 10–20° angle as illustrated in Fig. 1 (Culp *et al.*, 1991). After the agarose has solidified at room temperature, the glass slide is removed leaving a groove (Fig. 1). Multiple plates can be made each time and

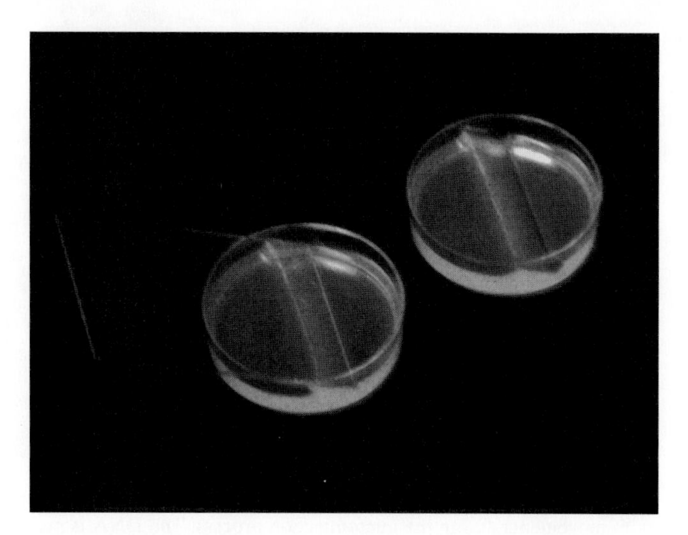

Fig. 1 A microinjection plate is made by inserting a glass microscope slide into warm agarose at a 10–20° angle and letting the agarose solidify. The slide is removed leaving a groove in the agarose. During microinjection, embryos are positioned along this groove so that they are supported when penetrated by the microinjection needle.

should be stored in ddH_2O at 4°C. A plate can be used many times if it is washed following each use with 70% ethanol and then with ddH_2O. A used plate is covered with plastic wrap and stored at 4°C.

4. Microinjection Setup

The microinjection system we use consists of a stereomicroscope, a micromanipulator, a magnetic stand, a syringe, a plastic two-way stopper, Teflon tubing, a needle holder, microinjection capillaries, and a microinjection plate. The micromanipulator is attached to the magnetic stand which is securely fastened to a metal surface. The syringe is filled with 3–4 ml of mineral oil and is directly connected to the two-way stopper. This stopper is connected to approximately 3 feet of 1-mm Teflon tubing that, at the other end, is connected to the microinjection needle holder. It takes considerable effort to attach the tubing to the needle holder. However, a tight fit is required to ensure that the microinjection system remains intact when under the pressure caused by pulling and pushing the syringe plunger. The needle holder is attached to the micromanipulator. Because an automatic microinjector is not needed, this system is quite inexpensive. Some individuals may prefer to microinject without the use of a micromanipulator. Although difficult, this would significantly reduce the cost of the system. The microinjection system is pictured in Fig. 2.

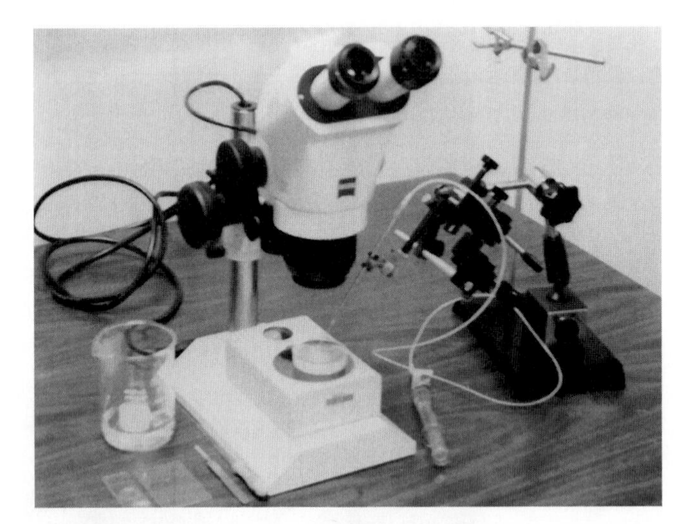

Fig. 2 A typical zebrafish microinjection system. A micromanipulator is used to control needle movements during the microinjection process. The DNA is pushed through the needle by a syringe containing mineral oil. Embryos, the needle, and the microinjected DNA are all visible under a stereomicroscope allowing for precise control of the microinjection process. The micromanipulator is controlled by the right hand, and the microinjection plate is moved on the *x* and *y* axis using the left hand. This system occupies little desktop space and is portable.

5. Preparation of the Microinjection Needle

Microinjection needles are made with a Flaming/Brown Micropipette Puller using fine-glass microinjection capillaries. The tip of the microinjection needle should be approximately 0.05–0.15 mm. Prior to microinjection, a needle is opened by breaking its tip under a stereomicroscope using a clean, sharp forcep. It will take practice to break the needle consistently at a position optimal for microinjection. A needle broken too much will be too large to puncture the embryos effectively and will likely destroy the embryos. Conversely, if a needle is broken too little, it will be difficult to load with DNA and will be more prone to clogging. The needle is attached to the needle holder and filled with mineral oil by pushing oil from the syringe, through the Teflon tubing, and into the needle. Effort should be made to avoid trapping air bubbles in the microinjection tube and needle.

The DNA prepared for microinjection is spun on high speed in a microcentrifuge for 2 min. to sediment any contaminating particulates. Depending on how many eggs one plans to inject, 2–10 μl of DNA is transferred onto a clean plastic cover slip and is overlaid with a few drops of mineral oil to prevent evaporation. We have been able to microinject 200–300 embryos per 1 μl of DNA. The DNA is loaded by placing the needle tip into the DNA solution using the micromanipulator and by gently pulling the 10cc syringe plunger. If loading occurs too slowly or the needle becomes clogged with debris, the needle can be broken again using forceps as described earlier and loading can be resumed. However, in most cases, it is better to start with a new needle. Because loading is a slow process, a clamp may be used to hold the syringe barrel during loading.

6. Collection and Preparation of Eggs

In the late afternoon of the day before microinjection, and at least 30 minutes after their last feeding, a male and one or two females are transferred from their fish tanks to a mouse cage containing a breeding trap. The male and female are separated using a divider (Fig. 3) and covered to prevent them from leaping out of the cage. One pair of fish can normally produce one to several hundred eggs, but sometimes no eggs or bad eggs are produced. Consequently, 10–15 pairs should be set up to ensure that a sufficient number of eggs will be obtained for microinjection. On the day of microinjection, after the lights have come on in the morning, the breeding trap (with the fish and the divider still inside) is moved to a new mouse cage filled halfway with fresh fish water. The fish are joined by removing the divider. How many mating pairs of fish to release at one time is determined by the speed of the microinjection process. For example, someone who microinjects quickly may need to release three pairs at once whereas a slower worker may need to release only one pair each time. Females will usually begin laying eggs after only a few minutes. Eggs are collected in a 50-ml beaker approximately 10 minutes after being laid. The fish water in the beaker is replaced with 15–20 ml of a diluted pronase solution (approximately 1 mg/ml in Holt-

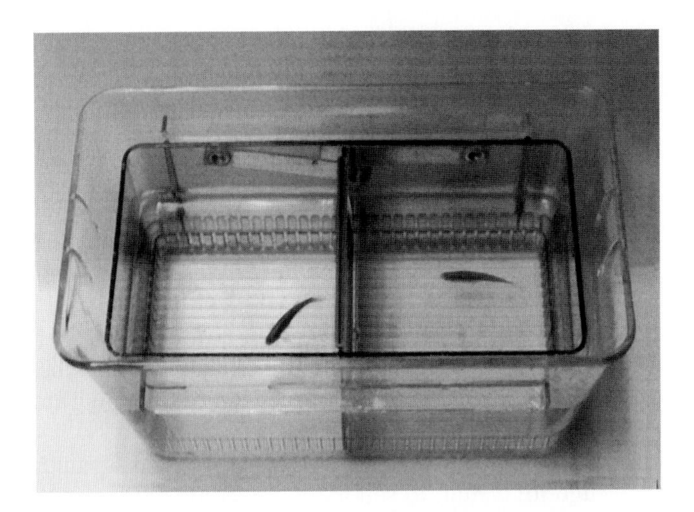

Fig. 3 The day prior to microinjection, male and female zebrafish are placed in a breeding trap inside of a mouse cage containing fish water. They are immediately separated by a divider and covered with a lid (not shown). The next morning, after the fish are placed in a new mouse cage filled halfway with fresh fish water, the divider is removed. This procedure makes it possible to control when eggs are laid so that they may be immediately collected and dechorionated. It is important to move quickly after the eggs have been laid because they will reach the two-cell stage in approximately 45 minutes.

freter's solution) prewarmed at 28–30°C, and the beaker is placed in a 28–30°C incubator for several minutes. The beaker can be gently swirled to help embryos fall out of their chorions. When approximately 5–10 chorions drop off the embryos (chorions can be seen floating in the solution), the eggs are washed 8–10 times with Holtfreter's solution to remove all traces of chorions and pronase. One should avoid treating the embryos with pronase too long because this will decrease the embryo survival rate.

7. Microinjection

A microinjection plate is filled with Holtfreter's solution. The dechorionated embryos are gently transferred into the middle groove of the microinjection plate using a glass Pasteur pipette. Approximately 200 embryos can be lined up in the groove of one plate. The microinjection needle should be positioned with the help of the micromanipulator. The needle goes into the cytoplasm of single-cell embryos from the top of the cell (Fig. 4A), and the DNA solution is microinjected by gently pushing the syringe plunger. The microinjected DNA normally occupies approximately one-fifth the volume of the single-cell mass which can be seen under the stereomicroscope (Fig. 4B). Because many of the embryos will not be in the correct orientation for microinjection, their position in the microinjection plate can be adjusted using the microinjection needle.

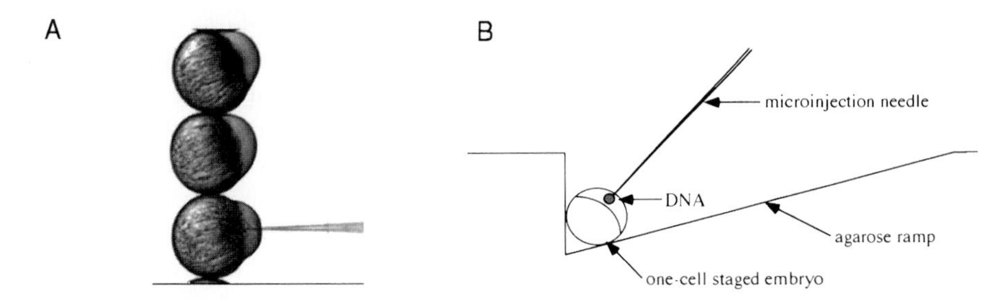

Fig. 4 The embryos are microinjected by first penetrating the cell wall with the microinjection needle (A) and then pushing gently on the syringe plunger until the microinjected DNA occupies approximately one-fifth of the cytoplasm volume (B). The needle is immediately pulled out of the cytoplasm, and the next egg is microinjected. It is possible to see the DNA being microinjected into the cytoplasm without a rhodamine dye, but this dye will increase the visibility of the DNA solution. It is important not to microinject too much DNA into the embryos because this can significantly decrease the survival rate.

Microinjected embryos are transferred into several Petri dishes containing fresh Holtfreter's solution and incubated at 28–30°C. Unfertilized and deformed eggs can be easily recognized and should be removed 4–5 hours after microinjection. Developed embryos (~50% epiboly) are transferred into a 24-well tissue culture plate in 1/2× Holtfreter's solution (5 embryos per well) and grown overnight at 28–30°C. The next day, dead and deformed embryos are removed, and healthy embryos are transferred into 1/4× Holtfreter's solution. After 48 hours of development, microinjected embryos are placed into regular fish water. The care and maintenance of microinjected embryos is a critical part of the microinjection process and will directly influence the number of surviving embryos.

8. Identification of Germline Transgenic Founder Fish by PCR

Although it is possible to screen germline transgenic zebrafish by examining expression of a reporter gene such as GFP or *lacZ,* polymerase chain reaction (PCR) still represents the most commonly used method to detect germline transmission of transgenes. To reduce the labor involved in this process, eggs from multiple pairs of fish are pooled as described later. Potential transgenic founder fish (2–3 months old) are mated to each other or to a nontransgenic wild-type fish. Because transmission of a transgene to the F_1 generation is often mosaic (ranging from 2 to 90%), it is necessary to collect approximately 100 eggs from each founder fish. F_1 embryos are grown 24 hours at 28–30°C and treated with pronase to remove their chorions as described earlier. Approximately 100 embryos are transferred to an Eppendorf tube containing 1 ml of DNA extraction buffer, vortexed briefly, and incubated with rotation at 55°C for 3–4 hours. The tubes are vortexed briefly, once per hour, during this incubation. The samples are then centrifuged at 14,000 rpm for 10 minutes at room temperature using a microcentrifuge. Ten samples are pooled by combining 20 μl of extraction solution from each sample in a new Eppendorf tube. The remaining DNA extraction solution from each sample should be saved at −20°C until further use. The combined DNA mixtures are precipitated with the addition of 300 μl of 100% ethanol followed by microcentrifugation. Preciptates are washed with 70% ethanol, vacuum dried, and dissolved in 100 μl of 1 × TE (pH 8.0).

Using the isolated genomic DNA, standard PCR can be performed to identify pools of DNA samples that contain a transgene. The primer set corresponding to the *GFP* reporter gene (see Materials) produces a 267-bp PCR product (Long *et al.,* 1997). A number of controls should be included in the PCRs. First, an internal control that detects an endogenous zebrafish gene should be included in each reaction to assess if the PCR reaction worked. The primer set corresponding to the zebrafish *Wnt5A* gene (see Materials) produces a 387-base pair PCR product (Lin *et al.,* 1994a). Second, a positive control that contains one transgenic embryo per 500 embryos should be included to ensure that the PCR conditions used are sensitive enough to detect such a low transgene concentration. This can be achieved by mixing a previously identified transgenic embryo with non-transgenic embryos or mixing transgene DNA with DNA isolated from non-transgenic fish at appropriate ratios. Finally, a negative control should be included to address possible contamination. A typical PCR result is shown in Fig. 5.

Once a positive pool is identified, PCR can be performed using the saved DNA extraction solutions from each individual sample to identify the transgenic founder fish. Using the pooling method, it is possible to screen 200 transgenic families per week. After a founder fish harboring a transgene is identified by PCR, other methods, such as Southern blotting and examination of transgene expression, should be performed on the progeny to confirm the identity of the founder fish.

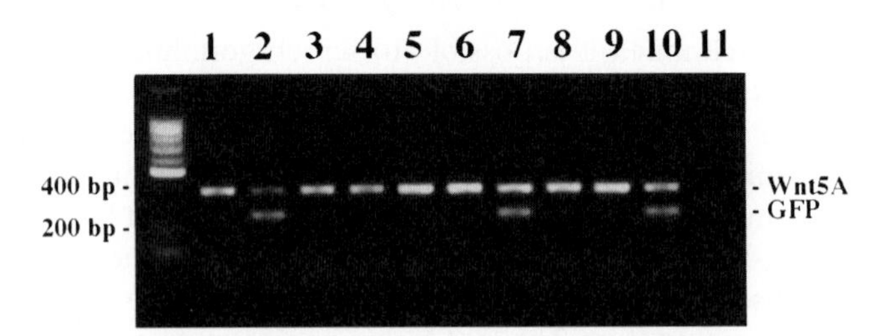

Fig. 5 PCR is commonly used to detect the presence of transgenes in the genome of microinjected zebrafish. This is an ethidium bromide stained gel from a typical PCR on genomic DNA isolated from 24-h-old F_1 embryos. Lanes 2, 7, and 10 represent embryos from transgenic founder fish as indicated by both Wnt5A and GFP PCR products. Lanes 1, 3, 4, 5, 6, 8, and 9 represent embryos from nontransgenic founder fish as indicated by a lack of the GFP PCR product. Lane 11 is a negative control PCR that was performed to assess possible contamination in the PCRs.

9. Identification of Germline Transgenic Founder Fish by Reporter Gene Expression

The founder transgenic fish can be identified based on the expression of a specific reporter gene in the founder's progeny. The luciferase, bacterial *lacZ* or GFP reporter genes can be expressed in zebrafish if they are driven by a proper promoter (Amsterdam *et al.*, 1995; Amsterdam *et al.*, 1996; Bayer and Campos-Ortega, 1992; Jessen *et al.*, 1998; Kuo *et al.*, 1995; Lin *et al.*, 1994b; Long *et al.*, 1997; Meng *et al.*, 1997; Moss *et al.*, 1996; Muller *et al.*, 1993, 1997; Rinder *et al.*, 1992; Verri *et al.*, 1997; Westerfield *et al.*, 1992; Williams *et al.*, 1996). When GFP is used as a reporter gene, embryos generated by mating a microinjected fish to a wild-type fish are observed for GFP expression under a fluorescence microscope. GFP expression in these embryos identifies the microinjected parent as a transgenic founder fish. Alternatively, embryos microinjected with the *lacZ* reporter gene can be mated to wild-type fish and their offspring analyzed for *lacZ* expression (Bayer and Campos-Ortega, 1992). About 100 dechorionated embryos are transferred into an Eppendorf tube, 1 ml of Fix Solution is added, and the embryos are incubated at room temperature for 8 min. The Fix Solution is replaced with Wash Solution and the embryos are incubated at room temperature for 2 minutes. This wash is repeated four times. The embryos are placed in Staining Solution for several hours to overnight at 37°C followed by observation under the microscope. Embryos displaying specific *lacZ* staining identify the microinjected parent as a transgenic founder fish.

10. Generation of Homozygous Transgenic Fish

The transgenic founder (F_O) fish can be bred to generate homozygous transgenic populations that can be maintained for generations without loss of the integrated transgenes. F_O fish are crossed with wild-type fish, and their progeny (F_1) are grown to sexual maturity. The presence of a transgene can be detected by analyzing GFP expression in F_1 fish or *lacZ* expression in F_2 fish as described earlier. Alternately, live transgenic F_1 fish can be identified by performing PCR on genomic DNA isolated from caudal fin clips. F_1 fish are anesthetized using a 0.16% tricaine stock solution diluted further in fish water by adding 4.2 ml stock solution per 100 ml fish water. The amount of tricaine to use varies with water volume and number of fish, but should be used sparingly so as not to kill the fish. With a razor blade, one-half to two-thirds of the tails are cut off and placed in separate Eppendorf tubes each containing 100 μl of DNA extraction buffer. The fish are placed in separate mouse cages numbered to correspond to the Eppendorf tube containing the tail cut. Each tube is vortexed thoroughly and then incubated with rotation at 55°C for 3–4 hours. The tubes are vortexed briefly once during every hour of this incubation. Following this incubation, the tubes are centrifuged at 14,000 rpm for 20 minutes at room temperature using a microcentrifuge. DNA samples are diluted 1:10 for PCR. If desired, multiple DNA samples can be pooled to minimize the number of PCR reactions.

Positively identified F_1 brothers and sisters are mated. An F_1 pair should normally produce 25% transgenic homozygote positives, 50% heterozygotes and 25% wild-type homozygotes among their progeny. F_2 embryos are grown to sexual maturity, and individual F_2 fish are mated to wild-type fish. The F_3 progeny are analyzed for the transgene either by assaying reporter gene expression or by analyzing individual F_3 embryos by PCR as described previously. A homozygous F_2 fish should pass the transgene to 100% of its offspring when mated with a wild-type fish. Identified homozygous F_2 fish can be mated to each other to produce a large homozygous population.

B. Specific Protocols for Promoter Analysis Using GFP in Living Embryos

Although tissue culture systems are useful for identifying many important noncoding *cis*-acting elements that regulate gene expression, transfection analysis in tissue culture cells cannot simulate the complex, rapidly changing microenvironment to which the promoter must respond during embryogenesis. Temporal and spatial analysis of promoter/enhancer activity can be only poorly mimicked *in vitro.* As noted in many chapters of this book, organ rudiments in zebrafish appear within 24 hours of fertilization. The development of these organs can be followed easily because development takes place outside the mother, and the early embryo is transparent. Therefore, the use of GFP in developing zebrafish embryos provides an excellent vertebrate whole animal system for the analysis of tissue specific promoter/enhancer activities.

By microinjecting DNA constructs containing a tissue specific promoter ligated to the GFP reporter gene into single cell zebrafish embryos, one can continuously examine the dynamic expression patterns of GFP. By generating deletions and point mutations in promoter/enhancer regions, one can identify the *cis*-acting elements responsible for spatial and temporal gene expression patterns. Using the injection method described earlier, two persons can generate more than 1000 microinjected embryos that transiently express GFP in a single day. This makes it possible to obtain statistically significant data for analyzing promoter/enhancer activity for each construct.

1. Generation of Promoter/Enhancer-GFP Constructs

The expression of a gene is controlled by transcription factors that recognize and bind to specific sequence motifs present in various locations (e.g., 5′ and 3′ flanking sequences) and intervening sequences of a gene. For analyzing promoter/enhancer activity, one can isolate a long stretch of sequence immediately upstream of the translation start codon and ligate it to the reporter gene. This ensures the preservation of basic promoter elements and potential regulatory elements in the 5′ untranslated region and also eliminates the laborious work of identifying the transcription start site for a particular gene.

To isolate promoter/enhancer sequences of a gene, a DNA fragment that contains the 5′ part of this gene is used as a probe to screen a genomic library.

This probe can be obtained by digesting a cDNA clone with proper restriction enzymes or by PCR using primers specific to the 5′ region of the gene. Although genomic libraries constructed using lambda phage vectors are often used for this purpose, bacterial artificial chromosome (BAC) or P_1-derived artificial chromosome (PAC) libraries are preferred. The identified positive clones are mapped by restriction analysis and a large fragment containing promoter/enhancer sequences is subcloned into a plasmid vector.

The promoter/enhancer region is then amplified by PCR using a specific primer complementary to the cDNA sequence just 5′ of the translation start codon and a primer complementary to vector sequences. A *Bam*HI site is usually introduced into the specific primer to facilitate subsequent cloning. The PCR conditions depend on the length of the promoter region, melting temperatures of the primers and other factors. If the promoter is over 6 kb long, we use the Expand Long Template PCR System (Boehringer Mannheim). After digestion with *Bam*HI and another enzyme suitable for cloning, the amplified fragment is gel-purified and ligated to the GFP reporter gene.

The promoter/enhancer-GFP constructs are prepared as described earlier and microinjected into the cytoplasm of single-cell zebrafish embryos. Depending on the nature of the promoter/enhancer elements, the injected embryos may express GFP as early as 4 hr postinjection (sphere stage) and may continue to express GFP for a few months. The GFP expression patterns in the microinjected embryos can be observed and recorded at a number of distinct developmental stages using fluorescence microscopy.

2. Generation of Deletion and Point Mutation Constructs

To identify small *cis*-acting elements that are responsible for regulating tissue-specific gene expression patterns, constructs containing deletions in the promoter/enhancer can be generated. Naturally occurring restriction sites may be used to create a series of gross promoter deletions. Each construct generated should be individually microinjected into single-cell embryos and examined.

PCR technology can also be used to create a deletion series within an enhancer/promoter region rapidly. This is achieved using a promoter/enhancer-GFP construct as a template and primers specific to the promoter region and a vector primer. The amplified fragments should all contain the GFP gene and a polyadenylation signal in addition to the promoter sequences. PCR reactions can be performed using Expand Long Template PCR System (Boehringer Mannheim) for 25 cycles (94°C, 30 sec; 55°C, 30 sec; 68°C, 6 minutes). The PCR products are gel-purified with GENECLEAN II Kit (Bio 101, Inc.) and directly used for microinjection without further subcloning.

The deletion assays using PCR can confine a potential regulatory sequence motif to 20–30 bp. To determine the core sequence necessary for the activity of a tissue-specific *cis*-acting element, site-directed base mutations can be generated by PCR. Mutant primers containing 2–3 mutant bases are used in conjunction with a vector primer for PCR amplification of the target sequence using a promoter/enhancer-GFP construct as the template. Again, the PCR products are used for microinjection without further subcloning.

To obtain statistically significant data, one should examine at least 200 injected embryos for each construct. When using rhodamine fluorescent dye as an injection marker, uninjected embryos are easily visualized and can be discarded. The original records should include the number of embryos that have or do not have specific GFP expression and the number of GFP-positive cells in each embryo.

3. Fluorescent Microscopic Observation and Imaging

Embryos are anesthetized using tricaine and examined under a FITC filter on a Zeiss or other type of microscope equipped with a color video camera or a digital camera (see Materials). Images of fluorescent embryos are collected separately under fluorescent and bright fields and stored in a computer. Adobe Photoshop software is used to superimpose the fluorescent image on the bright field image.

IV. Discussion

Transgenic zebrafish generated using plasmid DNA constructs are used for a variety of genetic analyses. We have generated transgenic zebrafish containing the GFP reporter gene ligated to 5.6 kb of zebrafish GATA-1 promoter sequence (Long *et al.,* 1997). These fish are used to obtain pure populations of hematopoietic progenitor cells by isolating fluorescent, GFP-expressing hematopoietic cells, from 16-hr-old embryos using fluorescence-activated cell sorting. These cells can be used to perform differential display PCR to identify genes uniquely expressed in the earliest hematopoietic progenitor cells. In addition to generating transgenic zebrafish, transient gene expression assays can be performed in living zebrafish embryos after microinjection of plasmid DNA. In contrast to *in situ* hybridization, these assays provide three-dimensional information of gene expression patterns in real time because living embryos can be observed for several days. We have used plasmid constructs containing GFP ligated to zebrafish GATA-2 promoter sequences to identify a regulatory element responsible for GATA-2 expression in the central nervous system (Meng *et al.,* 1997). Promoter analyses such as these are invaluable tools for dissecting the regulatory elements required for proper spatial and temporal gene expression patterns. However, for some genes, regulatory elements are located several kilobases upstream or downstream from the transcription start site. In these instances, microinjection of large DNA constructs such as bacterial, P_1-derived, or yeast artificial chromosomes may be required to identify these *cis*-acting sequences. We have developed a technique for modifying bacterial artificial chromosomes through homologous recombination to insert the GFP reporter gene (Jessen *et al.,* 1998). Using transient expression assays, we showed that embryos microinjected with the modified BAC clones were less mosaic and had improved GFP expression in hematopoietic progenitor cells compared with smaller plasmid constructs.

These types of studies exemplify some of the uses for gene transfer techniques in zebrafish. With the identification of zebrafish mutants carrying genetic defects in such diverse processes as hematopoiesis, somitogenesis, and neurogenesis (Driever *et al.,* 1996; Haffter *et al.,* 1996), the expansion of transgenic technologies in zebrafish will be invaluable for the analysis and perhaps identification of these genes.

Acknowledgments

We thank members of our laboratory for experimental contributions and comments. SL started to develop transgenic protocols while working as a postdoctoral fellow in Nancy Hopkins' laboratory at MIT and wishes to thank her for advice and support.

References

Amsterdam, A., Lin, S., and Hopkins, N. (1995). The Aequorea victoria green fluorescent protein can be used as a reporter in live zebrafish embryos. *Dev. Biol.* **171,** 123–129.

Amsterdam, A., Lin, S., Moss, L. G., and Hopkins, N. (1996). Requirements for green fluorescent protein detection in transgenic zebrafish embryos. *Gene* **173,** 99–103.

Antoch, M. P., Song, E. J., Chang, A. M., Vitaterna, M. H., Zhao, Y., Wilsbacher, L. D., Sangoram, A. M., King, D. P., Pinto, L. H., and Takahashi, J. S. (1997). Functional identification of the mouse circadian Clock gene by transgenic BAC rescue. *Cell* **89,** 655–667.

Bayer, T. A., and Campos-Ortega, J. A. (1992). A transgene containing lacZ is expressed in primary sensory neurons in zebrafish. *Development* **115,** 421–426.

Culp, P., Nüsslein-Volhard, C., and Hopkins, N. (1991). High-frequency germ-line transmission of plasmid DNA sequences injected into fertilized zebrafish eggs. *Proc. Natl. Acad. Sci. USA* **88,** 7953–7957.

Driever, W., Solnica-Krezel, L., Schier, A. F., Neuhauss, S. C., Malicki, J., Stemple, D. L., Stainier, D. Y., Zwartkruis, F., Abdelilah, S., Rangini, Z., Belak, J., and Boggs, C. (1996). A genetic screen for mutations affecting embryogenesis in zebrafish. *Development* **123,** 37–46.

Gibbs, P. D., Gray, A., and Thorgaard, G. (1994). Inheritance of P element and reporter gene sequences in zebrafish. *Mol. Mar. Biol. Biotechnol.* **3,** 317–326.

Haffter, P., Granato, M., Brand, M., Mullins, M. C., Hammerschmidt, M., Kane, D. A., Odenthal, J., van Eeden, F. J., Jiang, Y. J., Heisenberg, C. P., Kelsh, R. N., Furutani-Seiki, M., Vogelsang, E., Beuchle, D., Schach, U., Fabian, C., and Nüsslein-Volhard, C. (1996). The identification of genes with unique and essential functions in the development of the zebrafish, *Danio rerio. Development* **123,** 1–36.

Hanahan, D. (1989). Transgenic mice as probes into complex systems. *Science* **246,** 1265–1275.

Higashijima S, Okamoto, H., Ueno, N., Hotta, Y., and Eguchi G. (1997). High-frequency generation of transgenic zebrafish which reliably express GFP in whole muscles or the whole body by using promoters of zebrafish origin. *Dev. Biol.* **192,** 289–299.

Jaenisch, R. (1988). Transgenic animals. *Science* **240,** 1468–1474.

Jessen, J. R., Meng, A. M., McFarlane, R. J., Paw, B. H., Zon, L. I., Smith, G. R. and Lin, S. (1998). Modification of bacterial artificial chromosomes through chi-stimulated homologous recombination and its application in zebrafish transgenesis. *Proc. Natl. Acad. Sci. USA* **95,** 5121–5126.

Kuo, C. H., Uetsuki, T., Kim, C. H., Tanaka, H., Li, B. S., Taira, E., Higuchi, H., Okamoto, H., Yoshikawa, K., and Miki, N. (1995). Determination of a necdin cis-acting element required for neuron specific expression by using zebra fish. *Biochem. Biophys. Res. Commun.* **211,** 438–446.

Lin, S., Gaiano, N., Culp, P., Burns, J. C., Friedmann, T., Yee, J. K., and Hopkins, N. (1994a). Integration and germ-line transmission of a pseudotyped retroviral vector in zebrafish. *Science* **265,** 666–669.

Lin, S., Yang, S., and Hopkins, N. (1994b). lacZ expression in germline transgenic zebrafish can be detected in living embryos. *Dev. Biol.* **161,** 77–83.

Long, Q., Meng, A., Wang, H., Jessen, J. R., Farrell, M. J., and Lin, S. (1997). GATA-1 expression pattern can be recapitulated in living transgenic zebrafish using GFP reporter gene. *Development* **124,** 4105–4111.

Meng, A., Tang, H., Ong, B. A., Farrell, M. J., and Lin, S. (1997). Promoter analysis in living zebrafish embryos identifies a cis-acting motif required for neuronal expression of GATA-2. *Proc. Natl. Acad. Sci. U.S.A.* **94,** 6267–6272.

Moss, J. B., Price, A. L., Raz, E., Driever, W., and Rosenthal, N. (1996). Green fluorescent protein marks skeletal muscle in murine cell lines and zebrafish. *Gene* **173,** 89–98.

Muller, F., Lele, Z., Varadi, L., Menczel, L., and Orban, L. (1993). Efficient transient expression system based on square pulse electroporation and in vivo luciferase assay of fertilized fish eggs. *FEBS Lett.* **324,** 27–32.

Muller, F., Williams, D. W., Kobolak, J., Gauvry, L., Goldspink, G., Orban, L., and Maclean, N. (1997). Activator effect of coinjected enhancers on the muscle-specific expression of promoters in zebrafish embryos. *Mol. Reprod. Dev.* **47,** 404–412.

Powers, D. A., Hereford, L., Cole, T., Chen, T. T., Lin, C. M., Kight, K., Creech, K., and Dunham, R. (1992). Electroporation: A method for transferring genes into the gametes of zebrafish (*Brachydanio rerio*), channel catfish (*lctalurus punctatus*), and common carp (*Gyprinus carpio*). *Mol. Mar. Biol. Biotechnol.* **1,** 301–308.

Rinder, H., Bayer, T. A., Gertzen, E. M., and Hoffmann, W. (1992). Molecular analysis of the ependymin gene and functional test of its promoter region by transient expression in *Brachydanio rerio*. *DNA Cell Biol.* **11,** 425–432.

Stuart, G. W., McMurray, J. V., and Westerfield, M. (1988). Replication, integration and stable germ-line transmission of foreign sequences injected into early zebrafish embryos. *Development* **103,** 403–412.

Stuart, G. W., Vielkind, J. R., McMurray, J. V., and Westerfield, M. (1990). Stable lines of transgenic zebrafish exhibit reproducible patterns of transgene expression. *Development* **109,** 577–584.

Verri, T., Argenton, F., Tomanin, R., Scarpa, M., Storelli, C., Costa, R., Colombo, L., and Bortolussi, M. (1997). The bacteriophage T7 binary system activates transient transgene expression in zebrafish (*Danio rerio*) embryos. *Biochem. Biophys. Res. Commun.* **237,** 492–495.

Westerfield, M., Wegner, J., Jegalian, B. G., DeRobertis, E. M., and Puschel, A. W. (1992). Specific activation of mammalian Hox promoters in mosaic transgenic zebrafish. *Genes Dev.* **6,** 591–598.

Williams, D. W., Muller, F., Lavender, F. L., Orban, L., and Maclean, N. (1996). High transgene activity in the yolk syncytial layer affects quantitative transient expression assays in zebrafish (*Danio rerio*) embryos. *Transgenic Res.* **5,** 433–442.

Zelenin, A. V., Alimov, A. A., Barmintzev, V. A., Beniumov, A. O., Zelenina, I. A., Krasnov, A. M., and Kolesnikov, V. A. (1991). The delivery of foreign genes into fertilized fish eggs using high-velocity microprojectiles. *FEBS Lett.* **287,** 118–120.

CHAPTER 8

The Zebrafish Genome

John Postlethwait, Angel Amores, Allan Force, and Yi-Lin Yan

Institute of Neuroscience
University of Oregon
Eugene, Oregon 97403

I. Introduction

The genomes of various eutherian mammals are remarkably alike. Fluorescent *in situ* hybridization and comparative meiotic maps have revealed that entire chromosomes arms contain much the same set of genetic information in humans, cats, cows, pigs, and sheep (Lyons *et al.,* 1994; O'Brien *et al.,* 1997a,b; Rettenberger *et al.,* 1995a,b; Solinas-Toldo *et al.,* 1995). Thus, mammals have not only orthologous genes but also orthologous chromosomes. (Orthologues are genetic elements, such as genes or chromosome segments, in two different species that have been inherited from a single genetic element in the last common ancestor of the two species.) The order of genes within an orthologous pair of chromosomes is often more variable than their content (Watkins-Chow *et al.,* 1997), suggesting that inversions may become fixed in populations more frequently than translocations. The mouse genome appears to have suffered substantially more chromosome rearrangements than other well-studied mammalian genomes (Lyons *et al.,* 1994), perhaps because of the greater likelihood of fixation of translocations in small populations. The comparative mapping of various mammals makes it

METHODS IN CELL BIOLOGY, VOL. 60

0091-679X/99 $30.00

possible to predict the location of a gene in one mammal's genome from the location of that gene's orthologue in the chromosomes of another mammalian species. Can the same procedure be extended to zebrafish?

The last common ancestor of human and mouse lived about 100 million years ago, but the last common ancestor of human and zebrafish lived about 420 million years ago (Ahlberg and Milner, 1994). Would we expect four times the number of chromosome rearrangements to separate the human and zebrafish genome than separate the human and mouse genomes? This is an important question because of the potential power of comparative genomics to speed the analysis of zebrafish biology, and, reciprocally, to relate molecular genetic mechanisms discovered in zebrafish to human biology. Fortunately, recent comparative genetic mapping analysis has demonstrated substantial conservation of syntenies between zebrafish and mammalian genomes (Postlethwait *et al.,* 1998). (A pair of genes is syntenic if they are found on the same chromosome. In a conserved synteny, the orthologues of a pair of genes that is syntenic in one species are syntenic in another species.) Analysis of the zebrafish gene map further suggests that during the course of vertebrate evolution, the zebrafish and human lineages have shared two rounds of whole genome duplication, and a third whole genome duplication event probably occurred before the teleost radiation. As a consequence, for many genes in humans, there may be two copies in zebrafish, and for many human chromosome segments, zebrafish is likely to have two such segments.

II. Mapping the Zebrafish Genome

The haploid zebrafish genome has 25 chromosomes, most of which are difficult to distinguish (Daga *et al.,* 1996; Gornung *et al.,* 1997). These chromosomes contain about 1.7×10^9 base pairs of DNA (Hinegardner and Rosen, 1972), about half the mammalian genome size. The first genetic map of the zebrafish included about 400 genetic markers, mostly random amplified polymorphic DNAs (RAPDs), along with a few genes and mutations (Postlethwait *et al.,* 1994; Johnson *et al.,* 1996). The development of simple sequence length polymorphisms (SSLPs) (Knapik *et al.,* 1996, 1998) provided about 700 markers that were more consistent from laboratory to laboratory and from strain to strain. Anonymous DNA polymorphisms, including RAPDs (see Postlethwait *et al.,* Chapter 9, this volume), SSLPs RFLPs (restriction fragment length polymorphisms), and AFLPs (amplified fragment length polymorphisms, see Ransom and Zon, Chapter 12, this volume) provide a framework map for the positioning of biologically interesting genetic markers, including genes and mutations. Short interspersed repetitive elements can also provide markers for locus mapping (Shimoda *et al.,* 1996). Single strand conformation polymorphism (SSCP) is especially useful for the mapping of cloned genes (Beier *et al.,* 1992; Beier, 1998; Postlethwait *et al.,*

1998). The use of somatic cell hybrid panels and radiation hybrid panels has the advantage that genes can be mapped without the necessity of first identifying a polymorphism segregating in a cross (Ekker *et al.,* 1996).

Maps based on the rate of meiotic recombination in female zebrafish are about 2900 cM long (1 cM represents 1% recombinants) (Postlethwait *et al.,* 1994; Johnson *et al.,* 1996). Such maps employ haploid mapping panels constructed from the haploid progeny of individual females heterozygous for highly polymorphic genomes. Such mapping panels are of great utility when using markers with dominant/recessive alleles such as RAPDs and AFLPs because recessive alleles, whose phenotype is the absence of a band on the gel, are not obscured by dominant alleles, whose phenotype is the presence of the band. The sex-averaged maps based on an F_2 intercross is about 2300 cM (Knapik *et al.,* 1998), substantially shorter than the map based on female meiosis. Thus, the length of the male map should be about 1700 cM [$= 2(2300) - 2900$]. This suggests that in zebrafish as in mammals (Dietrich *et al.,* 1996; Jacob *et al.,* 1995; Donis-Keller *et al.,* 1987; Collins *et al.,* 1996) recombination in females is on average about 1.7 times as likely as in males. These map lengths give about 590 kb/cM for the female map, 720 kb/cM for the sex-averaged map, and, by inference, 850 kb/cM for the male map. The practical implications are that if one is performing a chromosome walk, where fine-structure mapping is critical, it is best to use female meiosis because fewer individuals will have to be genotyped to identify informative recombinants. On the other hand, when trying to obtain a rough linkage map location for a mutation rather quickly, following segregation in male meiosis would be more economical because the mutation is less likely to become separated from informative markers in male meiosis.

Because of the small size and similarity of zebrafish chromosomes, linkage groups (LG) identified by meiotic mapping have not yet been assigned to cytogenetically identifiable chromosomes. Nevertheless, the genetic maps consist of 25 linkage groups, each of which corresponds to a single chromosome. We know this because of the ability to perform tetrad analysis in zebrafish. George Streisinger developed methodologies to create fish that contain half-tetrad genomes. This involves fertilizing eggs *in vitro* with UV-irradiated sperm, followed by blocking the second meiotic division (Streisinger *et al.,* 1986; see Eisen *et al.,* this volume). The resulting diploid animal is homozygous for all loci between the centromere and the first recombination event. For a locus at the centromere, half of the half-tetrad embryos from a heterozygous mother are homozygous for one allele and half are homozygous for the other allele, with no heterozygotes. For loci farther away from the centromere, more and more heterozygotes are produced, diminishing the number of each type of homozygotes (e.g., see Postlethwait and Talbot, 1997). The scoring of loci along a chromosome in a population of half-tetrad animals thus provides the location of the centromere, and all loci linked to the same centromere are on the same chromosome (Kauffman *et al.,* 1995; Johnson *et al.,* 1995; 1996; Knapik *et al.,* 1998).

III. The Zebrafish Gene Map

Although anonymous DNA polymorphisms, such as SSLPs, AFLPs, and RAPDs, are highly useful for the construction of a framework map and for positional cloning projects (see Talbot and Schier, Ch. 15, this volume), those markers are usually species-specific and of limited use for comparative genome analysis. The location of cloned genes on genetic maps permits the comparison of genome arrangements between species. Figure 1 shows the location of about 150 cloned genes on a framework map of RAPDs and SSLPs. A number of conclusions can be drawn from the comparison of the zebrafish and human maps. [The location of mammalian genes was obtained from Online Mendelian Inheritance in Man (OMIM) (http://www3.ncbi.nlm.nih.gov/Omim/searchomim.html) and Mouse Genome Database (MGD) (http://www.informatics.jax.org/locus.html).]

1. Comparative mapping shows extensive conservation of syntenies between zebrafish and mammalian genomes. On nearly every chromosome, the apparent orthologues of genes that are syntenic in mammals are also syntenic in zebrafish. For example, as shown in Fig. 2D, 11 orthologous or highly homologous genes are syntenic on LG9 and the long arm of human chromosome two (Hsa2q). In contrast, this set of genes is on two chromosomes in mouse (Mmu1 and Mmu2). In general, the zebrafish genome seems to correspond more closely to the human genome than the mouse genome. This reflects the rapid chromosome evolution among rodents and the rather slow divergence of karyotypes in the primate lineage. Conservation of syntenies had already been detected between fish and mammals by Morizot (Morizot, 1990, 1994; Morizot and Siciliano, 1983; Morizot *et al.,* 1977, 1991) studying the cyprinodont *Xiphophorus* (swordtail). These results show that when a large number of genes has been mapped in zebrafish, it will be possible to infer from a gene's position in the zebrafish genome its likely location in the human genome. This conclusion is of enormous practical significance to zebrafish researchers.

2. The zebrafish genome has orthologues of paralogous chromosome segments in mammals. (Paralogues are genetic elements such as genes or chromosome segments within a species derived from a single ancestral genetic element.) Mammalian genomes have up to four copies of chromosome segments present in a vertebrate ancestor (Nadeau and Kosowsky, 1991; Lundin, 1993; Morizot, 1990; Ruddle *et al.,* 1994). Examples include much of Hsa4 and 5, segments of Hsa1, 6, 9, and 19 which contain the four mammalian Notch alleles (Katsanis *et al.,* 1996; Kasahara *et al.,* 1996), parts of Hsa11, 15, and 19, and portions of Hsa2, 7, 12, and 17, which contain the four mammalian *HOX* clusters and associated genes. Prince *et al.,* (1998) found three *hox* genes in zebrafish that did not map to the four previously identified *hox* clusters, and Amores *et al.* (1998) show that zebrafish has seven *hox* clusters (Amores *et al.,* 1998), two orthologues of human *HOXA* (*hoxaa* and *hoxab*), two of *HOXB* (*hoxba* and *hoxbb*), two of *HOXC*

(*hoxca* and *hoxcb*), and one of *HOXD* (*hoxda*). To illustrate the point that zebrafish have orthologues of individual mammalian *HOX*-bearing chromosomes, consider LG3, which contains seven genes whose apparent orthologues are on Mmu11 and/or Hsa17, not on the other mammalian *HOX*-bearing chromosomes (Fig. 2B). Three of the four mammalian *HOX*-bearing chromosomes have paralogues of *DLX* genes—*DLX1* on Hsa2 (*HOXD*), *DLX6* on Hsa7 (*HOXA*), and *DLX7* on Hsa17 (*HOXB*). LG3 has the *hoxbb* cluster, and *dlx8,* which phylogenetic analysis shows to be orthologus to *DLX7* (Ellies *et al.,* 1997). It is significant that this zebrafish *dlx* is related to the *DLX* that is found on the human *HOXB*-containing chromosome, not the paralogue found on the other *HOX*-bearing chromosomes. Likewise, mammals have two paralogues of the small neuropeptide genes *NPY* and *PPY* which are syntenic with *HOXA* and *HOXB* respectively. Zebrafish has clear orthologues of these genes (D. Larhammar, personal communication), with the *npy* gene being syntenic with *hoxab,* and the *ppy* gene being syntenic with *hoxbb,* in accord with syntenies in mammals. In addition, mammals have two retinoic acid receptor genes syntenic with *HOX* genes—*RARA* and *RARG,* which are syntenic with *HOXB* and *HOXC,* respectively. In zebrafish, *hoxbb* is syntenic on LG3 with *rara2b,* not *rarg,* which is syntenic with *hoxca* in zebrafish corresponding to its mammalian location. The principles derived from this analysis of LG3 are echoed by a similar analysis of other zebrafish chromosome segments. The conclusion is that the zebrafish genome has clear orthologues of paralogous chromosome segments found in mammals.

What was the origin of the paralogous chromosome segments found in mammals? Did they arise from tandem duplications of chromosome segments? Did they arise from duplications of individual chromosomes at many different times to create aneuploids of various types? Or did they arise from whole genome duplication events (polyploidization)? Because aneuploidy is so disastrous in vertebrate embryos—for example, Down Syndrome results from an extra copy of the smallest human chromosome—it is likely that duplication events tended to preserve euploidy. Because many portions of the mammalian genome appear to be present in duplicate paralogous copies, huge portions of the genome must have been duplicated. The most parsimonious explanation is that two whole genome duplication events occurred at some time in mammalian phylogeny producing four copies of each chromosome that was present in a vertebrate ancestor (Lundin, 1993). The four *HOX*-cluster-containing chromosomes and four *NOTCH* containing chromosome segments are the most obvious genetic fossils of those events yet excavated. An alternative explanation supported by phylogenetic analysis of collagen genes which are tightly linked to each of the four *HOX* clusters, is that these chromosomes arose sequentially, either by producing aneuploidy, or in three rounds of tetraploidization to give eight clusters followed by loss of four clusters (Bailey *et al.,* 1997). Because the zebrafish has clear orthologues of chromosome segments that exist as paralogous copies in mammalian genomes, the duplication events that produced those paralogous

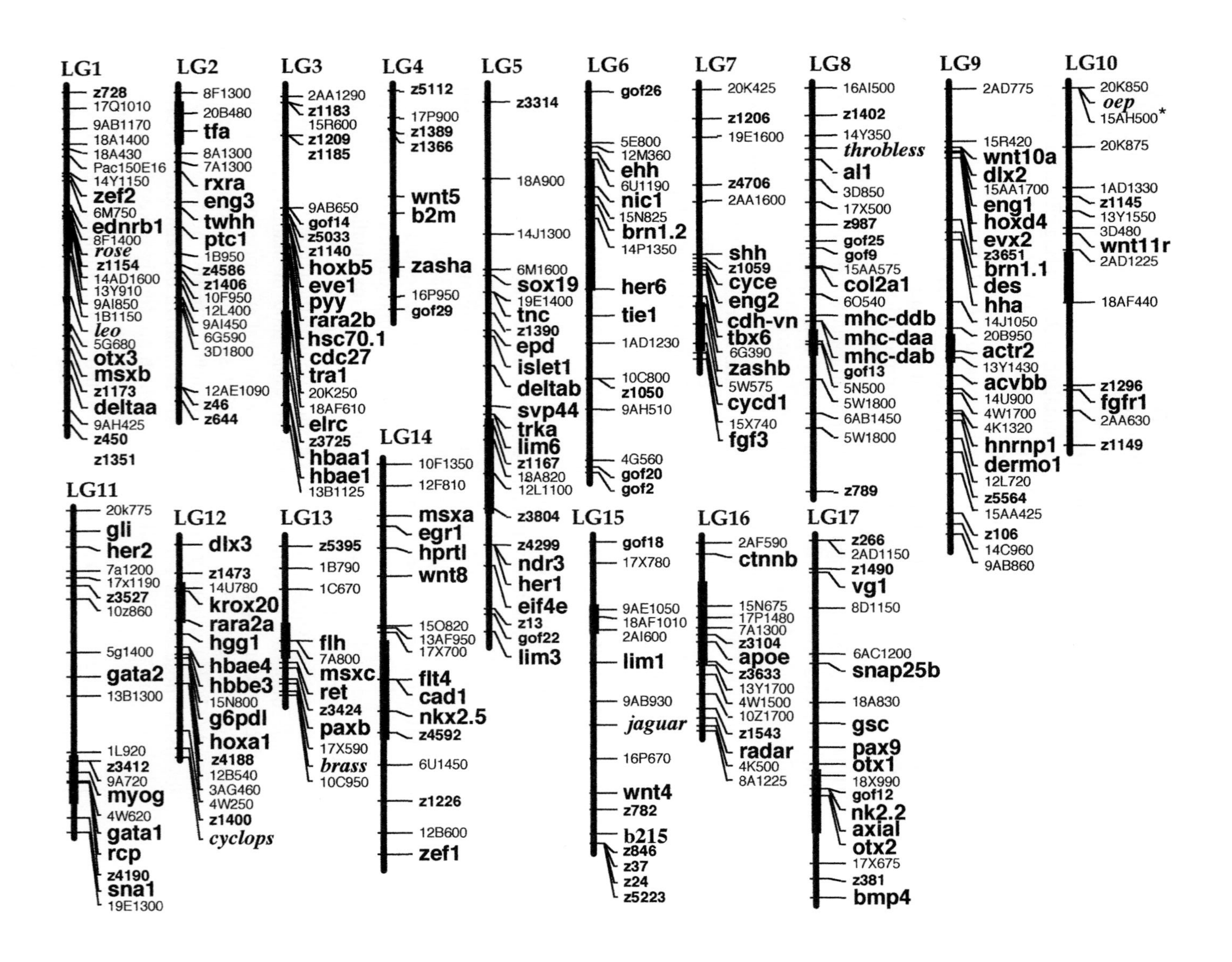

LG1
z728
17Q1010
9AB1170
18A1400
18A430
Pac150E16
14Y1150
zef2
6M750
ednrb1
8F1400
rose
z1154
14AD1600
13Y910
9AI850
1B1150
leo
5G680
otx3
msxb
z1173
deltaa
9AH425
z450
z1351
LG2
8F1300
20B480
tfa
8A1300
7A1300
rxra
eng3
twhh
ptc1
1B950
z4586
z1406
10F950
12L400
9AI450
6G590
3D1800
12AE1090
z46
z644
LG3
2AA1290
z1183
15R600
z1209
z1185
9AB650
gof14
z5033
z1140
hoxb5
eve1
pyy
rara2b
hsc70.1
cdc27
tra1
20K250
18AF610
elrc
z3725
hbaa1
hbae1
13B1125
LG4
z5112
17P900
z1389
z1366
wnt5
b2m
zasha
16P950
gof29
LG5
z3314
18A900
14J1300
6M1600
sox19
19E1400
tnc
z1390
epd
islet1
deltab
svp44
trka
lim6
z1167
18A820
12L1100
z3804
z4299
ndr3
her1
eif4e
z13
gof22
lim3
LG6
gof26
5E800
12M360
ehh
6U1190
nic1
15N825
brn1.2
14P1350
her6
tie1
1AD1230
10C800
z1050
9AH510
4G560
gof20
gof2
LG7
20K425
z1206
19E1600
z4706
2AA1600
shh
z1059
cyce
eng2
cdh-vn
tbx6
6G390
zashb
5W575
cycd1
15X740
fgf3
LG8
16AI500
z1402
14Y350
throbless
al1
3D850
17X500
z987
gof25
gof9
15AA575
col2a1
6O540
mhc-ddb
mhc-daa
mhc-dab
gof13
5N500
5W1800
6AB1450
5W1800
z789
LG9
2AD775
15R420
wnt10a
dlx2
15AA1700
eng1
hoxd4
evx2
z3651
brn1.1
des
hha
14J1050
20B950
actr2
13Y1430
acvbb
14U900
4W1700
4K1320
hnrnp1
dermo1
12L720
z5564
15AA425
z106
14C960
9AB860
LG10
20K850
oep
15AH500*
20K875
1AD1330
z1145
13Y1550
3D480
wnt11r
2AD1225
18AF440
z1296
fgfr1
2AA630
z1149
LG11
20k775
gli
her2
7a1200
17x1190
z3527
10z860
5g1400
gata2
13B1300
1L920
z3412
9A720
myog
4W620
gata1
rcp
z4190
sna1
19E1300
LG12
dlx3
z1473
14U780
krox20
rara2a
hgg1
hbae4
hbbe3
15N800
g6pdl
hoxa1
z4188
12B540
3AG460
4W250
z1400
cyclops
LG13
z5395
1B790
1C670
flh
7A800
msxc
ret
z3424
paxb
17X590
brass
10C950
LG14
10F1350
12F810
msxa
egr1
hprtl
wnt8
15O820
13AF950
17X700
flt4
cad1
nkx2.5
z4592
6U1450
z1226
12B600
zef1
LG15
gof18
17X780
9AE1050
18AF1010
2AI600
lim1
9AB930
jaguar
16P670
wnt4
z782
b215
z846
z37
z24
z5223
LG16
2AF590
ctnnb
15N675
17P1480
7A1300
z3104
apoe
z3633
13Y1700
4W1500
10Z1700
z1543
radar
4K500
8A1225
LG17
z266
2AD1150
z1490
vg1
8D1150
6AC1200
snap25b
18A830
gsc
pax9
otx1
18X990
gof12
nk2.2
axial
otx2
17X675
z381
bmp4

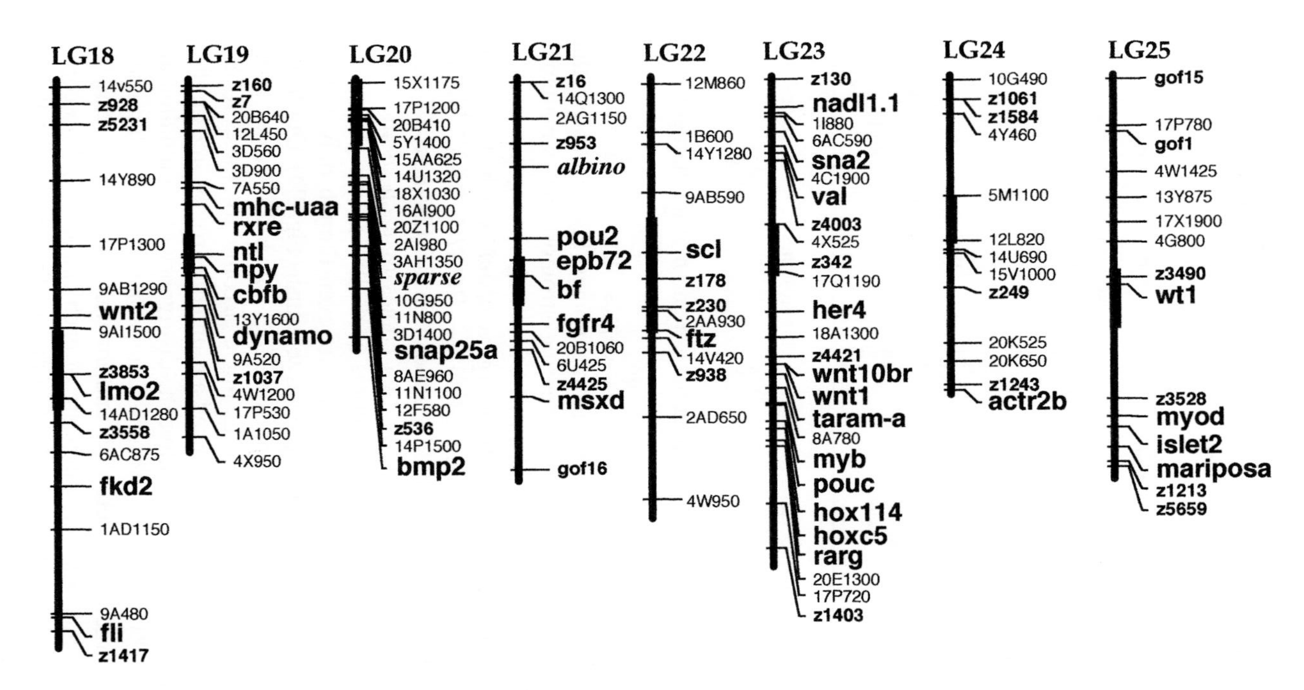

Fig. 1 A genetic map for the zebrafish. For references to genes, see ZFIN, the zebrafish database at http://zfish uoregon.edu/.

A HOXA

LG16	**Hsa7**	**LG19**
hoxaa	HOXA	hoxab
	EVX1	evx1
	DLX6	dlx6
	DLX5	dlx4
LG7	NPY	npy
eng2	EN2	eng3
shh	SHH	twhh
	INHBA	**LG2**
	TWIST	
	WNT2	
	GLI3	
cycd1	CCND1	
fgf3	GFGF3	

B HOXB

LG12	**Hsa17**	**Mmu11**	**LG3**
hoxba	HOXB	HOXB	hoxbb
			eve1
dlx7	DLX7		dlx8
dlx3	DLX3		
	PYY	Pyy	pyy
	WNT3	Wnt3	
rara2a	RARA	Rara	rara2b
hbae4		Hba	hbae1
	THRA1	Thra	tra1
	GFAP	Gfap	
	CDC27	Cdc27	cdc27
	CRNB1	Acrb	

C HOXC

LG23	**Hsa12**	**Mmu15**		**LG11**
hoxca	HOXC	Hoxc		hoxcb
			Mmu10	
dhh		Dhh		
	INHBC		Inhbc	
wnt1	WNT1	Wnt1		
	GLI		Gli	gli
rarg	RARG	Rarg		
	VDR			
	PRPH	Prph		
tara	ACVRLK1			
pouc		Emb		
myb			Myb	

D HOXD

LG9	**Hsa2**	**Mmu2**		
hoxda	HOXD	Hoxd		
evx2	EVX2	Evx2		
dlx1	DLX1	Dlx1		**LG1**
dlx2	DLX2	Dlx2	**Mmu1**	dlx5
eng1	EN1		En1	eng4
hha	IHH		Ihh	ehh
actbb	INHBB		Inhbb	**LG6**
dermo	DERMO			
	GLI2		Gli2	
		Rxra	Rxrg	
des	DES		Des	
	CRNA1	Acra		nic1
acvr2a		Acvr2a		
brn1.1			Brn1	brn1.2
		Gdf5		contact

Fig. 2 Conserved syntenies between duplicated *hox*-bearing chromosomes in zebrafish and their mammalian orthologues.

copies occurred before the divergence of the zebrafish and mammalian genomes 420 million years ago.

3. The zebrafish genome appears to have two copies of many chromosome segments that exist as single copies in mammalian genomes. Figure 2 shows that zebrafish have two copies of each of the *HOX*-bearing chromosomes of mammals (Amores *et al.,* 1998). For example, recall that LG3 contains the genes *hoxb5b, dlx8, rara2b,* and the embryonic alpha hemoglobin gene *hbae1,* which are orthologous to the mammalian genes on Hsa17 and Mmu11—*HOXB5, DLX7, RARA,* and *Hba.* Duplicates of these zebrafish loci appear on LG12—*hoxb5a, dlx7, rara2a,* and *hbae4.* Likewise, zebrafish have two copies of each of the other mammalian *HOX*-bearing chromosomes (Amores *et al.,* 1998). This generalization also goes for the *HOXD*-bearing chromosome, even though only a single *hoxd* paralogous cluster has been yet discovered in zebrafish (Fig. 2D). Duplicates of other human chromosome paralogy groups besides the *HOX*-cluster group are also apparent. At least three hypotheses might account for these results: (1) The zebrafish lineage may have experienced a round of chromosome duplications, probably a polyploidization event, after the divergence of mammalian and ray-finned fish lineages. (2) These chromosomes might have become duplicated shortly prior to the divergence of zebrafish and mammalian lineages, and the mammalian lineage has since lost one duplicate of each cluster. (3) Two copies of each *HOX* cluster might be present in human and mouse genomes but have not yet been discovered. These last two explanations seem less likely than the supposition that at least three of the *hox* clusters were duplicated independently in zebrafish evolution.

When did the fish-specific round of duplication occur? Is it a relatively recent event present only in *Danio rerio,* like the documented tetraploidy of salmonids (Allendorf and Thorgaard, 1984)? Or is it an ancient event shared by all fish of the order Cypriniformes (minnows) or all fish of the division Teleostei or all members of the class Actinopterygia (ray-finned fish)? Much more extensive genomic analysis of other fish must be conducted before an answer is known for sure. Some evidence, however, is available from current gene sequence data. Four *Hox* clusters have been isolated from the pufferfish, *Fugu rubripes,* and a phylogenetic analysis shows that each cluster is related to a specific one of the seven clusters in zebrafish (Amores *et al.,* 1998). For example, pufferfish *Hoxc* genes assort in a tree with zebrafish *hoxca,* not with zebrafish *hoxcb,* the sibling cluster of *hoxca.* If the duplication event had occurred after the divergence of zebrafish and pufferfish lineages, genes in the pufferfish *Hoxc* cluster would be equally related to the orthologue in zebrafish *hoxca* and *hoxcb.* For the small amount of data available, the same appears to be true for the *hox* genes of other teleost fish (Amores *et al.,* 1998). Thus, these results would be predicted from a tetraploidization event after the divergence of ray-finned and lobe-finned fishes (420 million years ago), but before the teleost radiation (about 100 million years ago). More data are necessary to confirm this supposition. In fish like goldfish and salmonids and in *Xenopus laevis,* there has been a still more recent independent tetraploidization.

IV. Two Genes in Zebrafish for One in Mammals

The significance of the conclusion that there was an extra tetraploidization very early in the zebrafish lineage is that zebrafish will often have two genes that are *both* orthologues of a single mammalian gene. Examples include *shh* and *twhh,* which are both about equally related to mammalian *SHH* (Ekker S *et al.,* 1995; Zardoya *et al.,* 1996), and *snap25a* and *snap25b,* which are both closely similar to *SNAP* in human (Risinger *et al.,* 1997) and which map on separate chromosomes (Postlethwait *et al.,* 1998); *trkb1* and *trkb2* are both orthologues of *TrkB* in rat, and *trkc1* and *trkc2* are both orthologues of *TrkC* in rats (Martin *et al.,* 1995); zebrafish has five *msx* genes, whereas mouse just has three. Even though assignment of orthologies to these genes by phylogenetic analysis is difficult (Ekker *et al.,* 1997), they map in paralogous chromosome segments (Postlethwait *et al.,* 1998); the phylogeny of *rxra* and *rxrd* suggests that these genes arose after a common precursor gene branched from the mammalian *RXR* family (Jones *et al.,* 1995); the human X-chromosome gene *L1* has two zebrafish orthologues, *nadl1.1* and *nadl1.2* (Tongiorgi *et al.,* 1995).

The suggestion that zebrafish may have two copies of many unique genes in mammals has implications for comparative molecular embryology. Suggestions about mammalian biology inferred from zebrafish developmental genetics will be more robust after the isolation and analysis of *both* zebrafish orthologues of the mammalian gene, when two exist. This makes the job of zebrafish gene isolation and analysis more difficult. Of course, in some instances, one of the duplicates may have been lost during zebrafish phylogeny. And in other cases, tandem duplications may have occurred independently in either the fish or mammalian lineages. Paralogues derived from tandem duplication can be distinguished from those originating in tetraploidization events by genetic or physical mapping because tandem duplicates will generally be located side by side in the genome (unless chromosome rearrangements have separated the two duplicates), and duplicates from tetraploidization will map in duplicated chromosomes along with a large suite of other duplicated genes.

The possibility of an extra, independent genome duplication in the zebrafish lineage that was not experienced by mammals does not diminish the utility of zebrafish as a superb model for vertebrate developmental genetics. The conservation of syntenies between mammals and zebrafish is remarkable and can be used to predict the map location of human genes whose orthologues are mutated in zebrafish. Because, however, two genes in zebrafish may often be sharing the role of a single gene in mammals, the functions of an individual gene in zebrafish may not be identical to its mammalian orthologue. Force *et al.* (1998) have suggested that duplicate genes, for example two gene copies resulting from a tetraploidization event, may become preserved during evolution by the degenerative loss of complementary gene subfunctions from the duplicate copies. This is called the duplication, degeneration, complementation model, or DDC. For example, if the precursor gene has essential functions in the hindbrain and in the fin bud, then after duplication, one copy might become mutated in the hindbrain enhancer and the duplicate copy in the fin bud enhancer. In such a case, both genes must be fully expressed for proper development, and both will be preserved in evolution. As theoretical analysis shows (Force *et al.,* 1998), this phenomenon can sometimes result in true orthologues having different functions, including different patterns of expression, in two different taxa. Thus, inferring the expression pattern of a mammalian gene from the expression of just one of its orthologues in zebrafish must be tempered by the realization that there will be more genetic redundancy in the zebrafish.

Because of the hypothesized tetraploidization event in ray-finned fish phylogeny, zebrafish mutations may often exhibit only a portion of the phenotype which might have been obtained if one had mutated the orthologous unduplicated gene in the latest common ancestor of zebrafish and mammals. Thus, if one observes a significant difference between the phenotype of a zebrafish null allele and a mouse knockout in orthologous genes, it could be due in part in some cases to the complementary sharing of gene functions by the two duplicate zebrafish copies of a single gene in mouse.

In some cases, the presence in zebrafish of two orthologues of a mammalian gene may provide an advantage for developmental genetic analysis. The duplication, degeneration, complementation model (Force *et al.,* 1998) suggests that the functions of a gene, including its temporal and spatial expression patterns, that were present in the common ancestor of zebrafish and mammals will frequently be shared complementarily between the two zebrafish gene duplicates. Thus, the zebrafish provides a special opportunity to investigate gene functions that might otherwise be obscured by other functions of those same genes. For example, a null mutation in a gene that acts both early and late in development might die or become developmentally abnormal due to the early essential function before the late role commences. In zebrafish, in some cases, the late and early functions may have sorted out independently between the two duplicate copies of the gene, thereby facilitating analysis.

The zebrafish, because its embryos are favorable for embryological investigation, also provides a special opportunity to investigate the evolution of gene duplicates and the role gene duplication plays in the origin of evolutionary innovations. It has long been thought that the redundancy occasioned by gene duplication provides the luxury of evolutionary experimentation because one of the redundant copies can spawn novel functions while the other retains essential ancestral functions (Ohno, 1970). In this way, it has been thought that old genes can assume new functions. It has been supposed that duplicate genes resulting from the tetraploidizations that occurred early in vertebrate phylogeny may have been the fodder for the evolution of vertebrate novelties, such as the neural crest, large brains, ectodermal placodes, and later, jaws, and paired appendages (Ohno, 1970; Holland *et al.,* 1994; Holland and Garcia-Fernàndez, 1996). The more recent genome duplication in the ray-finned fish lineage may provide gene copies that are more amenable to the comparative promotor analysis that can illuminate the evolutionary mechanisms that result in new gene functions related to gene duplication.

V. Gene Nomenclature in Zebrafish

If the hypothesis is supported that there was a tetraploidization event after the divergence of ray- and lobe-finned fishes but before the teleost radiation, then special care must be taken in the naming of zebrafish cloned genes to reflect gene affinities (Mullins, 1995). It is especially important to convey relationships of zebrafish genes to their human and mouse orthologues in a way that does not hinder the reading of zebrafish literature by mammalian biologists. Full sequences should be obtained for newly cloned genes rather than the small PCR fragments often published for some gene families, and rigorous phylogenetic analysis should be conducted so that names reflect the genes' evolutionary histories, if well documented, and don't mislead when phylogenies are not yet well understood. Where possible, zebrafish genes with clear orthologues to mammalian genes should be designated with the mammalian abbreviations found in OMIM and MGD—except, of course, that zebrafish genes are written all lowercase (e.g., *shh*) compared to human, with all uppercase (*SHH*), and mouse with just the first letter uppercase (*Shh*); in all three cases, of course, the genes are written in italics. If orthologies are ambiguous, names should not be given that imply mammalian orthology. For example, orthologies of the five zebrafish *msx* genes could not be assigned, and so they were called *msxa, msxb, msxc, msxd,* and *msxe* to indicate the uncertainty of their relationships to mammalian *Msx1, Msx2,* and *Msx3* (Ekker *et al.,* 1997). Consideration should be given to naming sibling genes in zebrafish with *a* and *b* or *1* and *2* suffixes where appropriate. For example, *shh* and *twhh* might have been called *shha* and *shhb* from their phylogenies (Ekker S *et al.,* 1995; Zardoya *et al.,* 1996). Mouse and human nomenclature forbids the use of punctuation, including hyphens and periods, in gene names, and the use of Greek symbols, because these characters interfere with automated searching algorithms; this convention would help the zebrafish community as well.

Care must be taken in the naming of zebrafish genes to not give overriding significance to expression patterns. Theoretical analysis suggests that the reciprocal, complementary loss of regulatory elements driving expression in certain body parts can contribute to the retention of duplicated genes during evolution (Force *et al.,* 1998). Thus, the sum of the expression patterns of gene duplicates may reflect the expression pattern of the ancestral gene just before the duplication event. Because conserved regulatory elements might become partitioned by chance differently and unequally between the two gene duplicates, expression patterns can sometimes be an inappropriate indictor of gene history. For example, the sequence available for the gene initially called *hoxa1* left open the possibility that it could be related to *HOXA1* or *HOXB1,* but because the expression pattern of the gene resembled more closely *HOXA1,* the gene was called *hoxa1* (Alexandre *et al.,* 1996). Subsequent complete sequence analysis of other zebrafish *hox* genes provided more phylogenetic information, and mapping data revealed that this gene is the first gene in one of the two zebrafish *hox* clusters phylogenetically related to *HOXB,* and the gene has subsequently been renamed *hoxb1b.*

VI. Conclusions

The zebrafish is a wonderful model for early embryonic development of vertebrates, and its genome shares substantial similarities to our own. The conservation of genome organization between humans and zebrafish will facilitate the transferal of information from one species to the other. The inference that zebrafish lineage appeared to have suffered an extra polyploidization step not experienced by mammals simply means that both zebrafish copies of a mammalian gene (when two copies have survived) should be studied for better connectivity of molecular genetic mechanisms of development in zebrafish to the tetrapod, especially human, condition.

Acknowledgments

This paper is dedicated to Don Morizot for his early keen insights into the evolutionary genetics of fishes. We thank G. Corley-Smith and E. Dodou for critical comments on the manuscript. Support was provided by R01RR10715 and P01HD22486.

References

Ahlberg, P., and Milner, A. (1994). The origin and early diversification of tetrapods. *Nature* **368,** 507–514.

Alexandre, D., Clarke, J. D. W., Oxtoby, E., Yan, Y.-L., Jowett, T., and Holder, N. (1996). Ectopic expression of Hoxa-1 in the zebrafish alters the fate of the mandibular arch neural crest and phenocopies a retinoic acid-induced phenotype. *Development* **122,** 735–746.

Allendorf, F., and Thorgaard, G. (1984). Tetraploidy and the evolution of salmonid fishes. *In* "Evolutionary Genetics of Fishes" (B. J. Turner, Ed.), pp. 1–46. Plenum Press, New York.

Amores, A., Force, A., Yan, Y.-L., Wang, Y.-L., Fritz, A., Joly, L., Prince, V., Ho, R., Amemiya, C., Ekker, M., and Postlethwait, J. H. (1998). *Hox* cluster duplication and the evolution of vertebrate morphological complexity. Submitted.

Bailey, W., Kim, J., Wagner, G., and Ruddle, F. (1997). Phylogenetic reconstruction of vertebrate Hox cluster duplications. *Mol. Biol. Evol.* **14,** 843–853.

Beier, D. R., Dushkin, H., and Sussman, D. J. (1992). Mapping genes in the mouse using single-strand conformation polymorphism analysis of recombinant inbred strains and interspecific crosses. *Proc. Natl. Acad. Sci. U.S.A.* **89,** 9102–9106.

Beier, D. R. (1998). Zebrafish: Genomics on the fast track. *Genome Res.* **8,** 9–17.

Collins, A., Frezal, J., Teague, J., and Morton, N. E. (1996). A metric map of humans: 23,500 loci in 850 bands. *Proc. Natl. Acad. Sci. U.S.A.* **93,** 14771–14775.

Daga, R. R., Thode, G., and Amores, A. (1996). Chromosome complement, C-banding, Ag-NOR and replication banding in the zebrafish *Danio rerio. Chromosome Res.* **4,** 29–32.

Dietrich, W. F., Miller, J., Steen, R., Merchant, M. A., Damron-Boles, D., Husain, Z., Dredge, R., Daly, M. J., Ingalls, K. A., O'Connor, T. J., *et al.* (1996). A comprehensive genetic map of the mouse genome. *Nature* **380,** 149–152.

Donis-Keller, H., Green, P., Helms, C., Cartinhour, S., Weiffenbach, B., Stephens, K., Keith, T. P., Bowden, D. W., Smith, D. R., Lander, E. S., Botstein, D., Akots, G., Rediker, K. S., Gravius, T., Brown, V. A., Rising, M. B., Parker, C., *et al.* (1987). A genetic linkage map of the human genome. *Cell* **51,** 319–337.

Ekker, M., Akimenko, M.-A., Allende, M. L., Smith, R., Drouin, G., Langille, R. M., Weinberg, E. S., and Westerfield, M. (1997). Relationships among *msx* gene structure and function in zebrafish and other vertebrates. *Mol. Biol. Evol.* **14,** 1008–1022.

Ekker, M., Speevak, M. D., Martin, C. C., Joly, L., Giroux, G., and Chevrette, M. (1996). Stable transfer of zebrafish chromosome segments into mouse cells. *Genomics* **33,** 57–64.

Ekker, S. C., Ungar, A. R., Greenstein, P., vonKessler, D. P., Porter, J. A., Moon, R. T., and Beachy, P. A. (1995). Patterning activities of vertebrate hedgehog proteins in the developing eye and brain. *Curr. Biol.* **5,** 944–955.

Ellies, D. L., Stock, D. W., Hatch, G., Giroux, G., Weiss, K. M., and Ekker, M. (1997). Relationship between the genomic organization and the overlapping embryonic expression patterns of the zebrafish *dlx* genes. *Genomics* **45,** 580–590.

Force, A., Lynch, M., Pickett, F. B., and Postlethwait, J. H. (1998). Evolution of gene functions and expression patterns in development: The duplication–degeneration–complementation model. Submitted.

Gornung, E., Gabrielli, I., Cataudella, S., and Sola, L. (1997). CMA3-banding pattern and fluorescence in situ hybridization with 18S rRNA genes in zebrafish chromosomes. *Chromosome Res.* **5,** 40–46.

Hinegardner, R., and Rosen, D. E. (1972). Cellular DNA content and the evolution of teleostean fishes. *Am. Nat.* **166,** 621–644.

Holland, P. W., Garcia-Fernandez, J., Williams, N., and Sidow, A. (1994). Gene duplications and the origins of vertebrate development. *Development* (Suppl.), 125–135.

Holland, P. W., and Garcia-Fernàndez, J. (1996). Hox genes and chordate evolution. *Devel. Biol.* **173,** 382–390.

Jacob, H. J., Brown, D. M., Bunker, R. K., Daly, M. J., Dzau, V. J., Goodman, A., Koike, G., Kren, V., Kurtz, T., Lernmark, A., *et al.* (1995). A genetic linkage map of the laboratory rat, *Rattus norvegicus. Nat. Genet.* **9,** 63–69.

Johnson, S. L., Gates, M. A., Johnson, M., Talbot, W. S., Horne, S., Baik, K., Rude, S., Wong, J. R., and Postlethwait, J. H. (1996). Centromere-linkage analysis and consolidation of the zebrafish genetic map. *Genetics* **142,** 1277–1288.

Johnson, S. L., Africa, D., Horne, S., and Postlethwait, J. H. (1995). Half-tetrad analysis in zebrafish: Mapping the ros mutation and the centromere of linkage group I. *Genetics* **139,** 1727–1735.

Jones, B. B., Ohno, C. K., Allenby, G., Boffa, M. B., Levin, A. A., Grippo, J. F., and Petkovich, M. (1995). New retinoid X receptor subtypes in zebra fish (*Danio rerio*) differentially modulate transcription and do not bind 9-cis retinoic acid. *Mol. Cell Biol.* **15,** 5226–5234.

Kasahara, M., Hayashi, M., Tanaka, K., Inoko, H., Sugaya, K., Ikemura, T., and Ishibashi, T. (1996). Chromosomal localization of the proteasome Z subunit gene reveals an ancient chromosomal duplication involving the major histocompatibility complex. *Proc. Natl. Acad. Sci. USA* **93,** 9096–9101.

Katsanis, N., Fitzgibbon, J., and Fisher, E. M. C. (1996). Paralogy mapping: Identification of a region in the human MHC triplicated onto human chromosomes 1 and 9 allows the prediction and isolation of novel *PBX* and *NOTCH* loci. *Genomics* **35,** 101–108.

Kauffman, E. J., Gestl, E. E., Kim, D. J., Walker, C., Hite, J. M., Yan, G., Rogan, P. K., Johnson, S. L., and Cheng, K. C. (1995). Microsatellite-centromere mapping in the zebrafish (*Danio rerio*). *Genomics* **30,** 337–341.

Knapik, E. W., Goodman, A., Atkinson, O. S., Roberts, C. T., Shiozawa, M., Sim, C. U., Weksler-Zangen, S., Trolliet, M. R., Futrell, C., Innes, B. A., Koike, G., McLaughlin, M. G., Pierre, L., Simon, J. S., Vilallonga, E., Roy, M., Chiang, P. W., Fishman, M. C., Driever, W., and Jacob, H. J. (1996). A reference cross DNA panel for zebrafish (*Danio rerio*) anchored with simple sequence length polymorphisms. *Development* **123,** 451–460.

Knapik, E., Goodman, A., Ekker, M., Chevrette, M., Delgado, J., Neuhauss, S., Shimoda, N., Driever, W., Fishman, M., and Jacob, H. (1998). *Nature Genet.,* **18,** 338–344.

Lundin, L. G. (1993). Evolution of the vertebrate genome as relected in paralogous chromosomal regions in man and the house mouse. *Genomics* **16,** 1–19.

Lyons, L. A., Raymond, M. M., and O'Brien, S. J. (1994). Comparative genomics: The next generation. *Anim. Biotech.* **5,** 103–111.

Martin, S. C., Marazzi, G., Sandell, J. H., and Heinrich, G. (1995). Five Trk receptors in the zebrafish. *Dev. Biol.* **169,** 745–758.

Morizot, D. C., Slaugenhaupt, S. A., Kallman, K. D., and Chakravarti, A. (1991). Genetic linkage map of fishes of the genus *Xiphophorus* (Teleostei: Poeciliidae). *Genetics* **127,** 399–410.

Morizot, D. C., Wright, D. A., and Siciliano, M. J. (1977). Three linked enzyme loci in fishes: Implications in the evolution of vertebrate chromosomes. *Genetics* **86,** 645–656.

Morizot, D. C., and Siciliano, M. J. (1983). Comparative gene mapping in fishes. *In* "Isozymes: Current Topics in Biological and Medical Research," Vol. 10, pp. 261–285. Alan R. Liss, New York.

Morizot, D. C. (1990). Use of fish gene maps to predict ancestral vertebrate genome organization. *In* "Isozymes: Structure, Function, and Use in Biology and Medicine," pp. 207–234. Wiley-Liss.

Morizot, D. C. (1994). Reconstructing the gene map of the vertebrate ancestor. *Anim. Biotech.* **5,** 113–122.

Mullins, M. (1995). Genetic nomenclature guide. Zebrafish. *Trends Genet.* (Suppl.), 31–32.

Nadeau, J., and Kosowsky, M. (1991). Mouse map of paralogous genes. *Mamm. Genome* **1,** S433–S460.

O'Brien, S. J., Cevario, S. J., Martenson, J. S., Thompson, M. A., Nash, W. G., Chang, E., Graves, J. A., Spencer, J. A., Cho, K. W., Tsujimoto, H., and Lyons, L. A. (1997a). Comparative gene mapping in the domestic cat (*Felis catus*). *J. Hered.* **88,** 408–414.

O'Brien, S. J., Weinberg, J., and Lyons, L. A. (1997b). Comparative genomics: Lessons from cats. *Trends Genet.* **13,** 393–399.

Ohno, S. (1970). *Evolution by Gene Duplication.* Springer Verlag, Heidelberg, Germany.

Postlethwait, J., Johnson, S., Midson, C. N., Talbot, W. S., Gates, M., Ballenger, E. W., Africa, D., Andrews, R., Carl, T., Eisen, J. S., Horne, S., Kimmel, C. B., Hutchinson, M., Johnson, M., and Rodriguez, A. (1994). A genetic linkage map for the zebrafish. *Science* **264,** 699–703.

Postlethwait, J. H., Yan, Y.-L., Gates, M. A., Horne, S., Amores, A., Brownlie, A., Donovan, A., Egan, E. S., Ekker, M., Force, A., Gong, Z., Goutel, C., Fritz, A., Kelsh, R., Knapik, E., Liao, E., Orr, M., O'Shea, S. A., Paw, B., Ransom, D., Singer, A., Thomson, T., Beier, D., Joly, J.-S., Larhammar, D., Rosa, F., Westerfield, M., Zon, L. I., Johnson, S. L., and Talbot, W. S. (1998). Vertebrate genome evolution and the zebrafish gene map. *Nature Genet.* **18,** 345–349.

Postlethwait, J. H., and Talbot, W. S. (1997). Zebrafish genomics: From mutants to genes. *Trends Genet.* **13,** 183–190.

Prince, V., Joly, L., Ekker, M., and Ho, R. (1998). Zebrafish hox genes: Genomic organization and modified colinear expression patterns in the trunk. *Development* **125,** 407–420.

Rettenberger, G., Klett, C., Zechner, U., Kunz, J., Vogel, W., and Hameister, H. (1995a). Visualization of the conservation of synteny between humans and pigs by heterologous chromosomal painting. *Genomics* **26,** 372–378.

Rettenberger, G., Klett, C., Zechner, U., Bruch, J., Just, W., Vogel, W., and Hameister, H. (1995b). ZOO-FISH analysis: Cat and human karyotypes closely resemble the putative ancestral mammalian karyotype. *Chromosome Res.* **3,** 479–486.

Risinger, C., Deitcher, D. L., Lundell, I., Schwarz, T. L., and Larhammar, D. (1997). Complex gene organization of synaptic protein SNAP-25 in *Drosophila melanogaster. Gene* **194,** 169–177.

Ruddle, F. H., Bentley, K. L., Murtha, M. T., and Risch, N. (1994). Gene loss and gain in the evolution of vertebrates. *Development* (Suppl.) 155–161.

Schuler, G. D., Boguski, M. S., Stewart, E. A., Stein, L. D., Gyapay, G., Rice, K., White, R. E., Rodriguez-Tomé, P., Aggarwal, A., Bajorek, E., *et al.* (1996). A gene map of the human genome. *Science* **274,** 540–546.

Shimoda, N., Chevrette, M., Ekker, M., Kikuchi, Y., Hotta, Y., and Okamoto, H. (1996). Mermaid, a family of short interspersed repetitive elements, is useful for zebrafish genome mapping. *Biochem. Biophys. Res. Comm.* **220,** 233–237.

Solinas-Toldo, S., Lengauer, C., and Fries, R. (1995). Comparative genome map of human and cattle. *Genomics* **27,** 489–496.

Streisinger, G., Singer, F., Walker, C., Knauber, D., and Dower, N. (1986). Segregation analysis and gene-centromere distances in zebrafish. *Genetics* **112,** 311–319.

Tongiorgi, E., Bernhardt, R. R., and Schachner, M. (1995). Zebrafish neurons express two L1-related molecules during early axonogenesis. *J. Neurosci. Res.* **42,** 547–561.

Watkins-Chow, D. E., Buckwalter, M. S., Newhouse, M. M., Lossie, A. C., Brinkmeier, M. L., and Camper, S. (1997). Genetic mapping of 21 genes on mouse chromosome 11 reveals disruptions in linkage conservation with human chromosome 5. *Genomics* **40,** 114–122.

Zardoya, R., Abouheif, E., and Meyer, A. (1996). Evolutionary analyses of hedgehog and Hoxd-10 genes in fish species closely related to the zebrafish. *Proc. Natl. Acad. Sci. USA* **93,** 13036–13041.

CHAPTER 9

Using Random Amplified Polymorphic DNAs in Zebrafish Genomic Analysis

John H. Postlethwait and Yi-Lin Yan
Institute of Neuroscience
University of Oregon
Eugene, Oregon 97403

Michael A. Gates
Developmental Genetics Program
Skirball Institute of Biomolecular Medicine
New York University Medical Center
New York, New York 10016

I. Introduction

A highly effective way to analyze the genetic mechanisms of development and organ function is to mutagenize randomly, uncover mutations that alter the phenotype of interest, determine the nature of the genetic change, identify the protein disrupted by the mutation, and investigate the role of the newly discovered gene. This strategy has been enormously successful in revealing molecular mechanisms of development in *Drosophila* and *Caenorhabditis elegans.* Large-scale screening for embryonic lethal mutations had not been applied to vertebrates until recently, when several mutagenesis screens have produced thousands

METHODS IN CELL BIOLOGY, VOL. 60

0091-679X/99 $30.00

of mutations in hundreds of genes essential for embryonic development in the zebrafish (Haffter *et al.,* 1996; Driever *et al.,* 1996; Riley and Grunwald 1995; Kimmel 1989; Henion *et al.,* 1996). The molecular genetic analysis of these mutations has begun (Schulte-Merker *et al.,* 1994; 1997; Sepich *et al.,* 1997; Talbot *et al.,* 1995; Kishimoto *et al.,* 1997; Zhang *et al.,* 1998). Molecular isolation of the defective genes is facilitated by a genetic map and rapid ways to locate mutated genes on the map (Postlethwait and Talbot, 1997). Here we explore how random amplified polymorphic DNAs (RAPDs) can help achieve these aims in zebrafish genomic analysis. We first review what RAPDs are and how to identify them; then we discuss various types of RAPDs and their advantages and disadvantages with respect to other genetic markers and, finally, the use of RAPDs in constructing maps and mapping mutations.

II. What Are RAPDs?

RAPDs have been used extensively as a method to generate a genetic map for a species in a short period of time (e.g., Al-Janabi *et al.,* 1993; Kiss *et al.,* 1992; Reiter *et al.,* Tanksley *et al.,* 1992; Rafalski and Tingey, 1993; Grattapaglia *et al.,* 1996). Figure 1 diagrams how a RAPD primer—an oligonucleotide of arbitrary sequence about 10 base pairs long—binds to many sites on a DNA molecule, in both orientations (Williams *et al.,* 1990; Welsh and McClelland, 1990; Tingey *et al.,* 1992). If a primer binds in opposite orientations at sites close enough to become amplified in a polymerase chain reaction (PCR), an amplified DNA fragment will result, for example, the band of 100 base pairs (bp) shown in Fig. 1A. A number of fragments of various size are usually produced from a single primer. Primers with different sequences amplify different patterns of bands from the same DNA; for example, Fig. 2 shows 15 different primers amplifying the same genomic DNA. The number of fragments formed depends on the length of the primer and the complexity of the DNA. Longer primers give fewer bands than shorter primers, with ten nucleotides being about optimal for organisms with genomes as complex as that of zebrafish. In zebrafish, most primers 10 nucleotides long give about 6 to 12 fragments of 100 to 2000 bp (Johnson *et al.,* 1994; Postlethwait *et al.,* 1994). Because different 10-mer primers, even if they differ by a single base, amplify distinct sets of DNA fragments, RAPDs can detect changes in sequence of a single base pair.

Different genetic strains often amplify fragments of different sizes. In some instances, a difference in genomic DNA sequence will create a primer binding site in one strain, say SJD (Fig. 1B), that is absent from another, for example C32 (Fig. 1A). As a result, strain-specific bands become amplified due to allelic differences between the strains. One allele amplifies a band of a specific size, for example, the 250 bp fragment in Fig. 1B, whereas the other allele fails to amplify any band from that site. Because a heterozygote between the band-present allele (Fig. 1D, lane 2) and the band-absent allele (Fig. 1D, lane 1) shows

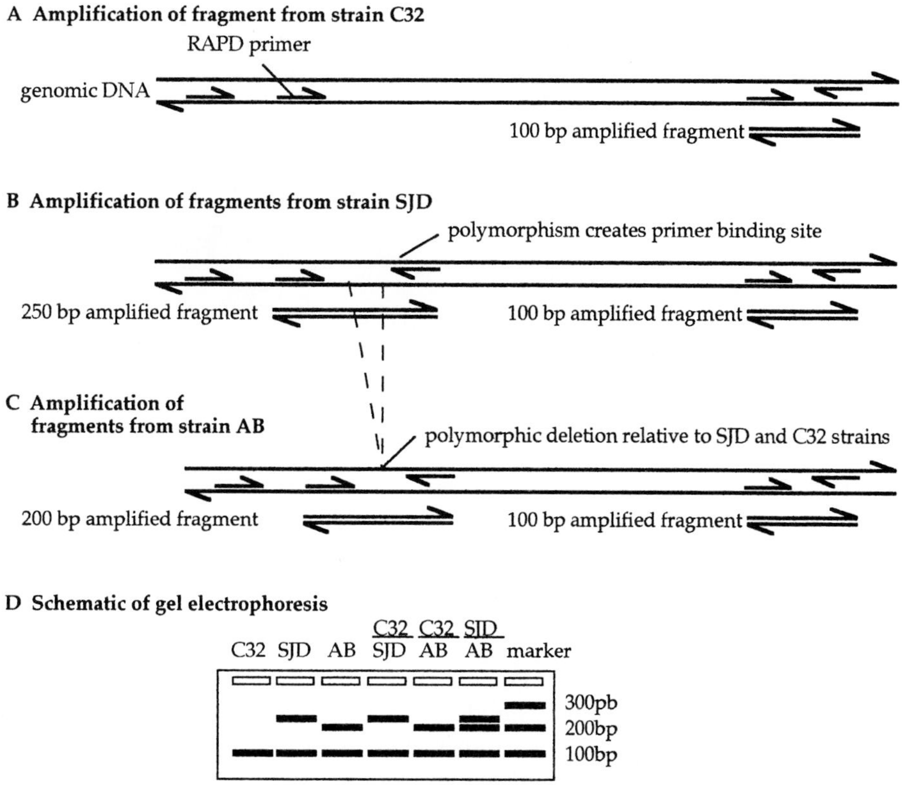

A Amplification of fragment from strain C32

RAPD primer

genomic DNA

100 bp amplified fragment

B Amplification of fragments from strain SJD

polymorphism creates primer binding site

250 bp amplified fragment

100 bp amplified fragment

C Amplification of fragments from strain AB

polymorphic deletion relative to SJD and C32 strains

200 bp amplified fragment

100 bp amplified fragment

D Schematic of gel electrophoresis

C32 SJD AB C32/SJD C32/AB SJD/AB marker

300pb

200bp

100bp

Fig. 1 How RAPDs work as genetic markers. (A) Because of their short length, RAPD primers bind at many sites along a chromosome, but amplify fragments only if they bind in opposite orientations and close enough to each other to copy the DNA between the fragments under the conditions used for the PCR. Here, a 100-bp fragment is diagrammed as being amplified from the C32 strain, which will show on the gel as a band indicated in part D of this figure, lane 1. (B) A polymorphism in the DNA at the same locus in a different strain, SJD, can create a new primer binding site that is fortuitously in the right orientation and distance for the amplification of a band, here shown as 250 bp in length on the gel in part D lane 2. A single-base-pair change is sufficient to create a new primer binding site. (C) Insertions or deletions between primer binding sites can also create polymorphic bands on the gel. Here a 50-bp deletion is indicated in strain AB, which results in a band of 200 bp on the gel in part D lane 3. (D) A schematic representation of a gel running the homozygotes of C32, SJD, and AB in the first three lanes, and all heterozygote combinations in the next three lanes. The 250-bp *SJD* allele is dominant to the absence of a fragment in the *C32* allele because the band forms from PCR using this primer on DNA of the C32/SJD heterozygote. Likewise, the 200-bp band from AB is dominant to the lack of a band in C32. The *SJD* and *AB* alleles of this locus, however, are codominant, because bands from both alleles are differentiated on the gel in a heterozygote.

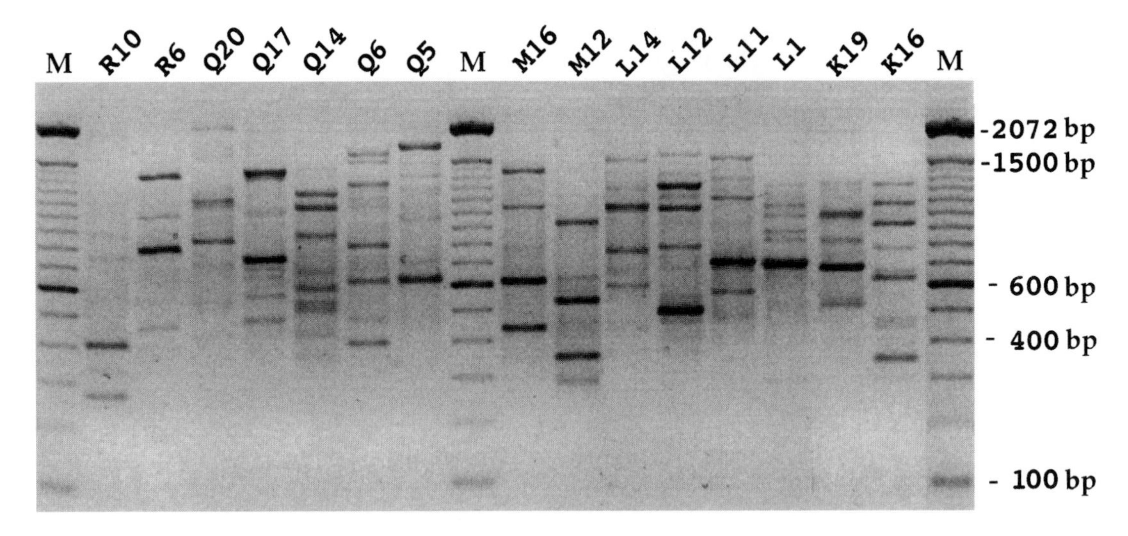

Fig. 2 Different RAPD primers amplify fragments of different sizes. Haploid zebrafish DNA was amplified with the 16 different RAPD primers listed. H, hundred base pair ladder.

the band (Fig. 1D, lane 4), the presence allele is dominant to the absence allele. The polymorphism identified by this allelic pair thus identifies a RAPD genetic marker. Figure 3 shows animals of different genotypes all amplified with the same primer. Bars at the left of the photograph indicate bands that are amplified from some genotypes and not from others.

Occasionally, RAPD markers have codominant alleles. These can occur if insertions or deletions arise between the two primer binding sites, thus causing bands of two different sizes to amplify and become visible after gel electrophore-

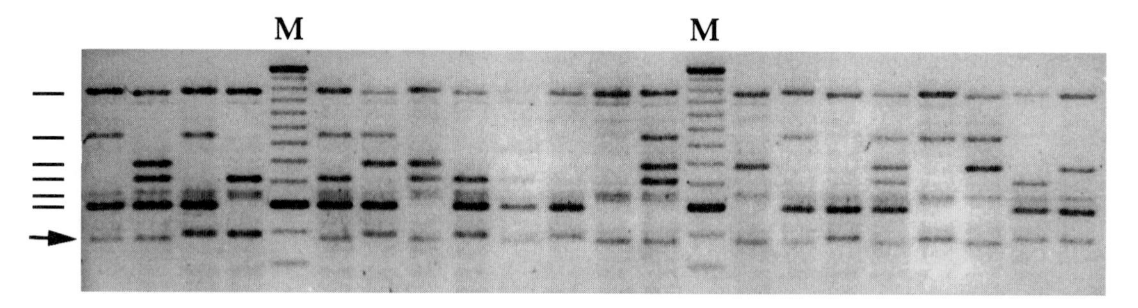

Fig. 3 Some RAPD loci have codominant alleles. The gel displays four dominant/recessive allele pairs (950, 800, 710, 600) (indicated with a dash at the left of the figure) that are unlinked, and a single codominant marker with alleles of 500 and 490 base pairs (indicated with an arrow).

sis. Relative to the genotype diagrammed in Fig. 1B, the genotype in Fig. 1C has a deletion between two of the primer binding sites. As a result, the band amplified at this locus will be smaller from the deletion allele (Fig. 1C, D, lane 3) than the band from the allele without the deletion (Fig. 1B, D, lane 2). Because a heterozygote between a short-band allele and a long-band allele will show both bands (Fig. 1D, lane 6), this type of allele is codominant. Only about 10% of RAPD markers are codominant (Johnson *et al.,* 1994, 1996). The arrow in Fig. 3 denotes a codominant RAPD marker with a size difference of about 10 bp. A very large insertion between two primer binding sites can prevent the amplification of a band, and thus also result in a recessive (no band) allele.

The presence of a particular band is a phenotype that is inherited as a dominant Mendelian genetic marker in a polymorphic cross. The haploid progeny of a heterozygote for two divergent genetic strains provides a particularly suitable family for investigating the inheritance of RAPD markers because, in haploids, dominant alleles do not obscure their recessive partners (Postlethwait *et al.,* 1994). Figure 3 shows the results of amplifying the DNA of 20 haploid progeny of a female hybrid between the AB and Darjeeling strains with a single RAPD primer. Each marker, indicated by the bar at the left of the figure, is present in about half the segregants and absent from the other half, as expected by Mendelian segregation in haploid gametes.

Genetic maps can be constructed from the segregation patterns of RAPD markers. Figure 4 shows the inheritance of three dominant RAPD markers using the commercially available RAPD primer A7 on the haploid progeny of

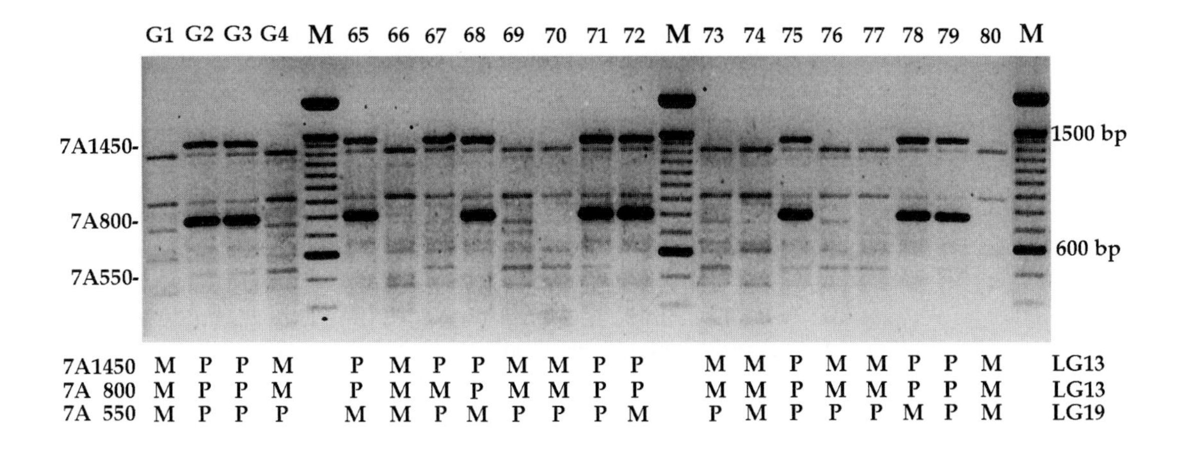

Fig. 4 Segregation of linked and unlinked RAPD markers. DNA from members of a haploid mapping panel was amplified with primer A7. Loci *7A1450* and *7A800* were tightly linked on LG13, with a single recombinant among 96 meiotic products (individual 67). In contrast, marker *7A550* was unlinked to the other two, and it maps to LG19.

a heterozygous female. Two of these loci, *7A1450* and *7A800,* are closely linked, with a single recombinant among 96 haploid progeny. Thus, these two markers are about a centiMorgan apart. The third marker, *7A550,* is unlinked to the other two. The genotypes inferred from the phenotypes on the gel are written at the bottom of the figure. The scoring of about 600 such markers, and the use of mapping software such as MapMaker (Lander *et al.,* 1987) or Map Manager (Manley and Cudmore, 1994) is sufficient to construct a closed map for zebrafish (Johnson *et al.,* 1996).

III. Advantages and Disadvantages of RAPDs as Polymorphic Markers

RAPDs are only one of the many types of genetic markers that are suitable for building genetic maps. Visible phenotypic markers, such as pigment patterns and bristle morphology, were used to make the early genetic maps of *Drosophila* and mouse, but only a few visible markers segregate in any given cross; this makes the accumulation of data painfully slow. The use of isozymes as genetic markers made many more loci available in single crosses. The best markers, however, are ones for which thousands can be followed in a single cross, and these are afforded by polymorphisms in DNA sequence. There are several ways to detect DNA polymorphisms that have been used with success in zebrafish research. These can be divided into two basic strategies—the first requires availability of a locus as a clone, which is either directly used as a probe, or which is sequenced to provide information for the design of PCR primers. This category includes classical restriction fragment length polymorphisms (RFLPs), microsatellite markers [also called simple sequence length polymorphisms (SSLPs)], and single strand conformation polymorphisms (SSCPs). Another category of DNA polymorphisms requires no previous knowledge of the loci identified; these include RAPDs and arbitrary fragment length polymorphisms (AFLPs). Each type has its own peculiar advantages, and different methods are more useful under different situations.

Prior sequence knowledge of a locus is expensive to obtain, but allows one to design PCR primers that amplify DNA from a single specific locus. Amplified regions that contain a simple sequence repeat (SSR), such as the dinucleotide sequence CA repeated many times, are often of different lengths in different strains because they have different numbers of repeats. These SSLPs are wonderfully useful because of their high degree of polymorphism, but they are very expensive to develop because they require the cloning and sequencing of hundreds of genomic DNAs containing CA repeats. Fortunately, over 700 of such markers have been developed for zebrafish (Goff *et al.,* 1992; Knapik *et al.,* 1996, 1998). The SSCPs also require cloning and sequence knowledge and can detect single-base-pair changes (Beier *et al.,* 1992). The SSCPs are useful for mapping sequenced genes, which often have polymorphisms in introns or 3′ untranslated regions in zebrafish crosses (Postlethwait *et al.,* 1998). An additional useful attribute of SSLPs and SSCPs is that they are codominant, allowing the heterozygote to be distinguished from both homozygotes. This property is highly useful in normal sexual crosses.

Whereas SSLPs and SSCPs are highly polymorphic but expensive to develop, RAPDs and AFLPs are cheap to develop, but tend to be strain-specific, and hence may not be informative in some crosses of interest. RAPD primers are available at comparatively low expense from commercial sources, and each primer provides about five polymorphisms between AB and strains derived from India, such as Darjeeling, India, and WIK (Johnson *et al.,* 1994). RAPD analysis is technically simple because the amplification products can be separated on an agarose gel and DNA polymorphisms can be directly visualized with ethidium bromide. These properties made the initial construction of the genetic map uncomplicated and safe enough for part-time undergraduate researchers, nine of whom were coauthors on the original mapping paper (Postlethwait *et al.,* 1994). In contrast, AFLPs can provide about 20 or more polymorphisms per primer pair. AFLPs employ radiolabeled DNAs and achieve superior resolution on sequencing-style acrylamide gels (see Ransom, this volume). The longer primers used for AFLPs make binding more specific than with the short RAPD primers; hence, the amplified fragments tend to be more reproducible and less dependent on exact experimental conditions than RAPDs. Because both RAPDs and AFLPs usually have dominant/recessive alleles, the heterozygous genotype cannot usually be distinguished from the homozygous dominant. This disadvantage can be overcome by using mapping panels consisting of haploids (Streisinger *et al.,* 1981; Postlethwait *et al.,* 1994) or diploidized haploids (Streisinger, 1981, 1986; Young *et al.,* 1998).

IV. Methods for Developing RAPD Markers

The following methodology for RAPDs was adapted from Williams *et al.* (1990). In a 25-μl reaction: 10 m*M* Tris–HCl pH 8.3, 50 m*M* KCl, 1.5 m*M* $MgCl_2$, 0.001% Gelatin, 100 μM each dNTP, 1.3 mg/ml BSA, 25 ng/μl primer, DNA (5–50 ng), 0.5 unit amplitaq DNA polymerase, Perkin Elmer, top with 40 μl mineral oil, if necessary. In a 96-well microtiter dish and MJ Research thermocycler PTC100 (M.J. Research, 24 Bridge Street, Watertown, MA 02172, tel: 617-924-2266 or 617-923-8000; Fax: 617-923-8080), use the following thermal cycling program: Step 1, 92°C, 1 min; step 2, 94°C, 5 sec; step 3, 92°C, 55 sec; step 4, 36°C, 60 sec; step 5, 72°C, 90 sec; step 6, repeat steps 2–5 36 times; step 7, 72°C, 5 min; step 8, 4°C, indefinite hold. Separate products on a 1% agarose gel with a 100-bp ladder as marker. RAPD primers can be obtained from Operon Technologies (1000 Atlantic Avenue, Alameda, CA 94501, tel: 510-865-8644 or 800-688-2248, fax: 510-865-5255). The 96-well U-bottom wells for PCR are from Falcon (3911 MicroTest III Flexible Assay Plate); Becton Dickinson Labware, 1950

Williams Drive, Oxnard, CA 93030). Sometimes extra BSA must be added to the reaction to counteract substances in the plastic reaction plates. The products of RAPD reactions are generally resolved on agarose gels after staining in ethidium bromide. RAPD primers can also be labeled with fluorescent labels, and the resulting fragments are separated and visualized on an automated DNA sequencer with an in-lane size marker added for standardization between lanes and among gels (Corley-Smith *et al.,* 1997). RAPD markers are named by the primer designation followed by the size of the band and the strain in which the band is found. For example, an 1100-bp band amplified by Operon primer B12 from AB strain DNA would be called 12B1100(A).

There are a number of ways to increase the number of polymorphisms identified by a set of RAPD primers. For example, the combination of two or more RAPD primers in a single reaction provides novel bands not present using either of the two primers separately. In another variation, the gel is blotted, probed with labeled microsatellite repeat probes, and autoradiographed (Ramser *et al.,* 1997). The result provides segregating bands different from the ethidium bromide-stained segregating bands.

V. Using RAPDs to Identify Markers Closely Linked to a Mutation

The RAPDs are useful for the genetic analysis of point mutations and in defining the nature of chromosome rearrangements (Talbot *et al.,* 1998). The positional cloning of mutations requires the identification of DNA polymorphisms closely linked and flanking the mutation of interest. A chromosome walk from one marker past the mutation to the other marker will contain DNA from the mutated locus (Zhang *et al.,* 1998). This use of RAPDs is quite powerful when combined with the DNA pooling strategy called bulked segregant analysis (Michelmore *et al.,* 1991; Kesseli *et al.,* 1992; Williams *et al.,* 1993; Paran and Michelmore, 1993; Churchill *et al.,* 1993). Figure 5 diagrams how bulked segregant analysis works. Consider a female fish that is heterozygous in coupling phase for a mutation *m,* a closely linked RAPD marker *1A150,* and a distantly linked (or unlinked) RAPD marker *2A200* (Fig. 5A). This female is homozygous for RAPD marker *3A280.* During meiosis, recombination will occur often between *m* and the distantly linked marker *2A200* but will hardly ever occur between *m* and *1A150,* the tightly linked marker (Fig. 5B). Wild-type haploid embryos will thus nearly always be $1A150^{+}$ in genotype and hence amplify the band, and mutant haploids will nearly always be $1A150^{-}$ and fail to amplify the band. If DNAs from about 20 wild-type embryos are bulked together into one DNA pool, and DNAs from about 20 mutant embryos are joined in a second pool, then the wild-type (Fig. 5C, lane 1), but not the mutant (Fig. 5C, lane 2), DNA pool will amplify the 150-bp band with primer A1. On the other hand, $2A200^{+}$ and $2A200^{-}$ will be represented at high frequencies in both mutant and wild-type DNA pools

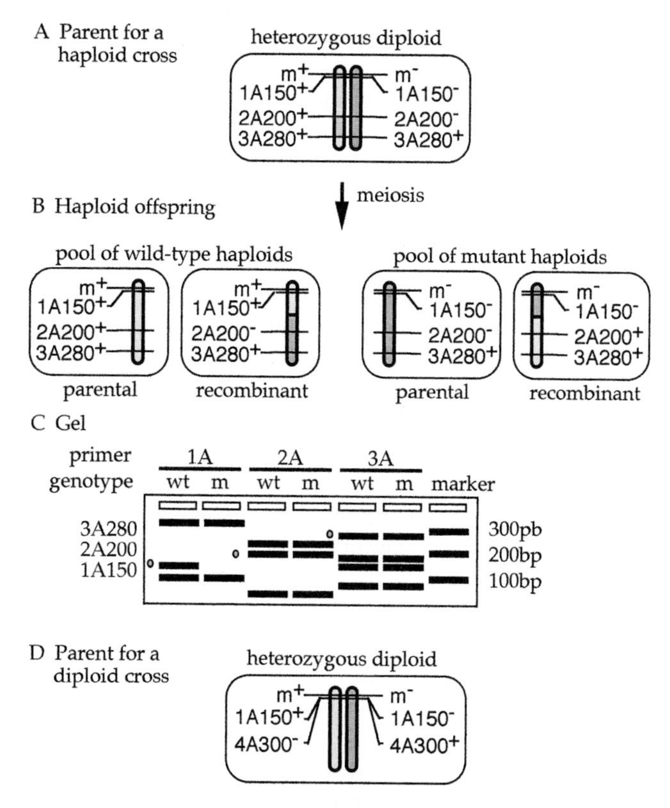

Fig. 5 The strategy of bulked segregant analysis. (A) Consider a female heterozygous in coupling phase for a mutation *m,* and three RAPD markers, *1A150, 2A200,* and *3A280.* (B) After meiosis and *in vitro* fertilization with UV–treated sperm, recombination will be frequent between the mutation and the distantly linked marker *A200* but will rarely occur between the mutation and the nearby marker *1A150.* If DNAs from mutant embryos are pooled together, the pool will almost never have the band-forming allele of the nearby marker *1A150* but will always have the band-forming allele of the distant marker *2A200,* and the reciprocal holds for the wild-type pool. (C) When the PCR products are separated on a gel, the nearby marker signifies its existence by showing up in the wild-type but not the mutant pool, while distantly linked and unlinked markers will be in all pools. (D) The method works for a diploid cross but detects only markers in coupling, not markers in repulsion phase with the recessive mutation.

as a result of frequent recombination between the mutation and the RAPD marker. If *2A200* were so distant it appeared to be unlinked to *m,* then there would be equal quantities of the plus and the minus alleles in both mutant and wild-type DNA pools, so the band will look the same in both gel lanes (Fig. 5C, lanes 3 and 4). Because the mother of this group of haploid segregants is homozygous for the locus-encoding band *3A280,* it will appear in both DNA pools (Fig. 5C, lanes 5 and 6). Experiments to identify RAPD loci closely linked to mutations are usually conducted using a variety of different RAPD primers to amplify DNAs from two pools (about 20 wild-type haploid segregants in one pool, and DNAs from 20 mutant haploid segregants in another pool), resulting in a gel similar to the one diagrammed in Fig. 5C. Markers closely linked to the mutation appear in one pool but are absent from the other. Unlinked loci and nonpolymorphic loci amplify bands in both pools, like the other bands shown on the gel.

Figure 6 shows bulked segregant analysis experiments conducted to identify RAPD markers closely linked to the mutation *silent heart,* which blocks the embryonic heart from beating (C. Walker and C. Kimmel, unpublished). Heterozygotes for the mutation on an AB genetic background were mated to Darjeeling males. From the offspring, heterozygous females were identified, and from them, haploid offspring were collected (Streisinger *et al.,* 1981). Haploid offspring were scored as either wild-type or mutant in phenotype. Two DNA pools, each consisting of 20 individuals, were assembled for each genotype. These four DNA pools then served as templates for the screening of several hundred commercially available RAPD primers, ten of which are shown in Fig. 6A. Bands that are present in one phenotypic pool but absent from the other are candidates for markers linked to the mutation. Such bands are visible for primer F6, I1, and AC6. The results from primers O6 and V8 demonstrate the value of running a duplicate pool consisting of different individuals for each phenotype because of false positives that may occur if a single pool is used for each phenotype. When a band appears to be linked to the mutation in two independent DNA pools, one scores the marker on individual segregants from the mapping cross (Fig. 6B). In the case shown, primer F6 amplified a band at 1800 that was about 7 cM away from the mutation: the band 6F1800 was absent from all but one of the wild-type haploids and present in all but two of the mutant haploids. The localization of the same RAPD band in a mapping cross scored for many loci identifies the position of the mutation (Postlethwait *et al.,* 1994).

Bulked segregant analysis works not only with haploid segregants but also with normal sexual crosses. Imagine that a fish is heterozygous for a RAPD marker *4A300* closely linked to *m* in repulsion rather than in coupling phase (Fig. 5D). The intercross of two such fish would provide a quarter of the progeny homozygous for the mutation, and most of these mutants would be homozygous for the recessive no-band allele of *1A150,* but they would be homozygous dominant for the band-forming allele of the closely linked locus *4A300.* The embryos homozygous for the wild-type allele of the mutation would usually be homozygous for the band-forming allele of *1A150,* but homozygous for the null allele of *4A300,* the marker linked in repulsion. If the homozygous wild-type embryos have a phenotype that is distinguishable from the heterozygotes, then the pooling protocol will allow the identification of both *1A150* and *4A300* as closely linked markers. But if homozygous wild-type embryos have the same phenotype as heterozygous individuals, which is the case with most zebrafish embryonic lethal mutations, then two thirds of the phenotypically wild-type pool will be heterozygous for marker *4A300* and, hence, show the band. Because the homozygous

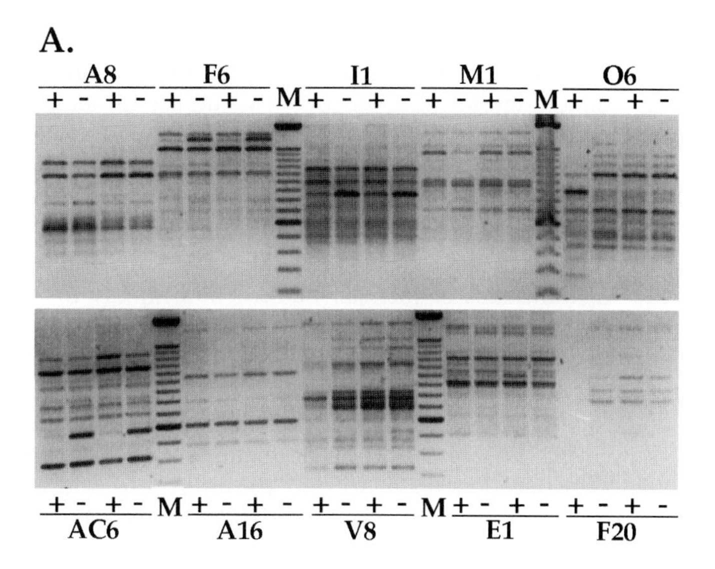

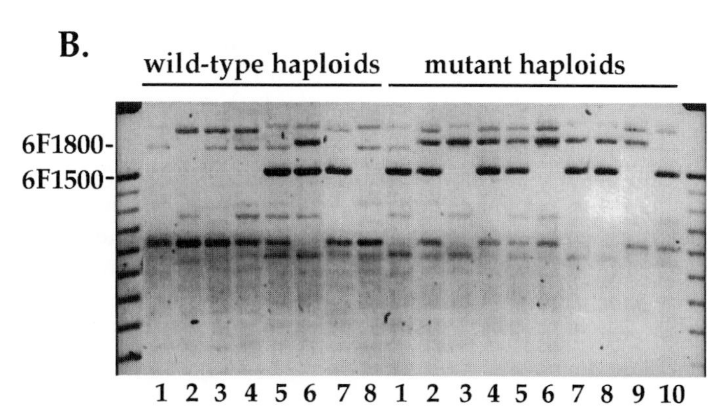

Fig. 6 Bulked segregant analysis for *silent heart*^{b109}. (A) Primer screen. Several hundred RAPD primers were screened for markers linked to *silent heart.* A female heterozygous for *silent heart* in the AB background was mated to a male from the Darjeeling strain. From a heterozygous daughter, haploid embryos were collected. DNA was extracted from 40 wild-type (+) and 40 mutant (−) haploid embryos. Two 20-individual DNA pools were created for each genotype and tested with the primers, 10 of which are shown here. Bands amplified by the F6 primer (6F1800) and the AC6 primer (6AC600) repeatedly differed in the two genotypes. (B) Individual animals from the mapping cross were checked to confirm linkage and determine map distances. The figure shows 8 wild-type and 10 mutant embryos checked individually with RAPD primer F6. Only one of the wild-type haploids (number 6) has the presence allele of the locus, whereas all but 2 of the mutant embryos have the presence allele. For locus 6F1800, there were a total of 7 recombinants among 95 haploids tested showing tight linkage. In contrast, for locus 6F1500 amplified by the same primer, there were 45 recombinants among 96 haploids, demonstrating that this locus is not linked to *silent heart.*

mutants which are homozygous for the band-forming allele of *4A300* also show the 300-bp band, this closely linked marker will exist but not be detected in a diploid intercross. Consequently, the use of dominant markers such as RAPDs or AFLPs for bulked segregant analysis works well for both haploid and sexual diploid crosses, but it is twice as efficient using haploid segregants because the use of haploids allows the detection of dominant/recessive markers whether they are linked in coupling or repulsion phase.

More than a thousand RAPD primers are commercially available, and so tens of thousands of loci can be screened rapidly for linkage to a mutation. If an average of five loci segregate per RAPD primer in a mapping cross, and one tests 1000 primers (2000 PCRs), one would screen 5000 loci for linkage to the mutation. Because there are about 3000 cM in the zebrafish female genetic map, this should generally yield a RAPD locus on average every 0.6 cM. At about 600 kilobase pairs/cM in the female map (Postlethwait *et al.,* 1994), this would provide a marker every 360 kb, on average. So this procedure would place a mutation somewhere within a 360-kb interval, on average, which can be spanned by a single YAC or a few PAC or BAC clones. Because one can screen through several hundred RAPDs much faster than doing a single step in a chromosome walk, it would seem prudent to find the very closest linked markers before embarking on a chromosome walk.

To make use of closely linked markers in a chromosome walk, it is necessary, of course, to have a mapping cross that is sufficiently large to resolve the markers from each other. A cross of 1000 meioses should resolve markers separated by about 0.1 cM (about 60 kb on average). The entire mapping panel has to be assayed only to find flanking markers located a few centimorgans from the mutation. After that, only the few recombinants need to be retested for markers suspected of being more closely linked. For an example, see Zhang *et al.,* (1998), which describes the use of RAPDs in the positional cloning of *one eyed pinhead.*

VI. Conversion of RAPD Markers to Sequence Tagged Sites

For RAPDs that are to be used repeatedly and for which there is no tolerance for error in genotyping, for example in a chromosome walk, it is useful to convert RAPDs to sequence tagged sites (STSs) called sequence tagged RAPDs, or STARs (see Johnson *et al.,* 1996; Talbot *et al.,* 1996; Comes *et al.,* 1997). To do this, use the appropriate RAPD primer to amplify the genotype containing the band-forming allele of the mutation, and cut the band from the gel. Isolate DNA from the band, and clone it into any of a number of vectors designed for cloning PCR fragments. After the band is sequenced, design robust primers for the band, and identify a restriction fragment polymorphism in the band that is segregating in the mapping cross. Amplification of the members of the mapping panel with the STAR primers followed by digestion with the appropriate restriction enzyme before gel electrophoresis allows the genotyping of the panel.

VII. Conclusions

RAPD markers are highly useful for the initial construction of a map because of their low up-front expense and are highly useful for finding loci closely linked to mutations because their high degree of polymorphism. On the other hand, RAPD methodology suffers from being strain-specific and subject to experimental conditions. For routine mapping, SSLPs have many advantages over RAPDs. However, for the identification of anonymous loci very close to mutations of high interest, RAPDs provide a convenient, simple, cost-effective, and, in fact, *rapid* way to meet the goal.

References

Al-Janabi, S. M., Honeycutt, R. J., McClelland, M., and Sobral, B. (1993). A genetic linkage map of *Saccharum spontaneum* L. "SES 208". *Genetics* **134,** 1249–1260.

Beier, D. R., Dushkin, H., and Sussman, D. J. (1992). Mapping genes in the mouse using single-strand conformation polymorphism analysis of recombinant inbred strains and interspecific crosses. *Proc. Natl. Acad. Sci. U.S.A.* **89,** 9102–9106.

Churchill, G., Giovannoni, J., and Tanksley, S. (1993). Pooled-sampling makes high-resolution mapping practical with DNA markers. *Proc. Natl. Acad. Sci. U.S.A.* **90,** 16–20.

Comes, A. M., Humbert, J. F., and Laurent, F. (1997). Rapid cloning of PCR-derived RAPD analysis. *Biotechniques* **23,** 210–212.

Corley-Smith, G. E., Lim, C. J., Kalmar, G. B., and Brandhorst, B. P. (1997). Efficient detection of DNA polymorphisms by fluorescent RAPD analysis. *Biotechniques* **22,** 690–692, 694, 696.

Driever, W., Solnica-Krezel, L., Schier, A. F., Neuhauss, S. C. F., Malicki, J., Stemple, D. L., Stainier, D. Y. R., Zwartkruis, F., Abdelilah, S., Rangini, Z., Belak, J., and Boggs, C. (1996). A genetic screen for mutations affecting embryogenesis in zebrafish. *Development* **123,** 37–46.

Goff, D., Galvin, K., Katz, H., Westerfield, M., Lander, E., and Tabin, C. (1992). Identification of polymorphic simple sequence repeats in the genome of the zebrafish. *Genomics* **14,** 200–202.

Grattapaglia, D., Bertolucci, F., Penchel, R., and Sederoff, R. (1996) Genetic mapping of quantitative trait loci controlling growth and wood quality traits in Eucalyptus grandis using a maternal half-sib family and RAPD markers. *Genetics* **144,** 1205–1214.

Haffter, P., Granato, M., Brand, M., Mullins, M. C., Hammerschmidt, M., Kane, D. A., Odenthal, J., Van Eeden, F. J. M., Jiang, Y. J., Heisenberg, C. P., Kelsh, R. N., Furutani-Seiki, M., Vogelsang, E., Beuchle, D., Schach, U., Fabian, C., and Nüsslein-Volhard, C. (1996). The identification of genes with unique and essential functions in the development of the zebrafish, *Danio rerio. Development* **123,** 1–36.

Henion, P. D., Raible, D. W., Beattie, C. E., Stoesser, K. L., Weston, J. A., and Eisen, J. S. (1996). Screen for mutations affecting development of zebrafish neural crest. *Dev. Genet.* **18,** 11–17.

Johnson, S. L., Midson, C. N., Ballinger, E. W., and Postlethwait, J. H. (1994). Identification of RAPD primers that reveal extensive polymorphisms between laboratory strains of zebrafish. *Genomics* **19,** 152–156.

Johnson, S. L., Gates, M. A., Johnson, M., Talbot, W. S., Horne, S., Baik, K., Rude, S., Wong, J. R., and Postlethwait, J. H. (1996). Centromere-linkage analysis and the consolidation of the zebrafish genetic map. *Genetics* **142,** 1277–1288.

Kesseli, R. V., Paran, I., and Michelmore, R. W. (1992). Efficient mapping of specifically targeted genomic regions and the tagging of these regions with reliable PCR-based genetic markers. *In* "Applications of RAPD Technology to Plant Breeding," pp. 31–36.

Kimmel, C. B. (1989). Genetics and early development of zebrafish. *Trends Genet.* **5,** 283–288.

Kishimoto, Y., Lee, K. H., Zon, L., Hammerschmidt, M., and Schulte-Merker, S. (1997). The molecular nature of zebrafish swirl: BMP2 function is essential during early dorsoventral patterning. *Development* **124,** 4457–4466.

Kiss, G. B., Csanadi, G., Kalman, K., Kalo, P., and Okresz, L. (1992). Construction of a basic genetic map for alfalfa using RFLP, RAPD, isozyme and morphological markers. *Mol. Genet. Gen.* **238,** 129–137.

Knapik, E. W., Goodman, A., Atkinson, O. S., Roberts, C. T., Shiozawa, M., Sim, C. U., Weksler-Zangen, S., Trolliet, M. R., Futrell, C., Innes, B. A., Koike, G., McLaughlin, M. G., Pierre, L., Simon, J. S., Vilallonga, E., Roy, M., Chiang, P. W., Fishman, M. C., Driever, W., and Jacob, H. J. (1996). A reference cross DNA panel for zebrafish (*Danio rerio*) anchored with simple sequence length polymorphisms. *Development* **123,** 451–460.

Knapik, E., Goodman, A., Ekker, M., Chevrette, M., Delgado, J., Neuhauss, S., Shimoda, N., Driever, W., Fishman, M., and Jacob, H. (1998). *Nat. Genet.*, **18,** 338–344.

Lander, E. S. (1987). MAPMAKER: An interactive computer package for constructing primary genetic linkage maps of experimental and natural populations. *Genomics* **1,** 174–181.

Manley, and Cudmore, R. (1994). http://mcbio.med.buffalo.edu/mapmgr.html.

Michelmore, R., Paran, I., and Kesseli, R. (1991). Identification of markers linked to disease-resistant genes by bulked segregant analysis: A rapid method to detect markers in specific genomic regions by using segregating populations. *Proc. Natl. Acad. Sci. U.S.A.* **88,** 9828–9832.

Paran, I., and Michelmore, R. W. (1993). Development of reliable PCR-based markers linked to downy mildew resistance genes in lettuce. *Theor. Appl. Genet.* **85,** 985–993.

Postlethwait, J., Johnson, S., Midson, C. N., Talbot, W. S., Gates, M., Ballenger, E. W., Africa, D., Andrews, R., Carl, T., Eisen, J. S., Horne, S., Kimmel, C. B., Hutchinson, M., Johnson, M., and Rodriguez, A. (1994). A genetic linkage map for the zebrafish. *Science* **264,** 699–703.

Postlethwait, J. H., Yan, Y.-L., Gates, M. A., Horne, S., Amores, A., Brownlie, A., Donovan, A., Egan, E. S., Ekker, M., Force, A., Gong, Z., Goutel, C., Fritz, A., Kelsh, R., Knapik, E., Liao, E., Orr, M., O'Shea, S. A., Paw, B., Ransom, D., Singer, A., Thomson, T., Beier, D., Joly, J.-S., Larhammar, D., Rosa, F., Westerfield, M., Zon, L. I., Johnson, S. L., and Talbot, W. S. (1998). Vertebrate genome evolution and the zebrafish gene map. *Nat. Genet.* **18,** 345–349.

Postlethwait, J. H., and Talbot, W. S. (1997). Zebrafish genomics: From mutants to genes. *Trends Genet.* **13,** 183–190.

Reiter, R., Williams, J., Feldmann, K., Rafalski, J., Tingey, S., and Scolnik, P. (1992). Global and local genome mapping in *Arabidopsis thaliana* by using recombinant inbred lines and random amplified polymorphic DNAs. *Proc. Natl. Acad. Sci. U.S.A.* **89,** 1477–1481.

Rafalski, J. A., and Tingey, S. B. (1993). Genetic diagnostics in plant breeding. *Trends Genet.* **9,** 275–280.

Ramser, J., Weising, K., Chikaleke, B., and Kahl, G. (1997). Increased informativeness of RAPD analysis by detection of microsatellite motifs. *Biotechniques* **23,** 285–290.

Riley, B. B., and Grunwald, D. J. (1995). Efficient induction of point mutations allowing recovery of specific locus mutations in zebrafish. *Proc. Natl. Acad. Sci. U.S.A.* **92,** 5997–6001.

Schulte-Merker, S., Lee, K. J., McMahon, A. P., and Hammerschmidt, M. (1997). The zebrafish organizer requires chordino. *Nature* **387,** 862–863.

Schulte-Merker, S., van Eeden, F., Halpern, M., Kimmel, C., Nusslein-Volhard, C. (1994) *no tail* (*ntl*) is the zebrafish homologue of the mouse T *Brachyury* gene. *Development* **120,** 1009–1015.

Sepich, D., Wegner, J., O'Shea, S., and Westerfield, M. (1997) An altered intron inhibits synthesis of the acetylcholine receptor alpha subunit in the paralyzed zebrafish mutant *nic1. Genetics* **184,** 361–372.

Streisinger, G., Walker, C., Dower, N., Knauber, D., and Singer, F. (1981). Production of clones of homozygous diploid zebrafish (*Brachydanio rerio*). *Nature* **291,** 293–296.

Streisinger, G., Singer, F., Walker, C., Knauber, D., and Dower, N. (1986). Segregation analysis and gene-centromere distances in zebrafish. *Genetics* **112,** 311–319.

Talbot, W. S., Egan, E. S., Gates, M. A., Walker, C., Ullmann, B., Neuhauss, S. C. F., Kimmel, C. B., and Postlethwait, J. H. (1998). Genetic analysis of chromosomal rearrangements in the cyclops region of the zebrafish genome. *Genetics* **148,** 373–380.

Talbot, W. S., Trevarrow, B., Halpern, M., Melby, A., Farr, G., Postlethwait, J., Jowett, T., Kimmel, C., and Kimelman, D. (1995). A homeobox gene essential for zebrafish notochord development. *Nature* **378,** 150–157.

Tanksley, S. D., Ganal, M. W., Prince, J., de Vicente, M., Bonierbale, M., Broun, P., Fulton, T., Pineda, J., Roder, M., Wing, R., Wu, W., and Young, N. (1992). High density molecular linkage maps of the tomato and potato genomes. *Genetics* **132,** 1141–1160.

Tingey, S. V., Rafalski, J. A., and Williams, J. G. K. (1992). Genetic analysis with RAPD markers. *In* "Applications of RAPD Technology to Plant Breeding," pp. 3–8.

Welsh, J., and McClelland, M. (1990). Fingerprinting genomes using PCR with arbitrary primers. *Nucl. Acids Res.* **18,** 7213–7218.

Williams, J. G. K., Kubelik, A. R., Livak, K. J., Rafalski, J. A., and Tingey, S. V. (1990). DNA polymorphisms amplified by arbitrary primers are useful as genetic markers. *Nucl. Acids Res.* **18,** 6531–6535.

Williams, J. G. K., Reiter, R. S., Young, R. M., and Scolnik, P. A. (1993). Genetic mapping of mutations using phenotypic pools and mapped RAPD markers. *Nucl. Acids Res.* **18,** 6531–6535.

Young, W., Wheeler, P., Coryell, V., Keim, and Thorgaard, G. (1998). A detailed linkage map of rainbow trout produced using doubled haploids. *Genetics* **148,** 839–850.

Zhang, J. J., Talbot, W. S., and Schier, A. F. (1998). Positional cloning identifies zebrafish one-eyed pinhead as a permissive EGF-related ligand required during gastrulation. *Cell* **92,** 241–251.

CHAPTER 10

Simple Sequence-Length Polymorphism Analysis

Eric C. Liao[*,†] and Leonard I. Zon[*]

[*] Division of Hematology/Oncology, Children's Hospital
Department of Pediatrics and Howard Hughes Medical Institute
Harvard Medical School
Boston, Massachusetts 02115

[†] Division of Health Sciences and Technology
Massachusetts Institute of Technology
Cambridge, Massachusetts 02139

The microsatellite genetic linkage map generated by Knapik and colleagues consists of 705 simple sequence-length polymorphism (SSLP) markers with an average resolution of 3.3 cM (Knapik *et al.*, 1998). The latest version of the map (July 1998) consists of 2000 markers and has been made available on the Internet by Drs. N. Shimoda and M. C. Fishman of the Massachusetts General Hospital. This latest map and its primer sequences can be accessed at http://zebrafish.mgh.harvard.edu/papers/unpublished/zf_map2k/. The primers for amplifying the SSLPs are commercially available from Research Genetics (1-800-533-4363). The following is a standard protocol that we find to be reliable in the amplification and visualization of these microsatellite markers.

METHODS IN CELL BIOLOGY, VOL. 60

0091-679X/99 $30.00

I. Analysis in DNA Sequencing Gels

A. Materials and Reagents

1. Polymerase chain reaction (PCR) buffer (1×): 2m*M* $MgCl_2$, 14m*M* Tris–HCl (pH 8.4), 68m*M* KCl, 0.001% gelatin, 0.14 mg BSA, and 0.14m*M* of each dNTP.
2. Standard 6% polyacrylamide, 8*M* urea sequencing gel.

B. Experimental Protocol

1. *PCR reaction mix.* We prefer labeling the SSLP-containing PCR product by direct incorporation of ^{32}P-labeled nucleotides, although 5′ end labeling of one or both PCR primers can also be employed (Sambrook *et al.*, 1989). For the former approach, we use the following PCR mixture: 30–50 ng of genomic DNA template, 15 ng each of 3′ and 5′ end primers, 8 μl 1× PCR buffer, 0.1 μl of 3000 Ci/mmol α-^{32}P-dCTP, and 0.1 μl (2 units) *Taq* DNA polymerase. Reactions are arrayed in 96-well microtiter plate formate. Because large numbers of reactions may be performed in a positional cloning project, we use a thermosealer (thermosealer, 96-well PCR plates, and sealer foils all available from Marsh; 1-800-445-2812) to expedite assembly of the reactions.

2. *PCR cycling.* Our PCR cycling is carried out in either Perkin Elmer 9600 or MJ PTC-100 thermocyclers. The reactions are denaturated at 94°C for 4 minutes, and then 30 cycles of amplification are performed using the following program: 94°C for 30 seconds, 55°C for 30 seconds, and 72°C for 60 seconds. Finally, terminal extension is performed at 72°C for 10 minutes, and the reactions are cooled to 4°C.

3. *Separation of SSLP-containing PCR product.* After PCR amplification, 4 μl of stop solution (90% formamide, 20m*M* EDTA, 01.% bromophenol blue, 0.1% xylene cyanol) is added to each reaction, and the reactions are denaturated at 94°C for 4 minutes. The samples are then placed on ice until electrophoresis. Each reaction (4 μl) is loaded in an individual gel lane, and the samples are electrophoresed at 65 W constant power for 1–2 hours. The tracking dyes, bromophenol blue and xylene cyanol, migrate with ssDNA fragments of approximately 80 nucleotides and 120–130 nucleotides, respectively. If the SSLP-containing PCR fragment is discrete and background bands from the reaction are minimal, one can double-load two different reactions per lane. In this case, the first set of samples are run an appropriate distance into the gel, the power supply is turned off, a second set of samples is applied to the gel, and electrophoresis is continued as before. Finally, gels are dried and exposed to X-ray film.

II. Alternative Analysis of SSLP-Containing PCR Product in High-Resolution Agarose Gels

1. PCR reactions are performed as described previously, but the reaction volume is increased to 25 μl, with the following components: 30–50 ng of genomic

DNA template, 15 ng each of 3′ and 5′ end primers, 18 μl 1× PCR buffer, and 0.1 μl (2 units) *Taq* DNA polymerase. Sterile, double-distilled water is added to bring the reaction volume to 25 μl.

2. Standard 2% agarose gels in 1× TBE are cast in a horizontal frame. Of ethidium bromide 5 μg are added per 10 ml of gel solution. Samples are loaded on the gel and electrophoresed until DNA fragments are sufficiently resolved. PCR products are visualized by use of a ultraviolet transilluminator.

III. Troubleshooting and Other Considerations

In our experience, the actual amplified marker size sometimes differs from the published "expected" size. We attribute such variability to allelic differences between different zebrafish genetic backgrounds. We also find that some marker polymorphisms do not segregate in the crosses of interest; to confirm lack of segregation, appropriate positive controls should be run on each gel. Although many SSLP-containing PCR products can be observed on polyacrylamide or agarose gels, some polymorphisms may be too subtle to score reliably. Finally, if the SSLP-containing PCR products are not detected, or their signals are weak, one should repeat the reactions with higher concentrations of the primers (e.g., 50 ng).

References

Knapik, E. W., Goodman, A., Ekker, M., Chevrette, M., Delgado, J., Neuhauss, S., Shimoda, N., Driever, W., Fishman, M. C., and Jacobs, H. J. (1998). A microsatellite genetic linkage map for zebrafish (*Danio rerio*). *Nat. Genet.* **18,** 338–343.

Sambrook, J., Fritsch, E. F., and Maniatis, T. (1989). "Molecular Cloning: A Laboratory Manual," 2nd ed. Cold Spring Harbor, NY: Cold Spring Harbor Laboratory.

CHAPTER 11

Gene Mapping in Zebrafish Using Single-Strand Conformation Polymorphism Analysis

Dorothee Foernzler and David R. Beier[1]

Genetics Division
Brigham and Women's Hospital
Harvard Medical School
Boston, Massachusetts 02115

I. Introduction

Positional cloning or a candidate gene analysis will be necessary to characterize the large number of genes causing developmental abnormalities in zebrafish generated by genome-wide mutagenesis screens (Haffter *et al.,* 1996; Driever *et al.,* 1996). A prerequisite for both of these strategies is a high-resolution genetic map of the zebrafish genome. Even though maps constructed using random amplified polymorphic DNAs (RAPDs) (Johnson *et al.,* 1996) and microsatellites (Knapik *et al.,* 1996) will facilitate positional cloning efforts, the success of a candidate gene approach will depend on the efficient localization of expressed

[1] Corresponding author: Genetics Division, Brigham and Women's Hospital, Harvard Medical School, 20 Shattuck Street, Boston, Massachusetts 02115

0091-679X/99 $30.00

sequences. In this chapter, we will introduce single-strand conformation polymorphism (SSCP) analysis as a powerful tool to map cDNAs in the zebrafish.

Originally developed for the detection of single-base-pair mutations at specific loci (Orita *et al.,* 1989), SSCP takes advantage of the fact that even single-nucleotide changes in short fragments of single-strand DNA may alter the intramolecular secondary conformation of these DNA stretches. If a short DNA fragment is denatured at high temperature and then quickly cooled, renaturation into double-stranded DNA is inhibited, and each single strand will assume its most-favored conformation according to its lowest free-energy state. This conformation is affected by changes in nucleotide content, which, under appropriate conditions, will often detectably shift its migration in a nondenaturing electrophoresis gel. Variant alleles identified by different migration patterns represent single-strand conformation polymorphisms.

Linkage studies using SSCP analyses have been successfully applied in the mouse using recombinant inbred strains or interspecific crosses (Beier *et al.,* 1992; Beier, 1993; Brady *et al.,* 1997). This technique has been extended for mapping genes in several other organisms including humans. It has been shown to be simple, rapid, and reproducible. No special equipment not common to any molecular biology laboratory is required. The PCR-based approach requires only small amounts of DNA, which makes it especially attractive for organisms for which the amount of DNA recoverable per animal is limited, as is the case for zebrafish. The polymorphisms detected using this technique are stably inherited, and SSCP can be applied to map any unique DNA region in a cross of sufficiently divergent strains. A disadvantage of this mapping strategy, in contrast to RFLP analysis by hybridization, is the requirement of species-specific sequence. Another limitation for the use of SSCP in linkage studies is the necessity for a sufficient polymorphism frequency. This, of course, is not specific for this form of analysis but is true for any sequence-dependent mapping technique.

SSCP applied for the purpose of gene mapping requires the PCR-mediated amplification of short DNA fragments derived from genomic sequences. Sequence templates for primer design ideally will correspond to regions less likely to be conserved to increase the chance of a polymorphic PCR product. For this purpose, 3′ untranslated regions and introns serve very well, although information regarding the latter is usually unavailable for cDNAs. To assess the polymorphism frequency of expressed sequences in commonly used zebrafish strains, we have tested over 150 primer pairs derived from 3′UTR. Every primer pair was initially tested for the presence of a polymorphism in the 3′UTR in parental DNA before mapping analyses with a reference cross DNA panel was performed. A maximal polymorphism frequency of 31% was detected for the AB and India strains used by Knapik *et al.,* (1996), and 42% for the SJD and C32 strains used by Johnson *et al.* (1996) (Foernzler *et al.,* 1998, submitted). As a comparison, the maximal polymorphism frequency in 3′ UTR sequences between inbred strains of mice was 20%, and in the two different species *Mus musculus* and *Mus spretus,* it was 70% (Brady *et al.,* 1997). The zebrafish lines used have been inbred for several generations or are clonal derivatives of a presumptive ancestor, but they cannot be considered as completely inbred. The number of genetically fixed loci in the genomes used for linkage analysis will influence the complexity of SSCP patterns.

We determined the strain distribution pattern of 17 polymorphic loci in zebrafish mapping crosses. On the MGH Reference Cross corresponding to the SSLP map (Knapik *et al.,* 1996), 6 loci polymorphic between AB/India were mapped, and 3 loci polymorphic between C32 and SJD were localized on the Oregon mapping cross (Johnson *et al.,* 1996). An example of this analysis is shown in Fig. 1. We included in our polymorphism screen 36 primers derived from partially characterized cDNA clones from kidney and embryonic cDNA libraries. In 13 of these expressed sequence tags, (ESTs) polymorphisms were detected, and 2 have been mapped. This demonstrates the feasibility of using ESTs for gene mapping in the zebrafish, which could be employed to generate a dense transcript map.

II. SSCP Methodology

For SSCP analysis, a polymerase chain reaction (PCR) is performed to amplify fragments between 150 and 350 bp. For best results, the template DNA should be diluted to a DNA concentration of 50 ng, although we successfully amplified PCR products with as little as 16 ng of zebrafish DNA per reaction on a regular basis. Excess template DNA will result in decreased SSCP sensitivity.

The choice of primers is often critical to the success of SSCP linkage studies and should be done carefully. To design PCR primers for SSCP analysis, at least

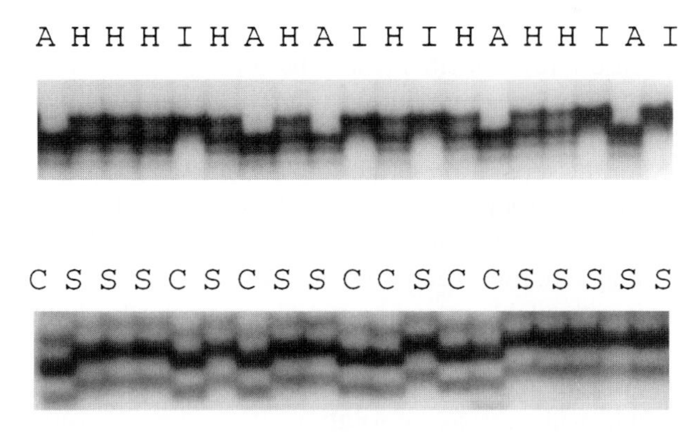

Fig. 1 SSCP analysis of expressed sequences. Genotype distributions of two different primer pairs corresponding to 3′UTR of expressed sequences are shown in progeny of two reference crosses. Top: Reference cross described by Knapik *et al.* (1996). Lanes correspond to diploid progeny of AB × India intercross. A: AB homozygote; H: heterozygote; I: India homozygote. Bottom: Reference cross described by Johnson *et al.* (1996). Lanes correspond to haploid progeny of SJD × C32 intercross. S: SJD; C: C32.

150 bp of unique sequence from less conserved regions such as 3′UTR, 5′UTR, or intron must be available. These sequences should be screened for possible vector contaminants, cloning artifacts, or repetitive regions using similarity search programs such as BLAST. A number of computer programs for optimized design of PCR primers are available. For our analyses, PCR primers were designed from 3′UTR of cDNAs using the Primer 0.5 program (Lincoln *et al.,* 1991) and the following criterion: primer length between 18 and 22 bp, primer GC% between 20 and 80%, primer melting temperature between 53°C and 60°C. A revised version of this program is Primer 3.0, which is available at http://www.genome.wi.mit.edu/genome_software/other/primer3.html.

Primers are 5′-end-labeled with $\gamma^{32}P$ using T4-polynucleotide kinase, which we have found to be more efficient and more sensitive than labeling using incorporation of radioactive nucleotides during the PCR reaction. Even though it is usually sufficient to label only one of the PCR primers, testing both primers is more efficient for polymorphism detection because, in some cases, the mobility of only one strand will be affected by a sequence change.

Most of the PCR primers designed using the criteria previously described will work under the standard PCR conditions that follow, but in some cases PCR conditions have to be optimized. If a primer pair does not amplify a product with the standard protocol, changing the Mg^{2+} concentration (within a range of 1.5–3 m*M* final concentration) or lowering the annealing temperature may result in more successful amplification. In contrast, an increase of the annealing temperature results in more specific PCR amplification in cases where the pattern complexity precludes unambigous scoring.

To obtain single-strand DNA, the fragments are denatured and electrophoresed on a nondenaturing polyacrylamide gel at 4°C. Standardized electrophoresis conditions are highly recommended and require special attention in order to ensure reproducible results because the pattern of single-strand migration is very sensitive to a number of different parameters including gel matrix composition, buffer concentration, and temperature. We use a 5% polyacrylamide gel matrix for the separation of the single-strand fragments. The SSCP electrophoresis runs for 2 hours at 40 W in a 4°C cold room. It is crucial to keep the temperature low because heating the gel will denature the intrastrand secondary structure that generates the polymorphic pattern of migration. If a means to cool the gel is unavailable, it can be run at low power for longer times (e.g., overnight). The gel is then transferred to a filter paper, dried, and autoradiographed overnight. Amplifying screens are *not* employed because they will result in diffuse bands and will decrease resolution. If maximal sensitivity is required (e.g., for very subtle polymorphisms), we have had success using a matrix sold by FMC BioProducts called MDE (Mutation Detection Enhancement) that is designed to enhance the separation of single-stranded products (Soto and Sukumar, 1992). However, the relative expense (compared with acrylamide gels) precludes its routine use in our lab. Additionally, the MDE matrix requires careful optimization of electrophoresis conditions because the results obtained are sensitive to even slight changes in the protocol.

A variety of strategies have been suggested for increasing the efficiency of detecting SSCPs, including varying the TBE buffer concentration, varying the concentration of cross-linker in the gel matrix, adding glycerol to the gel matrix, and other modifications. In our own experience, we have found that the most important variable for polyacrylamide gel analysis run in the cold is the buffer concentration, which should be 0.5× TBE. If increased resolution is required, analysis using MDE matrix has generally proven satisfactory.

III. Gene Mapping

This discussion of the utility of SSCP analysis for gene mapping in zebrafish presumes the availability of DNA from a mapping panel in which a sufficient number of markers have been typed such that assignment of a map position by linkage analysis is feasible. The general strategy for creating panels of this sort are illustrated by Johnson *et al.* (1996) and Knapik *et al.* (1996). The generation of diploid intercross or backcross progeny from a cross of divergent zebrafish strains is straightforward. The generation of mapping panels from haploid embryos is more complex; however, these data sets do not have heterozygous progeny, and the assignment of genotypes can, in many instances, be more conclusively scored (see Fig. 1).

Before a primer pair is selected for SSCP mapping analysis, it is tested for a polymorphism in parental DNA of the strains used to generate the mapping panel. It is often necessary to test several primer pairs (which ideally are nonoverlapping) to find one with an easily resolved polymorphism. If a scorable polymorphism is found, the mapping panel is tested, and the allele distribution for the gene locus is determined. The Mapmanager software (Manly, 1993; available at http//:mcbio.med.buffalo.edu/mapmgr.html) is highly recommended for recording genotype distribution data and for determining linkage. Originally developed for genetic mapping in the mouse, it is useful for the analysis of genetic mapping experiments using backcrosses, intercrosses, recombinant inbred (RI) strains, or radiation hybrids. It facilitates the storage, retrieval, and display of genotype data and includes tools for statistical analysis of the experimental results. Of particular utility is a simple interface for positioning new genes on a reference cross.

There are a number of potential difficulties in this procedure about which the investigator should be aware. In theory, a polymorphic fragment can be found in any expressed sequence as long as sufficient noncoding sequence is available. The efficiency of finding a polymorphic allele depends, to a considerable degree, on the size of the screened DNA region. If the 3′UTR is large enough, multiple primer pairs may be designed (which should ideally test minimally overlapping sequences) and tested for polymorphisms, or the use of intron or 5′UTR sequence can be employed for primer design.

Discerning whether an apparent SSCP is "real" can require some experience. It is not uncommon that minor bands will be variably detectable, and their presence may be incorrectly assessed as being strain-specific. A simple means to reduce this source of error is to perform duplicate reactions using the parental DNAs of the mapping cross and to load these samples in alternating sequence. A true strain-specific shift can generally be readily recognized. Complexity of the SSCP patterns may cause problems. In some cases, multiple independently segregating products are found for one primer pair. Again, testing several PCR primer combinations derived from a sequence of interest may clarify which polymorphism corresponds to the locus. Alternatively, it is possible to elute a band identified as polymorphic, reamplify the fragment, and determine its sequence.

In summary, SSCP mapping is an efficient tool to localize transcribed sequences in the zebrafish genome. Moreover, SSCP mapping is not restricted to fully characterized genes but can be used for localization of ESTs, similar to the efforts in human (Adams *et al.*, 1991) and mouse (Brady *et al.*, 1997). Because SSCP analysis can be used to map any unique genomic sequence, it is furthermore useful for analyzing sequences derived from the ends of large-insert clones, such as BACs, PACs, or YACs. For example, to assess possible chimerism in these clones, one can test that both ends colocalize. Additionally, in a chromosomal walk to a establish a contig over a region of interest, SSCP analyses may be helpful to determine the orientation of large-insert clones.

IV. Methods

A. Primer Labeling

1. Sufficient signal strength can obtained by labeling only one of the oligonucleotide primers, although maximal polymorphism detection is facilitated by labeling both. One microliter per reaction of a labeled primer mix prepared as follows will be used. The example is sufficient for 30 reactions:

24.2 μl dH_2O
3.0 μl 10× T_4 polynucleotide kinase buffer[2]
1.0 μl 20μM forward primer
1.0 μl 20μM reverse primer
1.2 μl γ^{32}P dATP, 6000Ci/mmole, 10mCi/μl
0.6 μl T_4 polynucleotide kinase (10 u/μl)

2. Incubate at 37°C for 35–45 minutes.

[2] 10× polynucleotide kinase buffer: 0.5M Tris–HCl pH 7.6; 0.1M $MgCl_2$; 0.05M Dithiothreitol; 1mM Spermidine–HCl; 1mM EDTA pH 8.0.

3. Heat inactivate the reaction at 68°C for 10minutes.

4. Keep reaction on ice until needed or store at −20°C.

B. PCR Reaction

1. All PCR reactions are carried out in 96-well plates with a final reaction volume of 12.5 μl, and 1μl of template DNA (16–50 ng) is added to each well. For mapping panels that are to be analyzed for many markers, it is convenient to aliquot template DNA into plates and allow it to air-dry. These plates can then be stored. The dried DNA is readily rehydrated during PCR.

2. Use 9.25 μl per reaction of a PCR master mix prepared as follows. The example is sufficient for 30 reactions:

205 μl	dH_2O
37.5 μl	10× PCR buffer[3]
2.5 μl	20μM forward primer
2.5 μl	20μM reverse primer
30 μl	χ^{32}P-labeled primer mix

3. Add 9.25 μl of the PCR master mix to each DNA template.

4. Mix well and overlay with 30 μl light mineral oil.

5. Mix on ice 0.5 μl of 10mM each dNTP (dATP, dCTP, dGTP, dTTP) and 1 unit of Taq polymerase for each PCR.

6. Perform hot-start PCR: Denature the PCR reaction at 94°C for 5 minutes; hold at 80°C.

7. Add 2.25 μl of dNTP/Taq mix to each reaction; mix well.

8. Amplify using appropriate thermocycling conditions for the primers used in the reaction. Our standard PCR conditions follows:

29–40 cycles:	annealing:	55°C for 2 minutes
	extension:	72°C for 3 minutes
	denaturing:	94°C for 1 minutes
final cycle:	annealing:	55°C for 2 minutes
	extension:	72°C for 7 minutes

9. Store PCR at −20°C or keep on ice until needed.

C. SSCP Gel Electrophoresis

1. Prepare gel mix for 5% polyacrylamide SSCP gel. Prepare 80 ml (sufficient for the Gibco BRL S2 sequencing gel apparatus used with a 0.4-mm spacer) as follows:

[3] 10× PCR buffer: 0.1M Tris–HCl pH 8.0; 0.5M *KCl; 0.02M* $MgCl_2$; 0.01% gelatin.

66 ml	dH_2O
4 ml	10× TBE (final concentration = 0.5×)
10 ml	acrylamide mix (38% acrylamide:2% Bis-acrylamide)
800 μl	10% APS (ammoniumperoxodisulfate) in H_2O, freshly prepared
30 μl	TEMED

2. Pour SSCP gel, insert comb, clamp well, and allow to polymerize for 2 hours or overnight. Avoid any bubbles in the gel.

3. Pre-equilibrate polymerized gel in gel apparatus with 0.5× TBE for at least 30 minutes at 4°C.

4. For each reaction, mix 2 μl PCR product and 4 μl USB stop solution solution (95% formamide; 0.05% bromphenol blue; 0.05% xylene cyanol; 0.02*M* EDTA pH 8.0) in a 96-well plate. Denature at 94°C for 5 minutes, transfer immediately onto ice.

5. Load 3–5 μl per reaction onto gel. Run the gel for 90–150 minutes (according to the size for PCR product) in 0.5× TBE at 4°C and 40 W with constant power.

6. Transfer the gel onto Whatman filter paper, cover with plastic wrap, and dry on a gel dryer for 30 minutes.

7. Expose the gel to X-ray film overnight; do not use amplifying screens.

References

Adams, M., Kelley, J., Gocayne, J., Dubnick, M., Polymeropoulos, M., Xiao, H., Merril, C., Wu, A., Olde, B. and Moreno, R. (1991). Complementary DNA sequencing: Expressed sequence tags and human genome project. *Science* **252,** 1651–1656.

Beier, D. R., Dushkin, H. and Sussman, D. J. (1992). Mapping genes in the mouse using single-strand conformation polymorphism analysis of recombinant inbred strains and interspecific crosses. *Proc. Natl. Acad. Sci. U.S.A.* **89,** 9102–9106.

Beier, D. R. (1993). Single-strand conformation polymorphism (SSCP) analysis as a tool for genetic mapping. *Mamm. Genome* **4,** 627–631.

Brady, K. P., Rowe, L. B., Her, H., Stevens, T. J., Eppig, J., Sussman, D. J., Sikela, J., and Beier, D. R. 1997. Genetic mapping of 262 loci derived from expressed sequences in a murine interspecific cross using single-strand conformational polymorphism analysis. *Genome Res.* **7,** 1085–1093.

Driever, W., Solnica-Krezel, L., Schier, A. F., Neuhauss, S. C. F., Malicki, J., Stemple, D. L., Stainier, D. Y. R., Zwartkruis, F., Abdelilah, S., Rangini, Z., Belak, J. and Boggs, C. (1996). A genetic screen for mutations affecting embryogenesis in zebrafish. *Development* **123,** 37–46.

Foernzler, D., Her, H., Knapik, E. W., Clark, M., Lehrach, H., Posthlethwait, J. H., Zon, L. I., and Beier, D. R. (1998). Gene mapping in zebrafish using single-strand conformation polymorphism analysis. *Genomics,* in press.

Haffter, P., Granato, M., Brand, M., Mullins, M. C., Hammerschmidt, M., Kane, D. A., Odenthal, J., van Eden, F. J. M., Jiang, Y.-J., Heisenberg, C.-P., Kelsh, R. N., Furutani-Seiki, M., Vogelsang, E., Beuchle, D., Schach, U., Fabian, C. and Nüsslein-Volhard, C. (1996). The identification of genes with unique and essential functions in the development of the zebrafish, *Danio rerio. Development* **123,** 1–36.

Johnson, S. L., Gates, M. A., Johnson, M., Talbot, W. S., Horne, S., Baik, K., Rude, S., Wong, J. R., and Posthlethwait, J. H. (1996). Centromere-linkage analysis and consolidation of the zebrafish genetic map. *Genetics* **142,** 1277–1288.

Knapik, E. W., Goodman, A., Atkinson, O. S., Roberts, C. T., Shiozawa, M., Sim, C. U., Weksler-Zangen, S., Trolliet, M. R., Futrell, C., Innes, B. A., Koike, G., McLaughlin, M. G., Pierre, L., Simon, J. S., Vilallonga, E., Roy, M., Chiang, P.-W., Fishman, M. C., Driever, W., and Jacob, H. J. (1996). A reference cross DNA panel for zebrafish (*Danio rerio*) anchored with simple sequence length polymorphisms. *Development* **123,** 451–460.

Lincoln, S. E., Daly, M. J., and Lander, E. S. (1991). PRIMER: A computer program for automatically selecting PCR primers. Available from primer@genome.wi.edu.

Manly, K. F. (1993). A Macintosh program for storage and analysis of experimental genetic mapping data. *Mamm. Genome* **4,** 303–313.

Orita, M., Suzuki, Y., Sekiya, T. and Hayashi, K. (1989). Rapid and sensitive detection of point mutations and DNA polymorphisms using the polymerase chain reaction. *Genomics* **5,** 74–879.

Soto, D., and Sukumar, S. (1992). Improved detection of mutations in the p53 gene in human tumors as single-stranded conformation polymorphs and double-stranded heteroduplex DNA. *PCR Methods Applic.* **2,** 96–98.

CHAPTER 12

Mapping Zebrafish Mutations by AFLP

David G. Ransom and Leonard I. Zon

Howard Hughes Medical Institute
Children's Hospital
Boston, Massachusetts 02115

I. Introduction

Amplified fragment length polymorphism (AFLP) is a robust and efficient method for selectively polymerase chain reaction (PCR)-amplifying genomic DNA restriction fragment length polymorphisms that exist between strains of organisms. This technology is a powerful way to identify markers that are closely linked to mutant loci and thereby facilitate positional cloning of the affected genes. AFLP has been used extensively to map mutations in plants (Rouppe van der Voort *et al.,* 1997; Simons *et al.,* 1997; Waugh *et al.,* 1997; Becker *et al.,* 1995; Thomas *et al.,* 1995), genotype bacteria (Janssen *et al.,* 1996; Lin *et al.,* 1996), and conduct quantitative trait analysis in plants (Nandi *et al.,* 1997; Quarrie *et al.,* 1997) and mammals (Otsen *et al.,* 1996). AFLP technology has also been adapted for use in differential display cloning of cDNAs (Money *et al.,* 1996;

METHODS IN CELL BIOLOGY, VOL. 60

0091-679X/99 $30.00

Habu *et al.,* 1997). We have had great success using AFLP in zebrafish to find markers that map within 0.3 cM of several mutant loci.

The AFLP methods described in this chapter are based primarily on the work of Vos *et al.* (1995). We also describe zebrafish DNA isolation methods developed in our laboratory. The patents for AFLP technology are owned by its developers at Keygene N.V. Both, Life Technologies and Perkin Elmer market kits licensed for AFLP of plant DNAs, and they maintain web sites describing the technology. These commercial kits also work well with zebrafish DNA. We chose to purchased reagents from several vendors to create our own kit.

AFLP is used to identify markers close to a mutation of interest after first testing the relevant markers that have been previously placed on maps of the zebrafish genome. The goal is to identify a marker within at least 1 cM of a mutation in order to facilitate the shortest possible walk to the gene using genomic clones. Examples of mapping and position cloning projects are described in detail by Schier and Talbot in Chapter 14 (this volume). The zebrafish genetic map contains random amplified polymorphic DNAs (RAPD) markers, genes (Johnson *et al.,* 1994, 1996; Postlethwait *et al.,* 1994), and a growing number of CA repeat markers (Knapik *et al.,* 1996, 1998). CA repeat markers or RAPDs can be use for bulk segregant analysis comparing pools of wild-type and mutant embryos using methods described by Postlethwait in Chapter 8 (this volume) and Knapik in Chapter 9 (this volume). Bulk segregant analysis of the progeny of AB/SJD or TU/wik hybrid mothers is used to map a mutation to a linkage group. If a candidate marker is not found well within 1 cM of the mutation, then pools of mutant haploids and wild-type haploids are collected and used for AFLP to identify closer markers. Production of haploid embryos is described by Walker in Chapter 3 (volume 59).

The AFLP technique consists of four basic steps: (1) complete digestion of genomic DNA from the offspring of AB/SJD hybrid females, (2) ligation of adapters, (3) selective amplification of a subset of the thousands of possible bands, and (4) gel analysis of the reaction products to detect bands that segregate with the mutation being studies. These steps are diagrammed in Fig. 1. These steps are briefly described as follows.

1. Genomic DNA from separate pools of wild-type and mutant embryos is digested to completion with two restriction enzymes, EcoRI (6-bp recognition site GAATTC) and MseI (4-bp recognition site TTAA). Based on the 1700-megabase size of the zebrafish genome and the average base composition (Postlethwait *et al.,* 1994), it is predicted that EcoRI cuts every 2 Kb, whereas MseI cuts every 300–400 bp to generate a very large number of fragments (>100,000). A fraction of these are MseI–EcoRI or EcoRI–EcoRI fragments.

2. Double-stranded oligonucleotide adapters are next ligated to the sticky ends of the pools of restricted DNA fragments.

3. Primers are then used to amplify a subset of the DNA fragments (steps 3 and 4 in Fig. 1). These primers are complementary to the MseI or EcoRI adapters

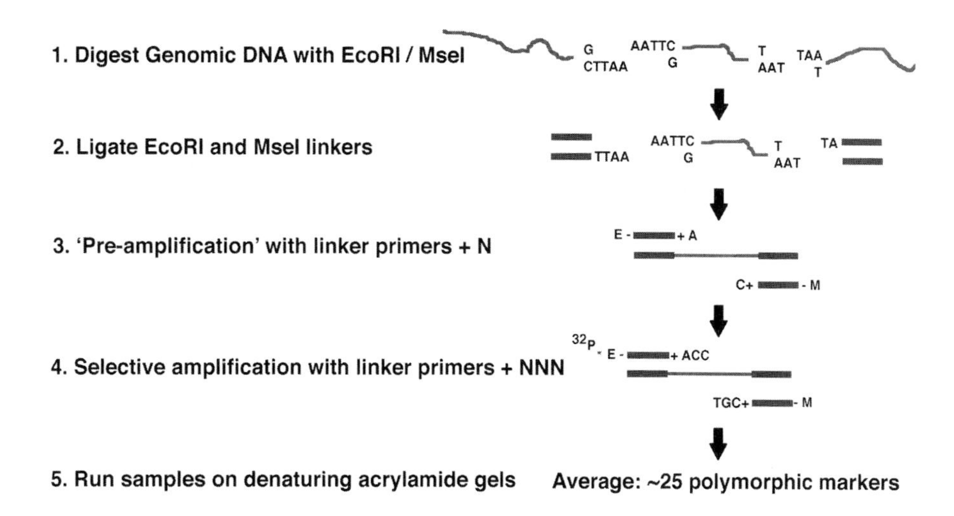

Fig. 1 The AFLP method of PCR-based DNA fingerprinting consists of major steps outlined here. Either haploid or diploid zebrafish DNA can be tested by AFLP. Separate pools of wild-type and mutant fish are processed in parallel, and the amplified bands are compared to identify bands that appear in only one pool.

and also contain three base extensions at the 3′-ends. The three base extensions specifically amplify the matching genomic DNA sequences. Taq DNA polymerase cannot extend DNAs if there are mismatches at the 3′-end of a primer. Thus, DNA fragments complementary to the 3′ extensions are selectively amplified, and a subset of the entire genome is tested in each reaction.

4. The amplified bands from PCR of each primer combination are then separated on a denaturing polyacrylamide gel-like sequencing reactions. The gels are dried and exposed to X-ray film to produce a genetic fingerprint of individual zebrafish or pools. Using radiolabeled EcoRI primers and three base extensions on both primers, we typically observe 50–100 bands per primer pair. Only a subset of these total bands are polymorphic between AB or TU and SJD strains of zebrafish. On average, 25 strain-specific bands are amplified per primer pair. Most segregating AFLP markers are dominant in one of the two mapping strains. Thus, 256 primer pairs can be used to rapidly test an estimated 6400 markers in haploids. Many-fold more loci are tested per reaction than are tested by RAPDs (Postlethwait *et al.*, 1994). If it is assumed that these markers are evenly distributed across the zebrafish genome, then 6400 markers tested on the 3000-cM genome allows identification of a markers at a density greater than 1 marker per centimorgan (Johnson *et al.*, 1996, Postlethwait *et al.*, 1994).

II. Extraction of Zebrafish Genomic DNA

The DNA extracted for AFLP must be clean enough for complete enzyme digestion. We have developed the following method to rapidly prepare dirty DNA from large numbers of individual embryos while retaining aliquots that can be stored and purified more extensively at a later time. Dirty prep DNA can be typed with CA repeats, RAPDs, or single strand conformation polymorphisms (SSCPs) to identify recombinant individuals. Samples saved from the recombinants can then be purified for AFLP.

Lysis Reagents

1. Lysis buffer (10m*M* Tris pH 8.3, 50m*M* KCl, 1.5m*M* $MgCl_2$, 0.3% NP40, 0.3% Tween-20, 1.8 mg/ml proteinase K)
2. DNAzol reagent (Gibco/BRL #10503)
3. 1*M* Hepes
4. 8m*M* NaOH
5. Glycogen 20 mg/ml (Boehringer Manheim #901393)

Extraction of Both Clean and Dirty Genomic DNA from the Same Samples

1. Sort embryos or larvae and collect up to 200 animals in 5-ml glass vial. Fix by washing twice in methanol and store in methanol at −20°C.

2. Put fixed animals into a Petri dish to separate individuals and transfer them into wells of 96-well plates using a transfer pipette. Remove methanol with a pulled pasteur pipette.

3. Dry the embryos for 10 min at 70°C by putting the uncapped plate in a thermocycler.

4. Add 50 μl of lysis buffer, and digest overnight at 55°C in an oven.

5. Remove debris by using a multichannel pipette set at 5 μl.

6. Split each lysate between dirty and clean preps. Transfer 25 μl of each lysate to a new plate. Add 50 μl of DNAzol to each well of the clean prep plate. Do not add DNAzol to the dirty preps. Store the DNAzol samples at 4°C. *Do not freeze.*

7. For the dirty prep portion, boil the embryos at 95°C for 10 minutes in a thermocycler and cool to 4°C. Dilute the sample by 1:30 for PCR. The dirty lysates can be stored at −20°C.

8. After recombinants have been identified, the appropriate stored DNAzol lysates can be processed. Transfer lysates to 2-ml microfuge tubes and add 1 μl of glycogen (20 μg/μl).

9. Precipitate DNA from the lysates by adding 25 μl of 100% ethanol. Mix samples by inversion, and incubate for 1–5 minutes at room temperature.

10. Centrifuge at 4K rpm for 2 minutes at room temperature. Remove the supernatant while avoiding the pellet of DNA.

11. Wash the pellet two times three with 1 ml of 95% ethanol. Suspend the DNA in the ethanol by inverting the tube three times. Respin the tubes at 4K rpm for 2 minutes to pellet the DNA.

12. Air-dry the DNA 15–30 minutes in an open tube. Do not overdry DNAs.

13. Dissolve DNA in 25 μl of 8mM NaOH. Pipette gently and do not vortex. Make up 8mM NaOH fresh each month.

14. Neutralize DNA solution by adding 1 μl of 1M HEPES.

15. Use as a stock for AFLP or store at 4°C. *Do not freeze.*

III. AFLP Methods

A. DNA Digestion

Assemble DNA digests on ice. We use thermocyclers (MJ Research #PTC100) and 200-μl tubes or 96-well PCR plates (Robbins Scientific #1055-00-0) for all steps. It is important to mix pooled DNAs together first and then to digest in order to ensure equal treatment of each individual sample that is pooled for bulk segregant analysis. We use lysates from 20 haploid animals each to form pools.

Digest Reagents

1. Eco RI/Mse I Mixture. 4 U MseI + 4 U EcoRI/μl. Add 10 μl stock Eco RI (NEB 101CS, 100 U/μl) to 250 μl stock MseI (NEB 525L, 4 U/μl).

2. One-Phor-All+ Buffer (Pharmacia #27-0901)

Digest Reaction

Reagents	Volumes
10 × One-Phor-All+	2.5 μl
DNA (from pools or individuals)	18 μl
EcoRI/MseI Mix	1 μl
ddH_2O	3.5 μl
Total	25 μl

Incubate in a thermocycler at 37°C for 2 hours, 72°C for 10 minutes, to 4°C > 5 minutes.

B. Ligation of Adapters

The adapter/ligation solution can be made in bulk and stored for later use. The final adapter/ligation solution is E-Adapter, 5 pmoles,/M-Adapter, 50 pmoles, 0.4mM ATP, 10mM TrisCl 7.5, 10mM MgOAc, 50mM KOAc in 1.44 ml. This is enough solution for 60 AFLP ligations. All primers were purchased as custom orders from Gibco/BRL.

Ligation Reagents

1. Linker Sequences

E-ADAPTER.1	5′-CTC GTA GAC TGC GTA CC-3′
E-ADAPTER.2	5′-AAT TGG TAC GCA GTC TAC-3′
M-ADAPTER.1	5′-GAC GAT GAG TCC TGA G-3′
M-ADAPTER.2	5′-TAC TCA GGA CTC AT-3′

2. Preparation of Adapter Linkers. Make one reaction each of E- and M-adapter per tube of adapter/ligation solution.

E-Adapter (5 pmoles/μl)		*M-Adapter* (50 pmoles/μl)	
Reagents	Volumes	Reagents	Volumes
E-Adapter.1 (0.32 μg/μl)	5 μl	M-Adapter.1 (1.56 μg/μl)	10 μl
E-Adapter.2 (0.34 μg/μl)	5 μl	M-Adapter.2 (1.33 μg/μl)	10 μl
ddH_2O	44 μl	ddH_2O	34 μl
10× One-Phor-All+ buffer	6 μl	10× One-Phor-All+ buffer	6 μl

Anneal primers by the following method: Cycle 94°C, 4 minutes; ramp 90 minutes to 45°C, 30 second; ramp 90 minutes to 4°C.

3. Preparation Adapter/Ligation Solution: 1.44 ml

Reagents	Volumes
E-Adapter	60 μl
M-Adapter	60 μl
ATP 100 mmol/l	6.0 μl
10× One-Phor-All	131 μl
ddH_2O	1183.0 μl
Store at −20°C	

Ligation Reaction

1. Add the following to digested DNA reaction:

Reagents	Volumes
Adapter ligation solution	24 μl
T4 DNA ligase (BRL #15224-017)	1 μl

2. Incubate in a thermocycler at 20°C for 2 hours.

3. Dilute samples 1:3 reaction by adding 100 μl of TE buffer (10m*M* TrisCl 8.0, 0.1m*M* EDTA) and mix. Note: This is 1/10 of the EDTA in standard TE. The diluted solution is sufficient for 30 pre-amplifications or about 7500 AFLPs!

4. Store at −20°C.

C. Preamplification

Preamplification with primers complementary to the linkers plus one base increases the specificity of the final specific amplification. We routinely use the four pre-amp primers that follow. Taq polymerase from any of the major manufacturers works well.

Pre-amp Reagents

1. Pre-amp primers

E-PRE-N 5′-GAC TGC GTA CCA ATT CN-3′
M-PRE-N 5′-GAT GAG TCC TGA GTA AN-3′

Our standard set of pre-amp primers follows:

E-PRE-A 5′-GAC TGC GTA CCA ATT CA-3′
E-PRE-T 5′-GAC TGC GTA CCA ATT CT-3′
M-PRE-C 5′-GAT GAG TCC TGA GTA AC-3′
M-PRE-G 5′-GAT GAG TCC TGA GTA AG-3′

2. Pre-amp Primer Mix. Primers are at a concentration of 28 ng/μl each, 0.26m*M* each dNTP in 1.080 ml (30 reactions).

Reagents	Concentration	Volumes
M-Pre-N	1 μg/μl	28 μl
E-Pre-N	1 μg/μl	28 μl
dNTPs	100m*M*	3 μl each
ddH_2O		1012 μl

3. 10× PCR buffer without Mg and 25 m*M* $MgCl_2$ (Boehringer Manheim #1699121). Reactions should be 1.5m*M* $MgCl_2$ final concentration.

Pre-amp Reaction. It is important NOT to preheat samples before PCR. Assemble reactions on ice; then put reactions in a room temperature thermocycler and start cycling. This will prevent loss of the nonphosphorylated and therefore unligated strands of the linkers.

1. Assemble reactions on ice.

Reagents	Volumes
Diluted template DNA	5 μl
Pre-amp primer mix	37 μl
10× PCR buffer	5 μl
25m*M* Mg	3 μl
Taq polymerase 5U/μl	0.1 μl
Total	50 μl

2. PCR for 20 cycles at: 94°C, 30 seconds; 56°C, 60 seconds; 72°C, 60 seconds; soak at 4°C.

3. Dilute 1:30 by adding 5 μl of pre-amplification reaction to 145 μl of TE. Make multiple tubes in a batch. Each 150-μl tube is sufficient for 30 specific amplifications. One 96-well plate will hold 24 sets. For four pools, the stocks should be arrayed wt1, mt1, wt2, mt2 to aid plate and gel loading.

4. Store at −20°C.

D. Primer Labeling

There are tens of thousands of potential enzyme and primer pair combinations that can be used for AFLP. We have identified 256 EcoRI/MseI primer pair combinations that give an average of 50–100 bands. These sequences of are shown in Fig. 2. They will test approximately 6400 loci when used with haploid DNAs or approximately 3200 loci if used with diploid embryos.

Primer Labeling Reagents

1. T4 Polynucleotide Kinase. 10 U/μl (BRL 18004-010).

2. 5X Kinase Buffer. (BRL 18004-010) 350m*M* TrisCl 7.6, 50m*M* MgC12, 500m*M* KCl, 5m*M* BMeOH.

3. EcoRI Primers. 27.8 ng/μl in H_2O. See Fig. 2.
EcoRI Primers. #E-NNN 5′-GAC TGC GTA CCA ATT CNN N-3′
EcoRI Primer. 0.5 ml (55.25 μl labelings, 2750 AFLP reactions)

Reagents	Volumes
EcoRI Primer (1 μg/μl)	28 μl
ddH_2O	972 μl

Primer Labeling Reaction

1. Assemble in PCR tubes. This enough for 50 specific amplifications.

Reagents	Volumes
EcoRI primer (pick one)	9 μl
5× kinase buffer	5 μl
Gamma-32P-ATP 6000Ci/m	10 μl
T4 polynucleotide kinase	1 μl
Total	25 μl

2. Incubate in a thermocycler at 37°C for 1 hour; 72°C, 10 minutes; soak at 4°C.

E. Selective Amplification

Selective amplification is performed using the three base extension primers on the preamplified DNA stocks. The mixes and volumes are designed to facilitate use of multichannel pipettes.

Selective Amplification Reagents

1. MseI Primers: 6.7 ng/μl, 0.225m*M* each dNTP.
   MseI primers: #5′-CAA GAT GAG TCC TGA GTA ANN N-3′
   MseI primer mix: 1.5 ml (330 AFLP reactions)

Reagents	Volumes
MseI Primer (1 μg/μl)	10 μl
dNTPs 100m*M*	3.5 μl each
ddH_2O	1453 μl

2. Mix 1. Add the following to a PCR tube per five reactions.

Reagents	Volumes
Labeled EcoRI primer	2.5 μl
MseI primer mix	22.5 μl
Total	25 μl

3. Mix 2. Add the following to a 1.5-ml tube per ten reactions.

Reagents	Volumes	X 11 for a 96-well plate
ddH_2O	67 μl	737 μl
10× PCR buffer	20 μl	220 μl
25m*M* Mg	12 μl	132 μl
Taq polymerase	1 μl	11 μl

Specific Amplification Reaction

1. Assemble AFLP reaction on ice then spin down. It is best to add DNAs in an array ready for the multichannel gel loader.

Reagents	Volumes
Diluted template DNA	5 μl
Mix 1 (primers/dNTPs)	5 μl
Mix 2 (Taq/buffer)	10 μl
Total	20 μl

A	1 M-CAA	2 M-CAC	3 M-CAG	4 M-CAT	5 M-CTA	6 M-CTC	7 M-CTG	8 M-CTT
1 E-AAC	A1	A2	A3	A4	A5	A6	A7	A8
2 E-AAG	A9	A10	A11	A12	A13	A14	A15	A16
3 E-ACA	A17	A18	A19	A20	A21	A22	A23	A24
4 E-ACC	A25	A26	A27	A28	A29	A30	A31	A32
5 E-ACG	A33	A34	A35	A36	A37	A38	A39	A40
6 E-ACT	A41	A42	A43	A44	A45	A46	A47	A48
7 E-AGC	A49	A50	A50	A52	A53	A54	A55	A56
8 E-AGG	A57	A58	A59	A60	A61	A62	A63	A64

E	17 M-GAA	18 M-GAC	19 M-GAG	20 M-GAT	21 M-GTA	22 M-GTC	23 M-GTG	24 M-GTT
1 E-AAC	E1	E2	E3	E4	E5	E6	E7	E8
2 E-AAG	E9	E10	E11	E12	E13	E14	E15	E16
3 E-ACA	E17	E18	E19	E20	E21	E22	E23	E24
4 E-ACC	E25	E26	E27	E28	E29	E30	E31	E32
5 E-ACG	E33	E34	E35	E36	E37	E38	E39	E40
6 E-ACT	E41	E42	E43	E44	E45	E46	E47	E48
7 E-AGC	E49	E50	E50	E52	E53	E54	E55	E56
8 E-AGG	E57	E58	E59	E60	E61	E62	E63	E64

Fig. 2 The grids show 256 primers pairs that amplify zebrafish DNA well and test an estimated 6400 loci. The primer pairs were given logical names noted in the boxes to facilitate record keeping. The pre-amp primer backbones are: E-Pre-N: 5′-GACTGCGTACCAATTCN -3′ and M-Pre-N: 5′-GATGAGTCCTGAGTAAN -3′. The specific amplification primer back bones are #-E-NNN: 5′-GACTGCGTACCAATTCNNN -3′ and #-M-NNN 5′-GATGAGTCCTGAGTAANNN -3′.

I	1 M-CAA	2 M-CAC	3 M-CAG	4 M-CAT	5 M-CTA	6 M-CTC	7 M-CTG	8 M-CTT
17 E-TAC	I1	I2	I3	I4	I5	I6	I7	I8
18 E-TAG	I9	I10	I11	I12	I13	I14	I15	I16
19 E-TCA	I17	I18	I19	I20	I21	I22	I23	I24
20 E-TCC	I25	I26	I27	I28	I29	I30	I31	I32
21 E-TCG	I33	I34	I35	I36	I37	I38	I39	I40
22 E-TCT	I41	I42	I43	I44	I45	I46	I47	I48
23 E-TGC	I49	I50	I50	I52	I53	I54	I55	I56
24 E-TGG	I57	I58	I59	I60	I61	I62	I63	I64

M	17 M-GAA	18 M-GAC	19 M-GAG	20 M-GAT	21 M-GTA	22 M-GTC	23 M-GTG	24 M-GTT
17 E-TAC	M1	M2	M3	M4	M5	M6	M7	M8
18 E-TAG	M9	M10	M11	M12	M13	M14	M15	M16
19 E-TCA	M17	M18	M19	M20	M21	M22	M23	M24
20 E-TCC	M25	M26	M27	M28	M29	M30	M31	M32
21 E-TCG	M33	M34	M35	M36	M37	M38	M39	M40
22 E-TCT	M41	M42	M43	M44	M45	M46	M47	M48
23 E-TGC	M49	M50	M50	M52	M53	M54	M55	M56
24 E-TGG	M57	M58	M59	M60	M61	M62	M63	M64

Fig. 2 (*Continued*)

4. PCR cycle profile: 13 cycles of [94°C, 30 seconds; 65°C − 0.7° per cycle, 30 seconds; 72°C, 60 seconds] then 23 cycles of [94°C, 30 seconds; 56°C, 30 seconds; 56°C, 30 seconds; 72°C, 60 seconds] for 23 cycles, then soak at 4°C.

F. Gel Analysis

We use radiolabeled primers and autoradiography for analysis of AFLP results. We find that denaturing polyacrylamide gels made with LongRanger Gel (FMC 50615) provide good resolution of bands from 50–500 bp and do not need to be fixed before drying. Dry gels give sharper banding patterns. It is also possible to use automated sequencer to perform nonradioactive AFLP. An example of results from a typical experiment mapping the *moonshine* mutant (Ransom *et al.*, 1996) are shown in Fig. 3.

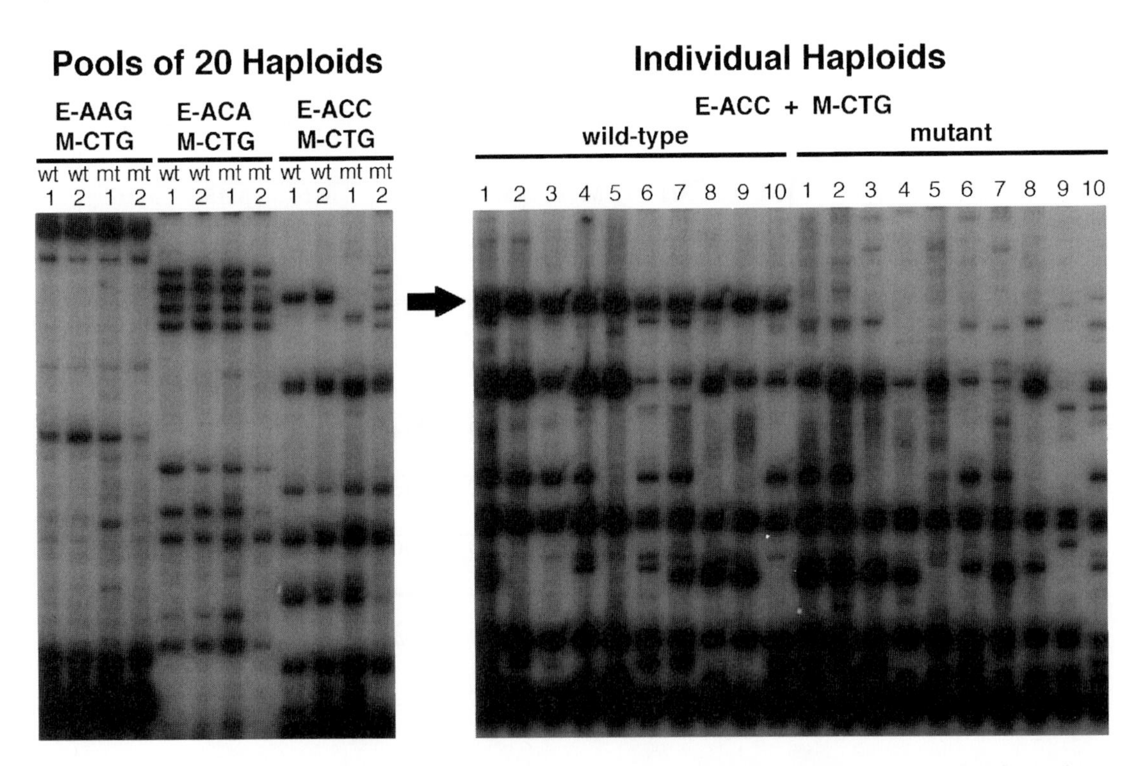

Fig. 3 An example of bulk segregant analysis of zebrafish *moonshine* mutants. This figure shows 20% of the length of a typical AFLP gel. In order to identify markers close to the *moonshine* gene, AFLP was performed on two pools of 20 haploid mutants and two pools of wild-type haploids. An arrow indicates a link band that was further tested on individual haploid embryos. Note that several unlinked polymorphic markers are visible in the individuals that are amplified by the same primer pair.

Gel Analysis

1. After PCR add 20 μl of squencing stop solution (95% formamide, 20 mM EDTA).

Reagents	Volumes
Formamide	38 ml
Bromphenol blue	0.02 g
Xylene cyanole	0.02 g
0.5*M* EDTA	1.6 ml

2. Before running on a gel, incubate in a thermocycler at 94°C for 4 minutes; soak at 4°C; place on ice.
3. Denaturing gel. Per LongRanger Gel.

Reagents	Volumes
Urea	21 g
10× TBE	6 ml
Longer Ranger Mix	5 ml
dH_2O	to 50 ml
TEMED	0.025 ml
Fresh 10% APS	0.25 ml

4. Prerun in 0.6× TBE for about 20 minutes. Load 3 μl of sample per lane. Run in 0.6× TBE about 2 hours at 55 W constant power until the bromphenol blue runs off the gel and xylene cyanole is about 10 cM from the bottom.
5. Do not fix. Dry. Expose O/N or about 12 hours without screens.

G. Band Recovery

After linked AFLP bands have been identified by bulk segregant analysis. AFLP should done on individuals to confirm the linkage. Bands can then be cut out of the gels and subcloned.

Recovery Reagents. Elution buffer:

0.5*M* ammonium acetate

1m*M* EDTA pH 8.00.2% SDS

Band Recovery

1. Punch holes through autorad into gel with a needle. Cut out four bands from individual AFLPs.
2. Add 500 μl of elution buffer and incubate at 55°C for 1 hour.

3. Spin down paper at 14K for 5 minutes. Remove 250 μl of sup. Respin to remove all debris.

4. Add an equal volume of EtOH and place on ice for 30 minutes. Glycogen may be used as a carrier if desired, but there is generally enough residual acrylamide to act as a carrier.

5. Spin 14,000 rpm for 15 minutes at 4°C.

6. Wash twice with 1 ml 70% EtOH.

7. Speedvac dry and resuspend in 50 μl TE.

8. Use 10 μl of DNA per 100 μl PCR with 100 pmol (~1 μg) each 3 base extension AFLP primer.

9. Run half of reamplified PCR product on an agarose TAE gel to cut out and subclone the band.

H. Conversion of AFLPs to SSCPs

It is often necessary to convert AFLP bands to SSCP assays. SSCPs can be performed on crude DNA preps and are useful for detecting polymorphisms in different families or strains of zebrafish. To design an SSCP assay, AFLP bands are subcloned and then sequenced, and 24-mer primers are made that amplify 200–300 bp of genomic DNA. The amplified products are run on cold nondenaturing gels to assess secondary structure conformations caused by differing sequences. SSCP mapping is described in detail by Beier and Foernzler in Chapter 10 (this volume).

SSCP Reagents

1. SSCP PCR Mix

Reagents	Volumes	
ddH_2O	98.5 ml	
1*M* $MgCl_2$	0.23 ml	
1*M* Tris pH 8.4	1.5 ml	
4*M* KCl	7.5 ml	
0.1% gelatin	1.5 ml	
100mg/ml BSA	0.150 ml	
100m*M* dATP	0.150 ml	(Boehringer Manheim #1051-440)
100m*M* dCTP	0.150 ml	(Boehringer Manheim #1051-458)
100m*M* dGTP	0.150 ml	(Boehringer Manheim #1051-466)
100m*M* dTTP	0.150 ml	(Boehringer Manheim #1051-482)

SSCP Reactions

1. Mix the following using stock solutions and a repeating pipettor.

Reagents	Volumes
DNA (1:33 to 1:50 dilution of haploids)	2 μl
RAPD PCR Mix	8 μl
Primers (50 ng per primer)	2 μl
*32*P dCTP (6000 Ci/mmol)	0.1 μl
Taq Polymerase 5U/μl	0.1 μl
Total	12 μl

2. Amplify for 40 cycles of (94°C for 3 minutes, 94°C for 30 seconds, anneal 10°C below primer Tm for 30 seconds, 72°C for 1 minute) then 72°C for 10 minutes and then hold at 4°C.
3. Add 5 μl of stop solution to PCR reaction.
4. Denature for 5 minutes at 98°C.

SSCP Gels

Reagents	Volumes
ddH_2O	66 ml
10× TBE	4 ml
Long Ranger Mix	10 ml
10% APS	800 μl
TEMED	30 μl

1. Load 3 μl of SSCP samples on nondenaturing gels.
2. Run in 1 × TBE running buffer. To run in the cold room, cool 20 minutes at 4°C and run at 50 W at 4°C for 2 hours. To run at room temperature, run at 4 W for 8–10 hours.

IV. General Considerations

AFLP is the fastest and most cost-effective currently available technique for identifying a high density of genetic markers linked to a mutant locus of interest. The method first described by Vos *et al.* (1995) has been successfully applied to mapping mutations in a broad range of organisms including zebrafish. Mapping AFLP markers does not require prior sequence characterization of the target genome, in contrast to the sequencing effort required for making CA markers (Knapik *et al.,* 1996, 1998). AFLP markers are also robust and reproducible within families. However, using specific AFLPs is not as consistently informative among different nonisogenic strains as using CA repeat markers (Knapik *et al.,* 1996, 1998). This problem can be overcome by subcloning the AFLP band of interest and developing a SSCP mapping assay. SSCPs are frequently informative in a wide range of zebrafish strains. AFLP PCR also tests tenfold more loci per reaction than RAPD primers (Johnson *et al.,* 1994, 1996; Postlethwait *et al.,* 1994). The relatively costly and labor-intensive clean DNA prep needed for AFLP is also a drawback in comparison to RAPD and CA mapping. We have adapted a extraction method that works around this problem by splitting samples for both clean and dirty preps.

A typical AFLP mapping project begins with a mutant that has been mapped to a linkage group by bulk segregant analysis of CA markers. Zebrafish are scored and sorted as mutant or wild-type based on phenotype. Flanking markers are then used to type animals and identify misscored haploids and true recombinants. Mapping with haploids doubles the number of informative AFLP markers that can be tested because greater than 90% of AFLP markers are dominant. Therefore, AB dominant markers are not informative in scoring the diploid progeny of a AB/SJD hybrid cross. Both diploid mutants and heterozygotes contain an AB dominant band, so no linkage can be detected. AFLP is performed on 80 total individuals in two pools of 20 wild-type phenotype zebrafish and two pools of 20 mutants. The use of multiple pools decreases the incidence of picking false positive bands that do not reproduce in two pools. Testing more pools with more recombinant individuals also increases the likelihood of distinguishing tightly linked markers. Using 256 primer pairs on four pools requires 1024 reactions or eleven, 96-well plates. These 1024 reactions can be run out on 16 sequencing gels. These 16 sequencing gels would test an estimated 6400 loci and identify markers at an average density of better than one marker per centimorgan. The resolution of this fine map depends on the number of animals tested. The DNA isolation methods described in this chapter allow a panel of 2000 animals to be screened efficiently. Linked markers are confirmed by AFLP on panels of DNAs from individuals that have recombinations near the mutation. Thus, AFLP mapping is a powerful and efficient method to create a fine map of a mutant locus and rapidly identify a marker that can be used to begin a position cloning project.

Acknowledgments

We thank Nathan Bahary, Adriana Donovan, and the other members of our laboratory for helpful discussions and technical advice. We also thank Chi-Bin Chien (MPI Tübingen) for suggestions to improve zebrafish DNA isolation methods.

References

Becker, J., Vos, P., Kuiper, M., Salamini, F. and Heun, M. (1995). Combined mapping of AFLP and RFLP markers in barley. *Mol. Gen. Genet.* **249,** 65–73.

Habu, Y., Fukada-Tanaka, S., Hisatomi, Y., and Iida, S. (1997). Amplified restriction fragment length polymorphism-based mRNA fingerprinting using a single restriction enzyme that recognizes a 4-bp sequence. *Biochem. Biophys. Res. Commun.* **234,** 516–521.

Janssen, P., Coopman, R., Huys, G., Swings, J., Bleeker, M., Vos, P., Zabeau, M., and Kersters, K. (1996). Evaluation of the DNA fingerprinting method AFLP as an new tool in bacterial taxonomy. *Microbiology* **142,** 1881–1893.

Johnson, S. L., Gates, M. A., Johnson, M., Talbot, W. S., Horne, S., Baik, K., Rude, S., Wong, J. R., and Postlethwait, J. H. (1996). Centromere-linkage analysis and consolidation of the zebrafish genetic map. *Genetics* **142,** 1277–1288.

Johnson, S. L., Midson, C. N., Ballinger, E. W. and Postlethwait, J. H. (1994). Identification of RAPD primers that reveal extensive polymorphisms between laboratory strains of zebrafish. *Genomics* **19,** 152–156.

Knapik, E. W., Goodman, A., Atkinson, O. S., Roberts, C. T., Shiozawa, M., Sim, C. U., Weksler-Zangen, S., Trolliet, M. R., Futrell, C., Innes, B. A., Koike, G., McLaughlin, M. G., Pierre, L., Simon, J. S., Vilallonga, E., Roy, M., Chiang, P. W., Fishman, M. C., Driever, W., and Jacob, H. J. (1996). A reference cross DNA panel for zebrafish (Danio rerio) anchored with simple sequence length polymorphisms. *Development* **123,** 451–460.

Knapik, E. W., Goodman, A., Ekker, M., Chevrette, M., Delgado, J., Neuhauss, S., Shimoda, N., Driever, W., Fishman, M. C. and Jacob, H. J. (1998). A microsatellite genetic linkage map for zebrafish (Danio rerio). *Nat. Genet.* **18,** 338–343.

Lin, J. J., Kuo, J., and Ma, J. (1996). A PCR-based DNA fingerprinting technique: AFLP for molecular typing of bacteria. *Nucleic Acids Res.* **24,** 3649–3650.

Money, T., Reader, S., Qu, L. J., Dunford, R. P. and Moore, G. (1996). AFLP-based mRNA fingerprinting. *Nucleic Acids Res.* **24,** 2616–2617.

Nandi, S., Subudhi, P. K., Senadhira, D., Manigbas, N. L., Sen-Mandi, S. and Huang, N. (1997). Mapping QTLs for submergence tolerance in rice by AFLP analysis and selective genotyping. *Mol. Gen. Genet.* **255,** 1–8.

Otsen, M., den Bieman, M., Kuiper, M. T., Pravenec, M., Kren, V., Kurtz, T. W., Jacob, H. J., Lankhorst, A., and van Zutphen, B. F. (1996). Use of AFLP markers for gene mapping and QTL detection in the rat. *Genomics* **37,** 289–294.

Postlethwait, J. H., Johnson, S. L., Midson, C. N., Talbot, W. S., Gates, M., Ballinger, E. W., Africa, D., Andrews, R., Carl, T., and Eisen, J. S. (1994). A genetic linkage map for the zebrafish. *Science* **264,** 699–703.

Quarrie, S. A., Laurie, D. A., Zhu, J., Lebreton, C., Semikhodskii, A., Steed, A., Witsenboer, H., and Calestani, C. (1997). QTL analysis to study the association between leaf size and abscisic acid accumulation in droughted rice leaves and comparisons across cereals. *Plant Mol. Biol.* **35,** 155–165.

Ransom, D. G., Haffter, P., Odenthal, J., Brownlie, A. J., Vogelsang, E., Kelsh, R. N., Brand, M., van Eeden, F. J. M., Furutani-Seiki, M., Granato, M., Hammerschmidt, M., Heisenberg, C.-P., Jiang, Y.-J., Kane, D. A., Mullins, M. C., and Nusslein-Volhard, C. (1996). Characterization of zebrafish mutants with defects in embryonic hematopoiesis. *Development* **123,** 311–319.

Rouppe van der Voort, J. N., van Zandvoort, P., van Eck, H. J., Folkertsma, R. T., Hutten, R. C., Draaistra, J., Gommers, F. J., Jacobsen, E., Helder, J., and Bakker, J. (1997). Use of allele specificity of comigrating AFLP markers to align genetic maps from different potato genotypes. *Mol. Gen. Genet.* **255,** 438–447.

Simons, G., van der Lee, T., Diergaarde, P., van Daelen, R., Groenendijk, J., Frijters, A., Buschges, R., Hollricher, K., Topsch, S., Schulze-Lefert, P., Salamini, F., Zabeau, M., and Vos, P. (1997). AFLP-based fine mapping of the Mlo gene to a 30-kb DNA segment of the barley genome. *Genomics* **44,** 61–70.

Thomas, C. M., Vos, P., Zabeau, M., Jones, D. A., Norcott, K. A., Chadwick, B. P., and Jones, J. D. (1995). Identification of amplified restriction fragment polymorphism (AFLP) markers tightly linked to the tomato Cf-9 gene for resistance to Cladosporium fulvum. *Plant J.* **8,** 785–794.

Vos, P., Hogers, R., Bleeker, M., Reijans, M., van de Lee, T., Hornes, M., Frijters, A., Pot, J., Peleman, J., and Kuiper, M. (1995). AFLP: A new technique for DNA fingerprinting. *Nucleic Acids Res.* **23,** 4407–4414.

Waugh, R., Bonar, N., Baird, E., Thomas, B., Graner, A., Hayes, P. and Powell, W. (1997). Homology of AFLP products in three mapping populations of barley. *Mol. Gen. Genet.* **255,** 311–321.

CHAPTER 13

Zebrafish Expressed Sequence Tags and Their Applications

Zhiyuan Gong

Department of Biological Sciences
National University of Singapore
Singapore 119260

I. Introduction

In recent years, the zebrafish, *Danio rerio,* has gained increasing popularity as it has become a new vertebrate model for developmental and genetic analyses. As a model organism, this species offers several major advantages. First, it is an

0091-679X/99 $30.00

excellent organism for embryological studies because of the availability of a large number of eggs, rapid and external development, and transparency during most of the embryogenesis. Second, its relatively short life cycle makes it suitable for genetic analysis. Third, the cost of maintaining a zebrafish facility is considerably lower than those for other vertebrate model organisms. Recently, the generation of thousands of zebrafish mutants has further enhanced the position of the zebrafish as a new vertebrate model in developmental biology (Driever *et al.*, 1996; Haffter *et al.*, 1996) and the characterization of these mutants will undoubtedly shed further light on the mechanism of development.

Despite these advances, at the present time, the zebrafish still suffers from a major drawback: the number of genes that have been isolated from this species is rather limited, as compared to other model organisms such as the mouse, *Drosophila,* and *Caenorhabditis elegans.* Currently, only a handful of zebrafish DNA sequences are available in DNA databases (about 500 non-EST entries were available as of June 1997), whereas hundreds of thousands of human and *C. elegans* sequences, tens of thousands of mouse and *Drosophila* sequences, and a few thousand *Xenopus* sequence entries have been deposited in DNA databases. In order to improve the availability of zebrafish genes, we have adapted a rapid method to identify and isolate zebrafish cDNA clones (Gong *et al.*, 1997). Our approach is to partially sequence randomly selected clones from zebrafish cDNA libraries so that the partial sequences can be used as tags for identifying the clones through sequence homology search in DNA databases. I describe here our method known as cDNA clone tagging, which should be applicable to any small research group. This protocol has also been used successfully at the undergraduate level so that students can practice several basic molecular techniques including polymerase chain reaction (PCR), plasmid DNA preparation, DNA sequencing, and sequence analysis.

II. Procedure of cDNA Clone Tagging

A. Outline

The approach we used is, in principle, similar to the common approach to generating expressed sequenced tags (ESTs) (Adams *et al.*, 1991). However, in most EST projects, selection of cDNA clones is on a completely random basis. Consequently, a large number of redundant clones and short-insert clones without the coding sequence are unavoidably accumulated. Because the objective of our study is to identify and isolate a maximal number of zebrafish genes with minimal resources, we have modified the experimental protocol to include the following. First, the cDNA libraries are constructed by a directional cloning approach using Stratagene's Lambda ZAP Uni-XR cloning system. Each selected clone is sequenced only once from the 5′ end to access the coding region because the coding sequence is more conserved than the untranslated regions and should provide a better chance to identify a gene. Second, only a clone harboring a relatively large insert (>0.7 kb) is selected to ensure that every clone that is sequenced contains at least part of the coding region. Third, plasmid DNA is prepared from each selected clone and is kept for future reference and application after sequencing. The basic procedure of this approach is summarized in Fig. 1.

B. Construction of cDNA Libraries

Directional cloning is a prerequisite for 5′-end sequencing. The cloning system we use is Stratagene's Lambda ZAP Uni-XR cloning system. The advantages of this system include directional cloning with the 5′ end at the EcoR I site and the 3′ end at the Xho I site, *in vivo* excision to conveniently convert the lambda phage DNA vector into the pBluescript plasmid vector in a bacterial host, and versatility of the pBluescript such as *in vitro* transcription for both strands of RNA and expression of fusion proteins. We prefer a phage-based library to a plasmid-based library because the former is easier to store and can be used for other common purposes. Mass *in vivo* excision can be performed by coinfection

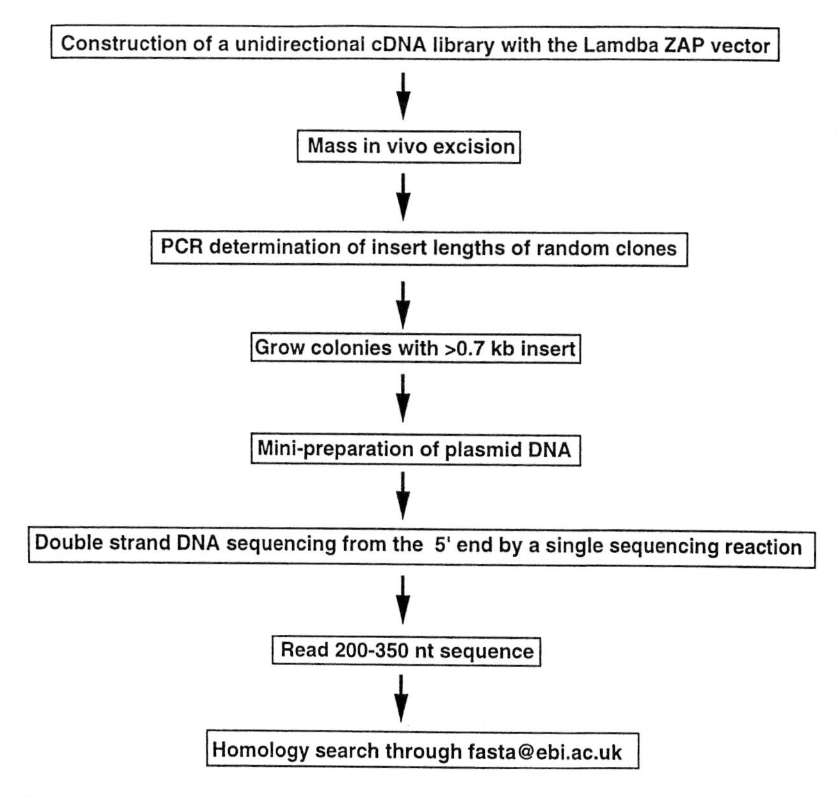

Fig. 1 Flowchart of cDNA clone tagging. Detailed description of the procedure is given in the text.

of phage library and a helper phage. Usually, $\sim 10^7$ pfu infectious phages or 10× of the number of original clones in the library are used for each mass *in vivo* excision. The phagemid phages after *in vivo* excision can be stored at −70°C for a long term by supplementing with 7% DMSO (dimethylsulfoxide). Prior to selection of cDNA clones, phagemid phages either freshly prepared or from a long-term storage can be used to infect host bacteria, and the infected bacteria can be plated on an ampicillin LB (Luria Bertani) agar plate.

C. Determination of Insert Length by PCR

There are two main purposes for determining insert length. First, the information on insert length helps us to avoid sequencing small insert or empty clones which have little or no use for identification of a cDNA clone. Second, it is useful to evaluate whether the clone is full length and how much is missing from the 5′ end. To determine the insert length for each clone, a bacterial colony can be used directly as a template in PCR. Briefly, a sterile pipette tip is used to gently touch a single colony and to transfer some of it into a PCR mix. Prior lysis of bacterial cells is not necessary because repeated denaturations of the DNA template during PCR effectively lyse the cells and release DNA for amplification. Primers SK (5′-CGCTCTAGAACTAGGATC) and T7 (5′-GTAATACGACTCACTATAGGGC) should be used because they are closest to the cloning sites. PCR is carried out for 30 cycles of 94°C, 1 minute; 68°C, 1 minute; and 70°C, 1.5 minute. The PCR products is examined by agarose gel electrophoresis using the 100-bp DNA ladder (Life/Technologies) as a molecular marker.

D. Plasmid DNA Preparation

Many methods are available for preparing high-quality plasmid DNA for double-strand DNA sequencing. Generally, a minipreparation from 1.5 ml of enriched TB (Terrific Broth) medium or 2–3 ml of LB medium can produce 10–30 μg of plasmid DNA (high copy number plasmids such as pBluescript), which is sufficient for DNA sequencing and backup copy storage. In our study, two methods for minipreparation of plasmid DNA are used: one is a manual method by alkaline lysis, and the other uses commercially available kits. Usually, the commercial kit produces reproducible high-quality DNA for sequencing.

For the manual method, a bacterial pellet is suspended in 100 μl of glucose solution (50m*M* glucose/10m*M* EDTA/25m*M* Tris, pH 8.0) and lysed with 200 μl of freshly made alkaline solution (0.2*N* NaOH/1% SDS) by gentle inversion, followed by addition of 150 μl of 4*M* KOAc (pH 4.8). After centrifugation, the supernatant is incubated with 40 μg of RNase A at 37°C for 20 minutes, extracted by phenol/chloroform, and precipitated with 2.5 volume of ethanol. The DNA pellet is then resuspended in 20 μl of 0.8*M* NaCl and precipitated by equal volume of 13% PEG (polyethylene glycol; MW 8000). The PEG precipitation is important to generate high-quality DNA for sequencing.

Several manufacturers produce mini-DNA preparation kits using silica-based resin. DNA is selectively bound to the resin and eluted with water. The procedure is fast, and plasmid DNA can be purified from bacteria in as short as 15 minutes. Tedious RNase digestion and hazardous organic extraction are eliminated. The quality of DNA is generally better than the manually prepared DNA for the purpose of sequencing. We routinely use the Wizard Minipreps Kit from Promega.

E. DNA Sequencing

Because of the relative ease of preparation of double-stranded plasmid DNA, double-stranded DNA sequencing is the method of choice for cDNA clone tagging. In our cDNA identification project, most of the sequencing is performed by manual sequencing using Pharmacia's T7 Sequencing Kit. Following the standard protocol provided by the manufacturer, each sequencing reaction is sufficient to load at least four times for gel electrophoresis. Because only one loading is required for identification of cDNA clones in most cases, the reaction can be scaled-down to 50–70% of the suggested volume to reduce cost. Usually, 1.5–2 μg of plasmid DNA is used for each reaction. Rapid denaturation and annealing can be performed using accurately calibrated 1.00*M* NaOH and 1.00*M* HCl. ^{35}S-dNTP is preferred to ^{32}P-dNTP for the manual sequencing because a better gel resolution can be achieved and thus a longer sequence can be read.

A more convenient method for performing DNA sequencing reaction is to use a cycle-sequencing method. It requires less DNA and usually generates better sequencing ladders. But this method depends on the availability of a PCR machine. We use Perkin Elmer's AmpliCycle Sequencing Kit. With this kit, we usually use purified plasmid DNA as template to a maximal amount (150 fmol, about 0.5 μg of 4 kb DNA) and minimal ^{33}P-dATP (<5 μCi per reaction). With 25 cycles of reaction using the thermal cycler, GeneAmp PCR system 9600 (Perkin Elmer), clearly readable signals can be produced after an overnight autoradiography.

For these manual sequencing methods, usually a sequence of 200–350 nucleotides can be obtained from a single loading using Model S2 Sequencing Gel Electrophoresis Apparatus from Life/Technologies. The accuracy of our sequence is $>95\%$ based on sequence match from a few selected clones whose sequences from zebrafish are available in databases, and most of the mismatches occur beyond 200 bp of an EST sequence.

F. Sequence Analysis

The partial sequence is submitted to Mail-FASTA via email (fasta@ebi.ac.uk) for sequence homology search (Pearson and Lipman, 1988), using a format based on the instruction manual from the server, which can be obtained by typing HELP in a single line in an email message to Mail-FASTA. An example of the sequence format sending to Mail-FASTA follows:

```
TITLE ZF-E72
LIB GBALL
LIST 100
Align 30
ONE
SEQ
TTGATCTTCTTAGACTTCACACATACCGTCTCGACATGGCACCCAAGAAGGCCAAA
AGGAGGGCAGCAGGAGGAGAGGGTTCCTCCAACGTCTTCTCCATGTTTGAGCAGAG
CATTGACCAGAACAGAGAC GGTATAAT
END
```

The clone name is ZF-E72 (TITLE ZF-E72) and the search is against all DNA entries in the EMBL database (recent release + new entries) (LIB EMALL). Up to 100 of the most homologous sequences can be listed (LIST 100) and 30 of the most homologous sequence alignments can be shown (Align 30); the two commands can be omitted if default values are chosen (i.e. 50 most homologous sequences and 10 alignments). If the sequence is the 5′ sequence (i.e., sense strand), only ONE sequence is needed for search. If the sequence contains an apparent poly T (complementary to a poly A tail of mRNA) at the beginning, it is most likely that the 3′ sequence and two strands (normal + complementary) are needed to search for homology; thus, ONE in line 5 should be replaced by TWO or omitted because the default is TWO in the Mail-FASTA program. The actual sequence should be placed between the SEQ and END commands.

We prefer the FASTA program to the BLAST program because the former is capable of introducing gaps (deletions and insertions) for maximal alignment. This is particularly important for comparing a sequence generated by a single-run sequencing, which may contain some reading errors. In our experience, the two programs produce the same result in most cases, but FASTA generally produces a more reliable alignment, whereas BLAST tends to break homologous domains into several segments. Recently a new version of BLAST, which allows the introduction of gaps, is available and may be a good substitute for the FASTA program (Altschul *et al.,* 1997).

The identification of a zebrafish homologous cDNA is generally based on a high-sequence identity (usually >65%) over a relatively long range (usually >150 bp). The probability for random match is generally smaller than 10^{-5}. For the majority of the clones we sequenced, there is a clear match or nonmatch. For example, if a sequence is from an actin gene, this sequence will pick up all actin DNA sequences from various species in the first 100 listed sequences of highest homology. If a sequence does not match any sequence in the database, the first 100 sequences listed will be unrelated. For some clones, their match may be impressive because of the presence of repetitive sequences. If this is the case, the repetitive sequence should be removed before homology search is performed.

However, for some clones whose homology is marginal to the match criteria, the following should be considered.

1. Is the match global (i.e., over the entire length of an EST sequence)? The real match is generally global unless there is an intron in the matched sequence. Potential introns can be quickly recognized based on the GT-AG rule (Sharp, 1987). However, sometimes the sequence may be matched only for a particular protein domain of high conservation and thus represents a member of super family.

2. Do they have the same reading frame? A quick glimpse of the reading frame from a sequence alignment is to examine whether they have a regular mismatch every three nucleotides. Because of the wobble of genetic codons, the third nucleotide in many codons are under neutral selection in evolution.

3. The amino acid sequences can be examined for all three potential reading frames to ascertain their homology.

G. Prescreening Hybridization

A common problem arising from sequencing of randomly selected cDNA clones in an unnormalized cDNA library is the accumulation of redundant clones because of the repeated selection of some abundantly expressed clones. For example, among the 480 random clones sequenced from the adult cDNA library we used, 20 are vitellogenin clones and represent 4.2% of the adult library. Therefore, it is preferable to avoid selecting the clones that have already been sequenced. An effective way of getting rid of these clones is to perform a hybridization with the probes from most abundant clones which have already been picked. Previously we performed a such prescreening hybridization with a mixed probe of 12 most abundantly represented clones in an embryonic cDNA library and 16% of the clones showed positive and thus were eliminated (Gong *et al.*, 1997).

To perform a prescreening hybridization, insert DNAs can be amplified individually by PCR from the plasmids of selected clones using two vector primers, SK and T7. The PCR products are purified from agarose gel by the Gel Extraction Kit (Qiagen) and mixed in approximately equal amounts and labeled with ^{32}P-dNTP by nick translation or random priming. Bacterial colonies after infection with phagemid phages are spotted manually onto an ampicillin LB plate and each plate of 150 mm diameter can be spotted easily with 400–500 well-separated colonies. Colony *in situ* hybridization is carried out, and it is important to include 1 μg/ml of denatured pBluescript DNA in the hybridization buffer to reduce hybridization background because the purified insert DNAs may contain a trace amount of pBluescript vector DNA which will produce false positive signals. To more effectively use pBluescript DNA to reduce hybridization background, the

DNA should be broken into small fragments by acid and alkaline treatment. Briefly, 10 μl of 3*M* HCl is added to 100 μl of DNA solution (10–100 μg DNA) and incubated at room temperature for 5 minutes. Then 30 μl of NaOH is added and incubated at 37°C for 15 minutes, followed by the addition of 20 μl of 3*M* NaOH to neutralize the DNA sample. Hybridization is usually performed at a modest stringent condition, and posthybridization wash should be carried out at a stringent condition with the final wash at 0.2 × SET (1× = 0.15*M* NaCl; 1m*M* EDTA; 20m*M* Tris, pH 7.8) at 65°C. After autoradiograph, only the hybridization negative colonies are selected for cDNA clone tagging as described earlier.

III. Management of Tagged cDNA Clones

It is important to generate a large number of tagged cDNA clones or EST clones. It is equally important to manage these clones properly. With a large number of tagged cDNA clones collected, these clones can be organized into a new type of gene library, termed tagged gene library. This is literally closer to the meaning of "library" because it can be more organized and controllable than a traditional gene library. For this purpose, we have maintained our tagged clones in two forms: plasmid DNAs (hardware) and DNA sequences (software). For every clone sequenced, we store the plasmid DNA at −70°C. In case that the clone is needed in the future, a trace amount of DNA can be used for transformation into a bacterial host, and thus the plasmid DNA can be amplified and purified. For transportation of cDNA clones, 1–2 μl of plasmid DNA (~0.5 μg) is usually spotted onto a Whatman paper and sent by regular mail; the receiver can recover the DNA by soaking in sterile water, followed by bacterial transformation. For every sequence obtained, we store it in a DNASIS file. A directory summarizes all information available, such as closest gene and species, EST length, percent match, range of identity, insert length, full length of coding region, and cellular location of encoded protein. A hardcopy of the FASTA search result is also filed. For all identified clones, a catalog classifies them into the following categories: cytosolic proteins, cytoskeleton, membrane proteins, mitochondrial proteins, nuclear proteins, secreted and extracellular proteins, translation machinery, and unknown location. So far, the majority of the sequences we obtained have also been submitted to the EST database with the access numbers of H56779-H56890, AA549765-AA549840, AA555439-AA555625, and AA566216-AA566921 (Gong *et al.*, 1997).

To collect as many different zebrafish cDNA clones as possible for a tagged zebrafish gene library, we have carried out the initial work in two zebrafish cDNA libraries: an embryonic library made from mixed stages of embryos from 6 hpf (gastrulation) to 72 hpf (hatched fry) and an adult library made from a combination of both female and male adult fish. The rationale for choosing the two libraries is that they conceivably cover the most, if not all, expressed genes, including those expressed only at certain embryonic stages or in certain adult tissues. So far we have sequenced 725 clones from the embryonic library and 480 clones from the adult library (Table I). In addition, we have also sequenced 105 clones from an eye cDNA library and 57 clones from a brain cDNA library. Among these tagged clones, we have identified 287 distinct zebrafish cDNA clones, of which only 7 preexist in the public databases. Thus we have isolated 279 new zebrafish cDNA clones with some known functions in other species.

For a tagged cDNA library, ideally, each clone is collected only once to generate a unique set of all expressed genes. However, with an expanding number of random clones sequenced, there is a dramatic increase of redundant clones and thus a decrease of tagging efficiency. As shown in Fig. 2, the number of embryonic clones which are identified for every 100 random clones remains about the same: 41–56. However, the number of *new* clones, counted by the first appearance of an identified clone, decreases from 42 to 14 for the first six 100-blocks because of the increase (from 12 to 39) in the number of *redundant* clones which are counted by the second and subsequent appearances of the same clone. Therefore, to avoid selection of redundant clones, a prescreening hybridization is necessary after sequencing a few hundred clones. In a preliminary prescreening hybridization, the 12 most abundant embryonic clones have been used as probes (Table II). These clones appear for a total of 94 times out of 570 embryonic clones and thus represent 16.5% of the clones in the embryonic cDNA library. There have been sequenced 155 hybridization negative clones, of these, 63 of them (or 41%) have been identified, and 34 (or 22 per 100) new clones have been added. This is a significant increase (57%) compared to the number of new clones identified from the last 100 unsubtracted clones (14 per 100) (Fig. 2). Furthermore, the number of redundant clones is reduced by half from 39 to 10

Table I
Summary of Tagged Zebrafish cDNA Clones

		Identified clones		
cDNA libraries	Sequenced[a]	New[b]	Redundant[c]	Total[d]
Embryonic[e]	725	186	150	336 (46%)
Adult	480	124	66	190 (47%)
Eye	105	6	4	10 (9.5%)
Brain	57	1	0	1 (1.8%)
Total	1367	287[f]	260	546 (40%)

[a] The number of clones sequenced and filed.
[b] The identified clones first appear.
[c] The identified clones which have been picked previously.
[d] Total number of identified clones excluding the clones which match only ESTs or repetitive sequences.
[e] Including 155 embryonic clones after a prescreening hybridization to subtract the 12 most abundant clones.
[f] The same gene repeatedly selected in different categories is counted only once.

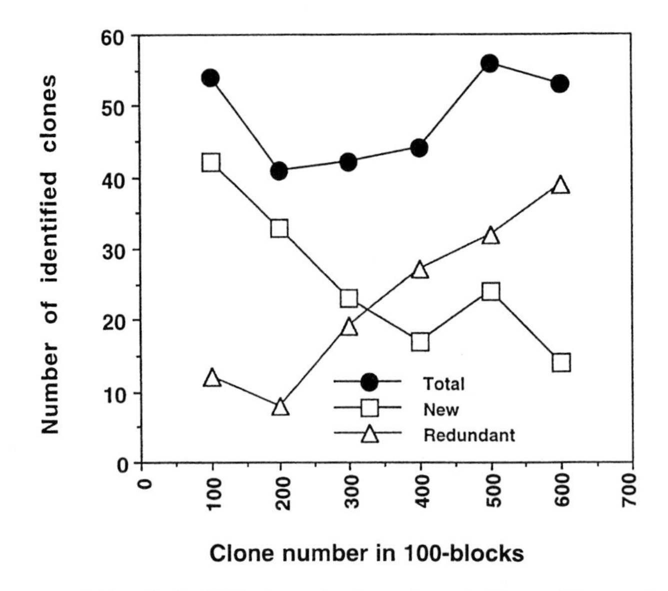

Fig. 2 Summary of identified cDNA clones in the embryonic library. The number of clones is counted for every 100 tagged clones in a sequence from clone #1 to clone #570. New, the identified clones which first appear; Redundant, the same clones which have been picked previously; Total, sum of the new and redundant clones. The last 100-block (600) is based on the statistics from clone #571–570 after normalization to 100.

per 100 clones sequenced. To maximally benefit from a cDNA clone tagging project, it is important to avoid redundant clones rather than simply to pursue the number of identified new clones because unidentified clones may be identified in the future when more DNA sequences are available from other species.

To perform a more thorough prescreening hybridization, we have recently amplified and purified the inserts individually from the 50 most abundantly expressed clones based on our tagged cDNA clones from the embryonic library and the adult library. These clones constitute 47% of the embryonic library and 26% of the adult library. The names of these clones and estimated representations in the two libraries are shown in Table II. Obviously, many of these clones are ubiquitously expressed and thus are useful to other zebrafish cDNA libraries including tissue specific libraries for a prescreening hybridization. We store the 50 insert DNAs separately, and a mixture of selected inserts can be made for specific purposes such as eliminating unwanted clones in a library prior to cDNA clone tagging. When more information is available in the future, more inserts can be purified and added to the list; thus, a more aggressive prescreening hybridization can be performed.

An alternative method to avoid sequencing redundant clones is to construct an equalized (or normalized) cDNA library in which all cDNA clones, abundant or rare, have about equal representation. The approaches to constructing such an equalized library include kenetic hybridization (Takahashi and Ko, 1994) and oligonucleotide hybridization (Meier-Ewert *et al.,* 1993). However, because of normalization, many abundantly expressed clones which may be important at this stage of developmental analysis are not easily isolated. Also, some closely related genes and splicing variants may be removed during the process of normalization. Furthermore, some of the procedures to generating such a normalized library are tedious and require tremendous work and sophisticated facilities which are not easily accessible to a small research group. Therefore, tagging with unnormalized cDNA libraries remains important for rapid isolation of useful cDNA clones, particularly in the initial stage when many of the abundantly expressed cDNA clones have not been isolated.

It is also important to tag different tissue-specific or cell-type-specific cDNA libraries. Because different tissues contain different sets of abundant clones, it is possible to obtain more diversified cDNA clones from these specific libraries and thus to reduce the number of redundant clones from the same cDNA library. A disadvantage in using zebrafish to construct tissue-specific libraries is the difficulty of collecting sufficient materials because zebrafish are so tiny. Another disadvantage is that few zebrafish cell lines are available for cell-type-specific cDNA libraries. To overcome the difficulty of working on limited materials, an amplification of total mRNA by PCR may be carried out to increase the starting materials for library construction (Froussard, 1993).

As demonstrated in Fig. 2, the number of the clones which should be sequenced from an unsubtracted library should preferably be small, and a prescreening hybridization to remove the most abundant clones can be carried out when the redundant clones become a problem. For example, in the embryonic library, the number of clones which should be sequenced is around 200 where the frequency for identification of a new gene remains relatively high (33–42%) and redundant rate is low (<12%). Different libraries may have different frequencies of abundant clones, and thus the number of clones which should be sequenced initially may vary. In the eye and brain cDNA libraries, the identification rates are very low (1.8–9.5%) compared to the embryonic and adult cDNA libraries (nearly 50%). However, the percentage of clones containing repetitive sequences such as di-, tri-, or tetra-nucleotide repeats and interspersed repetitive DNA families including Alu, DANA, and Tcl transposon families are much higher in the eye (6.7%) and brain (10.7%) libraries than those in the whole adult (0.8%) and embryonic (2.8%) libraries. The reason for these differences is not clear, but it is likely that the majority of expressed genes in these two tissues remain uncharacterized. Because redundancy is not a problem, obviously the number of clones which should be sequenced could be much larger than the number for the embryonic library.

IV. Applications of Tagged cDNA Clones

A. Expression Profile

Large-scale cDNA clone tagging has become an important approach to analyzing gene expression. By this approach, two types of information can be obtained

Table II
Most Abundantly Expressed Zebrafish Genes in Embryos and Adults

Gene name	Representations (%)	
	Embryonic	Adult
*mt 16S rRNA	3.86	0.63
*actin	2.98	1.25
*elongation factor 1α	1.75	0.63
*myosin light chain 2	1.40	0.63
*acidic ribosomal protein PO	1.05	1.25
*mt cytochrome b	0.88	0.21
*ADP/ATP carrier protein	0.88	0.42
α-tubulin	0.79	0.21
*mt cytochrome C oxidase subunit I	0.70	1.04
*elongation factor 2	0.70	0.42
*mt NADH dehydrogenase subunit I	0.70	0.21
polyubiquitin	0.53	0.63
ribosomal protein L7a	0.53	0.63
mt cytochrome C oxidase subunit II	0.53	0.42
*c-myc purine-binding transcription factor	0.53	0.42
ribosomal protein S6	0.53	0.42
*laminin receptor	0.53	<0.21
ribosomal protein S20	0.53	<0.21
creatine kinase M	0.40	1.25
parvalbumin	0.40	0.83
mt ATPase subunit 6/8	0.40	0.63
ribosomal protein S4, X isoform	0.40	0.63
G protein β subunit homolog	0.40	0.42
myosin heavy chain	0.40	0.42
ribosomal protein S9	0.40	0.42
cytokeratin I	0.40	0.21
mt cytochrome C oxidase subunit III	0.40	0.21
mt NADH dehydrogenase subunit II	0.40	0.21
mt NADH dehydrogenase subunit 4	0.40	0.21
ribosomal protein L3, J1 protein	0.40	0.21
ribosomal protein L8	0.40	0.21
ribosomal protein S3	0.40	0.21
cardiac myosin alkali light chain	0.40	<0.21
cytokeratin S	0.40	<0.21
α-tropomyosin	0.40	<0.21
calcium ATPase	0.40	<0.21
glyceraldehyde 3-phosphate dehydrogenase	0.26	1.04
ribosomal protein L6	0.26	0.42
ribosomal protein L11	0.26	0.42
poly A binding protein	0.26	0.21
troponin T isoform	0.26	0.21
ribosomal protein L9	0.26	0.21
Csa-19	0.26	0.21
mt NADH dehydrogenase subunit 4L	0.13	0.42

(*continues*)

Table II *(continued)*

Gene name	Representations (%)	
	Embryonic	Adult
ribosomal protein S3a	0.13	0.42
ribosomal protein S5	0.13	0.42
vitellogenin	<0.13	4.17
ZP3, egg membrane protein	<0.13	1.46
β-tubulin	<0.13	0.63
translationally controlled tumor protein	<0.13	0.63

Note: The 12 most abundant clones in the embryonic library used for a preliminary prescreening hybridization are labeled with asterisks; mt, mitochondrial.

from an unbiased cDNA library: the profile of expressed genes in the source tissue and the relative abundance of these transcripts (Adams *et al.,* 1995). Both types of information are important to the understanding of the function of the source tissue or cell type.

In a previous study on fish, we found that at least 27% of the random clones in a fish pituitary cDNA library encode pituitary hormones and at least 23% of the random clones in a fish liver cDNA library are for serum proteins, consistent with the major functions of the two organs (Gong *et al.,* 1994). For zebrafish, we have compared the profile of the expressed genes based on their cellular location. The distribution of the identified clones, as shown in Fig. 3, also provides preliminary information about the energy distribution of the organism. In both the embryo and the adult, the clones encoding translational machinery proteins constitute a large fraction. However, the portion in the embryo (24.4%) is significantly higher than in the adult (17.3%), indicating a higher activity of protein synthesis in the embryo. Similarly, the cytoskeletal protein mRNAs are more prevalent in the embryo, suggesting an important role of these proteins during early development. In contrast, the percentages of mRNAs for secreted proteins and cytosolic proteins are significantly higher in the adult, indicating that the adult fish shifts more energy to protein secretory and housekeeping cellular functions.

B. Identification of Interesting Clones

So far, we have generated tags for over 1300 cDNA clones, but about 60% of them remain unidentified. These unidentified clones provide a rich resource for identifying interesting zebrafish genes in future. In addition to a traditional hybridization screening approach, the gene of interest can also be retrieved by electronic screening against the sequence database. So far, all of our partial sequences have been used to search for a homology against all existing sequences in the public DNA databases at least once. Some sequence homology may be

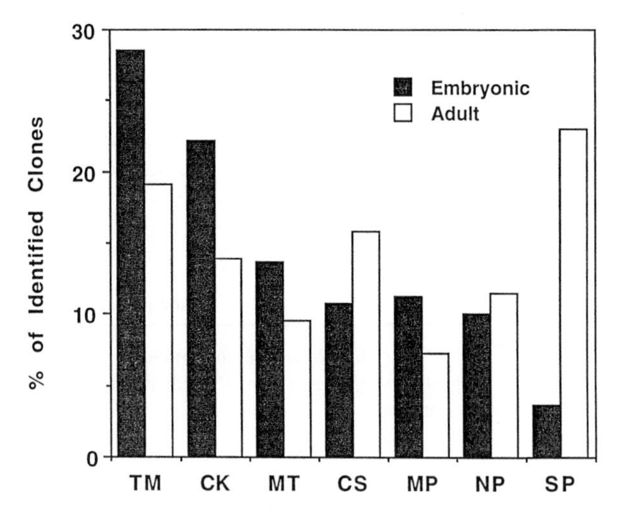

Fig. 3 Distribution of identified cDNA clones in the embryonic (dark bars) and adult (light bars) libraries. The percentages of identified clones for different categories in their respective libraries are shown. The clones isolated from the embryonic library after subtractive hybridization are not included. TM, translation machinery; CK, cytoskeleton; MT, mitochondria (excluding rRNA clones); CS, cytosolic; MP, membrane proteins; NP, nuclear proteins; and SP, secreted and extracellular proteins.

overlooked by the preliminary search if the homology is relatively low and does not appear on the list of the first 100 most homologous sequences. To identify a zebrafish homolog corresponding to a known gene in other species, an alternative approach is to use the known gene sequence to search against all of our partial sequences, and then to concentrate on the first few most homologous clones. An example of cloning the human homolog of yeast *CDC27* gene by electronic screening of human EST database has been reported (Tugendreich *et al.,* 1993). Recently, a large-scale identification of human EST clones homologous to *Drosophila* mutant genes has also been reported (Banfield *et al.,* 1996).

C. Isolation of Full-Length cDNA Clones

So far, 287 distinct zebrafish cDNA clones have been isolated through cDNA clone tagging, but only 98 of them (34%) have a coding region of full length. Thus, for many of the identified cDNAs, full-length clones remain to be isolated for other studies such as functional analysis and expression of recombinant proteins. To recover the full-length clone, a conventional method is to use the tagged partial clone to screen a cDNA library. Here I describe a rapid PCR approach to obtain a full-length clone.

As shown in Fig. 4, the gene specific primer A can be designed based on the EST sequence, and PCR is performed using the primer A and a vector primer,

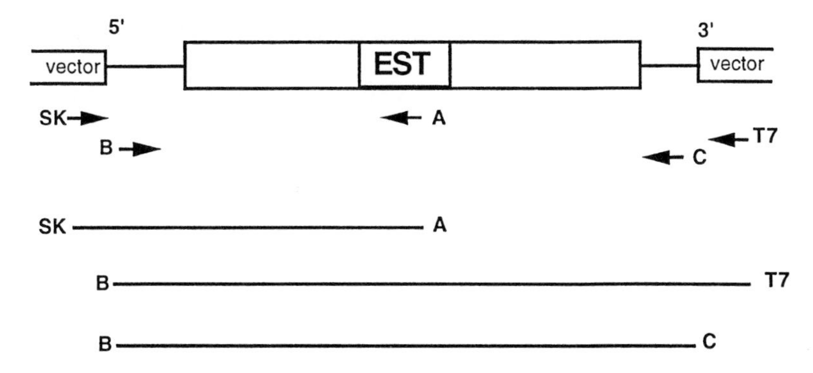

Fig. 4 Strategy of isolation of a full-length cDNA clone based on an EST sequence. The complete coding region is represented by a bar, and 5′ and 3′ UTRs are shown as thin lines. Arrows, PCR primers.

SK. After amplification, the PCR fragment can be cloned directly into a TA vector (e.g., pT7Blue from Novegen). The identity of the PCR fragment can be confirmed by DNA sequencing using the gene-specific primer or vector primers. It is important to have enough 5′ nucleotide sequence (>20 nucleotides) before the gene-specific primer A and thus the overlapping region can be used to confirm the identity of the amplified PCR fragment. To obtain a full-length clone in a single DNA fragment, the 5′ end of the newly amplified clone should be sequenced and thus a 5′ primer, B, can be designed. PCR can be performed by primers B and T7, a vector primer from the other end. An alternative and more efficient amplification is to use two gene-specific primers, B and C. The primer C can be obtained based on the 3′ sequence from the original EST clone. Another advantage to using the two gene-specific primers (B and C) is the possibility of performing RT-PCR directly when there is a difficulty obtaining a full-length cDNA clone from a cDNA library.

The gene-specific primers are usually 20 nucleotide long and sometimes six to eight additional nucleotides are introduced at the 5′ end of the primers to introduce a restriction site to facilitate cloning. The cDNA library (phages) is directly used as template for PCR amplification, which is generally performed for 30–35 cycles at 60°C for annealing and 65°C for extension. A common problem in amplifying the 5′ missing end is the presence of multiple fragments because of the heterogeneity of 5′ cDNA ends. The length of the missing 5′ end can be estimated from the alignment of the EST sequence and a homologous sequence. Thus, a fragment sufficiently long to cover the first ATG codon can be identified and purified from the gel for cloning. Sometime the long fragment is difficult to amplify because of the presence of smaller insert clones which out-compete the long fragment during PCR. If this is the case, the long fragment, if visible, can be cut from the agarose gel, frozen at −20°C for >20 minutes and thawed at 37°C for 5 minutes. There will be some liquid with DNA eluted from the gel, and the liquid can be used for another round of PCR amplification to generate enough DNA for cloning.

An alternative way to avoid the 5′ heterogeneity is to use a series of diluted libraries or sublibraries as a template for PCR amplification. For example, our embryonic cDNA library contains 2.6×10^6 primary clones. To reduce the complexity of the library, the library can be plated onto several Petri dishes with about 5×10^4 clones each dish and eluted individually as a sublibrary. The sublibrary usually produces a better PCR result because of the reduced complexity. However, a disadvantage is that many sublibraries have to be tested before an interesting clone can be amplified.

D. Isolation of Gene Promoters

Isolation of a gene promoter is important to study gene regulation through dissection of *cis*-elements and examination of their interaction with *trans*-acting factors. The availability of a promoter will also be useful to drive foreign gene expression in transgenic research. Currently, very few zebrafish gene promoters have been isolated and characterized. The availability of a large number of tagged zebrafish cDNA clones provides a vast opportunity to isolate zebrafish gene promoters. A conventional approach for this purpose is to screen a genomic DNA library with a cDNA probe. The screening and postscreening characterization procedures are generally tedious and labors intensive. Recently, we have developed a rapid method for isolating a gene promoter based on the EST information (Liao *et al.,* 1997). It is basically a linkers mediated PCR approach. The strategy of amplification, including the designing of linker DNA and PCR primers, is diagrammed in Fig. 5, as exemplified by the isolation of a myosin light chain (MLC) gene promoter. The zebrafish genomic DNA is first digested by selected restriction enzymes and modified by T4 DNA polymerase (or Klenow DNA polymerase) to generate blunt ends (if necessary). Because distribution of restriction sites is more or less random, it is advisable to use more than one restriction enzyme to isolate a promoter with a reasonable size. The digested genomic DNA is ligated with a partially double-stranded linker DNA which consists of a long (Oligo 1, 51mer) and a short oligonucleotide (Oligo 2, 11mer). Two linker-specific primers—L1 (21mer) and L2 (20mer)—are designed based

---

Fig. 5 (A) Schematic representation of the linker-mediated PCR approach to isolating a gene promoter based on an EST sequence. A detailed description of the procedure is given in the text. MLC, myosin light chain. (B) Schematic representation of the linker and primers used in isolation of a myosin light chain gene promoter. The 5′ EST sequence from the myosin light chain gene is shown for both strands, and the initiation ATG codon is boxed. All four primer sequences—L1, L2, E1, and E2—are shown and aligned with the linker DNA sequence or the 5′ sequence of the myosin light chain clone. Restriction sites (Hind III and EcoR I) introduced by the linker DNA are underlined.

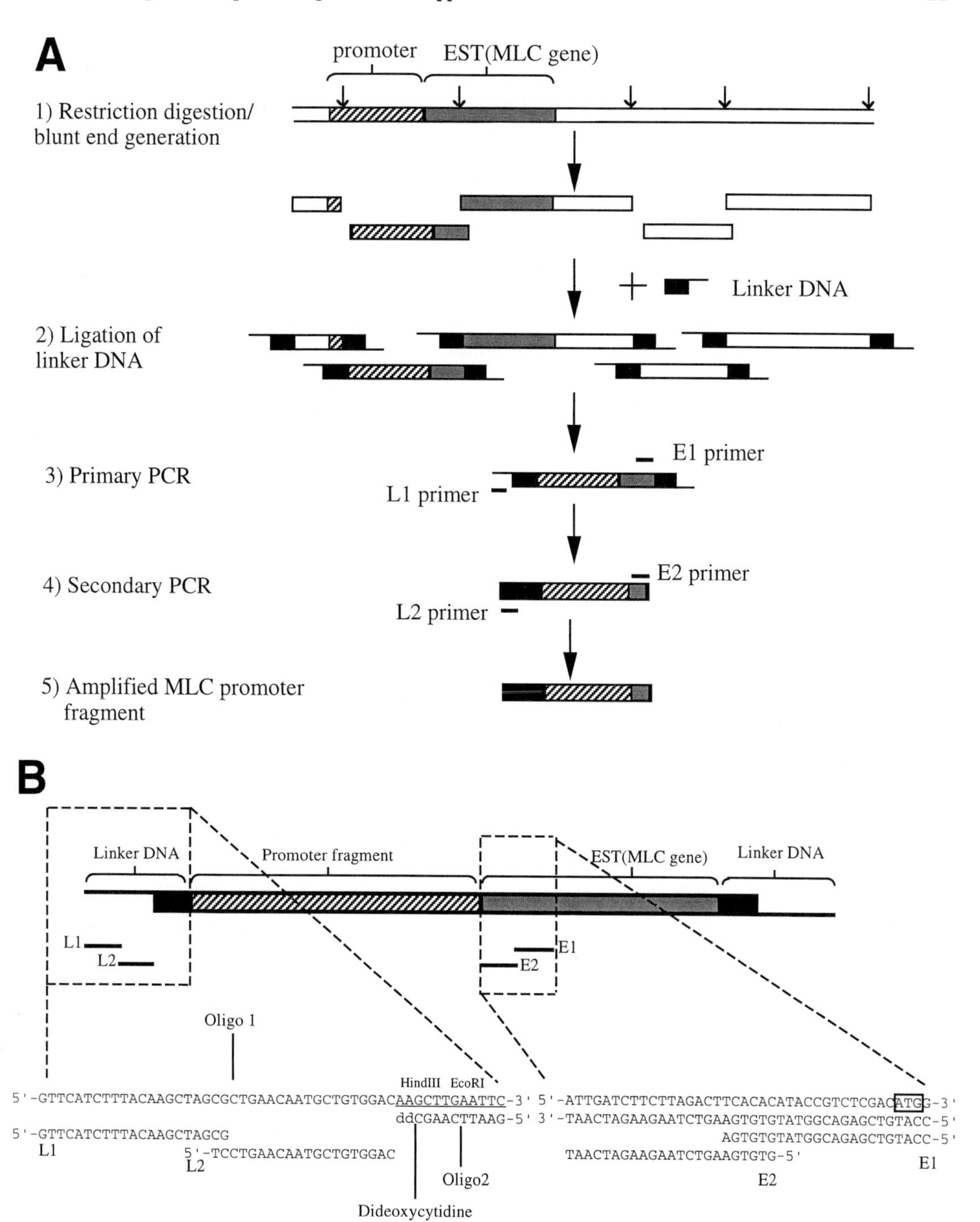
A
promoter
EST(MLC gene)
1) Restriction digestion/ blunt end generation
Linker DNA
2) Ligation of linker DNA
E1 primer
3) Primary PCR
L1 primer
E2 primer
4) Secondary PCR
L2 primer
5) Amplified MLC promoter fragment
B
Linker DNA
Promoter fragment
EST(MLC gene)
Linker DNA
L1
L2
E1
E2
Oligo 1
HindIII EcoRI
5'-GTTCATCTTTACAAGCTAGCGCTGAACAATGCTGTGGACAAGCTTGAATTC-3'
ddCGAACTTAAG-5'
5'-GTTCATCTTTACAAGCTAGCG
L1
5'-TCCTGAACAATGCTGTGGAC
L2
Oligo2
Dideoxycytidine
5'-ATTGATCTTCTTAGACTTCACACATACCGTCTCGACATGG-3'
3'-TAACTAGAAGAATCTGAAGTGTGTATGGCAGAGCTGTACC-5'
AGTGTGTATGGCAGAGCTGTACC-5'
TAACTAGAAGAATCTGAAGTGTG-5'
E1
E2

on the overhang portion of the long oligonucleotide sequence. Two gene-specific primers—E1 (23mer) and E2 (23mer)—are designed based on the EST sequence. The first ATG codon is not included in the second primer (E2) because, in most cases, the analysis of the promoter requires no coding region; thus, the final PCR product can be directly cloned into a reporter gene vector. Sufficient 5′ sequence is left upstream of the PCR primer E2 so that the identity of the amplified promoter region can be confirmed by sequence alignment.

As shown in Fig. 5A, two consecutive rounds of PCR are performed using two respective sets of primers—L1 and E1, and L2 and E2—to ensure the high specificity of the PCR product. The two linker-specific primers correspond to only the upper strand (Oligo 1) but not to the lower strand (Oligo 2), and the lower strand has a dideoxythymidine at the 3′ end. During PCR, L1 and L2 anneal only to the strand extended from gene-specific primer E1 or E2, whereas the extension of the lower strand, which could happen from intrastrand annealing of the two linker DNA ends or from heterostrand annealing of the linker DNA, is blocked by the presence of a dideoxynucleotide. Therefore, the presence of a dideoxynucleotide and the lack of the priming site in the linker DNA will greatly reduce background amplification. Similarly, the strategy described here can also be used for genome walking in positional cloning.

To amplify a long promoter region, it is important to use a long-distance PCR mix such as the Advantage Tth Polymerase Mix (Clontech). The first round of PCR is performed using primers L1 and E1. Each reaction (50 μl) contains 5 μl of 10× Tth PCR reaction buffer (1× = 15m*M* KOAc, 40m*M* Tris, pH 9.3), 2.2 μl of 25m*M* $Mg(OAc)_2$, 5 μl of 2m*M* dNTP, 1 μl of L1 (0.2 μg/μl), 1 μl of El (0.2 μg/μl), 33.8 μl of H_2O, 1 μl (50 ng) of linker ligated genomic DNA, and 1 μl of 50× Tth polymerase mix (Clontech). The cycling conditions are as follows: 94°C/1 minute, 35 cycles of 94°C/30 seconds and 68°C/6 minutes, and finally 68°C/8 minutes. After the primary round of PCR is completed, the PCR product is diluted 100-fold. There is 1 μl of diluted PCR product used as the template for the second round of PCR (nested PCR) with primers L2 and E2, as described for the primary PCR but with the following modification: 94°C/1 minute, 25 cycles of 94°C/30 seconds and 68°C/6 minutes, and finally 68°C/8 minutes.

E. Tissue and Cell-Type Markers

The identified zebrafish cDNA clones provide a valuable resource to select molecular markers for different tissues or cell types. For some of the clones, tissue specificity can be predicted from the gene name and confirmed by *in situ* hybridization and northern blot hybridization. These clones can also be used to express recombinant proteins to raise antibodies for expression analysis. Some of the potential clones which can be used as molecular markers for various tissues and cell types are summarized in Table III.

F. Genome Mapping

Another important application of these tagged clones is to construct dense genetic and physical maps for the zebrafish genome, which are valuable to the

Table III
Some Potential Molecular Markers from the Tagged cDNA Clones

Tissues/cells	Clones[a]
Skeletal muscle	Myosin light chain 2, creatine kinase M, myosin heavy chain
Slow axial muscle	Slow myosin binding protein C
Heart	Myosin heavy chain, slow myosin binding protein C
Erythrocytes	Globin
Liver	Transferrin, vitellogenin
Intestine	Cystein-rich intestinal protein, intestinal fatty acid binding protein
Pancrease	Elastase, trypsin, chymotrypsin
Cartilage	Cartilage matrix protein
Skin	Keratin
Lens	Crystallin
Early CNS	Sox 19
Neurons	Vimentin
Ovary	ZP2, ZP3
Rapid dividind cells	Cyclin B

[a] The suitability of some of the clones as molecular markers has been tested by *in situ* hybridization and Northern blot hybridization.

analysis of zebrafish mutants and a zebrafish genome project. Currently, several approaches are available in zebrafish to construct a genome map including RAPD (random amplified polymorphic DNA) (Postlethwait *et al.*, 1994), SSLP (single sequence length polymorphism) (Knapik *et al.*, 1996), SSCP (single strand conformation polymorphism) (Beier, 1993), somatic cell hybrids (Ekker *et al.*, 1996), and radiation hybrids (Cox *et al.*, 1990; C. Kwok, personal communication). The information generated from the EST project can be used in genome mapping by all of the means. For the approaches based on polymorphic DNA such as RAPD, SSLP, and SSCP, usually the 3′ untranslated sequence, which is more polymorphic than the 5′ coding region, is desired. The 3′ sequencing can be easily carried out because all our sequenced plasmid DNAs are retained. The advantage of 5′ sequencing first is the possibility to avoid repeated sequencing redundant clones from the 3′ end. Sometimes the 3′ sequencing reaction is difficult to perform because of the presence of a long polyA tail. To overcome the difficulty, we have designed an anchored oligo dT primer which contains 15T residues followed by two degenerated nucleotides (Liao and Gong, 1997). The sequence is 5′-TTTTTTTTTTTTTTTVN; where V = A, C, and G; N = A, C, G, and T. This anchored oligo dT primer can be used to sequence all oligo dT primed cDNA clones from the 3′ end irrespective of their cloning vectors. For genome mapping with somatic cell hybrids and radiation hybrids, no DNA polymorphism is required and the 5′ EST sequences can be used directly to design primers for PCR. The EST clones can also be used for hybridization to somatic cell hybrid DNAs and radiation hybrid panel DNAs. Similarly, these tagged clones can be used in zebrafish YAC, BAC, and PAC libraries for construction of a physical map. Finally, these tagged clones can also be used for chromosome *in situ* hybridization.

G. Candidate Cloning of Genes for Mutants

Because of the lack of a dense genome map for zebrafish at the present time, the candidate cloning approach remains the most effective way to isolate mutant genes in zebrafish. The accumulation of a large number of identified zebrafish cDNA clones is likely to provide some clues to the mutated genes and thus some candidate EST clones may be used for isolation of mutant genes. To test the candidacy of an EST clone, it is necessary to demonstrate the colocalization of the mutant gene and the gene for the candidate EST clone on a genome map, and then the candidate EST clone can be used to isolate the corresponding genomic DNA from the mutant zebrafish. The final confirmation of its authenticity for the mutant requires a transgenic approach to rescue the mutant. In addition, the EST sequences are also a useful resource to identify the genes interrupted by a transgenic insertion. A recent example has been reported in zebrafish by Allende *et al.* (1996) who have isolated a novel gene which is interrupted by insertion mutagenesis; the sequence flanking the insertion is similar to a human EST. With this clue, the complete zebrafish gene is identified and isolated.

Acknowledgments

I thank Tie Yan, Ji Liao, Jiangyan He, Yaling Guo, and Sze Ern Lee in my laboratory for their contributions to the zebrafish EST project and Dr. V. Korzh for comments on the manuscript. Our work is supported by NUS academic research grants from Singapore.

References

Adams, M. D., Kelley, J. M., Gocayne, J. D., Dubnick, M., Polymeropoulos, M. H., Xiao, H., Merril, C. R., Wu, A., Olde, B., Moreno, R. F., Kerlavage, A. R., McCombie, W. R., and Venter, J. C. (1991). Complementary DNA sequencing: Expressed sequence tags and human genome project. *Science* **252,** 1651–1656.

Adams, M. D., and 84 co-authors (1995). Initial assessment of human gene diversity and expression patterns based upon 83 million nucleotides of cDNA sequence. *Nature* (Suppl.) **377,** 3–17.

Allende, M. L., Amsterdam, A., Becker, T., Kawakami, K., Gaiano, N., and Hopkins, N. (1996). Insertional mutagenesis in zebrafish identifies two novel genes, pescadillo and dead eye, essential for embryonic development. *Genes Dev.* **10,** 3141–3155.

Altschul, S. F., Madden, T. L., Schaffer, A. A., Zhang, J., Zhang, Z., Miller, W., and Lipman, D. L. (1997). Gapped BLAST and PSI-BLAST: A new generation of protein database search programs. *Nucl. Acids Res.* **25,** 3389–3402.

Beier, D. R. (1993). Single-strand conformation polymorphism (SSCP) analysis as a tool for genetic mapping. *Mamm. Genome* **4,** 627–31.

Banfield, S., Borsani, G., Rossi, E., Bernard, L., Guffanti, A., Rubboli, F., Marchitiello, A., Giglio, S., Coluccia, E., Zollo, M., Zuffard, O., and Ballabio, A. (1996). Identification and mapping of human cDNA homologous to *Drosophila* mutant genes through EST database searching. *Nat. Genet.* **13,** 167–174.

Cox, D. R., Burmeister, M., Price, E. R., Kim, S., and Myers, R. M. (1990). Radiation hybrid mapping: A somatic cell genetic method for constructing high-resolution maps of mammalian chromosomes. *Science* **250,** 245–250.

Driever, W., Solnica-Krezel, L., Schier, A. F., Neuhauss, S. C. F., Malicki, J., Stemple, D. L., Stainier, D. Y. R., Zwartkruis, F., Abdelilah, S., Rangini, Z., Belak, J., and Boggs, C. (1996). A genetic screen for mutations affecting embryogenesis in zebrafish. *Development* **123,** 37–46.

Ekker, M., Speevak, M. D., Martin, C. C., Joly, L., Giroux, G., and Chevrette, M. (1996). Stable transfer of zebrafish chromosome segments into mouse cells. *Genomics* **33,** 57–64.

Froussard, P. (1993). rPCR: A powerful tool for random amplification of whole RNA sequences. *PCR Methods Applic.* **3,** 185–190.

Gong, Z., Hu, Z., Gong, Z. Q., Kitching, R., and Hew, C. L. (1994). Bulk isolation and identification of fish genes by cDNA clone tagging. *Mol. Mar. Biol. Biotech.* **3,** 243–251.

Gong, Z., Yan, T., Liao, J., Lee, S. E., He, J., and Hew, C. L. (1997). Rapid identification and isolation of zebrafish cDNA clones. *Gene.* **201,** 87–98.

Haffter, P., Granato, M., Brand, M., Mullins, M. C., Hammerschmidt, M., Kane, D. A., Odenthal, J., van Eeden, F. J. M., Jiang, Y.-J., Heisenberg, C.-P., Kelsh, R. N., Furutani-Seiki, M., Vogelsang, E., Beuchle, D., Schach, U., Fabian, C., and Nüsslein-Volhard, C. (1996), The identification of genes with unique and essential functions in the development of the zebrafish, *Danio rerio. Development* **123,** 1–36.

Knapik, E. W., Goodman, A., Atkinson, O. S., Roberts, C. T., Shiozawa, M., Sim, C. U., Weksler-Zangen, S., Trolliet, M. R., Futrell, C., Innes, B. A., Koike, G., McLaughlin, M. G., Pierre, L., Simon, J. S., Vilallonga, E., Roy, M., Chiang, P.-W., Fishman, M. C., Driever, W., and Jacob, H. J. (1996). A reference cross DNA panel for zebrafish (*Danio rerio*) anchored with simple sequence length polymorphisms. *Development* **123,** 451–460.

Liao, J., and Gong, Z. (1997). Sequencing of 3′ cDNA clones using anchored oligo dT primers. *BioTechniques* **23,** 368–370.

Liao, J., Chan, C. H., and Gong, Z. (1997). An alternative linker-mediated polymerase chain reaction method using a dideoxynucleotide to reduce amplification background. *Anal. Biochem.* **253,** 137–139.

Lipman, D. L. (1997). Gapped BLAST and PSI-BLAST: a new generation of protein database search programs. *Nucl. Acids Res.* **25,** 3389–3402.

Meier-Ewert, S., Maier, E., Ahmadi, A., Curtis, J., and Lehrach, H. (1993). An automatic approach to generating expressed sequence catalogues. *Nature* **361,** 375–376.

Pearson, W. R., and Lipman, D. J. (1988). Improved tools for biological sequence comparison. *Proc. Natl. Acad. Sci. U.S.A.* **85,** 2444–2448.

Postlethwait, J. H., Johnson, S. L., Midson, C. N., Talbot, W. S., Gates, M., Ballinger, E. W., Africa, D., Andrews, R., Carl, T., Eisen, J. S., Horne, S., Kimmel, C. B., Hutchinson, M., Johnson, M., and Rodriguez, A. (1994). A genetic linkage map for the zebrafish. *Science* **264,** 699–703.

Sharp, P. A. (1987). Splicing of messenger RNA precursors. *Science* **235,** 766–771.

Takahashi, N., and Ko, M. S. (1994). Toward a whole cDNA catalog: construction of an equalized cDNA library from mouse embryos. *Genomics* **23,** 202–210.

Tugendreich, S., Boguski, M. S., Seldin, M. S., and Hieter, P. (1993). Linking yeast genetics to mammalian genomes: Identification and mapping of the human homolog of CDC27 via the expressed sequence tag (EST) data base. *Proc. Natl. Acad. Sci. U.S.A.* **90,** 10031–10035.

CHAPTER 14

Zebrafish YAC, BAC, and PAC Genomic Libraries

Chris T. Amemiya,[*] Tao P. Zhong,[†] Gary A. Silverman,[‡] Mark C. Fishman,[†] and Leonard I. Zon[||]

[*]Center for Human Genetics
Boston University School of Medicine
Boston, Massachusetts 02118-2394

[†]Cardiovascular Institute
Massachusetts General Hospital East
Charlestown, Massachusetts 02129-2060

[‡]Department of Medicine
Harvard Medical School
Cambridge, Massachusetts

[||]Howard Hughes Medical Institute
Children's Hospital
Boston, Massachusetts 02115

0091-679X/99 $30.00

I. Introduction

The number of mutations affecting zebrafish development thus far uncovered via mutagenesis screens (Kahn, 1994; Driever *et al.,* 1996; Hafter *et al.,* 1996) is indeed impressive and exciting. However, if we are to use these mutants effectively for studying the underlying functional biology of development, it will be necessary to "positionally clone" (Collins, 1992) the genes responsible for these developmental defects. This is a nontrivial and labor-intensive task that requires an arsenal of both genetic [e.g., simple sequence repeat markers, random amplified polymorphic DNA (RAPD) markers] and genomic reagents (e.g., somatic cell/radiation hybrids, genomic libraries). In this chapter, we discuss the zebrafish large-insert genomic libraries constructed to date in yeast artificial chromosome (YAC), bacterial artificial chromosome (BAC), and P1-derived artificial chromosome (PAC) vectors. Notably, it is not the purpose of this chapter to discuss the actual methodologies in constructing large-insert genomic libraries; readers interested in these aspects should consult the primary literature on this subject. Rather, we review and discuss: (1) the different vector systems employed; (2) the characteristics and public availability of the respective libraries; (3) how the libraries are screened and used for positional cloning; and (4) future applications of these large-insert libraries to zebrafish developmental biology. Specific details regarding the technical aspects of library construction can be found in separate reports describing the respective libraries.

II. Cloning Systems

Genomics efforts in nonmammalian species have benefited extensively from the Human Genome Initiative, especially with regard to the types of genomics reagents that can be developed and in the overall approaches used for positional cloning (reviewed in Birren and Lai, 1996). The three cloning systems to be discussed here (YACs, BACs, and PACs) have all had a major impact on positional cloning in human and mouse systems and are anticipated to have a comparable impact on zebrafish genetics and genomics. A comparison of the characteristics of the three vector systems is provided in Table I, and an overview of the general method for genomic cloning using these vector systems is given in Fig. 1 (see color plate). The most notable feature of these systems relative to conventional lambda and cosmid systems is their greatly increased insert sizes, a substantial advantage for chromosomal walking and positional cloning efforts. The large insert sizes are also advantageous in that they can better encompass full-length genes (complete with their regulatory sequences), which allows for expedient characterization of gene structure and organization as well as better reagents for functional genomics experiments.

A. The YAC System

The YAC cloning system (Burke *et al.,* 1987; Nelson and Brownstein, 1994) uses two vector arms, both containing yeast selectable markers and telomere

Table I
Comparison of the YAC, BAC, and PAC Cloning Systems[a]

	YAC	BAC	PAC
Host cells	*Saccharomyces cerevisiae* AB1380, J57D	*Escherichia coli* DH10B	*Escherichia coli* DH10B
Transformation method	Spheroplast transformation	Electroporation	Electroporation
DNA topology of recombinants	Linear	Circular, supercoiled	Circular, supercoiled
Maximum insert size	⟩⟩ 1 mb	~300 kb	~300 kb
Selection for recombinants	*ade2* supF red–white color selection	LacZ blue–white color selection	*Sac*IIB selective growth
Selection for vector	Dropout media (lacking tryptophan and uracil)	Chloramphenicol	Kanamycin
Enzyme used for partial digestions	*Eco*RI	*Hin*dIII	*Mbo*I or *Sau*3A
Stability[b]	Varies from clone to clone but can be very unstable	Very stable	Very stable
Degree of chimerism[c]	Varies from library to library but can be higher than 50% of clones in library	Very low	Very low
Degree of co-cloning[d]	Occasionally	Undetectable	Undetectable
Purification of intact insert	Relatively difficult	Easy	Easy
Direct sequencing of insert	Difficult	Relatively easy	Relatively easy
Clone mating[e]	Yes	No	No

[a] This table compares the vector systems that have been most commonly used by the genomics community: pYAC4, pBeloBAC11, and pCYPAC2. It should be noted that many recent variations on these vectors have been generated such that some of the features in this table do not hold true (e.g., P. de Jong and coworkers have generated a BAC vector, pBACe3.6, that uses the *Sac*IIB selective growth system of P1-based clones in lieu of blue-white color selection; *cf.* http://bacpac.med.buffalo.edu/vector.shtml).

[b] Stability is defined as the ability of inserts to remain intact after repeated cell generations.

[c] Chimerism is defined as the cloning of noncontiguous genomic DNA into the same clone (Green *et al.,* 1991).

[d] Cocloning is defined as the presence of multiple, independent clones per cell.

[e] Clone mating is where two overlapping clones are combined into a single yeast cell, where they undergo homologous recombination to produce larger clones (Green and Olson, 1990a; Silverman *et al.,* 1990; Silverman, 1996b).

sequences, and one containing a centromere and a yeast autonomously replicating sequence (ARS). The (dephosphorylated) vector arms are ligated to the size-selected genomic DNA to be cloned (usually *Eco*RI partial digests) and subsequently transformed into yeast cells which have had their cell walls enzymatically removed (spheroplasts). To ensure that only clones containing both vector arms are propagated, the transformed spheroplasts are plated in top agar prepared with selective media; clones with inserts are further detected via a red–white color selection scheme. Transformants comprise features necessary for mitotic propagation in the yeast cells: namely, two telomeres, an ARS, a centromere, and presumably a genomic insert. Colonies are then picked (arrayed) into wells of a microtiter dish which contain media supplemented with a cryoprotectant. The YAC system is highly advantageous in that it is capable of propagating extremely large (megabase-sized) fragments (Nelson and Brownstein, 1994). The disadvantages of the system are many and include their relatively high rate of insert chimerism, inherent instability, slow growth rate, and overall difficulty in handling (Table I). Moreover, as YACs are linear chromosomes, they cannot be routinely isolated away from the endogenous yeast chromosomes except via separation by pulsed field gel electrophoresis (PFGE; Schwartz and Cantor, 1984; Chu *et al.,* 1986; Lee *et al.,* 1996), a relatively tedious procedure. Despite these problems YACs have had an integral role in physical mapping and positional cloning in the human and mouse (e.g., Green and Olson, 1990a; Burke, 1991; Larin *et al.,* 1991), and have been used for constructing transgenic cell lines and animals for functional studies (e.g., Jakobovits *et al.,* 1993; Maas *et al.,* 1997; Mendez *et al.,* 1997; Peterson *et al.,* 1997; Smith *et al.,* 1997).

B. The BAC System

The BAC system (Shizuya *et al.,* 1992) is bacterial plasmid-based. The system operates under the principle that the *Escherichia coli* F-factor plasmid replicon is capable of replicating, as observed in Hfr conjugation, DNA molecules as large as the entire *E. coli* chromosome (Kornberg and Baker, 1991). The vectors used for BAC cloning have been stripped of all accessory components and are essentially an F-factor replicon with a chloramphenicol resistance gene (Hosoda *et al.,* 1990; Leonardo and Sedivy, 1990; Shizuya *et al.,* 1992); the latest BAC vectors also possess a LacZ blue–white color selection system (Kim *et al.,* 1996). As with the YAC system, partially digested genomic DNA (usually *Hin*dIII) is extensively size-selected and ligated to dephosphorylated BAC vector (7.5 kb). Ligation products are then electrotransformed into an *E. coli* host strain that possesses a cell wall defect (DH10B; Life Technologies) and recombinants isolated by plating on selective media. The selected clones are arrayed into microtiter dishes using media containing a cryoprotectant. The BAC system can propagate molecules up to 300 kb, considerably smaller than what is possible in the YAC system. Nonetheless, BACs are highly advantageous in that they essentially are large plasmids and, as such, can be handled and characterized using simple modifications of routine methods used in all molecular genetics laboratories. BACs also appear to be comparatively stable because of their single copy F-factor origin of replication and do not exhibit the high levels of chimerism observed in YACs.

C. The PAC System

The PAC cloning system (Ioannou *et al.,* 1994; Amemiya *et al.,* 1996) is very similar to the BAC system, except that it uses a (single copy) P1 bacteriophage replicon and a kanamycin resistance gene (Sternberg, 1990). As with the BAC system, size-selected partial digests (usually *Mbo*I or *Sau*3A) are ligated to dephosphorylated vector and electrotransformed into DH10B cells. The PAC vector (e.g., pCYPAC2, pCYPAC6; ~16 kb) additionally contains a *Sac*BII selection system against noninsert-containing clones, whereby only recombinant clones grow vigorously in the presence of 5% sucrose (Pierce *et al.,* 1992). Selected PACs are arrayed in microtiter dishes containing a cryoprotectant. The PAC system can accommodate inserts of up to 300 kb and exhibits essentially all the advantages of BACs. In addition, the PAC system contains an inducible P1 lytic replicon that permits multicopy propagation of PAC plasmids when the clones are grown in the presence of 1m*M* IPTG (Sternberg, 1990). The disadvantage of the PAC system relative to BACs, is the larger vector size, especially for whole-clone shotgun sequencing projects; however, recent studies (Amemiya, unpublished) have shown that the PAC vector can be reduced in size to ~8 kb and function without compromising efficiency. Both PACs and BACs have been used extensively for physical mapping and functional analysis in humans (Brown *et al.,* 1995; Constantinou-Deltas *et al.,* 1996; Antoch *et al.,* 1997; Cooper *et al.,* 1997; Hubert *et al.,* 1997; Matsumoto *et al.,* 1997; Mullins *et al.,* 1997; Nechiporuk *et al.,* 1997; Nielson *et al.,* 1997) and are being implemented in large-scale DNA sequencing efforts (Venter *et al.,* 1996; Ansorge *et al.,* 1997; Boysen *et al.,* 1997a).

III. Available Zebrafish Genomic Libraries and the Need for Multiple Libraries

To date, two zebrafish YAC libraries, one BAC library, and one PAC library have been constructed. With regard to YACs, one library has been generated at Boston Children's Hospital and the other has been generated at Massachusetts General Hospital; these libraries are hereafter referred to as BCH-YAC and MGH-YAC, respectively. The MGH-YAC library is described in Zhong *et al.* (1998). The BAC library was constructed by the University of California—Berkeley and GenomeSystems (St. Louis, MO) and is briefly described in Barth *et al.* (1997). The PAC library was constructed at the Boston University School of Medicine. Relevant details regarding the respective libraries are provided in Table II. Notably, all libraries are publicly available or accessible through the primary contacts and/or commercial distributors (see Table II).

All the libraries encompass from four to five haploid genome equivalents, and all have been used successfully to isolate known genes and/or genomic regions (Barth *et al.,* 1997; Graser *et al.,* 1998; Zhang *et al.,* 1998, unpublished data). The average insert sizes of the four libraries range from 470 kb down to 90 kb and in the order: MGH-YAC > BCH-YAC > PAC > BAC. *In toto,* the genomic coverage encompassed in all four libraries is 3.5×10^{10} bp, or roughly 20× coverage. It is important to note that all three kinds of resources—YACs, BACs, and PACs—are complementary to one another. That is, none of the three vector systems employed is completely devoid of cloning biases, and there will be regions in each respective library that are under- and/or overrepresented. Also, because three different restriction enzymes were used for the generation of partial digests

Table II
Description of Current Zebrafish YAC, BAC, and PAC Libraries

	BCH-YAC	MGH-YAC	BAC	PAC
Zebrafish strain used	AB	AB	AB	AB
Tissue used	Adult whole blood	5-day embryos	Adult whole blood	Adult whole blood
Number of animals used	~200	25,000	1	~200
Method for agarose embedding	Amemiya *et al.*, 1996	Zhong *et al.*, 1998	Barth *et al.*, 1997	Amemiya *et al.*, 1996
Enzyme used for partial digests	*Eco*RI	*Eco*RI	*Hin*dIII	*MboI*
Vector used	pYAC4 (Burke *et al.*, 1987)	pRML1/pRML2 (Spencer *et al.*, 1993)	pBeloBac11 (Wang *et al.*, 1997)	pCYPAC6 (http://med-humgen 14.bu.edu)
Host strain	AB1380	J57D	DH10B™	DH10B™
Total clones in library	36,000	17,000	100,000	100,000
Estimated insert size[a]	240 kb	470 kb	90 kb	115 kb
Percent of clones with no insert	<10%	<1%	?	<10%
Estimated genomic representation[b]	5×	4.7×	4×	4–5×
Primary contact	Gary Silverman (silverman__g@a1.tch.harvard.edu)	Tao Zhong (zhong@cvrc-taco.mgh.harvard.edu)	David Smoller (dave@genomesystems.com)	Chris Amemiya (camemiya@bu.edu)
Commercial distributor of library	GenomeSystems	Research Genetics	GenomeSystems	GenomeSystems
Primary mode of library screening	High-density filters	PCR on Pools	High-density filters	High-density filters PCR on pools

[a] Based on analysis of over 100 individual recombinants by PFGE

[b] Based on: (1) the number of recombinants in the respective libraries and assuming a genome size of ca. 1700 mb and (2) average number of "hits" in limited screening of the libraries with single-copy probes.

(Table II), this increases the overall probability that all genomic fragments are represented by the available libraries. The success in generating these first-generation libraries is highly encouraging; consequently, the next generation of libraries is expected to be of far greater quality and representation.

IV. Screening Methods

There are two basic methods for screening arrayed genomic libraries: PCR analysis on pooled samples and hybridization of high-density colony filters. Both methods are applicable to the zebrafish libraries discussed here (Table II). For PCR screening, the library is subdivided into "smart" pools, and PCR is performed on DNAs isolated from the pools using primers unique to the genomic region being targeted. Each pool can consist of thousands of individual clones; various hierarchical or multidimensional pooling strategies have been published for this purpose, with the objective of detecting clone "intersections" (Green and Olson, 1990b; Amemiya *et al.*, 1990, 1991; Barillot *et al.*, 1991). PCR screening is usually a two- or three-tiered approach, the last round of screening generally requiring PCR on rows and columns of a microtiter dish (Heard *et al.*, 1989) or direct colony hybridization of positive microtiter dishes (Nizetic *et al.*, 1991). The MGH-YAC library has been pooled and is currently screenable only using such a PCR strategy (Zhong *et al.*, 1998); the PAC library has been also been pooled for PCR screening (S. Johnson and J. Postlethwait, personal communication). The advantage of the pool-PCR approach is its sensitivity, ease of implementation, and reliability. The general method is very robust and can be successfully used by relatively inexperienced personnel as well as by large, high-throughput (robotics-driven) genomics laboratories. The disadvantage of this method, particularly when done manually, is that it is relatively labor-intensive.

The hybridization approach relies on robotically produced high-density colony filters of the entire YAC, BAC, or PAC libraries (Nizetic *et al.*, 1991; Olsen *et al.*, 1993; Hoheisel *et al.*, 1996). These membranes (usually 22 × 22 cm) may contain from 18,432 to 28,800 individual clones and are processed so that colony DNAs are affixed to the membranes; positive clones are detected using hybridization with labeled probes (DNA fragments, riboprobes, oligonucleotides). Moreover, because the colonies have been applied to the filters in duplicate and in diagnostic grid patterns, the true identity of the positive hybridizing clones are immediately discernible. An example of such a hybridization is given in Fig. 2. The advantage of this screening method is that it is very fast, allowing for identification of specific clones after a single round of screening. The chief disadvantages involve the high cost of producing/purchasing the filters and the limitations on their reuse for repeated probings (up to ten times or so).

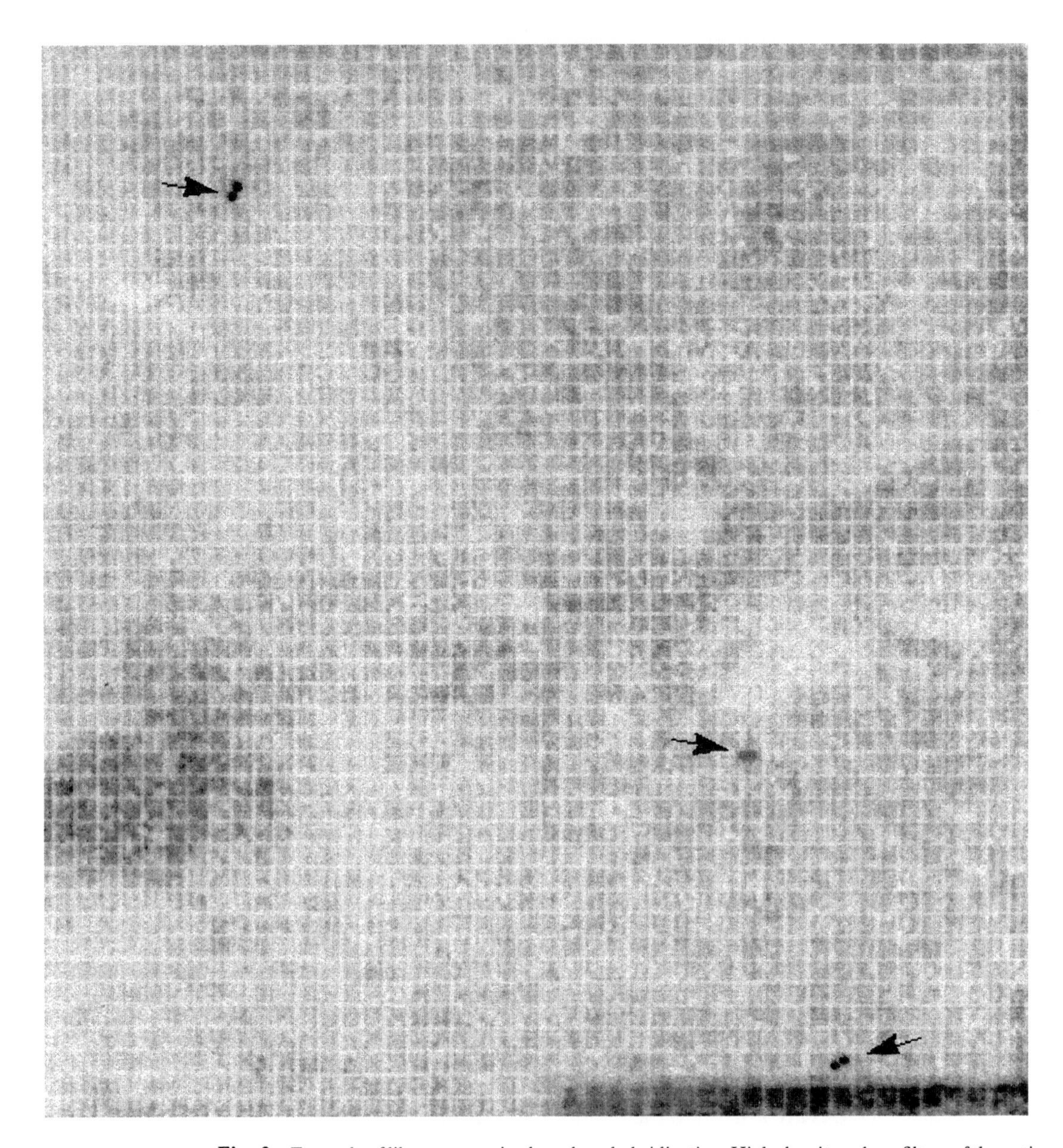

Fig. 2 Example of library screening by colony hybridization. High-density colony filters of the entire zebrafish PAC library (GenomeSystems) were screened with a single-copy probe to exon 1 of the zebrafish *Hoxa*-11 gene. Each filter contains up to 18,432 individual clones, which were spotted in a duplicate pattern that is diagnostic (i.e., allows for exact clone numbers to be immediately inferred from their hybridization patterns). The 400-bp probe was labeled with α-^{32}PdCTP in a random priming reaction (Feinberg and Vogelstein, 1983) and hybridized to the filters essentially following the manufacturer's recommendations. A portion of one 22×22 cm filter is shown. The duplicate positive signals are denoted by arrows and are clearly visible above the background hybridization level. A minimal background hybridization pattern is advantageous for orienting the filters and positive signals. The three probe-positive clones were identified, grown up from their respective microtiter well positions within the library, and authenticated via restriction analysis and hybridization.

V. Positional Cloning Using YAC, BAC, and PAC Libraries

Positional cloning (Collins, 1992) can be broken down into four successive steps: (1) genetic mapping, or localization of the locus to a specific chromosomal region via linkage analysis; (2) cloning (coverage) of that region in large-insert genomic clones; (3) identification of candidate genes (transcripts) that are encoded in the region; and (4) detection of lesions in the candidate genes that may give rise to the mutant phenotype. As is evident, each successive step results in increasingly higher resolving power. This strategy has been very successful for human disease loci, with the number of genes positionally cloned to date totaling close to 100 (ongoing tally from the National Human Genome Research Institute). The ability to identify any unknown human gene is further expedited by "positional candidates" (Collins, 1995), gene sequences such as expressed sequenced tags (ESTs) which have been physically mapped or anchored to defined regions of the genome. Thus, if one has genetically narrowed the region to which a particular gene is thought to occur, the publicly available sequence databases can be searched for transcripts emanating from that region, and these candidates can subsequently be examined.

With respect to zebrafish, the mapping resources and genomic reagents thus far available are comparatively depauperate to that for the human and mouse. Nonetheless, a pure positional cloning approach has recently been used to identify genes encoding the gastrulation defect, *one-eyed pinhead* (Zhang *et al.,* 1998), as well as the hematopoiesis defect, *sauternes* (Brownlie *et al.,* submitted). Notably, these studies both used the large-insert libraries discussed here and served as paradigms for the many ongoing or planned positional cloning projects in zebrafish.

A. Genetic Mapping

As discussed earlier, the first step in cloning a mutation-causing gene involves genetic mapping: specifically, to localize the gene to a linkage (chromosomal) group then to use previously placed (polymorphic) markers to increasingly narrow the region of interest. This requires setting up requisite crosses with genetically disparate strains so that the meiotic segregation of the mutant phenotype (relative to the polymorphisms) can be followed. Discussion into the kinds of genetic markers available in zebrafish, the unique mapping strategies that can be employed in this species, and the current status of the linkage map may be found in Postlethwait *et al.* (1994, 1998), Johnson *et al.* (1994, 1995), Kauffman *et al.* (1995), Knapik *et al.* (1996, 1998), Postlethwait and Talbot (1997), and Beier (1998). Linkage mapping will be necessary until the region between a flanking marker and the mutant gene is probably less than 1 cM, where 1 cM $\cong$ 625 kb in the zebrafish genome (Postlethwait *et al.,* 1994). In many cases, there may be insufficient markers to obtain such a narrow genetic interval and new markers may need to be generated "on the fly" (see Postlethwait and Talbot,

1997; Beier, 1998; and Zhang *et al.,* 1998 for discussion as to the generation of new polymorphic markers).

B. Chromosomal Walking

The next step in positional cloning involves isolating the region of interest from large-insert genomic libraries by systematic chromosomal walking. This iterative process involves screening libraries (see Section IV) with the most proximal flanking markers, aligning the resultant clones via restriction mapping or by sequence-tagged-site (STS)-based analysis (Ota and Amemiya, 1996; Amemiya *et al.,* 1996), generating new markers (probes) from the ends of the clones, and then rescreening the genomic libraries with the probes until a "walk" is established that encompasses the genomic region of interest. There is no "one way" to do a chromosomal walk because the logistics of each walk will vary with the specifics and requirements of each individual project. However, because chromosomal walking is a relatively tedious undertaking, the narrower the genetic interval (from VA) the faster the walk will be completed. The direction (orientation) of the walk relative to the target gene also needs to be established. This is done by generating new (polymorphic) markers from the walk (usually the ends of YACs, BACs, or PACs) and testing these against recombinant meioses: the nearer one gets to the gene of interest, the lower the frequency of recombination will be; and ultimately, there are no detectable crossovers in all meioses tested.

The decision to use YACs, BACs, or PACs will be based, in part, on the magnitude of the anticipated walk, as well as from personal preference. The increased sizes of YACs allow for much larger steps to be taken; however, the ease of handling of BACs and PACs allows for faster walking and characterization. It is conceivable that a combination of all three cloning resources will be required for a given project.

1. End-Isolation of YAC, BAC, and PAC Inserts

A crucial step in establishing a chromosomal walk is to generate probes and/or PCR primers for rescreening the genomic library. In the simplest case scenario, quick and dirty hybridization probes can be generated by using whole inserts, by single-primer amplification reactions from vector arms, or by T7/SP6 riboprobe runoff reactions (BACs and PACs only) (Evans and Lewis, 1989; Kim *et al.,* 1994, 1995; Ashworth *et al.,* 1995; Olsen *et al.,* 1996). These probes are used in hybridizations to high-density colony filters using large amounts of sheared and denatured genomic DNA in order to block repeats before and during the hybridizations (Evans and Lewis, 1989; Kim *et al.,* 1994, 1995; Ashworth *et al.,* 1995; Olsen *et al.,* 1996). Even under these suppression–hybridization conditions, however, these types of probes may result in high background signals when hybridized to genomic libraries owing to the presence of repeat sequences such as *Alu*I,

DANA, and mermaid elements (He *et al.,* 1992; Izsvak *et al.,* 1996; Shimoda *et al.,* 1996); and thus the level of false-positive clones may be significant, requiring considerable effort in subsequent clone confirmation and error-checking.

The preferred (more reliable) method for generating probes and/or polymerase chain reaction (PCR) primers for library screening is to obtain DNA sequences from the ends of the large-insert clones. This method allows selection of unique sequences to design oligonucleotides for direct use as hybridization probes or for PCR primers, which in turn may be used for screening pooled libraries (Green and Olson, 1990b; Amemiya *et al.,* 1992; Zhong *et al.,* 1998) or for generating amplicons for use as hybridization probes. The end-sequences are also important in that they can serve as STS anchors, thus enriching the genomic map. Direct sequencing using vector-specific primers can be performed, albeit with some difficulty, on BAC and PAC clones (Benes *et al.,* 1997; Fajas *et al.,* 1997; Boysen *et al.,* 1997b; Voss *et al.,* 1997). The key to generating usable sequence data from these reagents is isolating very good quality DNA, and, in the case of automated sequencing with dye-terminators, the use of a large excess of sequencing primer (Boysen *et al.,* 1997b). After sequence determination, the data can be compared to public databases in order to detect possible coding sequences and known repeats. Resulting sequences can then be input into a primer analysis algorithm to design oligonucleotide primers for probes and/or for PCR amplification.

An alternative to direct sequence determination involves the physical isolation of the ends of genomic inserts. This is necessary when using YACs because direct sequencing from these clones is difficult. Isolation of YAC ends can be accomplished using either PCR-based methods or by cloning the ends by plasmid rescue methods (Silverman, 1993, 1996a; Spencer *et al.,* 1993; Wu *et al.,* 1996; Zhong *et al.,* 1998). The latter is a method whereby the YAC DNA is digested with a judiciously chosen restriction enzyme, allowed to cyclize (via self-ligation), and transformed into a bacterial host. Notably, the vector arms contain a bacterial replicon and antibiotic resistance marker that allow bacterial propagation of end sequences. DNA sequencing is performed using primers that flank the insert cloning site.

The plasmid end-rescue method can also be used for BACs and PACs; Figure 3 illustrates a simple scheme for isolating the ends from PAC inserts employing of a pUC19 rescue vector. As with the direct sequencing methods, data are generated from the T7 and SP6 primer sites in the PAC (BAC) vector. The advantages of this method over direct sequencing of BACs and PACs are that the sequences generated are usually of better quality because of the higher purity of the pUC19-rescued DNA and that additional (internal) sequences can be obtained using pUC/M13 "universal" primers. Specific protocols for isolation of insert ends from PAC and BAC clones are available from C. Amemiya's website (http://med-humgen14.bu.edu).

Finally, it is important to note that there is a possibility that the ends of the inserts within large genomic clones are not from contiguous regions in the genome. These "chimeric" clones are much more of a problem with YAC libraries

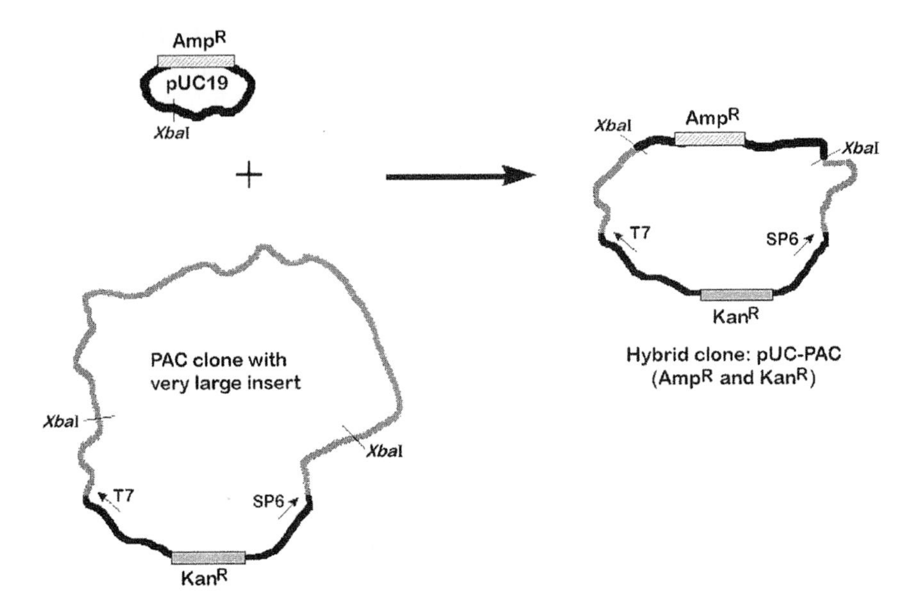

Fig. 3 A simple cloning scheme to rescue the ends of PAC clones. PAC DNA is digested with *Xba*I, which cleaves within the insert DNA but not within the vector. After heat-inactivation, the DNA fragments are ligated *en masse* with *Xba*I-digested pUC19. The ligation is then transformed into competent DH10B *E. coli* cells and plated onto agar media containing both ampicillin and kanamycin, thereby selecting only pUC/PAC hybrid molecules. DNA sequencing is performed using the T7 or SP6 promoter sites in the PAC vector; internal sequences may additionally be obtained using pUC/M13-forward and -reverse primers. For rescuing BAC inserts, BAC DNA is cleaved with *Nsi*I, ligated to *Pst*I-digested pUC19, transformed into *E. coli,* and plated onto media containing ampicillin and chloramphenicol. This general method results in efficient rescue of the ends of PAC (and BAC) inserts and can be accomplished in a relatively short time. Other restriction enzymes may be used as long as they cleave the insert but not vector DNA. Specific protocols for this and other PAC/BAC methods can be found at C. Amemiya's web site (http://med-humgen14-bu.edu).

because of *de novo* recombinations during transformation (Larionov *et al.,* 1994a, 1994b). Thus, after determining the sequence from the ends of a YAC clone, it is often recommended that these ends be checked to ensure that the fidelity of the insert DNA has not been compromised. In the mouse or human systems, this is generally done by PCR mapping to a somatic cell hybrid panel and detecting whether or not the ends emanate from the same chromosomal region. Somatic cell hybrid panels are now being developed for the zebrafish. Alternatively, clone ends can be hybridized to a Southern blot of zebrafish DNA digested with a panel of macrorestriction enzymes (and fractionated by PFGE); shared bands between probes would indicate physical linkage.

2. Characterization of Large-Insert Clones

After large-insert clones are isolated, they should be oriented with respect to one another within the contig. This is necessary to determine the extent of overlap between the clones and to determine the "minimal spanning path" across the contig. This ensures that each subsequent walking step encompasses the largest genomic coverage. This characterization need not be of high resolution as in "sequence-ready maps" (Ashworth *et al.,* 1995; Olsen *et al.,* 1996; Cooper *et al.,* 1997); rather, one simply would like to generate a rough framework map. Clones are easily oriented via a combination of PCR and restriction analysis. Large-insert clones are aligned on the basis of the presence or absence of PCR (STS) markers. To better orient the clones, crude DNAs (from minipreps) can be digested with relatively infrequent six- or eight-cutter restriction enzymes and electrophosed in an agarose gel in order to "fingerprint" the respective clones (Ota and Amemiya, 1996). If necessary, these gels can be blotted and hybridized with various markers from throughout the walk to confirm directionality. To estimate the insert lengths, it is necessary to electrophorese the clones through pulsed field gels; YACs can be electrophoresed intact because of their linear nature, however, BACs and PACs must first be digested with *Not*I (which flanks the cloning site of the vectors).

An attractive alternative to restriction fingerprinting in the zebrafish may be the use of interspersed repeat PCR (IR-PCR) (Nelson *et al.,* 1989; Ledbetter *et al.,* 1990; Amemiya *et al.,* 1992). In this strategy, universal primers are generated in order to recognize the most conserved sequences of respective interspersed repeat elements; thus PCR can be used with these primers in order to amplify the DNA in-between these repeat elements and to generate an IR-PCR profile. This strategy, employing *Alu* and L1 sequences, has been used extensively in human genome mapping and positional cloning (e.g., Ledbetter *et al.,* 1990; Aslanidis *et al.,* 1992; Cohen *et al.,* 1993; Higgins *et al.,* 1998). Because the zebrafish genome has interspersed repeats such as DANA and mermaid elements, universal IR-PCR primers can be designed and applied to PAC, BAC, or YAC clones, each of which should theoretically harbor at least a few of these sequences (Izsvak *et al.,* 1996; Shimoda *et al.,* 1996).

C. Isolation of Candidate Genes from Large-Insert Clones

Given a haploid genome size of 1.7×10^9 bp and a crude estimate of 100,000+ genes in the zebrafish (Postlethwait *et al.,* 1994, 1998), on average, this species should possess one gene (or more) every ~17 kb. By extrapolation, a contig of 470 kb (the size of one clone from the MGH-YAC library) may contain over 25 genes, and isolation of the specific transcriptional unit being sought may be a somewhat daunting task. It is therefore imperative that the essential genomic region be reduced down to as narrow a region as possible; it is also advantageous that the region be contained in BACs, PACs, and/or cosmids (Zhang *et al.,* 1998) because these reagents allow isolation of transcriptional units much more efficiently than from YACs. Methods commonly employed for isolating coding sequences from large genomic regions include (1) direct cDNA selection, (2) exon trapping, (3) hybridization of large-insert clones to cDNA libraries, and (4) direct sequencing of BAC and PAC clones.

1. Direct cDNA Selection

Direct cDNA selection is a method in which the genomic clones (inserts) are denatured and hybridized with a PCR-amplifiable population of cDNAs (e.g., cDNA that has been digested with a frequent cutting restriction enzyme and linkers added); bound cDNAs are then re-amplified via PCR and either taken through another round of selection or cloned (Del Mastro *et al.,* 1995; Del Mastro and Lovett, 1997; Guimera *et al.,* 1997). Note, before hybridization, the genomic insert is either biotin-labeled or covalently affixed to a small strip of nylon filter; then it is preblocked with excess sheared and denatured genomic DNA to prevent repeat sequences in the cDNA population from being isolated. The genomic population is then hybridized with the cDNAs, after which it is washed extensively to remove unbound material (this is possible because the genomic DNAs are affixed to either streptavidin–paramagnetic beads or a nylon filter strip). The bound cDNAs are then released via denaturation and re-amplified by PCR. This process is repeated once or twice to increase the fidelity and sensitivity of the capture procedure with the products being cloned and sequenced after the final round of selection. After sequencing of the short cDNA fragments, quality control procedures on potential candidates should be conducted (i.e., homology searches and rehybridization to the genomic clone). Candidates that are unique sequences and that emanate from the genomic contig should be used as probes to isolate their full-length counterparts via cDNA library screening; alternatively, primers can be generated from the candidates in order to isolate cDNAs by reverse transcription PCR (RT-PCR). The caveats of the direct cDNA selection method are that it is relatively labor-intensive and will work only if the cDNA population being used contains the gene (transcript) of interest.

2. Exon Trapping

Exon trapping is a method in which genomic inserts are subcloned into a plasmid vector that, when transfected into a mammalian cell line, permits mRNA splicing if the accompanying genomic DNA contains the appropriate (consensus) splice signals (Duyk *et al.,* 1990; Hamaguchi *et al.,* 1992; Krizman, 1997). After transfection, RNA is isolated from the cells, reverse transcribed using a 3′ vector primer, and then subjected to PCR using strategically placed primers in order to amplify trapped exons. The resultant products are then subcloned and sequenced. As with direct cDNA selection (discussed earlier), quality control measures need to be taken with the captured exons, and full-length cDNAs need to be isolated and characterized. Exon trapping is a very sensitive method that, unlike direct cDNA selection, does not rely on the inherent expression level of a given gene. There is, however, a requirement that the gene of interest contains multiple exons and that the splice sites are sufficiently conserved to undergo splicing in an exogenous (mammalian) system.

3. Direct Hybridization of Large-Insert Clones to cDNA Libraries

This method uses large-insert clones as probes on cDNA libraries (Elvin *et al.,* 1990; Marchak and Collins, 1993; Kern and Hampton, 1997). The pulsed field gel-separated insert DNAs are labeled to very high specific activity and hybridized to cDNA library filters using extensive blocking of repeat and vector sequences. This is a very straightforward and expedient method, particularly if the cDNA library being screened is in a plasmid vector, and arrayed and spotted on high-density filters (Kern and Hampton, 1997). However, most of the available zebrafish cDNA libraries are high-titer lambda libraries and are not yet available in arrayed format. For these conventional lambda cDNA libraries, this screening mode could result in comparatively high background levels (Marchak and Collins, 1993) and may be difficult to implement if numerous positive clones are detected in the initial round of screening. To increase the likelihood of success, the library being screened should be derived from the specific stage or tissue in which the anticipated gene would be expressed. For example, in the *oep* project, BAC and PAC clones from the critical genomic region were further subcloned into cosmid vectors, and these in turn were used for direct hybridization to a blastula/gastrula lambda cDNA library (Zhang *et al.,* 1998). Likewise, in the *sauterne* project, whole BAC and PAC inserts were used to screen an arrayed hematopoietic cDNA library (Brownlie *et al.,* submitted). In both of these cases, numerous candidate clones were identified, sequenced, and characterized.

4. Direct Sequencing of BAC or PAC Clones

Genomic BAC or PAC clones that likely contain the gene of interest can be subjected to DNA sequencing (Martin-Gallardo *et al.,* 1992; Boysen *et al.,* 1997a; Higgins *et al.,* 1998). This is generally done by subcloning the BACs or PACs via random shotgun methods, sequencing large numbers of subclones with an automated sequencer and then using computer algorithms to reassemble the DNA sequences. For good accuracy, the sequencing redundancy needs to be at least six- or sevenfold coverage. However, for gene identification, the stringency of sequence coverage can be much lower (one- to twofold coverage; "sequence skimming"). The genomic sequences are then compared to all known sequences in the public databases using BLAST (Altschul *et al.,* 1990, 1997) or other alignment algorithms in order to identify potential homologs. Additionally, several on-line algorithms exist for identifying putative coding and regulatory sequences in genomic DNA (Kamb *et al.,* 1995; Matis *et al.,* 1996; Claverie, 1997; Rawlings and Searls, 1997).[1] After genes or potential coding regions are identified, one still needs to procure the corresponding cDNAs via library screening

[1] Note that these programs have been developed and evaluated primarily using mammalian sequences. As such, they may or may not work sufficiently with *Danio* (or other teleost) DNA. As an example, G. Litman and coworkers have had only limited success in identifying coding sequences in a project involving ~100 kb of contiguous DNA from *Spheroides nephelus* (Southern pufferfish) in spite of the fact that several known genes are encoded in the region.

or RT-PCR methods. This method for gene identification is both labor-intensive and costly and can realistically be done only in large sequencing centers or via commercial sequencing companies.

5. Authenticity of Candidate Genes

After candidate genes have been isolated and confirmed to derive from the essential genomic region, it is prudent to determine which candidates should be pursued further. As mentioned previously, it is likely that multiple cDNAs will have been isolated up to this point. DNA sequence data and accompanying homology searches may or may not shed light on the potential functions of the isolated cDNAs. The tractability of the zebrafish embryo for whole-mount *in situ* hybridization can be used to assess *in vivo* expression patterns of the candidates. This is important because the true candidates will be predicted to exhibit some defined expression pattern, whereas genes showing very disparate patterns may be provisionally ruled out. This criterion, in fact, was used for further delimiting the candidate genes for *oep* (Zhang *et al.,* 1998).

D. Detection of Mutations in Candidate Genes

In order to determine if a candidate gene is involved in etiology of a particular mutant condition, it is necessary to sequence the gene from that mutant animal. This involves isolating the cDNA from a (cDNA) library from the mutant animal or, using PCR, to amplify the gene from either mRNA (RT-PCR) or directly from genomic DNA [as in the case of *oep* (Zhang *et al.,* 1998)]. If differences are observed between wild-type sequences and those from the mutants, this may indicate a phenotype-altering mutation or a merely a polymorphism. If the mutation is the only difference and if this change is consistent among all mutant animals, the change is probably responsible for the mutant phenotype. Likewise, if a particular nucleotide substitution results in an obvious nonpermissive amino acid change, the mutation is probably phenotype altering. Detection of a mutation which affects the integrity of the reading frame (deletion, insertion, frameshift) would be more compelling evidence than that the lesion is phenotype altering. Finally, functional assays such as mutant rescue provide strong evidence that a candidate gene is associated with a given phenotype or developmental process (Strahle *et al.,* 1997; Kishimoto *et al.,* 1997; Liao *et al.,* 1998). For example, in the case of *oep* (Zhang *et al.,* 1998), the investigators used microinjected mRNA from the *oep* wild-type gene to rescue definitively the mutant phenotype, whereas the mutant allele mRNAs could not rescue the phenotype.

VI. Further Applications of YAC, BAC, and PAC Resources to Zebrafish Genetics

A. Exploitation of the Yeast Homologous Recombination System

Saccharomyces cerevisiae has an inherently high level of homologous recombination, which, in part, explains the high chimerism frequency in YAC libraries

(Larionov *et al.,* 1994a, b). However, homologous recombination has also proven to be an invaluable tool for yeast geneticists in that it is used extensively to study and manipulate genes and the genome. Insofar as mammalian systems are concerned, it has recently been demonstrated that one can cotransfect yeast cells with HMW DNA together with a linearized targeting vector (whose ends contain homologous sequences to the HMW DNA, i.e., are recombinogenic); this results in the targeting vector specifically rescuing (cloning) the sequences being sought. Moreover, if this targeting vector contains requisite markers and replicons for both yeast and bacterial propagation, the entire construct can be shuttled out of yeast and into *E. coli.* Such vectors have been generated for rescuing multiple- as well as single-copy genes from the human and mouse genome (Bradshaw *et al.,* 1995, 1996; Larionov *et al.,* 1996a,b; Kouprina *et al.,* 1997, 1998). It is conceivable that this method also will be useful in zebrafish genomics for everything from chromosomal walking to rescuing rearranged alleles to generating chromosome-specific libraries from somatic cell hybrids.

Another application of homologous recombination in yeast is clone mating. In some instances, zebrafish genes may be too large to be encompassed in a single BAC, PAC, or YAC clone. If one wishes to study overall gene structure/organization as well as functional expression in a transgenic animal, it may be necessary to isolate the gene as an intact entity. This can be done by crossing yeast of different mating types that harbor, respectively, overlapping YAC clones (Silverman, 1996b). This general methodology has been used in human genetics to characterize very large genes, such as the cystic fibrosis and BCL2 loci (Green and Olson, 1990a; Silverman *et al.,* 1990).

B. Transgenesis and Gene Expression Studies

Large-insert clones are extremely useful for functional genomics experiments because there is an increased probability that they will contain *cis*-regulatory elements (enhancers) that may be located a considerable distance from coding sequences. Because these regulatory elements are necessary for precise recapitulation of the spatial and temporal expression patterns of the gene, transgene-reporter constructs using larger clones are clearly advantageous (Bradshaw *et al.,* 1996). Further, the possibility of introducing a green fluorescence protein cassette (Chalfie *et al.,* 1994) just downstream of the endogenous promoter will enable regulatory patterns of developmentally relevant genes to be observed *in vivo* without sacrificing the animals (Long *et al.,* 1997; Higashijima *et al.,* 1997). Even though targeted mutagenesis in large-insert clones is comparatively more difficult to accomplish than in small plasmids, methods are being developed with this exact purpose in mind (Bradshaw *et al.,* 1995, 1996; Yang *et al.,* 1997; Jessen *et al.,* 1998).

VII. Summary

Numerous positional cloning projects directed at isolating genes responsible for the myriads of observed developmental defects in the zebrafish are anticipated in the very near future. In this chapter, we have reviewed the YAC, BAC, and PAC large-insert genomic resources available to the zebrafish community. We have discussed how these resources are screened and used in a positional cloning scheme and have pointed out frequently formidable logistical considerations in the approach. Despite being extremely tedious, positional cloning projects in the zebrafish will be comparatively easier to accomplish than in human and mouse, because of unique biological advantages of the zebrafish system. Moreover, the ease and speed at which genes are identified and cloned should rapidly increase as more mapping reagents and information become available, thereby paving the way for meaningful biological studies.

Acknowledgments

We thank the following individuals for help and/or suggestions: Asaf Halevi, Tatsuya Ota, Gary W. Litman, David Smoller, and Eric S. Lander. The zebrafish Hoxa11 exon-1 probe was obtained from Chiu-hua Chen (Yale University, New Haven, CT). This work was funded, in part, by grants from NIH (NIAID-AI39008 to C.T.A.; NICHD-HD284 and NCI-CA69331 to G.A.S.; NHLBI-HL49579, NCRR-RR08888 to M.C.F.; NIDDK-DK49216 and NIDDK-DK53298 to L.I.Z.) and NSF (IBN-9614940 to C.T.A.). T.A.P. was supported by NIH fellowship HNLBI-HL07208.

Internet Web Sites Relevant to YAC, BAC, and PAC Cloning

http://med-humgen14.bu.edu—Chris Amemiya's laboratory site at Boston University School of Medicine; contains vector sequences and PAC/BAC protocols.

http://bacpac.med.buffalo.edu—Pieter de Jong's laboratory site at Roswell Park Cancer Institute; contains information about existing PAC and BAC libraries constructed in his laboratory, vector sequences and PAC/BAC protocols.

http://www.tree.caltech.edu—Caltech Genome Research Laboratory (Mel Simon, Director); contains information about BAC vectors, libraries, and BAC sequencing.

http://gc.bcm.tmc.edu:8088/centers.html—A list of all human genome centers.

http://www-seq.wi.mit.edu—MIT Genome Sequencing Project (Eric Lander, Director).

http://hubcap.clemson.edu/~schoi/BAC.html—Clemson University BAC Center (Rod Wing, Director).

References

Altschul, S. F., Gish, W., Miller, W., Myers, E. W., and Lipman, D. J. (1990). Basic local alignment search tool. *J. Mol. Biol.* **215,** 403–410.

Altschul, S. F., Madden, T. L., Schaffer, A. A., Zhang, J., Zhang, Z., Miller, W., and Lipman, D. J. (1997). Gapped BLAST and PSI-BLAST: A new generation of protein database search programs. *Nucl. Acids Res.* **25,** 3389–3402.

Amemiya, C. T., Aslanidis, C., Alleman, J. A., Chen, C., and de Jong, P. J. (1990). Use of a multidimensional pooling scheme and ALU-PCR for cosmid contig-mapping in the myotonic dystrophy region (19q13.2–3). *Am. J. Hum. Genet.* **47,** 958.

Amemiya, C. T., Alegria-Hartman, M. J., Aslanidis, C., Chen, C., Nikolic, J., Gingrich, J. C., and de Jong, P. J. (1992). A two-dimensional YAC pooling strategy for library screening via STS and Alu-PCR methods. *Nucl. Acids Res.* **20,** 2559–2563.

Amemiya, C. T., Ota, T., and Litman, G. (1996). Construction of P1 artificial chromosome (PAC) libraries from lower vertebrates. *In* "Nonmammalian Genomic Analysis" (B. W. Birren and E. Lai, eds), pp. 223–256. Academic Press, San Diego.

Ansorge, W., Voss, H., and Zimmermann, J. (1997). "DNA Sequencing Strategies." Wiley, New York.

Antoch, M. P., Song, E. J., Chang, A. M., Vitaterna, M. H., Zhao, Y., Wilsbacher, L. D., Sangoram, A. M., King, D. P., Pinto L. H., and Takahashi, J. S. (1997). Functional identification of the mouse circadian Clock gene by transgenic BAC rescue. *Cell* **89,** 655–667.

Ashworth, L. K., Alegria-Hartman, M., Burgin, M., Devlin, L., Carrano, A. V., and Batzer, M. A. (1995). Assembly of high-resolution bacterial artificial chromosome, P1-derived artificial chromosome, and cosmid contigs. *Anal. Biochem.* **224,** 564–571.

Aslanidis, C., Jansen, G., Amemiya, C., Shutler, G., Mahadevan, M., Tsilfidis, C., Chen, C., Alleman, J., Wormskamp, N. G. M., Vooijs, M., Buxton, J., Johnson, K., Smeets, H. J. M., Lennon, G. G., Carrano, A. V., Korneluk, R. G., Wieringa, B., and de Jong, P. J. (1992). Cloning of the essential myotonic dystrophy region and mapping of the putative defect. *Nature* **355,** 548–551.

Barillot, E., Lacroix, B., and Cohen, D. (1991). Theoretical analysis of library screening using a *N*-dimensional pooling strategy. *Nucl. Acids Res.* **19,** 6241–6247.

Barth, A. L., Dugas, J. C., and Ngai, J. (1997). Noncoordinate expression of odorant receptor genes tightly linked in the zebrafish genome. *Neuron* **19,** 359–369.

Beier, D. R. (1998). Zebrafish: Genomics on the fast track. *Genome Res.* **8,** 9–17.

Benes, V., Kilger, C., Voss, H., Paabo, C., and Ansorge, W. (1997). Direct primer walking on P1 plasmid DNA. *BioTechniques* **23,** 98–100.

Birren, B. and Lai, E. (1996). "Nonmammalian Genomic Analysis: A Practical Guide." Academic Press, San Diego.

Boysen, C., Simon, M. I., and Hood, L. (1997a). Analysis of the 1.1-Mb human alpha/delta T-cell receptor locus with bacterial artificial chromosome clones. *Genome Res.* **7,** 330–338.

Boysen, C., Simon, M. I., and Hood, L. (1997b). Fluorescence-based sequencing directly from bacterial and P1-derived artificial chromosomes. *BioTechniques* **23,** 978–982.

Bradshaw, M. S., Bollekens, J. A., and Ruddle, F. H. (1995). A new vector for recombination-based cloning of large DNA fragments from yeast artificial chromosomes. *Nucl. Acids Res.* **11,** 4850–4856.

Bradshaw, M. S., Shashikant, C. S., Belting, H. G., Bollekens, J. A., and Ruddle, F. H. (1996). A long-range regulatory element of Hoxc8 identified by using the pClasper vector. *Proc. Natl. Acad. Sci. U.S.A.* **93,** 2426–2430.

Brown, M. A., Jones, K. A., Nicolai, H., Bonjardim, M., Black, D., McFarlene, R., de Jong, P., Quirk, J. P., Lehrach, H., and Solomon, E. (1995). Physical mapping, cloning, and identification of genes within a 500-kb region containing BRCA1. *Proc. Natl. Acad. Sci. U.S.A.* **92,** 4362–4366.

Burke, D. T. (1991). The role of yeast artificial chromosome clones in generating genome maps. *Curr. Opin. Genet. Dev.* **1,** 69–74.

Burke, D. T., Carle, G. F., and Olson, M. V. (1987). Cloning of large segments of exogenous DNA into yeast by means of artificial chromosome vectors. *Science* **236,** 806–812.

Chalfie, M., Tu, Y., Euskirchen, G., Ward, W. W., and Prasher, D. C. (1994). Green fluorescent protein as a marker for gene expression. *Science* **263,** 802–805.

Chu, G., Vollrath, D., and Davis, R. W. (1986). Separation of large DNA molecules by contour-clamped homogeneous electric fields. *Science* **234,** 1582–1585.

Claverie, J. M. (1997). Computational methods for the identification of genes in vertebrate genomic sequences. *Hum. Mol. Genet.* **6,** 1735–1744.

Cohen, D., Chumakov, I., and Weissenbach, J. (1993). A first-generation physical map of the human genome. *Nature* **366,** 698–701.

Collins, F. S. (1992). Positional cloning: Let's not call it reverse anymore. *Nat. Genet.* **1,** 3–6.

Collins, F. S. (1995). Positional cloning moves from perditional to traditional. *Nat. Genet.* **9,** 347–350.

Constantinou-Deltas, C., Bashiardes, E., Patsalis, P. C., Hadjimarcou, M., Kroisel, P. M., Ioannou, P. A., Roses, A. D., and Lee, J. E. (1996). Complete coding sequence, exon/intron arrangement and chromosome location of ZNF45, a KRAB-domain-containing gene. *Cytogenet. Cell Genet.* **75,** 230–233.

Cooper, P. R., Nowak, N. J., Higgins, M. J., Simpson, S. A., Marquardt, A., Stoehr, H., Weber, B. H., Gerhard, D. S., de Jong P. J., and Shows., T. B. (1997). A sequence-ready high-resolution physical map of the best macular dystrophy gene region in 11q12–q13. *Genomics* **41,** 185–192.

Del Mastro, R. G., Wang, L., Simmons, A. D., Gallardo, T. D., Clines, G. A., Ashley, J. A., Hilliard, C. J., Wasmuth, J. J., McPherson, J. D., and Lovett, M. 1995. Human chromosome-specific cDNA libraries: New tools for gene identification and genome annotation. *Genome Res.* **5,** 185–194.

Del Mastro, R. G., and Lovett, M. (1997). Isolation of coding sequences from genomic regions using direct selection. *Methods Mol. Biol.* **68,** 183–199.

Driever, W., Solnica-Krezel, L., Shier, A. F., Neuhauss, S. C. F., Malicki, J., Stemple, D. L., Stainier, D. Y. R., Zwartkruis, F., Abdelilah, S., Rangini, Z., Belak, J., and Boggs, C. (1996). A genetic screen for mutations affecting embryogenesis in zebrafish. *Development* **123,** 37–46.

Duyk, G. M., Kim, S. W., Myers, R. M., and Cox, D. R. (1990). Exon trapping: A genetic screen to identify candidate transcribed sequences in cloned mammalian genomic DNA. *Proc. Natl. Acad. Sci. U.S.A.* **87,** 8995–8999.

Elvin, P., Slynn, G., Black, D., Graham, A., Butler, R., Riley, J., Anand, R., and Markham, A. F. (1990). Isolation of cDNA clones using yeast artificial chromosome probes. *Nucl. Acids Res.* **18,** 3913–3917.

Evans, G. A., and Lewis, K. A. (1989). Physical mapping of complex genomes by cosmid multiplex analysis. *Proc. Natl. Acad. Sci. U.S.A.* **86,** 5030–5034.

Fajas, L., Stacks, B., and Anwerx, J. (1997). Cycle sequencing of large DNA templates. *BioTechniques* **23,** 1034–1036.

Feinberg, A. P., and Vogelstein, B. (1983). A technique for radiolabeling DNA restriction endonuclease fragments to high specific activity. *Anal. Biochem.* **132,** 6–13.

Graser, R., Vincek, V., Takami, K., and Klein, J. 1998. Analysis of zebrafish Mhc using BAC clones. *Immunogenet.* **47,** 318–325.

Green, E. D., and Olson, M. V. (1990a). Chromosomal region of the cystic fibrosis gene in yeast artificial chromosomes: a model for human genome mapping. *Science* **250,** 94–98.

Green, E. D., and Olson, M. V. (1990b). Systematic screening of yeast artificial-chromosome libraries by use of the polymerase chain reaction. *Proc. Natl. Acad. Sci. U.S.A.* **87,** 1213–1217.

Green, E. D., Riethman, H. C., Dutchik, J. E., and Olson, M. V. (1991). Detection and characterization of chimeric yeast artificial-chromosome clones. *Genomics* **11,** 658–669.

Guimera, J., Pucharcos, C., Domenech, A., Casas, C., Solans, A., Gallardo, T., Ashley, J., Lovett, M., Estivill, X., and Pritchard, M. (1997). Cosmid contig and transcriptional map of three regions of human chromosome 21q22: Identification of 37 novel transcripts by direct selection. *Genomics* **45,** 59–67.

Hafter, P., Granato, M., Brand, M., Mullins, M. C., Hammerschmidt, M., Kane, D. A., Odenthal, J., van Eeden, F., Jiang, Y.-J., Heisenberg, C.-P., Kelsch, R. N., Furutani-Seiki, M., Warga, R., Vogelsand, E., Beuchle, D., Schack, U., Fabian, D., and Nüsslein-Volhard, C. (1996). The identification of genes with unique and essential functions in the development of the zebrafish, *Danio rerio. Development* **123,** 1–36.

Hamaguchi, M., Sakamoto, H., Tsuruta, H., Sasaki, H., Muto, T., Sugimura, T., and Terada, M. (1992). Establishment of a highly sensitive and specific exon-trapping system. *Proc. Natl. Acad. Sci. U.S.A.* **89,** 9779–9783.

He, L., Zhu, Z., Faras, A. J., Guise, K. S., Hackett, P. R., and Kupuschinski, A. R. (1992). Characterization of *Alu*I repeats of zebrafish *(Brachydanio rerio). Mol. Mar. Biotechnol.* **1,** 125–135.

Heard, E., Davies, B., Feo, S., and Fried, M. (1989). An improved method for the screening of YAC libraries. *Nucl. Acids Res.* **17,** 5861.

Higashijima, S., Okamoto, H., Ueno, N., Hotta, Y., and Eguchi, G. (1997). High-frequency generation of transgenic zebrafish which reliably express GFP in whole muscles or the whole body by using promoters of zebrafish origin. *Dev. Biol.* **192,** 289–299.

Higgins, M. J., Day, C. D., Smilinich, N. J., Ni, L., Cooper, P. R., Nowak, N. J., Davies, C., de Jong, P. J., Hejtmancik, F., Evans, G. A., Smith, R. J., and Shows, T. B. (1998). Contig maps and genomic sequencing identify candidate genes in the Usher 1C locus. *Genome Res.* **8,** 57–68.

Hoheisel, J. D., Maier, E., Mott, R., and Lehrach, H. (1996). Integrated genome mapping by hybridization techniques. *In* "Nonmammalian Genomic Analysis" (B. W. Birren and E. Lai, eds), pp. 319–346. Academic Press, San Diego.

Hosoda, F., Nishimura, S., Uchida, H., and Ohki, M. (1990). An F factor based cloning system for large DNA fragments. *Nucl. Acids Res.* **18,** 3863–3869.

Hubert, R. S., Mitchell, S., Chen, X. N., Ekmekji, K., Gadomski, C., Sun, Z., Noya, D., Kim, U. J., Chen, C., Shizuya H., Simon, M., de Jong, P. J., and Korenberg, J. R. (1997). BAC and PAC contigs covering 3.5 Mb of the Down syndrome congenital heart disease region between D21S55 and MX1 on chromosome 21. *Genomics* **41,** 218–226.

Ioannou, P. A., Amemiya, C. T., Garnes, J., Kroisel, P. M., Shizuya, H., Chen, C., Batzer, M. A., and de Jong, P. J. (1994). A new bacteriophage P1-derived vector for the propagation of large human DNA fragments. *Nature Genet.* **6,** 84–89.

Izsvak, Z., Ivics, Z., Garcia-Estefania, D., Fahrenkrug, S. C., and Hackett, P. B. (1996). DANA elements: A family of composite, tRNA-derived short interspersed DNA elements associated with mutational activities in zebrafish (*Danio rerio*). Proc. Natl. Acad. Sci. U.S.A. **93,** 1077–1081.

Jakobovits, A., Moore, A. L., Green, L. L., Vergara, G. J., Maynard-Currie, C. E., Austin, H. A., and Klapholz, S. (1993). Germ-line transmission and expression of a human-derived yeast artificial chromosome. *Nature* **362,** 255–258.

Jessen, J. R., Meng, A., McFarlane, R. J., Paw, B. H., Zon, L. I., Smith, G. R., and Lin, S. (1998). Modification of bacterial artificial chromosomes through chi-stimulated homologous recombination and its application in zebrafish transgenesis. *Proc. Natl. Acad. Sci. U.S.A.* **95,** 5121–5126.

Johnson, S. L., Midson, C. N., Ballinger, E. W., and Postlethwait, J. H. (1994). Identification of RAPD primers that reveal extensive polymorphisms between laboratory strains of zebrafish. *Genomics* **19,** 152–156.

Johnson, S. L., Africa, D., Horne, S., and Postlethwait, J. H. (1995). Half-tetrad analysis in zebrafish: Mapping the *ros* mutation and the centromere of linkage group I. *Genetics* **139,** 1727–1735.

Kahn, P. (1994). Zebrafish hit the big time. *Science* **264,** 904–905.

Kamb, A., Wang, C., Thomas, A., DeHoff, B. S., Norris, F. H., Richardson, K., Rine, J., Skolnick, M. H., and Rosteck, P. R., Jr. (1995). Software trapping: a strategy for finding genes in large genomic regions. *Comput. Biomed. Res.* **28,** 140–153.

Kauffman, E. J., Gestl, E. E., Kim, D. J., Walker, C., Hite, J. M., Yan, G., Rogan, P. K., Johnson, S. L., and Cheng, K. C. (1995). Microsatellite-centromere mapping in the zebrafish (*Danio rerio*). *Genomics* **30,** 337–341.

Kern, S., and Hampton, G. M. (1997). Direct hybridization of large-insert genomic clones on high-density gridded cDNA filter arrays. *Bio Techniques* **23,** 120–124.

Kim, U. J., Shizuya, H., Birren, B., Slepak, T., de Jong, P., and Simon, M. I. (1994). Selection of chromosome 22-specific clones from human genomic BAC library using a chromosome 22-specific cosmid library pool. *Genomics* **22,** 336–339.

Kim, U. J., Shizuya, H., Chen, X. N., Deaven, L., Speicher, S., Solomon, J., Korenberg, J., and Simon, M. I. (1995). Characterization of a human chromosome 22 enriched bacterial artificial chromosome sublibrary. *Genet. Anal.* **12,** 73–79.

Kim, U. J., Birren, B. W., Slepak, T., Mancino, V., Boysen, C., Kang, H. L., Simon, M. I., and Shizuya, H. (1996). Construction and characterization of a human bacterial artificial chromosome library. *Genomics* **34,** 213–218.

Kishimoto, Y., Lee, K. H., Zon, L., Hammerschmidt, M., and Schulte-Merker, S. (1997). The molecular nature of zebrafish swirl: BMP2 function is essential during early dorsoventral patterning. *Development* **124,** 4457–4466.

Knapik, E. W., Goodman, A., Atkinson, O. S., Roberts, C. T., Shiozawa, M., Sim, C. U., Weksler-Zangen, S., Trolliet, M. R., Futrell, C., Innes, B. A., Koike, G., McLaughlin, M. G., Pierre, L., Simon, J. S., Vilallonga, E., Roy, M., Chiang, P. W., Fishman, M. C., Driever, W., and Jacob, H. J. (1996). A reference cross DNA panel for zebrafish *(Danio rerio)* anchored with simple sequence length polymorphisms. *Development* **123,** 451–460.

Knapik, E. W., Goodman, A., Ekker, M., Chevrette, M., Delgado, J., Neuhauss, S., Shimoda, N., Driever, W., Fishman, M. C., and Jacob, H. J. (1998). A microsatellite genetic linkage map for zebrafish (*Danio rerio*). *Nat. Genet.* **18,** 338–343.

Kornberg, A., and Baker, T. A. (1991). "DNA Replication." W. H. Freeman, New York.

Kouprina, N., Graves, J., Cancilla, M. R., Resnick, M. A., and Larionov, V. (1997). Specific isolation of human rDNA genes by TAR cloning. *Gene* **197,** 269–276.

Kouprina, N., Graves, A. L., Afshari, C., Barrett, J. C., Resnick, M. A., and Larionov, V. (1998). Functional copies of a human gene can be directly isolated by transformation-associated recombination cloning with a small 3′ end target sequence. *Proc. Natl. Acad. Sci. U.S.A.* **95,** 4469–4474.

Krizman, D. B. (1997). Gene isolation by exon trapping. *Meth. Mol. Biol.* **68,** 167–182.

Larin, Z., Monaco, A. P., and Lehrach, H. (1991). Yeast artificial chromosome libraries containing large inserts from mouse and human DNA. *Proc. Natl. Acad. Sci. U.S.A.* **88,** 4123–4127.

Larionov, V., Kouprina, N., Nikolaishvili, N., and Resnick, M. A. (1994a). Recombination during transformation as a source of chimeric mammalian artificial chromosomes in yeast (YACs). *Nucl. Acids Res.* **22,** 4154–4162.

Larionov, V., Graves, J., Kouprina, N., and Resnick, M. A. (1994b). The role of recombination and RAD52 in mutation of chromosomal DNA transformed into yeast. *Nucl. Acids Res.* **22,** 4234–4241.

Larionov, V., Kouprina, N., Graves, J., Chen, X. N., Korenberg, J. R., and Resnick, M. A. (1996a). Specific cloning of human DNA as yeast artificial chromosomes by transformation-associated recombination. *Proc. Natl. Acad. Sci. U.S.A.* **93,** 491–496.

Larionov, V., Kouprina, N., Graves, J., and Resnick, M. A. (1996b). Highly selective isolation of human DNAs from rodent-human hybrid cells as circular yeast artificial chromosomes by transformation-associated recombination cloning. *Proc. Natl. Acad. Sci. U.S.A.* **93,** 13925–13930.

Ledbetter, S. A., Nelson, D. L., Warren, S. T., and Ledbetter, D. H. (1990). Rapid isolation of DNA probes within specific chromosome regions by interspersed repetitive sequence polymerase chain reaction. *Genomics* **6,** 475–481.

Lee, J. S., Birren, B., and Lai, E. (1996). Introduction to pulsed-field gels and preparation and analysis of large DNA. *In* "Nonmammalian Genomic Analysis" (B. W. Birren and E. Lai, eds.), pp. 1–24. Academic Press, San Diego.

Leonardo, E. D., and Sedivy, J. M. (1990). A new vector for cloning large eukaryotic DNA segments in *Escherichia coli. Biotechnology* **8,** 841–844.

Liao, E. C., Paw, B. H., Oates, A. C., Pratt, S. J., Postlethwait, J. H., and Zon, L. I. (1998). SCL/Tal-1 transcription factor acts downstream of cloche to specify hematopoietic and vascular progenitors in zebrafish. *Genes Dev.* **12,** 621–626.

Long, Q., Meng, A., Weng, H., Jessen, J. R., Farrell, M. J., and Lin, S. (1997). GATA-1 expression pattern can be recapitulated in living transgenic zebrafish using GFP reporter gene. *Development* **124,** 4105–4111.

Maas, A., Dingjan, G. M., Savelkoul, H. F., Kinnon, C., Grosveld, F., and Hendriks, R. W. (1997). The X-linked immunodeficiency defect in the mouse is corrected by expression of human Bruton's tyrosine kinase from a yeast artificial chromosome transgene. *Eur. J. Immunol.* **27,** 2180–2187.

Marchak, D. A., and Collins, F. S. (1993). The use of YACs to identify expressed sequences: cDNA screening using total YAC insert. *In* "YAC Libraries: A User's Guide" (D. L. Nelson and B. H. Brownstein, eds), pp. 113–126. W. H. Freeman, New York.

Martin-Gallardo, A., McCombie, W. R., Gocayne, J., FitzGerald, M., Wallace, S., Lee, B. M., Lamerdin, J., Trapp, S., Kelley, J., Liu, L.-I., Dubnick, M., Dow, L., Kerlavage, A., DeJong, P., Carrano, A., Fields, C., and Venter, J. C. (1992). Automated DNA sequence analysis of 106 kilobases from human chromosome 19q13.3. *Nat. Genet.* **1,** 34–39.

Matis, S., Xu, Y., Shah, M., Guan, X., Einstein, J. R., Mural, R., and Uberbacher, E. (1996). Detection of RNA polymerase II promoters and polyadenylation sites in human DNA sequence. *Comput. Chem.* **20,** 135–140.

Matsumoto, N., Soeda, E., Ohashi, H., Fujimoto, M., Kato, R., Tsujita, T., Tomita, H., Kondo, S., Fukushima, Y., and Niikawa, N. (1997). A 1.2-megabase BAC/PAC contig spanning the 14q13 breakpoint of t(2; 14) in a mirror-image polydactyly patient. *Genomics* **45,** 11–16.

Mendez, M. J., Green, L. L., Corvalan, J. R., Jia, X. C., Maynard-Currie, C. E., Yang, X. D., Gallo, M. L., Louie, D. M., Lee, D. V., Erickson, K. L., Luna, J., Roy, C. M., Abderrahim, H., Kirschenbaum, F., Noguchi, M., Smith, D. H., Fukushima, A., Hales, J. F., Klapholz, S., Finer, M. H., Davis, C. G., Zsebo, K. M., and Jakobovits, A. (1997). Functional transplant of megabase human immunoglobulin loci recapitulates human antibody response in mice. *Nat. Genet.* **15,** 146–156.

Mullins, L. J., Kotelevtseva, N., Boyd, A. C., and Mullins, J. J. (1997). Efficient Cre-lox linearization of BACs: Applications to physical mapping and generation of transgenic animals. *Nucl. Acids Res.* **25,** 2539–2540.

Nechiporuk, T., Nechiporuk, A., Sahba, S., Figueroa, K., Shibata, H., Chen, X. N., Korenberg, J. R., de Jong, P., and Pulst, S. M. (1997). A high-resolution PAC and BAC map of the SCA2 region. *Genomics* **44,** 321–329.

Nelson, D. L., Ledbetter, S. A., Corbo, L., Victoria, M. F., Ramirez-Solis, R., Webster, T. D., Ledbetter, D. H., and Caskey, C. T. (1989). *Alu* polymerase chain reaction: a method for rapid isolation of human-specific sequences from complex DNA sources. *Proc. Natl. Acad. Sci. U.S.A.* **86,** 6686–6690.

Nelson, D. L., and Brownstein, B. H. (1994). "YAC Libraries: A User's Guide." W. H. Freeman, New York.

Nielsen, L. B., McCormick, S. P. A., Pierotti, V., Tam, C., Gunn, M. D., Shizuya, H., and Young, S. G. (1997). Human apolipoprotein B transgenic mice generated with 207- and 145-kilobase pair bacterial artificial chromosomes. Evidence that a distant 5′-element confers appropriate transgene expression in the intestine. *J. Biol. Chem.* **272,** 29752–29758.

Nizetic, D., Drmanac, R., and Lehrach, H. (1991). An improved bacterial colony lysis procedure enables direct DNA hybridization using short (10, 11 bases) oligonucleotides to cosmids. *Nucl. Acids Res.* **19,** 182.

Olsen, A. S., Combs, J., Garcia, E., Elliot, J., Amemiya, C., de Jong, P. J., and Threadgill, G. (1993). Automated production of high density cosmid and YAC colony filters using a robotic workstation. *Bio Techniques* **14,** 116–123.

Olsen, A. S., Georgescu, A., Johnson, S., and Carrano, A. V. (1996). Assembly of a 1-Mb restriction-mapped cosmid contig spanning the candidate region for Finnish congenital nephrosis (NPHS1) in 19q13.1. *Genomics* **34,** 223–225.

Ota, T., and Amemiya, C. T. (1996). A nonradioactive method for improved restriction analysis and fingerprinting of large P1 artificial chromosome clones. *Genet. Anal.* **12,** 173–178.

Peterson, K. R., Clegg, C. H., Li, Q., and Stamatoyannopoulos, G. (1997). Production of transgenic mice with yeast artificial chromosomes. *Trends Genet.* **13,** 61–66.

Pierce, J. C., Sauer, B., and Sternberg, N. (1992). A positive selection vector for cloning high molecular weight DNA by the bacteriophage P1 system: Improved cloning efficacy. *Proc. Natl. Acad. Sci. U.S.A.* **89,** 2056–2060.

Postlethwait, J., Johnson, S., Midson, C. N., Talbot, W. S., Gates, M., Ballinger, E. W., Africa, D., Andrews, R., Carl, T., Eisen, J. S., Horne, S., Kimmel, C. B., Hutchinson, M., Johnson, M., and Rodriguez, A. (1994). A genetic map for the zebrafish. *Science* **264,** 699–703.

Postlethwait, J. H., and Talbot, W. S. (1997). Zebrafish genomics: from mutants to genes. *Trends Genet.* **13,** 183–190.

Postlethwait, J. H., Yan, Y.-L., Gates, M. A., Horne, S., Amores, A., Brownlie, A., Donovan, A., Egan, E. S., Force, A., Gong, Z., Goutel, C., Fritz, A., Kelsh, R., Knapik, E., Liao, E., Paw, B., Ransom, D., Singer, A., Thomson, M., Abduljabbar, T. S., Yelick, P., Beier, D., Joly, J.-S., Larhammar, D., Rosa, F., Westerfield, M., Zon, L. I., Johnson, S. L., and Talbot, W. S. (1998). Vertebrate genome evolution and the zebrafish gene map. *Nat. Genet.* **18,** 345–349.

Rawlings, C. J., and Searls, D. B. (1997). Computational gene discovery and human disease. *Curr. Opin. Genet. Dev.* **7,** 416–423.

Schwartz, D. C., and Cantor, C. R. (1984). Separation of yeast chromosome-sized DNAs by pulsed field gel electrophoresis. *Cell* **37,** 67–75.

Shepherd, N. S., Pfrogner, B. D., Coulby, J. N., Ackerman, S. L., Vaidnyanathan, G., Sauer, R. H., Balkenhol, T. C., and Sternberg, N. (1994). Preparation and screening of an arrayed human genomic library generated with the P1 cloning system. *Proc. Natl. Acad. Sci. U.S.A.* **91,** 2629–2633.

Shimoda, N., Chevrette, M., Ekker, M., Kikuchi, Y., Hotta, Y., and Okamoto, H. (1996). Mermaid, a family of short interspersed repetitive elements, is useful in zebrafish genome mapping. *Biochem. Biophys. Res. Comm.* **220,** 233–237.

Shizuya, H., Birren, B., Kim, U. J., Mancino, V., Slepak, T., Tachiiri, Y., and Simon, M. (1992). Cloning and stable maintenance of 300-kilobase-pair fragments of human DNA in Escherichia coli using an F-factor-based vector. *Proc. Natl. Acad. Sci. U.S.A.* **89,** 8794–8797.

Silverman, G. A., Green, E. D., Young, R. L., Jockel, J. I., Domer, P. H., and Korsmeyer, S. J. (1990). Meiotic recombination between yeast artificial chromosomes yields a single clone containing the entire BCL2 protooncogene. *Proc. Natl. Acad. Sci. U.S.A.* **87,** 9913–9917.

Silverman, G. A. (1993). Isolating vector-insert junctions from yeast artificial chromosomes. *PCR Meth. Appl.* **3,** 141–150.

Silverman, G. A. (1996a). End-rescue of YAC clone inserts by inverse PCR. *Methods Mol. Biol.* **54,** 145–155.

Silverman, G. A. (1996b). Reconstruction of large genomic segments of DNA by meiotic recombination between YACs. *Methods Mol. Biol.* **54,** 199–216.

Smith, D. J., Stevens, M. E., Sudanagunta, S. P., Bronson R. T., Makhinson, M., Watabe, A. M., O'Dell, T. J., Fung, J., Weier, H. U., Cheng, J. F., and Rubin, E. M. (1997). Functional screening of 2 Mb of human chromosome 21q22.2 in transgenic mice implicates minibrain in learning defects associated with Down syndrome. *Nat. Genet.* **16,** 28–36.

Spencer, F., Kenter, G., Connelly, C., and Hieter, P. (1993). Targeted recombination-based cloning and manipulation of large DNA segments in yeast. *Methods: A Companion to Meth. Enzymol.* **5,** 161–175.

Sternberg, N. (1990). Bacteriophage P1 cloning system for the isolation, amplification, and recovery of DNA fragments as large as 100 kilobase pairs. *Proc. Natl. Acad. Sci. U.S.A.* **87,** 103–107.

Strahle, U., Fischer, N., and Blader, P. (1997). Expression and regulation of a netrin homologue in the zebrafish embryo. *Mech. Dev.* **62,** 147–160.

Strong, S. J., Ohta, Y, Litman, G. W., and Amemiya, C. T. (1997). Marked improvement of PAC and BAC cloning is achieved using electroelution of pulsed-field gel-separated partial digests of genomic DNA. *Nucl. Acids Res.* **25,** 3959–3961.

Venter, J. C., Smith, H. O., and Hood, L. (1996). A new strategy for genome sequencing. *Nature* **381,** 364–366.

Voss, H., Nentwich, U., Duthie, S., Wiemann, S., Benes, V., Zimmermann, J., and Ansorge, W. (1997). Automated cycle sequencing with Taquenase: protocols for internal labeling, dye primer and "doublex" simultaneous sequencing. *Biotechniques* **23,** 312–318.

Wang, K., Boysen, C., Shizuya, H., Simon, M. I., and Hood, L. (1997). Complete nucleotide sequences of two generations of a bacterial artificial chromosome cloning vector. *BioTechniques* **23,** 992–994.

Wu, C., Zhu, S., Simpson, S., and de Jong, P. J. (1996). DOP-vector PCR: A method for rapid isolation and sequencing of insert termini from PAC clones. *Nucl. Acids Res.* **24,** 2614–2615.

Yang, X. W., Model, P., and Heintz, N. (1997). Homologous recombination based modification in *Escherichia coli* and germline transmission in transgenic mice of a bacterial artificial chromosome. *Nat. Biotechnol.* **15,** 859–865.

Zhang, J., Talbot, W. S., and Schier, A. F. (1998). Positional cloning identifies zebrafish one-eyed pinhead as a permissive EGF-related ligand required during gastrulation. *Cell* **92,** 241–251.

Zhong, T. P., Kaphingst, K., Akella, U., Haldi, M., Lander, E. S., and Fishman, M. C. (1998). Zebrafish genomic library in yeast artificial chromosomes. *Genomics* **48,** 136–138.

CHAPTER 15

Positional Cloning of Mutated Zebrafish Genes

William S. Talbot and Alexander F. Schier

Developmental Genetics Program
Skirball Institute of Biomolecular Medicine and
Department of Cell Biology
New York University School of Medicine
New York City, New York 10016

0091-679X/99 $30.00

I. Introduction

Zebrafish mutations provide an invaluable resource for dissecting genetic pathways that regulate vertebrate development, physiology, and behavior (Driever *et al.,* 1996; Haffter *et al.,* 1996). The optical clarity and accessibility of the zebrafish embryo allow exquisite analysis of mutant phenotype and a detailed understanding of gene function at the organismal and cellular level. To understand the molecular mechanisms underlying these cellular functions, it is essential to clone the mutated genes (reviewed in Postlethwait and Talbot, 1997). Three approaches have been used successfully to clone genes defined by mutations in zebrafish: retroviral tagging, testing candidate genes, and positional cloning.

Insertional mutagenesis relies on the ability to induce mutations by insertion of a DNA sequence that can be used as a tag to clone the mutated gene. Pseudotyped retroviral vectors have been successfully used to induce mutations in zebrafish (Gaiano *et al.,* 1996; reviewed in Schier *et al.,* 1996 and Amsterdam and Hopkins, Ch. 5, this volume). Mutated genes can then be cloned by virtue of the inserted proviral genome. Current protocols for retroviral mutagenesis are relatively inefficient when compared to chemical mutagenesis. Furthermore, there is no established protocol for isolating retrovirally tagged alleles of loci initially identified with other mutagens, such as the alkylating agent *N*-ethyl-*N*-nitrosourea (ENU). Further technical advances, such as the use of transposons as insertional agents (Ivics *et al.,* 1997; Raz *et al.,* 1998; Ivics *et al.,* Ch. 6, this volume), might improve the efficiency of isolating mutants, thus making insertional mutagenesis more widely applicable.

In the *candidate gene approach,* one assembles a collection of cloned genes that have some properties expected of the mutated locus and tests these genes as candidates (reviewed in Postlethwait and Talbot, 1997). In some successful applications of this approach, a gene's expression pattern (Talbot *et al.,* 1995) or mutant phenotype in another species (Schulte-Merker *et al.,* 1994) is the basis for candidate selection. Linkage analysis then provides a test to determine whether the candidate gene might correspond to the locus. In other cases, the systematic mapping of genes and mutations identifies candidate genes directly by linkage (reviewed in Collins, 1995). The principal drawbacks of the candidate approach are twofold. First, most zebrafish genes have not yet been isolated and are thus not immediately available as candidates. Second, the criteria for candidate selection are often subjective, and the correct candidate might be overlooked. As genomic resources for the zebrafish advance, the candidate gene approach will become more systematic, as many genes will be accessible through expressed sequence tag (EST) data bases. Furthermore, large-scale mapping efforts will allow many potential candidate genes to be tested by simply mapping a mutation of interest.

At present, neither candidate genes nor insertional alleles are available for the majority of interesting zebrafish mutants. In these cases, *positional cloning* can be used to identify the affected genes (Bender *et al.,* 1983). Positional cloning is an unbiased approach that is applicable to any mutation whose inheritance can be traced, even if nothing is known about the gene or biochemical pathways affected by the mutation. Despite a genome size that is two-thirds that of mouse and human, zebrafish has two features that are advantageous for positional cloning projects. First, high fecundity allows the analysis of several thousand meioses and fine mapping of a mutation to a small interval. Second, the accessibility of the embryo allows for RNA or DNA injection to rescue mutant phenotypes and rapid analysis of the expression pattern of candidate cDNAs. The recent isolation of the embryonic patterning gene *one-eyed pinhead* (*oep*) as a novel EGF-related factor highlights the potential of the positional cloning approach in zebrafish (Zhang *et al.,* 1998). Neither candidate cDNAs or ESTs nor insertional alleles were available for *oep,* making positional cloning the only current avenue to identify this gene.

Positional cloning projects consist of three steps (Fig. 1). The first phase involves the identification of a DNA segment (often referred to as a "marker")

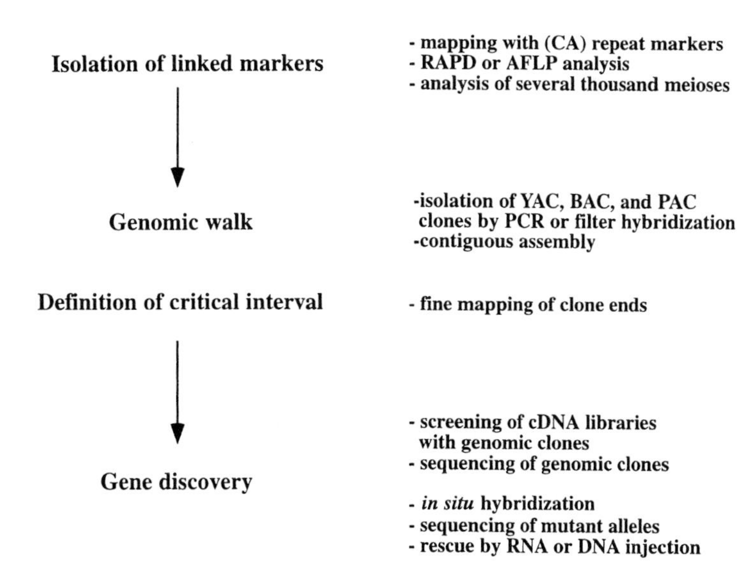

Fig. 1 Positional cloning. Overview of steps involved in positional cloning of genes identified by mutation.

that is near the mutant locus as judged by linkage analysis. In the second phase, the marker is used to isolate clones from genomic libraries. If the marker and the mutant locus are closely linked, the first clones isolated may contain the gene. Otherwise, the process is repeated to carry out a genomic walk until linkage analysis demonstrates that the mutant locus is encompassed in a contiguous stretch of genomic DNA (often referred to as a "contig"). The third phase involves the identification of the gene within the contig region. This may be accomplished in a number of ways, ranging from transgenic rescue of the mutation to sequence analysis of the genomic region. In the following sections, we discuss in more detail these major steps in positional cloning projects in zebrafish and provide the corresponding protocols.

II. Initiating a Positional Cloning Project

A. Mutant Alleles

Before embarking on a positional cloning project, it is important to consider the nature of the available mutant alleles. Most loci likely to be the focus of positional cloning projects are identified in screens performed with the chemical muagen ENU (Mullins *et al.,* 1994; Solnica-Krezel *et al.,* 1994). ENU-induced alleles are useful for genetic mapping experiments because they are predominantly intragenic lesions (mainly point mutations) that do not disrupt meiotic recombination in the vicinity of the mutant locus. By contrast, gamma-rays and X-rays can induce chromosomal rearrangements that obscure the position of the mutant locus, and these mutations should not be used for high-resolution mapping experiments (Talbot *et al.,* 1998). However, gamma-ray-induced alleles, particularly deletions, are quite useful for other purposes. For example, a deletion allele provides a quick test to determine whether a candidate gene or marker maps to the vicinity of the mutant locus: sequences that are present in the genomic DNA of deletion mutants (as determined by PCR assay or Southern hybridization) are either not located in the deleted interval or are repeated elsewhere in the genome (Schier *et al.,* 1997; Zhang *et al.,* 1998). Thus sequences failing this "deletion test" should be avoided when selecting probes to continue a genomic walk.

In the gene discovery phase of the project, it is a great advantage to have multiple ENU-induced alleles. This accelerates the discovery of molecular lesions in the locus and increases the likelihood of identifying an obvious "molecular null" allele, such as a nonsense or frameshift mutation at the beginning of the open reading frame. The potential peril of relying on only one allele is exemplified by *oep,* for which two ENU-induced alleles were characterized (Zhang *et al.,* 1998). Whereas one of the alleles (oep^{tz57}; Haffter *et al.,* 1996) is a point mutation that destroys the initiator codon and is hence a null allele, the other (oep^{m134}; Schier *et al.,* 1997) is a stop codon that deletes only the last 16 residues of the oep protein (Zhang *et al.,* 1998). An analysis resting solely on the latter allele might not have been sufficient proof of the identity of the mutant locus, particularly when one considers that the C-terminally truncated protein encoded by the *oep*^{m134} allele can rescue *oep* mutants when expressed at high levels in the early embryo.

B. Isolation of Tightly Linked Genetic Markers

Positional cloning projects are initiated by genetic mapping experiments that identify DNA polymorphisms (markers) that are tightly linked to the mutation of interest. It is important that the linked markers be as close to the mutation as possible because the assembly of a genomic walk composed of more than a few steps is a substantial undertaking. Considering that available zebrafish genomic libraries have average insert sizes ranging from 100 to 500 kb and that 1 cM of genetic distance represents an average of 600 kb of DNA, it is reasonable to initiate a positional cloning project with a marker located within 1 cM of the mutation (Postlethwait *et al.,* 1994; Postlethwait and Talbot, 1997). Note, however, that the rate of recombination per unit physical distance varies across the genome, so that genetic distance is only a rough indicator of the length of the genomic walk needed to reach any particular mutation. Thus the assumption that a tightly linked genetic marker is in close physical proximity does not necessarily hold in a region where recombination is suppressed, as often occurs near the centromere. In the case of the telomere-proximal *oep* genomic region, a physical distance of ~400 kb corresponded to a genetic distance of ~0.3 cM.

In species such as mouse and human with extremely dense genetic maps, one often identifies suitable entry points for positional cloning by mapping a mutation and then determining which are the closest markers in the region. The zebrafish genetic map currently is made up of more than 1000 PCR-based polymorphisms, including RAPDs (randomly amplified polymorphic DNA; see the following discussion and Postlethwait *et al.,* Ch. 9, this volume), SSLPs (simple sequence length polymorphisms; see the following discussion and Liao and Zon, Ch. 10, this volume), and cloned genes (Postlethwait *et al.,* 1998). These markers are spaced at an average interval of 3 cM (1.8 Mb). Therefore the average distance between a mutation and its closest mapped marker will be 1.5 cM or 0.9 Mb. Thus, in favorable cases, mapping a mutation will identify a previously known marker useful for positional cloning, and this will occur more frequently as more markers are added to the genetic map. However, there will be many cases when mutations are not within 1 cM of any previously mapped marker, and additional markers must be sought. This is doubly so because it is advisable to initiate genomic walks from markers on both sides of the mutation as a safeguard against encountering unclonable or highly repetitive regions that can hinder positional cloning projects.

Genomic regions containing mutations of interest can be saturated with genetic markers using bulked segregant analysis (BSA), a method first developed for plant positional cloning projects (Michelmore *et al.,* 1991; Giovannoni *et al.,* 1991). BSA compares PCR-based markers derived from two pooled genomic

DNA samples, one from wild-type individuals and the other from mutant siblings (Figs. 2 and 3). Markers linked to the mutation are differentially amplified between the two pools, whereas unlinked marker fragments are present in both pools because of independent assortment of markers on different chromosomes and crossing-over of distant markers on the same chromosome. Putatively linked markers identified by differential amplification from pooled samples are scored in individuals to verify linkage. Confirmed markers are then scored in larger families (composed of ~100–200 individuals) to determine which are candidates to be tightly linked to the mutant locus. Markers that still appear to be tightly linked to the mutation are then tested in larger mapping panels to measure the genetic distance more precisely and to resolve which of the identified markers is closest to the mutation. The markers used to clone *oep* were initially identified as RAPDs in a BSA screen using 10-mer primers of arbitrary sequence (Schier *et al.,* 1997), demonstrating the utility of this approach in zebrafish positional cloning projects (Fig. 3).

1. Marker Technologies

Genetic mapping experiments rely on PCR-based methods for detecting naturally occurring DNA polymorphisms between different strains of zebrafish. As described by Postlethwait *et al.* (Ch. 9, this volume), Liao and Zon (Ch. 10, this volume), and Johnson and Zon (App. 1, this volume), F_1 fish from a number of strain combinations can be used to generate crosses suitable for the PCR-based markers commonly examined by bulked segregant analysis.

a. Simple Sequence Length Polymorphisms

SSLPs are present in the zebrafish genome as tracts of $(CA)_n$ and other low-complexity sequences whose length varies in different strains (Knapik *et al.,* 1998; see Chapter 20 for in-depth review). Primer pairs amplifying more than 500 SSLPs were commercially available as of February 1998 from Research Genetics. These primers can detect polymorphic fragments linked to mutations in BSA. SSLPs have the advantage that they can be used to screen genomic libraries by PCR or oligonucleotide hybridization and to assign a mutation to a linkage group in maps with frameworks of SSLPs. However, the relatively small number of SSLPs currently available limits their usefulness because the average interval between SSLP markers is ~3.5 Mb. Hence, it is often necessary to screen other markers because, only in occasional cases, will one find SSLPs closely flanking a mutation of interest.

b. RAPD and AFLP Markers

As described in detail by Postlethwait *et al.* (Ch. 9, this volume) and Ransom and Zon (Ch. 12, this volume), randomly amplified polymorphic DNA (RAPD; Williams *et al.,* 1990); and amplified fragment length polymorphism (AFLP; Vos *et al.,* 1995); markers are generated with primers of arbitrary sequence, so they are available in practically unlimited numbers. This is a key advantage for positional

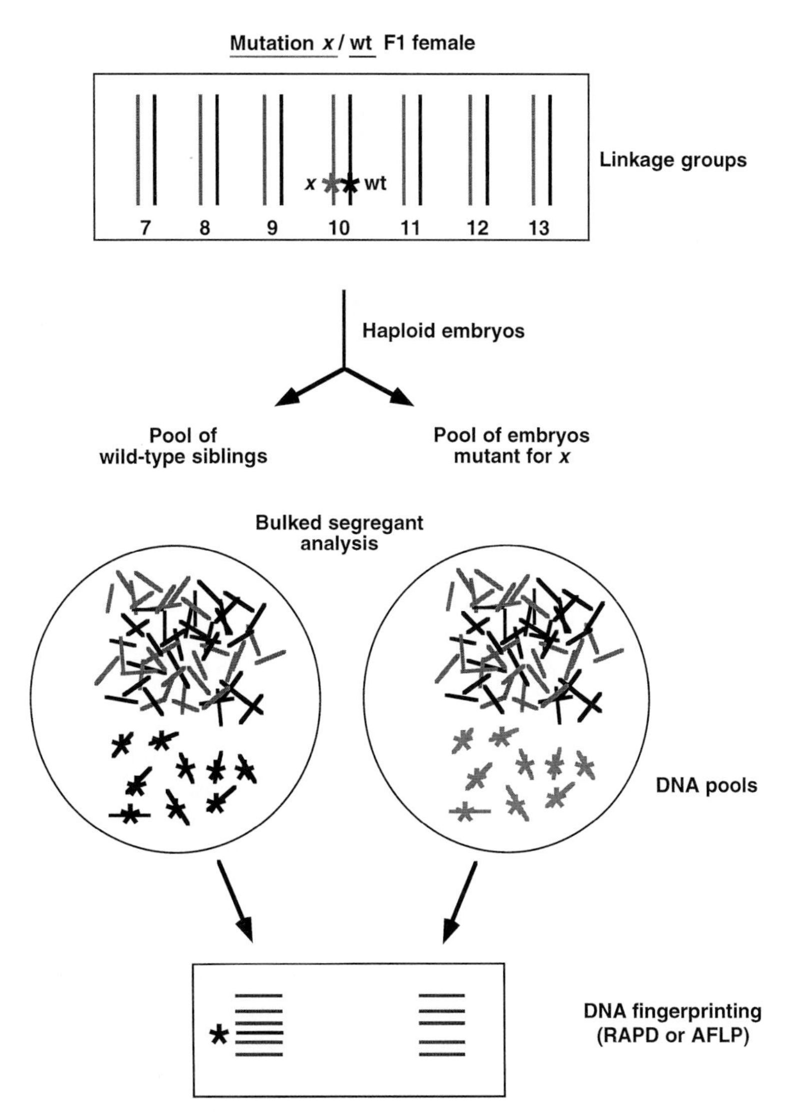

Fig. 2 Bulked segregant analysis and isolation of linked markers. In this example, haploid embryos are derived from a single heterozygous female (gray and black chromosomes). DNA from pools of phenotypically wild-type and mutant embryos is prepared. Mutation x is located on linkage group 10. Markers on other linkage groups and markers on linkage group 10 that are at a distance from mutation x are present in both pools because of independent assortment and crossing-over. The only difference between the wild-type and mutant DNA pools is the region harboring the locus of interest. DNA fingerprinting methods (see text for details) amplify marker fragments from genomic DNA in the two pools. Most marker fragments (seperated by gel electrophoresis) are shared between the pools, either because unlinked chromosomal regions are present in both pools or because the DNA marker is nonpolymorphic in the region linked to wild-type and mutant locus. Occasionally, a band amplifies in one pool but not the other, constituting a strong candidate for a linked marker.

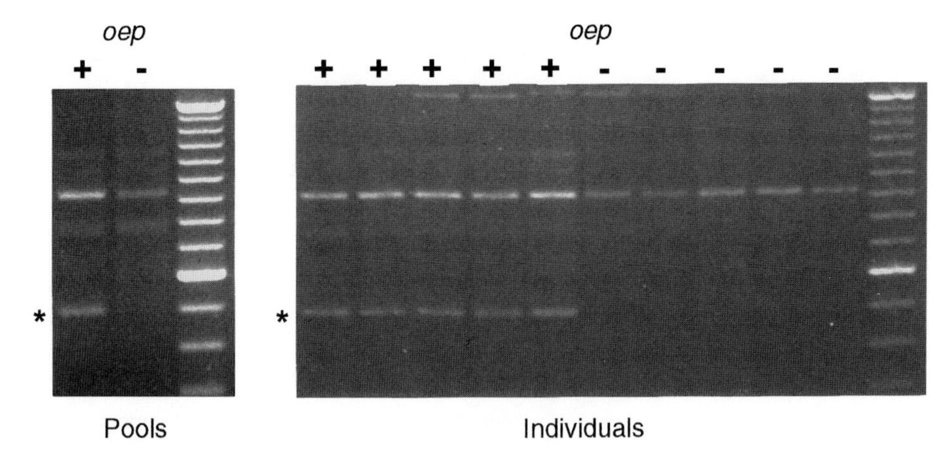

Fig. 3 Isolation of RAPD marker linked to *one-eyed pinhead (oep)*. DNA pools of 20 mutant (−) or wild-type (+) embryos were subjected to RAPD analysis (Schier *et al.*, 1997). Marker 15.AH500 (*) amplifies in wild-type but not mutant pools. Analysis of individual embryos confirms linkage. Further analysis established that this marker is located 0.03 cM from *oep* (Zhang *et al.*, 1998).

cloning, and BSA can assay thousands of AFLPs and RAPDs to identify those residing near the mutant locus. In an ideal case, one would identify markers within 0.5 cM of the mutant locus, so the goal of BSA is to find a marker in a 1-cM region of the ~3000-cM zebrafish genome. Thus a BSA screen should be designed to assay at least several thousand markers. Because AFLP markers allow higher throughput than RAPDs, an AFLP screen is currently the best way to identify markers suitable for positional cloning. A single AFLP assay using a unique primer pair generates more than ten polymorphic markers in a typical zebrafish mapping cross (Ransom and Zon, Ch. 12, this volume). Therefore, a screen of wild-type and mutant genomic DNA pools with more than 300 AFLP primer pairs, which a single investigator can achieve in a month or less, provides a good likelihood of finding a tightly linked marker.

AFLPs and RAPDs are typically scored as dominant markers, meaning that one allele is scored as a result of the presence of an amplified fragment, whereas the other is scored as a result of its absence (e.g., see Figs. 2, 3 and 4). This presents little difficulty in analyzing haploid embryos, but it is less efficient for diploid crosses because the homozygous dominant genotype cannot be distinguished from the heterozygote. In this case, dominant markers are informative if the dominant allele is linked to the wild-type but not the mutant chromosome.

It is sometimes difficult to screen a genomic library with RAPD markers directly because they are amplified by short primers that might hybridize to multiple sites in the genome, and these marker fragments often contain repetitive sequences. Thus, for markers demonstrated to be tightly linked to the mutation in initial experiments (e.g., 0–2 recombinants among 200 meioses), the best course for further analysis is to derive primers specific for the linked polymorphism

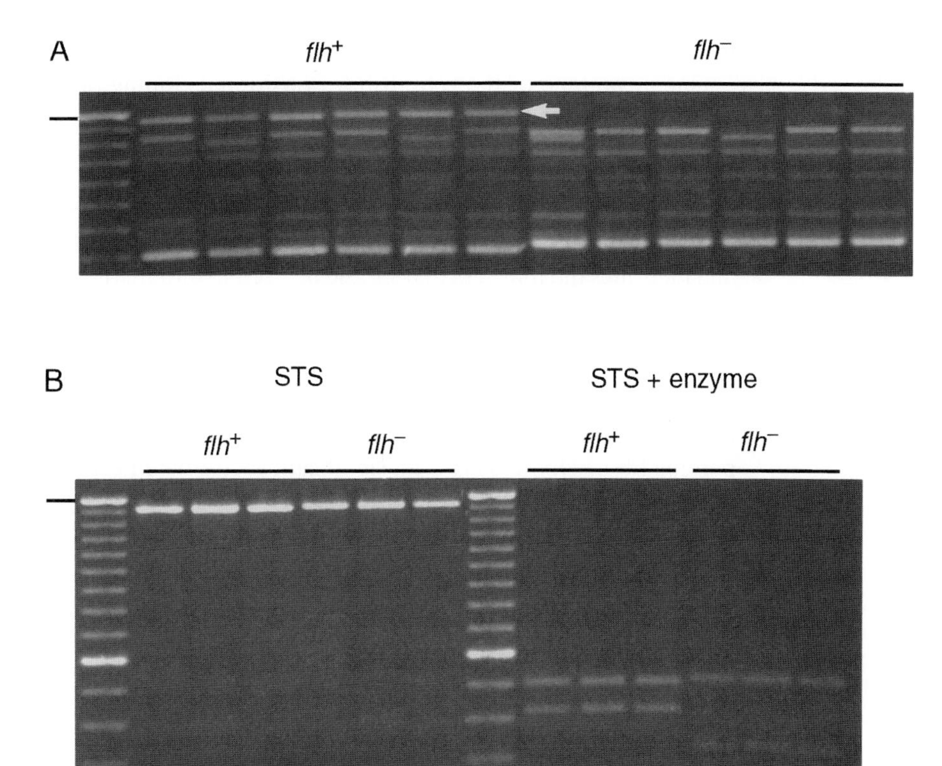

Fig. 4 Construction of an STS marker from a RAPD fragment. (A) The RAPD marker 7A.1450 is linked to the *floating head (flh)* locus. A panel comprised of 12 individual haploid siblings was assayed in RAPD PCR reactions with the primer 5′-GAAACGGGTG-3′. One of the amplified fragments, termed 7A.1450 (arrow), is linked to *flh,* as indicated by the presence of the marker fragment in the wild-type but not the mutant samples. This marker was initially identified in a BSA screen of wild-type and mutant genomic DNA pools. (B) The 7A.1450 STS is linked to *flh.* A pair of 21-mer primers derived from the sequence of the cloned 7A.1450 RAPD (Protocol 15.3) amplifies a fragment from genomic DNA of both wild-type and mutant haploid individuals (left panel). A polymorphism linked to *flh* is revealed when the STS is cleaved with the restriction enzyme RsaI (right panel), confirming that the STS was derived from the 7A.1450 RAPD. No recombinants between *flh* and the 7A.1450 locus were identified among 1332 haploid individuals scored with the RAPD, the STS, or both. Lines denote the 1500-bp fragment of the 100-bp Ladder (Life Technologies) that serves as a size standard.

from the sequence of the cloned marker fragment (Protocol 15.3). Locus-specific markers generated in this way are referred to as sequence-tagged sites (STS; Figure 4). The marker of interest may comigrate with other unlinked fragments that may be mistakenly cloned, so it is imperative to confirm linkage of the isolated STS and the mutant locus. Polymorphisms in STSs may be detected with restriction enzymes or by single-strand conformational analysis (Protocol 15.4, which follows, and Foernzler and Beier, Ch. 11, this volume).

C. High-Resolution Mapping

One advantage of the zebrafish for positional cloning projects is the ability to use high-resolution mapping to resolve DNA sequences that are in close physical proximity. Before a walk is initiated, it is essential to create a high-resolution map of the region containing the mutant locus by scoring large numbers of meiotic products with markers tightly linked to the mutation (Fig. 5). This gives a precise measurement of the genetic distance between markers and the mutation and identifies the markers that should be used to initiate the walk. High-resolution mapping also allows one to orient a contig assembled in a genomic walk, or even a single clone, by determining which end lies genetically closer to the mutation (Fig. 6). Moreover, mapping the ends of clones near the mutation allows one to determine whether the frequency of genetic recombination in the region of the mutation deviates from the genome-wide average of about 1% recombination per 600 kb of DNA. This is an significant factor when estimating the number of steps that will be required to traverse the distance between a particular marker and the mutant locus. Finally, and importantly, genetic mapping experiments determine which clone or clones within a contig contain the locus of interest (often referred to as "critical interval"). Thus mapping experiments greatly expedite a positional cloning project by directing the walk and by limiting the search for the gene after the genomic walk is completed.

To gain the most advantage from mapping experiments, it is important to score all genetic markers of interest in the same mapping panel. This obviates the

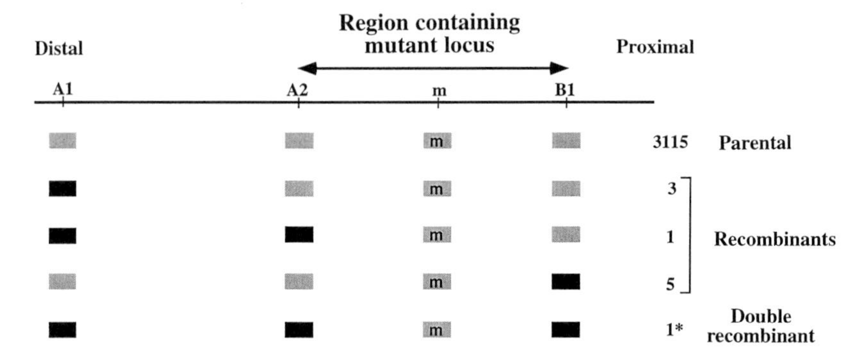

Fig. 5 Mapping of linked markers and identification of recombinants. DNA markers A1, A2, and B1 are closely linked to mutation m and are polymorphic between wild-type and mutant chromosomes. 3125 haploid mutant embryos derived from m/+ heterozygous F_1 females are scored. The chromosomal region defined by the three markers is derived from the mutant chromosome (gray boxes) in 3115 embryos (parental). Three embryos are wild-type (black box) for marker A1 but mutant for the other two markers, indicating recombination events between A1 and A2. One and five embryos are recombinants between A2 and the mutation m and B1 and m, respectively. One embryo (*) that was scored as phenotypically mutant is wild type for markers A1, A2, and B1. This embryo could be a double recombinant. However, it is also possible that the embryo is actually wild-type and has been misclassified as mutant because double crossovers in a small interval are rare.

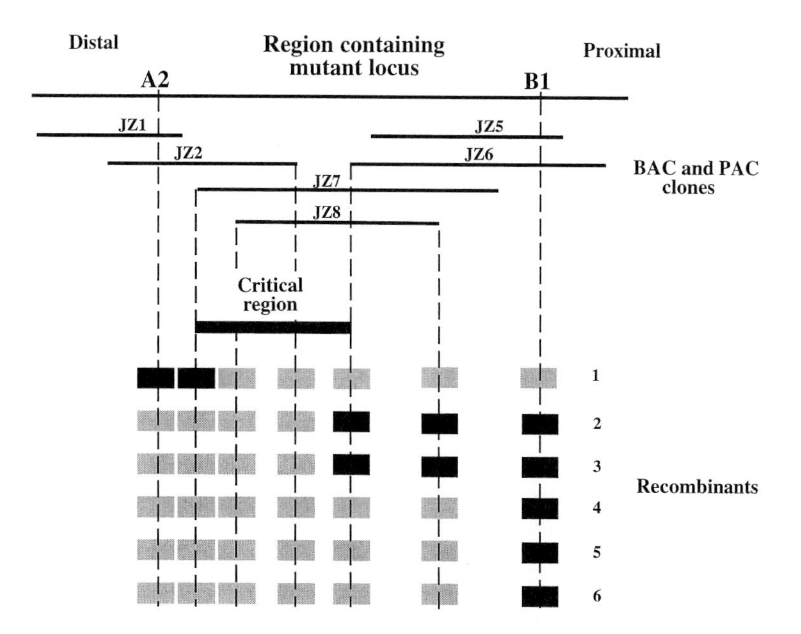

Fig. 6 Assembly of genomic walk, fine mapping, and identification of critical region. Marker A2 identified clones JZ1 and JZ2 and marker B1 identified JZ5 and JZ6. The proximal and distal ends of JZ2 and JZ6, respectively, then identified JZ7 and JZ8. For fine mapping, end sequences were used to generate PCR markers that were scored in the recombinant embryos identified in Fig. 5. The distal end of JZ7 is still one recombinant away from the mutant locus, whereas the distal end of JZ6 is two recombinants away from the mutant locus. The distal end of JZ8 and the proximal end of JZ2 are in the 0 recombinant interval. This places the mutation in a critical interval between the distal end of JZ7 and the distal end of JZ6. Black and gray boxes depict wild-type and mutant genomic regions, respectively.

need to conduct repeatedly thousands of genotype assays because only individuals previously shown to carry a crossover in the region of interest need to be analyzed (Fig. 6). One might begin a typical set of experiments by scoring 2000–3000 haploid individuals from mutant mapping crosses with markers flanking the mutation. Producing 1000 or more zebrafish embryos is a matter of a few days and scoring their genotypes is straigthforward if one develops the appropriate markers. Markers that can be scored on agarose gels, such as STS size polymorphisms revealed by enzyme digestion (Fig. 4), are convenient for these purposes. For example, STS polymorphisms derived from RAPDs linked to *floating head* (*flh*) and *oep* were scored using agarose gels on panels of 1300 and 3100 individuals, respectively (Talbot *et al.*, 1995; Schier *et al.*, 1997; Zhang *et al.*, 1998). To minimize errors, the phenotypes of each individual added to the mapping panel should be double-checked, and no individuals with ambiguous phenotypes should be used for mapping experiments. Mis-scored embryos will appear to be recombinants, leading to an overestimate of the genetic distance between the marker and the mutation (Fig. 5). For example, a wild-type haploid embryo that is misclassified as mutant will be scored as a recombinant. It would be concluded that this embryo contains the mutant allele of the locus of interest as infered from the phenotype but the wild-type allele of the linked marker. To detect such errors, the panel should be scored with markers on both sides of the mutant locus. This will reveal individuals that are putative double recombinants (Fig. 5). In the previous example of an embryo misclassified as mutant, both flanking marker alleles would be from the wild-type chromosome, indicative of a potential double recombination event. Provided the flanking markers are tightly linked to the mutation, these animals are very likely to be phenotypically mis-scored rather than being bona fide double recombinants because double crossovers in a short interval are rare. These putatively misclassified individuals should be excluded from initial analysis but can be analyzed later with markers from the walk that are more closely linked to the mutation to determine if these embryos might indeed be double recombinants.

These initial mapping experiments identify the relatively rare recombinants that will be useful in resolving any additional markers located in the vicinity. Nonrecombinant embryos need not be analyzed further because individuals that carry no crossovers in the region will not be useful for resolving new markers lying within it. Thus, after scoring a large mapping panel for flanking markers, one may obtain the genetic map resolution of the entire panel by scoring only a few key recombinants.

In summary, we recommend initiating a positional cloning project by testing several thousand AFLP markers (corresponding to more than 200 AFLP primer pair combinations) on pools of 40 wild-type and 40 mutant embryos. To identify flanking markers closest to the mutant locus, 200 individual embryos should then be tested. Ideally, an additional 2000–3000 embryos are then tested for fine-mapping using an agarose-scorable STS. These steps can be carried out within a few months.

D. PROTOCOL 15.1
Rapid Isolation of Genomic DNA from Embryos

This procedure generates DNA of sufficient quality for most PCR assays but cannot be used for other purposes [e.g., restriction enzyme digestion, amplified fragment length polymorphism (AFLP)] that demand high-quality samples of double-stranded DNA. One embryo provides enough genomic DNA for up to several thousand PCR assays.

1. Dechorionate individual embryos in embryo medium and transfer to 96-well PCR plates (MJ Research).
2. Add 50 μl lysis buffer (10m*M* Tris–HCl pH 8, 1.0m*M* EDTA, 0.3% Tween, 0.3% NP40).

3. Incubate for 10 minutes at 98°C.
4. Digest by addition of 5 μl proteinase K (10 mg/ml; Boehringer) at 55°C o/n.
5. Inactivate proteinase K by incubation at 98°C for 10 minutes.
6. Store genomic DNA at −80°C.

E. PROTOCOL 15.2
PCR with Genomic DNA Templates

1. Dilute genomic DNA prepared as above 25- to 250-fold in 0.1 × TE (1m*M* Tris, pH 8.0, 0.1m*M* EDTA). Use 5 μl of diluted genomic DNA in a 25-μl reaction.

2. Design primers 20–24 nucleotides in length, with T_m of 55–60°C.

3. For PCR analysis in 25-μl reactions, use the following final reaction conditions: 10m*M* Tris–HCl pH 8.4, 50m*M* KCl, 1.5m*M* $MgCl_2$, 0.001% gelatin, 100μ*M* each dNTP, 0.1 mg/ml BSA, 100n*M* each forward and reverse primer, 0.3 units Taq DNA polymerase. Cycling parameters: 3 minutes at 94°C followed by 35 cycles of 30 seconds at 94°C, 30 seconds at 55°C, and 30 seconds at 72°C, followed by 3 minutes at 72°C.

F. PROTOCOL 15.3
Constructing a Sequence-Tagged Site from RAPD and AFLP Markers

1. Perform PCR with appropriate RAPD or AFLP primers to amplify linked marker.

2. Excise gel band containing marker fragment.

3. Reamplify marker using excised fragment as template, and electrophorese products in low-melting-point agarose gel.

4. Excise and clone appropriate fragment using the TA cloning kit (Invitrogen) or a similar system.

5. Identify clones with the appropriate insert by restriction enzyme and PCR tests.

6. Sequence the marker fragments. Design 20 to 24-bp primers with T_m of 55–60°C, taking care to avoid repeated sequences identified by similarity searches.

7. Confirm that the STS is genetically linked to the mutation by scoring a polymorphism in a mutant mapping cross. Single strand confirmation polymorphisms SSCPs; see Protocol 15.4 and Foernzler and Beier, Ch. 11, this volume) can often be identified in STS markers, and restriction site polymorphisms are also very useful (Fig. 4), particularly when large numbers of individuals are to be scored.

G. PROTOCOL 15.4
Single-Strand Conformational Polymorphism Analysis

1. Generate labeled fragments (100–300 bp) by including 3 μCi of [α-^{32}P]dATP per PCR using the conditions described in Protocol 15.2.
2. Add 30 μl sample running buffer (80% formamide, 0.1% bromophenol blue; 0.1% xylene cyanol; 1m*M* EDTA; 10m*M* NaOH).
3. Denature samples at 98°C for 10 minutes and immediately cool on ice.
4. Separate 4 μl of single stranded amplification product on a nondenaturing 4.5% polyacrylamide gel (acrylamide : bisacrylamide 39 : 1) in 1× TBE at 4°C at 25–40 W.

III. Isolating Genomic Clones and Defining the Critical Region

The second phase of a positional cloning project involves the application of isolated markers to identify genomic clones in the region and the assembly of a contig containing the mutant gene (Fig. 6). Here we focus on two aspects of this step: (1) choosing and screening libraries of large-insert genomic clones and (2) monitoring the progress of the walk to ensure that clones containing the mutant locus are identified.

A. Large-Insert Genomic Libraries

Genomic libraries constructed in bacterial artificial chromosome (BAC), P_1 artificial chromosome (PAC), and yeast artificial chromosome (YAC) vectors are available for zebrafish positional cloning projects. These vector systems and libraries are described in more detail elsewhere in this volume (Amemiya *et al.*, Ch. 14, this volume), so we will provide only a brief overview. BAC and PAC clones (available from Genome Systems) are propagated in *E. coli*, which simplifies manipulation of these clones compared to YACs, which grow in yeast. YACs have the advantage, however, that they can contain larger inserts: zebrafish YAC libraries (available from Research Genetics and Genome Systems) have average insert sizes of 240–480 kb, compared to 80 and 110 kb for the BAC and PAC libraries, respectively. Thus the choice of library depends on the physical distance to be traversed. For example, in the cloning of *oep*, the genetic distance between the nearest marker and the mutation was 0.03 cM (1 recombinant among 3122 meioses), corresponding to an estimated physical distance of 20 kb. Therefore, BAC and PAC libraries were screened, and, according with the estimate, the *oep* gene was contained in the same BAC clone as the genetic marker. In cases where the marker and the mutation recombine more frequently (>0.5 cM), it is advisable to initiate the walk with YAC clones and then to isolate BACs and

PACs in the vicinity of the mutation for detailed analysis of the region. We recommend the use of BAC or PAC clones in the final phase of the walk and for gene discovery because these are less treacherous than YAC clones, which are often chimeric or internally deleted.

To initiate a genomic walk, the closest markers on either side of the mutant locus are used to isolate genomic clones. Libraries can be screened by PCR or by hybridization to filters with arrayed clones (Baxendale *et al.,* 1991; Kern and Hampton, 1997). Both methods are convenient, although we typically use PCR if robust, locus-specific primers are available. This is done efficiently using PCR on DNA samples from pooled clones. In this approach, upper pools containing DNA from clones on multiple plates, single plate pools, and column and row pools are screened for amplification of a specific STS. This method allows the identification of clones containing a given STS within a few days. The oligonucleotides that serve as primers for a marker can also be used to screen a library by hybridization to arrayed filters (see Protocol 15.5 and Fig. 7), which is a useful method for libraries for which PCR pools are not available.

To characterize newly identified clones, end sequences are determined by direct sequencing of BAC or PAC clones or by sequencing the rescued ends of YAC clones (Protocols 15.7 and 15.9). In addition, we routinely determine the insert size of BAC and PAC clones by pulsed field gel electrophoresis. Primer pairs derived from end sequences are tested for amplification on the following

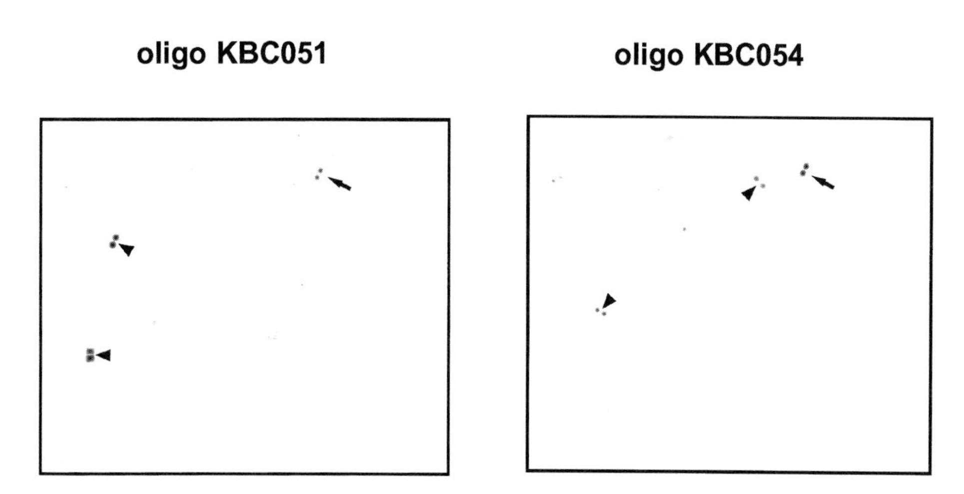

Fig. 7 Identification of genomic clones by hybridization of oligonucleotide probes to arrayed PAC filters. Oligonucleotides KBC051 and KBC054 were derived from the end of a BAC clone (97A17) from the *one-eyed pinhead* walk (Zhang *et al.,* 1998). End-labeled oligos were seperately hybridized to duplicate PAC filter sets (Protocol 15.5). Note that each oligo hybridizes to two unique clones (arrowheads) that are unlikely to be linked to *oep* and one overlapping clone that is located in the *oep* region (arrow; corresponding to PAC 134F10 in Zhang *et al.,* 1998). Each clone is spotted on the filters twice, allowing convenient identification of positive clones.

templates: (1) the original genomic clone to verify that primers amplify the sequenced end, (2) any additional genomic clones from the walk to determine overlap between clones, and (3) genomic DNA to determine if primers amplify a single band. We have found that end sequences often contain repetitive DNA, making the identification of primer pairs that amplify a single, genetically linked segment difficult. It is thus advisable to test several primer combinations for a given end. It is particularly useful to analyze primer pair combinations in deletion alleles. Primer pairs that amplify in wild-type embryos but not deletion mutants can then be used for testing the overlap among clones, continuing the walk and fine-mapping. Because of high chimerism of YAC clones, it is particularly important to determine if the isolated YAC ends are linked to the mutant locus (Green *et al.,* 1991). This can be done by the "deletion test" or by genetic mapping using the amplified ends as STSs. After overlap has been established among the isolated clones, terminal STSs are used on the mapping panel to monitor if any crossovers have been passed and assess the progress of the walk. To carry out the next step of the walk, the primer pair that amplifies sequences from the end of the contig (i.e., the end that does not amplify previously identified clones) is used to screen the library again by PCR.

The major time-consuming step in the PCR-based approach is the requirement to identify a new STS for continuing the walk. This demands that the ends of identified genomic clones are sequenced and that specific primers are synthesized and then tested for overlap and suitability to serve as STSs. It can be faster to isolate clones for a new step by hybridization with clones from the previous step (Fig. 8). This eliminates the need for end-sequence determination and STS development. However, this approach can be hindered by the repetitive sequences in genomic clones, which cause cross-hybridization between clones that are in different regions of the genome. This problem can be offset by preannealing (suppressing) the probe with unlabeled zebrafish genomic DNA (Protocol 15.6). Note that this method is not perfect and that it is essential to confirm the map location of novel clones by STS linkage analysis.

B. Defining the Critical Region

1. High-Resolution Mapping

Once a contig has been assembled, it is necessary to delimit the critical region where the mutant locus resides. At this step, high-resolution genetic mapping can be used to great benefit (Fig. 6). As described earlier, STSs derived from the genomic clones of the contig are tested on the recombinants in the mapping panel. The closest flanking markers that are still separated by one recombinant from the mutant locus define the borders of the critical interval. In the case of *oep,* for example, analysis of STSs derived from PAC, BAC, and cosmid clones in 3122 meioses limited the *oep* genomic region to a ~100-kb interval.

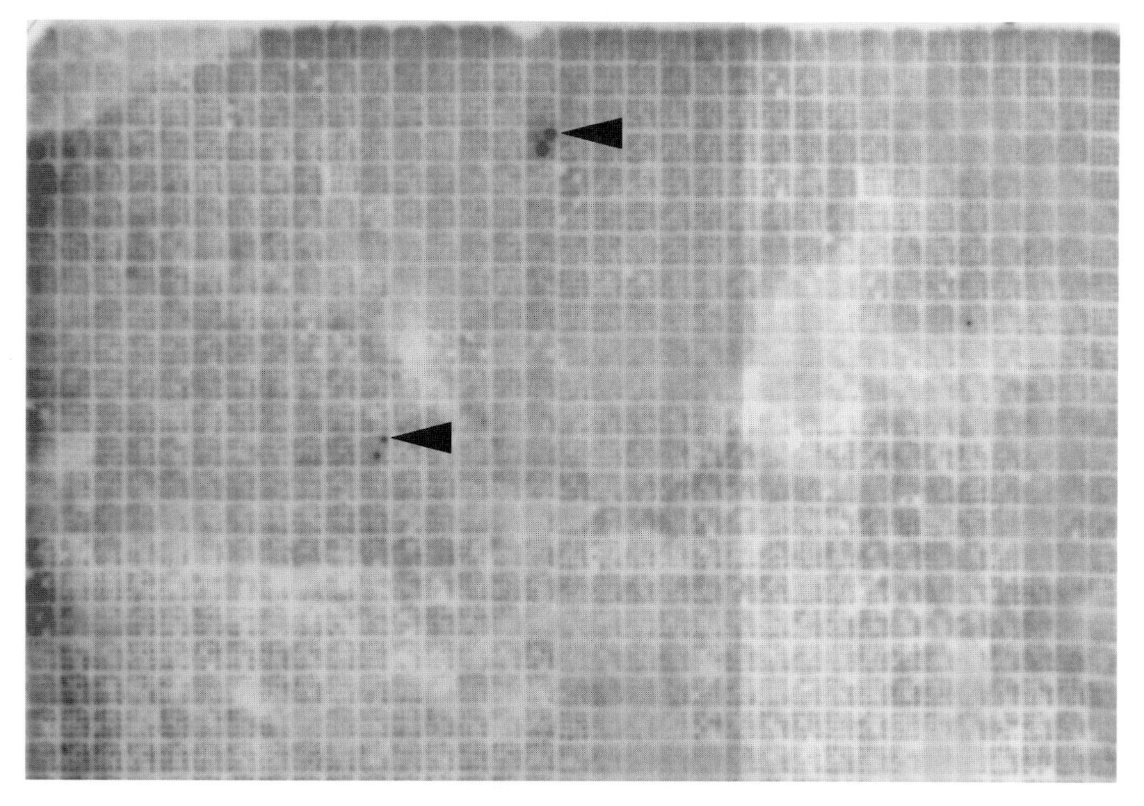

Fig. 8 Screening a genomic library by hybridization with a BAC clone insert. A probe (prepared as in Protocol 15.6) from the 120-kb genomic clone BAC 8P19 was hybridized to a zebrafish BAC library arrayed on high-density filters. A portion of a filter with two hybridizing clones (arrowheads) is shown. Each clone is spotted on the filters twice, allowing convenient identification of positive clones.

2. Complementation by Injection of Genomic Clones

Another potentially powerful method to delimit the critical interval containing the mutant locus involves the injection of genomic clones and assay for (partial) complementation of the mutation. Recent experiments show that *flh* mutants can be partially rescued by injection of BAC clones that contain the wild-type gene (Yan *et al.,* 1998). Phenotypic rescue in these experiments was only partial, presumably reflecting the mosaic inheritance and expression of the injected transgene, which has been documented for other clones injected into one-cell embryos (Westerfield *et al.,* 1992). This suggests that transient transgenic rescue will be useful for gene discovery in some positional cloning projects. It is too soon to say how widely this method will be useful because the partial nature of the rescue might hinder identification of genes acting in a small number of cells.

Furthermore, late-acting genes may not be scorable in transient assays. Despite these concerns, microinjection rescue assays may provide a rapid approach to characterize loci, such as *flh,* that produce easily scorable, early phenotypes.

C. PROTOCOL 15.5
Screening BAC and PAC Libraries by Hybridization Using Oligonucleotide Probes

1. Screen filters with arrayed BAC and PAC clones (Genome Systems) using oligonucleotides end-labeled with [γ-^{32}P]ATP and T4 polynucleotide kinase. Use oligonucleotides from primer pairs that amplify a single band in genomic PCR and hybridize them individually to separate filter sets.
2. Prehybridize fresh filter sets in Hyb buffer (6× SSC, 5× Denhardt's Solution, 0.5% SDS, 0.1 mg/ml salmon sperm DNA) for 2 hours at 55°C.
3. Hybridize at T = (T_m of the oligonucleotide − 4 to 6°C) in Hyb buffer at 10^6 cpm/ml for 4 hours.
4. Wash filters 2 × 7.5 minutes in 5× SSC/0.5% SDS at hybridization temperature.
5. Rinse briefly with 5× SSC/0.5% SDS at room temperature.
6. Expose to film and determine coordinates of double-positive BAC or PAC clones. By selecting clones hybridizing to both oligonucleotides, repetitive sequences can be excluded in most cases.
7. Filters can be stripped by washing in 0.1× SSC/0.1%SDS at 68°C for 2 hours.

D. PROTOCOL 15.6
Screening Genomic Libraries by Hybridization with BAC/PAC Probes

1. Gel purify cosmid or BAC or PAC inserts (~100 ng) after NotI digestion.
2. Digest with EcoRI to reduce the DNA fragment size.
3. Label by random priming in the presence of [α-^{32}P]dATP and [α-^{32}P]dCTP.
4. For competition of repetitive sequences, incubate labeled probes with 200 μg of sheared zebrafish genomic DNA and 0.5 μg/μl poly(CA)/poly(GT) (Sigma-0307) in a total of 600 μl.
5. Boil for 8 minutes.
6. Chill on ice.
7. Add 1*M* NaPO4 pH 7.4 to a final concentration of 0.12*M*.
8. Allow DNA to anneal for 2.5 hours at 65°C.
9. Use suppressed probe at 10^6 cpm/ml for hybridization to arrayed filters that have been prehybridized in 6× SSC, 5× Denhardt's, 0.5% SDS, 100 μg/ml salmon sperm DNA at 68°C for 2 hours.

10. Hybridize o/n at 68°C.
11. Wash filters for 15 minutes in 1× SSC/0.5% SDS at room temperature.
12. Wash filters for 15 minutes 1× SSC/0.5% SDS at 65°C.
13. Wash filters for 15 minutes 0.1× SSC/0.5% SDS at 65°C.
14. Rinse in 1× SSC/0.5% SDS at room temperature.
15. Expose to film and determine coordinates of positive clones.

E. PROTOCOL 15.7
YAC End Rescue

Isolation of Yeast Genomic DNA (from Kaiser *et al.*, 1994)

1. Grow 10 ml yeast cultures to saturation in YPD at 30°C

2. Collect the cells by centrifugation for 2 minutes in a clinical centrifuge. Remove the supernatant and resuspend the cells in 0.5 ml of distilled water. Transfer the cells to a 1.5 ml microfuge tube and collect them by centrifugation for 5 seconds in a microfuge.

3. Decant the supernatant and briefly vortex the tube to resuspend the pellet in the residual liquid.

4. Add 0.2 ml of 2% Triton X-100, 1% SDS, 100m*M* NaCl, 10m*M* Tris–Cl (pH 8), and 1m*M* Na_2EDTA. Add 0.2 ml of phenol:chloroform:isoamyl alcohol (25:24:1). Add 0.3 g of acid-washed glass beads [0.45–0.5 mm beads (Sigma) soaked in nitric acid and washed in copious amounts of distilled water].

5. Vortex for 3–4 minutes. Add 0.2 ml of TE (pH 8).

6. Centrifuge for 5 minutes in microfuge. Transfer the aqueous layer to a fresh tube. Add 1 ml of 100% ethanol. Mix by inversion.

7. Centrifuge for 2 minutes in microfuge. Discard the supernatant. Resuspend the pellet in 0.4 ml of TE plus 3 μl of a 10 mg/ml solution of RNase A (in 50m*M* potassium acetate, pH5.5; boiled for 10 minutes). Incubate for 5 minutes at 37°C. Add 10 μl of 4*M* ammonium acetate plus 1 ml of 100% ethanol. Mix by inversion.

8. Centrifuge for 2 minutes in a microfuge. Discard the supernatant. Air-dry the pellet (~10-20 μg DNA) and resuspend in 50 μl of TE.

Recovery of YAC Ends in E. coli, for Clones in YAC Vector pRML1/2 (from Zhong *et al.*, 1998)

1. Digest 1 μl (~0.2-0.4μg) of yeast DNA with BamH1 or Spe1 in 100 μl.

2. Phenol extract (phenol, phenol/chloroform, chloroform, 1× each). Add 1 μl tRNA (10 μg/μl stock) and ethanol precipitate.

3. Centrifuge 15 minutes in Eppendorf centrifuge at 12,000 *g* at 4°C.

4. Discard supernatant and resuspend pellet in 25 μl TE.

5. Add 10 μl ligase buffer, 5 μl ligase (400 units/μl New England Biolabs), 60 μl H_2O. Incubate at 16°C overnight.

6. Ethanol precipitate and spin for 15 minutes at 12,000 *g* at 4°C.

7. Resuspend in H_2O and electroporate into *E. coli* DH10B.

F. PROTOCOL 15.8
Preparation of BAC/PAC DNA for Direct Sequence Analysis

This protocol employs QIAGEN DNA purification columns and reagents, with some modifications of the standard QIAGEN plasmid protocol that improve the yield for large clones.

1. Resuspend pellet from 100 ml culture in 50 ml buffer P1 (50m*M* Tris–HCl, pH 8.0, 10m*M* EDTA) with 200 μg/ml RNase A.

2. Lyse cells by addition of 50 ml buffer P2 (200m*M* NaOH, 1% SDS). Incubate 10 minutes at room temperature.

3. Add 50 ml of chilled buffer P3; incubate on ice for 15 minutes.

4. Centrifuge at 20,000 *g* for 30 minutes, and discard pellet. Filter or recentrifuge supernatant to remove any remaining debris.

5. Add 105 ml isopropanol to supernatant. Centrifuge 15,000 *g* for 30 minutes to collect DNA precipitate.

6. Resuspend pellet in 0.5 ml dH_2O, add 5 ml buffer QBT (750m*M* NaCl, 50m*M* MOPS, pH 7.0, 15% ethanol, 0.15% Triton X-100), and apply to QIAGEN-tip 500 that has been equilibrated with 10 ml buffer QBT.

7. Wash column twice with 30 ml buffer QC (1*M* NaCl, 50m*M* MOPS, pH 7.0, 15% ethanol). Elute BAC/PAC DNA in 15 ml buffer QF (1.25*M* NaCl, 50m*M* Tris, pH 8.5, 15% ethanol) that has been warmed to 65°C.

8. Precipitate DNA with 10.5 ml isopropanol. Centrifuge at 15,000 *g* for 30 minutes, and wash pellet with 70% ethanol. Resuspend DNA in 50 μl.

G. PROTOCOL 15.9
Sequence Analysis with BAC and PAC Templates

1. Use 1–2 μg of BAC/PAC DNA in 10 μl H_2O as template for sequencing reaction.

2. Add 2 μl of primer (12.5 pmol/μl). For terminal sequence analysis of PAC and BAC clones, the appropriate primers are:

PACSP6: 5′-GGCCGTCGACATTTAGGTGACAC-3′
PACT7: 5′-CCGCTAATACGACTCACTATAGGG-3′

BACSP6: 5′-CGCCAAGCTATTTAGGTGACAC-3′
BACT7: 5′-GTAATACGACTCACTATAGGG-3′

3. Add 8 μl ABI Ready Reaction mix.
4. Cycle with following program:
   1. 95°C, 5 minutes
   2. 95°C, 30 seconds
   3. 50°C, 20 seconds
   4. 60°C, 4 minutes
   5. Go to step 2 within this cycle, 49 more times.
   6. 4°C

5. Remove unincorporated nucleotides and electrophorese as for standard plasmid templates (see, for example Wilson and Mardis, 1997a).

IV. Gene Discovery

A. Identifying Candidate Genes

1. Screening cDNA Libraries with Large-Insert Genomic Clones

The final phase of a positional cloning project is discovering the mutated gene within the genomic interval defined in the previous step. The most straightforward way to identify transcription units in an interval of genomic DNA is to directly screen an appropriate cDNA library with BAC, PAC, and cosmid clones that span the region. As with screening genomic libraries, the problem of repetitive sequences (which are present in untranslated sequences of some cDNAs) can be partly offset by prehybridizing the probe with unlabeled total zebrafish genomic DNA and simple repeat sequences (Baxendale *et al.,* 1991; Kern and Hampton, 1997). Despite these precautions, spurious clones are isolated by this method, and care must be used to ensure that cDNAs chosen for further analysis do, in fact, map to the region of the mutation. This can be achieved by assaying nonrepetitive STS markers derived from the cDNAs on genomic clones defining the interval or by genetically mapping cDNA-derived STSs in the recombinant panel used to orient and monitor the genomic walk.

In most cases multiple, potentially different, cDNA clones will be isolated. To classify cDNAs, we recommend fingerprinting the clones by digestion with frequently cutting restriction endonucleases such as Dde1. Clones can then be further subgrouped by end sequencing.

2. Sequence Analysis of Critical Region

In some cases, genetic mapping experiments may limit the locus to an interval that is small enough to be completely sequenced (Wilson and Mardis, 1997b).

This approach has been used successfully in a number of human positional cloning projects (e.g., Yu *et al.,* 1996), and there is nothing that would prevent its use in zebrafish as well. With the advent and continuing improvement of automated sequencing technology, the effort required to sequence a BAC or PAC clone might not be substantially greater than that required to sequence and map all the cDNAs identified by direct hybridization, particularly if the genomic region contains many genes. As the database of zebrafish ESTs becomes more comprehensive, this approach will become more compelling for the discovery of zebrafish genes because partial "sample squencing" of subclones from the BAC/PAC will quickly identify genes in the region.

B. Testing Candidate Genes

Once identified, candidate genes may be evaluated further by a number of criteria, including sequence, expression pattern, and ability to rescue the mutant phenotype.

1. *In Situ* Hybridization

Whole mount *in situ* hybridization is a routine method in the zebrafish (Jowett, Ch. 6, vol. 59), and it is likely that this will be informative in searches for many genes whose activities are developmentally restricted. However, prioritizing candidates on the basis of expression pattern involves assumptions based on mutant phenotype, and these can be misleading. For example, some genes required for the development of limited cell populations may be ubiquitously expressed. Thus *in situ* hybridization provides a convenient initial assay to prioritize candidates, but it is important not to eliminate candidates from further consideration solely on the basis of expression pattern.

2. Mutant Rescue by RNA or DNA Injection

Mutant rescue assays provide another means of evaluating candidates. Candidates identified as cDNAs may be quickly tested in a rescue assay by microinjection of synthetic RNA transcribed from plasmid templates (Schulte-Merker *et al.,* 1997; Kishimoto *et al.,* 1997; Zhang *et al.,* 1998; Ivics *et al.,* Ch. 6, this volume). Clones with rescuing activity become very strong candidates for the mutant locus. Negative results are much harder to interpret because it is difficult to compare early ubiquitous expression of a synthetic transcript to the endogenous situation in which the gene's promoter controls its expression. Thus RNA injection could lead to false negatives and also to dominant effects at early stages that obscure the ability to assay for rescue at later stages. In addition, this approach requires the effort to ensure that the injected clones encode a complete open reading frame. Rescue by genomic DNA injection (Yan *et al.,* 1998) does not suffer concerns about ectopic expression because the endogenous promoter drives expression. If rescue is achieved with a genomic DNA fragment, testing subcloned fragments should be an extremely effective way to identify the gene.

3. Sequencing of Mutant Alleles

Final proof of the identity of the mutated locus necessitates discovery of molecular lesions in mutant alleles of the gene. We view sequence analysis of mutant alleles as a final test of candidates supported by other assays such as expression pattern or rescuing activity, but one could elect to use this as the primary means of evaluating candidates if automated sequencing resources are available. Most ENU-induced mutations that have been characterized in zebrafish to date are nonsense mutations or missense mutations in conserved residues, which can be easily discovered by sequencing cDNA or genomic DNA corresponding to the open reading frame (ORF) (e.g. Talbot *et al.*, 1995; Brand *et al.*, 1996; Kishimoto *et al.*, 1997; Schulte-Merker *et al.*, 1997; Zhang *et al.*, 1998). However, there are examples of apparently ENU-inuduced mutations that disrupt essential noncoding elements, such as control regions, which can be much more difficult to identify. This underscores the importance of analyzing multiple alleles in the gene discovery phase. Mutant cDNA sequences can be conveniently obtained by RT-PCR from the appropriate RNA samples (Grinblat and Sive, Ch. 9, vol. 59), an approach that is not complicated by large introns. In cases where the mutant RNA is not synthesized or is difficult to obtain, genomic DNA may be amplified by PCR and sequenced. When genomic DNA amplification is hindered by large introns, one can define exon boundaries by directly sequencing PAC or BAC clones containing the gene using exon-specific primers.

C. PROTOCOL 15.10
Screening cDNA Libraries by Hybridization with BAC/PAC Probes

1. Follow steps 1–8 in Protocol 15.6.
2. Use suppressed probe for hybridization at 10^6 cpm/ml to cDNA filters (50,000 plaques/filter) that have been prehybridized in 6× SSC, 5× Denhardt's, 0.5% SDS, 100 μg/ml salmon sperm DNA at 68°C for 2 hours.
3. Hybridize o/n at 68°C to duplicate filters.
4. Wash filters for 15 minutes in 1× SSC/0.5% SDS at room temperature.
5. Wash filters for 15 minutes 1× SSC/0.5% SDS at 65°C.
6. Wash filters for 15 minutes 0.1× SSC/0.5% SDS at 65°C.
7. Rinse in 1× SSC/0.5% SDS at room temperature.
8. Expose to film and pick positive plaques from primary screen for secondary screen under the same conditions (use freshly suppressed probe, do not reuse previously suppressed probes).

9. For lambda ZAP libraries, excise phagemids using ExAssist helper phage (Stratagene).

10. Group clones by Dde1 restriction enzyme analysis and end-sequence determination as necessary.

11. Use representative cDNA clones to synthesize antisense and sense probes for whole mount *in situ* hybridization on wild-type embryos.

D. PROTOCOL 15.11
Mutation Detection by RT-PCR and Direct Sequencing

Point mutations often do not affect the presence of the mRNA of a mutant gene. To determine if the RNA of a mutant gene is present, it is advisable to first perform whole mount *in situ* hybridization on mutant embryos. In the cases where the candidate mutant mRNA is present, RT-PCR (Grinblat and Sive, Ch. 9, vol. 59) is performed and cDNAs are directly sequenced.

RNA Isolation

1. Homogenize embryos in TRIzol Reagent (Life Technologies). Use 200 μl for ten 30-hour-old embryos. One embryo allows isolation of ~500 ng total RNA.
2. Incubate for 5 minutes at room temperature.
3. Add 40 μl of chloroform per 200 μl TRIzol Reagent.
4. Shake for 15 seconds and let stand for 3 minutes at room temperature.
5. Centrifuge at 10,000 *g* for 15 minutes at 4°C.
6. Transfer the aqueous (upper) phase to a fresh tube and precipitate the RNA by addition of 100 μl of isopropanol per 200 μl TRIzol used for homogenization.
7. Incubate for 10 minutes at room temperature.
8. Centrifuge at 10,000 *g* for 10 minutes at 4°C.
9. Discard supernatant and wash pellet in 200 μl 70% ethanol per 200 μl TRIzol used for homogenization.
10. Centrifuge at 7500 *g* for 5 minutes at 4°C.
11. Air dry for 5 minutes.
12. Dissolve pellet in 10 μl H_2O per 200 μl TRIzol used for homogenization.

Reverse Transcription

1. For cDNA synthesis mix

total RNA	2 μl (~1 μg)
3 μg/ul random hexamers	1 μl
H_2O	7 μl

2. Incubate at 70°C for 10 minutes.
3. Incubate on ice for 10 minutes.
4. Add

5× RT buffer	4μl
(250mM Tris-HCl (pH 8.3), 375 mM KCl, 15 mM $MgCl_2$; GibcoBRL)	
0.1*M* DTT	2 μl
10m*M* each dNTP	1 μl
40 μ/μl RNase inhibitor	1 μl
H_2O	1 μl

5. Incubate at 37°C for 2 minutes.
6. Add 1 μl Superscript II RNaseH-reverse transcriptase (200 μ/μl; GibcoBRL).
7. Incubate at 37°C for 90 minutes.
8. Incubate at 95°C for 10 minutes.

For *PCR reaction,* follow Protocol 2 using 5 μl (1/4) of the cDNA reaction and primers that amplify 200- to 800-bp fragments. Separate on agarose gel and isolate amplified fragment using the QIAEX II gel extraction kit according to the manufacturer's instructions (Qiagen).

Sequence directly with the primers used earlier and/or internal primers. For manual sequencing, use the CYCLIST Taq DNA sequencing kit according to the manufacturer's instructions (Stratagene). For analysis with an ABI377 automated sequencer, use the Dye Terminator cycle sequencing kit (ABI) according to the manufacturer's instructions.

E. PROTOCOL 15.12
Mutation Detection in Genomic DNA

The intron/exon boundaries of candidate genes may be identified by using primers derived from cDNAs for PCR on genomic DNA and sequence analysis of BAC and PAC clone templates.

1. Use primers specific for segments of candidate cDNAs for PCR on genomic DNA. Primers that amplify the same fragment from cDNA clones and genomic DNA lie within the same exon. Introns lie between primers that amplify larger fragments from genomic DNA and large introns may lie between primer pairs that fail to amplify.
2. Sequence BAC or PAC clone templates with primers adjacent to the sites of possible introns.
3. Use PCR with primers derived from intron sequences to amplify individual exons or groups of exons separated by small introns from genomic DNA of wild-type and mutant siblings.

4. These fragments may be sequenced directly (as in Protocol 15.11), or they may be subcloned using the TA cloning system (Invitrogen) and sequenced. It is important to confirm that observed sequence changes are not errors introduced during the amplification by analyzing independently amplified fragments, particularly when individual subclones are sequenced.

Acknowledgments

We thank our colleagues and the members of our laboratories for sharing their protocols and expertise; Jiaojiao Zhang and Howard Sirotkin for providing the data shown in Figures 7 and 8, respectively; and Liliana Solnica-Krezel and Dirk Meyer for critical comments on the manuscript. Work in our laboratories is supported by NIH grants 1R01 RR12349-01 (W.S.T.), 1R01 GM57825-01 (W.S.T.), 1R01 GM56211-01 (A.F.S.), and 1R21 HG01704-01 (A.F.S.).

References

Amemiya, C. T., Zhong, T. P., Silverman, G. A., Fishman, M. C., and Zon, L. I. (1999). Zebrafish YAC, BAC, and PAC genomic libraries. *In* "Methods in Cell Biology, Volume 60" (H. W. Detrich, III, M. Westerfield, and L. I. Zon, Eds.). San Diego: Academic Press.

Amsterdam, A., and Hopkins, N. (1999). Retrovirus-mediated insertional mutagenesis in zebrafish. *In* "Methods in Cell Biology, Volume 60" (H. W. Detrich, III, M. Westerfield, and L. I. Zon, Eds.). San Diego: Academic Press.

Baxendale, S., Bates, G. P., MacDonald, M. E., Gusella, J. F., and Lehrach, H. (1991). The direct screening of cosmid libraries with YAC clones. *Nucl. Acids Res.* **19,** 6651.

Bender, W., Spierer, P., and Hogness, D. S. (1983). Chromosomal walking and jumping to isolate DNA from the Ace and rosy loci and the bithorax complex in *Drosophila melanogaster. J. Mol. Biol.* **168,** 17–33.

Brand, M., Heisenberg, C. P., Jiang, Y. J., Beuchle, D., Lun, K., Furutani-Seiki, M., Granato, M., Haffter, P., Hammerschmidt, M., Kane, D. A., Kelsh, R. N., Mullins, M. C., Odenthal, J., van Eeden, F. J. M., and Nüsslein-Volhard, C. (1996). Mutations in zebrafish genes affecting the formation of the boundary between midbrain and hindbrain. *Development* **123,** 179–190.

Collins, F. S. (1995). Positional cloning moves from perditional to traditional. *Nat. Genet.* **9,** 347–350.

Driever, W., Solnica, K. L., Schier, A. F., Neuhauss, S. C., Malicki, J., Stemple, D. L., Stainier, D. Y., Zwartkruis, F., Abdelilah, S., Rangini, Z., Belak, J., and Boggs, C. (1996). A genetic screen for mutations affecting embryogenesis in zebrafish. *Development* **123,** 37–46.

Foernzler, D., and Beier, D. R. (1999). Gene mapping in zebrafish using single-strand conformation polymorphism analysis. *In* "Methods in Cell Biology, Volume 60" (H. W. Detrich, III, M. Westerfield, and L. I. Zon, Eds.). San Diego: Academic Press.

Gaiano, N., Amsterdam, A., Kawakami, K., Allende, M., Becker, T., and Hopkins, N. (1996). Insertional mutagenesis and rapid cloning of essential genes in zebrafish. *Nature* **383,** 829–832.

Giovannoni, J. J., Wing, R. A., Ganal, M. W., and Tanksley, S. D. (1991). Isolation of molecular markers from specific chromosomal intervals using DNA pools from existing mapping populations. *Nucl. Acids Res.* **19,** 6553–6558.

Green, E. D., Riethman, H. C., Dutchik, J. E., and Olson, M. V. (1991). Detection and characterization of chimeric yeast artificial chromosome clones. *Genomics* **21,** 7–17.

Grinblat, Y., Lane, M. E., Sagerström, C., and Sive, H. (1999). Analysis of zebrafish development using explant culture assays. *In* "Methods in Cell Biology, Volume 59" (H. W. Detrich, III, M. Westerfield, and L. I. Zon, Eds.). San Diego: Academic Press.

Haffter, P., Granato, M., Brand, M., Mullins, M. C., Hammerschmidt, M., Kane, D. A., Odenthal, J., van Eeden, F. J. M., Jiang, Y. J., Heisenberg, C. P., Kelsh, R. N., Furutani-Seiki, M., Vogelsang,

E., Beuchle, D., Schach, U., Fabian, C., and Nüsslein-Volhard, C. (1996). The identification of genes with unique and essential functions in the development of the zebrafish, *Danio rerio*. *Development* **123,** 1–36.

Hammerschmidt, M., Blader, P., and Strähle, U. (1999). Strategies to perturb zebrafish development. *In* "Methods in Cell Biology, Volume 59" (H. W. Detrich, III, M. Westerfield, and L. I. Zon, Eds.). San Diego: Academic Press.

Ivics, Z., Hackett, P. B., Plasterk, R. H., and Izsvak, Z. (1997). Molecular reconstruction of Sleeping Beauty, a Tc1-like transposon from fish, and its transposition in human cells. *Cell* **91,** 501–510.

Ivics, Z., Izsvák, Z., and Hackett, P. B. (1999). Genetic applications of transposons and other repetitive elements in zebrafish. *In* "Methods in Cell Biology, Volume 60" (H. W. Detrich, III, M. Westerfield, and L. I. Zon, Eds.). San Diego: Academic Press.

Johnson, S. L., and Zon, L. I. (1999). Genetic backgrounds and some standard stocks and strains used in zebrafish developmental biology and genetics. *In* "Methods in Cell Biology, Volume 60" (H. W. Detrich, III, M. Westerfield, and L. I. Zon, Eds.). San Diego; Academic Press.

Jowett, T. (1999). Analysis of protein and gene expression. *In* "Methods in Cell Biology, Volume 59" (H. W. Detrich, III, M. Westerfield, and L. I. Zon, Eds.). San Diego: Academic Press.

Kaiser, C., Michaelis, S., and Mitchell, A. (1994). "Methods in Yeast Genetics." Cold Spring Harbor Laboratory Press, Cold Spring Harbor, New York, p. 142.

Kern, S., and Hampton, G. M. (1997). Direct hybridization of large-insert genomic clones on high-density gridded cDNA filter arrays. *Biotechniques* **23,** 120–124.

Kishimoto, Y., Lee, K.-H., Zon, L., Hammerschmidt, M., and Schulte-Merker, S. (1997). The molecular nature of zebrafish swirl: BMP2 function is essential during dorsoventral patterning. *Development* **124,** 4457–4466.

Knapik, E. W., Goodman, A., Ekker, M., Chevrette, M., Delgado, J., Neuhauss, S., Nobuyoshi, S., Driever, W., Fishman, M. C., and Jacob, H. J. (1998). A Microsatellite genetic linkage map for zebrafish (*Danio rerio*). *Nat. Genet.* **18,** 338–343.

Liao, E. C., and Zon, L. I. (1999). Simple sequence-length polymorphism analysis. *In* "Methods in Cell Biology, Volume 60" (H. W. Detrich, III, M. Westerfield, and L. I. Zon, Eds.). San Diego: Academic Press.

Michelmore, R. W., Paran, I., and Kesseli, R. V. (1991). Identification of markers linked to disease-resistance genes by bulked segregant analysis: a rapid method to detect markers in specific genomic regions by using segregating populations. *Proc. Natl. Acad. Sci. U.S.A.* **88,** 9828–9832.

Mullins, M. C., Hammerschmidt, M., Haffter, P., and Nüsslein-Volhard, C. (1994). Large-scale mutagenesis in the zebrafish: In search of genes controlling development in a vertebrate. *Curr. Biol.* **4,** 189–202.

Postlethwait, J. H., Johnson, S., Midson, C. N., Talbot, W. S., Gates, M., Ballinger, E. W., Africa, D., Andrews, R., Carl, T., Eisen, J. S., Horne, S., Kimmel, C. B., Hutchinson, M., Johnson, M., and Rodriguez, A. (1994). A genetic map for the zebrafish. *Science* **264,** 699–703.

Postlethwait, J. H., and Talbot, W. S. (1997). Zebrafish genomics: From mutants to genes. *Trends Genet* **13,** 183–190.

Postlethwait, J. H., Yan, Y.-L., and Gates, M. A. (1999). Using random amplified polymorphic DNAs in zebrafish genomic analysis. *In* "Methods in Cell Biology, Volume 60" (H. W. Detrich, III, M. Westerfield, and L. I. Zon, Eds.). San Diego: Academic Press.

Postlethwait, J. H., Yan, Y.-L., Gates, M. A., Horne, S., Amores, A., Brownlie, A., Donovan, A., Egan, E. S., Force, A., Gong, Z., Goutel, C., Fritz, F., Kelsh, R., Knapik, E., Liao, E., Paw, P., Ransom, D., Singer, A., Thomson, T., Yelick, P., Beier, D., Joly, J.-S., Larhammar, D., Rosa, F., Westerfield, M., Zon, L. I., Johnson, S. L., and Talbot, W. S. (1998). Vertebrate genome evolution and the zebrafish gene map. *Nat. Genet.* **18,** 345–349.

Ransom, D. G., and Zon, L. A. (1999). Mapping zebrafish mutations by AFLP. *In* "Methods in Cell Biology, Volume 60" (H. W. Detrich, III, M. Westerfield, and L. I. Zon, Eds.). San Diego: Academic Press.

Raz, E., van Luenen, H. G. A. M., Schaerringer, B., Plasterk, R. H. A., and Driever, W. (1998). Transposition of the nematode *Caenorhabditis elegans* Tc3 element in the zebrafish *Danio rerio*. *Curr. Biol.* **15,** 82–88.

Schier, A. F., Joyner, A. L., Lehmann, R., and Talbot, W. S. (1996). From screens to genes: Prospects for insertional mutagenesis in zebrafish. *Genes Dev.* **10,** 3077–3080.

Schier, A. F., Neuhauss, S. C., Helde, K. A., Talbot, W. S., and Driever, W. (1997). The one-eyed pinhead gene functions in mesoderm and endoderm formation in zebrafish and interacts with no tail. *Development* **124,** 327–342.

Schulte-Merker, S., Lee, K. J., McMahon, A. P., and Hammerschmidt, M. (1997). The zebrafish organizer requires chordino. *Nature* **387,** 862–863.

Schulte-Merker, S., VanEeden, F. J. M., Halpern, M. E., Kimmel, C. B., and Nüsslein-Volhard, C. (1994). No tail (Ntl) is the zebrafish homologue of the mouse T (brachyury) gene. *Development* **120,** 1009–1015.

Solnica-Krezel, L., Schier, A. F., and Driever, W. (1994). Efficient recovery of ENU-induced mutations from the zebrafish germline. *Genetics* **136,** 1401–1420.

Talbot, W. S., Egan, E. S., Gates, M. A., Walker, C., Ullmann, B., Neuhauss, S. C., Kimmel, C. B., and Postlethwait, J. H. (1998). Genetic analysis of chromosomal rearrangements in the cyclops region of the zebrafish genome. *Genetics* **148,** 373–380.

Talbot, W. S., Trevarrow, B., Halpern, M. E., Melby, A. E., Farr, G., Postlethwait, J. H., Jowett, T., Kimmel, C. B., and Kimelman, D. (1995). A homeobox gene essential for zebrafish notochord development. *Nature* **378,** 150–157.

Vos, P., Hogers, R., Bleeker, M., Reijans, M., van de Lee, T., Hornes, M., Frijters, A., Pot, J., Peleman, J., Kuiper, M., and Zabeau, M. (1995). AFLP: A new technique for DNA fingerprinting. *Nucl. Acid Res.* **23,** 4407–4414.

Westerfield, M., Wegner, J., Jegalian, B. G., DeRobertis, E. M., and Puschel, A. W. (1992). Specific activation of mammalian Hox promoters in mosaic transgenic zebrafish. *Genes Dev.* **6,** 591–598.

Williams, J. G., Kubelik, A. R., Livak, K. J., Rafalski, J. A., and Tingey, S. V. (1990). DNA polymorphisms amplified by arbitrary primers are useful as genetic markers. *Nucl. Acids Res.* **18,** 6531–6535.

Wilson, R. K., and Mardis, E. R. (1997a). Fluorescence-based DNA sequencing. *In* "Genome Analysis. A Laboratory Manual," Vol. 1, pp. 301–395. Cold Spring Harbor, New York: Cold Spring Harbor Laboratory Press.

Wilson, R. K., and Mardis, E. R. (1997b). Shotgun Sequencing. *In* "Genome Analysis. A Laboratory Manual," Vol. 1, pp. 397–454. Cold Spring Harbor, New York: Cold Spring Harbor Laboratory Press.

Yan, Y.-L., Talbot, W. S., Egan, E. S., and Postlethwait, J. H. (1998). Mutant rescue by BAC clone injection in zebrafish. *Genomics* **50,** 287–289.

Yu, C. E., Oshima, J., Fu, Y. H., Wijsman, E. M., Hisama, F., Alisch, R., Matthews, S., Nakura, J., Miki, T., Ouais, S., Martin, G. M., Mulligan, J., and Schellenberg, G. D. (1996). Positional cloning of Werner's syndrome. *Science* **272,** 258–262.

Zhang, J., Talbot, W. S., and Schier, A. F. (1998). Positional cloning identifies zebrafish one-eyed pinhead as a permissive EGF-related ligand required during gastrulation. *Cell* **92,** 241–251.

Zhong, T. P., Kaphingst, K., Akella, U., Haldi, M., Lander, E. S., Fishman, M. C. (1998). Zebrafish genomic library in yeast artificial chromosomes. *Genomics* **48,** 136–138.

CHAPTER 16

Construction and Characterization of Zebrafish Whole Genome Radiation Hybrids

Cheni Kwok,[1] Ricky Critcher, and Karin Schmitt[2]

Department of Genetics
University of Cambridge
Cambridge CB2 3EH, United Kingdom

[1] Present address: Genetic Technologies, SmithKline Beecham Pharmaceuticals, New Frontiers Science Park (North), Third Avenue, Harlow, Essex CM19 5AW, United Kingdom

[2] Present address: Millennium Pharmaceuticals, 640 Memorial Drive, 5th Floor, Cambridge, Massachusetts 02139

METHODS IN CELL BIOLOGY, VOL. 60

0091-679X/99 $30.00

I. Introduction

Zebrafish whole genome radiation hybrids are produced by fusing irradiated zebrafish cells to a hamster cell line that is hypoxanthine phosphoribosyl transferase (HPRT) or thymidine kinase (TK) deficient. The irradiation procedure randomly breaks the zebrafish genome, and the DNA fragments generated are rescued via the fusion process. Surviving cells are grown in hypoxanthine-aminopterine-thymidine (HAT) medium, which selects for hybrids containing the zebrafish HPRT gene or TK gene. In addition, unselected DNA fragments from the zebrafish are also retained. Experience has shown that a dose of 3000 rads produces fragments in a size range required for map continuity and optimal resolution for cloning. After picking the surviving hybrid clones and DNA extraction, the panel of hybrids is tested by polymerase chain reaction (PCR) for retention of genomic fragments by typing several microsatellite markers across the genome. In addition, retention can also be evaluated by FISH methods. For large-scale mapping or positional cloning projects, statistical programs are available to calculate map order and distances based on PCR typing of many markers across the hybrid panel. The advantages of whole genome radiation hybrids are (1) the ability to map any marker as long as it differs between rodent and zebrafish and (2) tailoring panels to different mapping resolutions depending on the application by adjusting the radiation dose.

II. Basic Protocol: Construction of Radiation Hybrids

The following protocol describes the construction of a panel of radiation hybrids. The donor cells are irradiated with a lethal dose of X rays, which results in double-strand breaks in the donor DNA. Fusion of irradiated cells with unirradiated recipient cell is mediated by using polyethylene glycol. After the fusion, the cells are subjected to HAT selection, which selects for the complementing gene (i.e., hypoxanthine phosphoribosyl transferase or thymidine kinase) from the irradiated donor cells. In a given hybrid, multiple fragments of donor chromosomes are present, and most are retained independent of selection.

A. Materials

Donor cells: Diploid zebrafish cell line with normal karyotype, for example AB9; a diploid fibroblast-like cell line derived from fin amputation of zebrafish

(AB strain). The AB9 cell strain can be obtained from American Type Culture Collection (Rockville, Maryland).

Recipient cells: Somatic TK^- or $HPRT^-$ hamster cell line. For example, Wg3H, a hypoxanthine-guanine phosphoribosyltransferase deficient ($HPRT^-$) hamster line.

Tissue culture: Tissue culture media optimized for donor and recipient cells. For example, AB9 was incubated in polyantibiotic solution, and cells were maintained in DMEM with 15–20% fetal calf serum, penicillin/streptomycin, and glutamine at 28°C in 5% CO_2; Wg3H was grown in E4 media with 10% fetal calf serum at 37°C

Transition medium: Bromo-2′ deoxyuridine, 100×; 6-thioguanine (2-amino-6-mercaptopurine). Trypsin/EDTA solution, 37°C.

HAT medium (selective medium): 50 × HAT supplement in complete medium with serum appropriate for recipient cell, 37°C. Dilute with media to 1× HAT.

PBSA: phosphate-buffered saline without calcium or magnesium.

PEG: 50% (v/v): 5.5 g of PEG4000 (Sigma) and 5 ml of serum-free growth medium, mixed and autoclaved. The pH is adjusted to 8.2.

25-, 75-, and 175-cm^2 tissue culture flasks

6-well tissue culture plates

Pasteur pipettes

Hemocytometer

Torrex X-ray machine

Swinging bucket rotor

Cryotubes

Microscope

B. Preparation of Donor and Recipient Cell Lines

1a. Donor cells: Grow a confluent 175-cm^2 flask of donor cells (i.e., about 3×10^7 cells) in the appropriate medium and temperature.

1b. Recipient cells: Grow the recipient cell lines for two passages with complete medium, serum and supplemented with transition medium (such as 6-thioguanine for Wg3H cell lines). Allow an additional passage without the transition medium prior to fusion. Grow the cells to near-confluency in a 175 cm^2 flask (i.e., about 3×10^7 cells).

The addition of transition medium allows the elimination of any spontaneously arisen drug-resistant cells. For example, addition of 6-thioguanine to Wg3H cells kills off any $HPRT^+$ revertants; similarly, the addition of BUDR to A23 cell lines eliminates TK^- revertants.

2. Harvest the cell lines from flasks using 37°C trypsin/EDTA solution. Add medium with serum to terminate the trypsinization.

Caution: Care should be taken not to overtrypsinize the cells. If the cells are trypsin sensitive, use dilute trypsin.

3. Take an aliquot of tissue suspension and perform a cell count by viewing it on a hemocytometer under the light microscope. Dilute as necessary such that both cell lines are of equal concentration.

4. Transfer 1×10^7 donor cell line in a 75-cm^2 flask. For recipient cells, transfer 1×10^7 cells in HAT medium in a second 75-cm^2 flask. Incubate both flasks on ice before fusion.

To ensure that the recipient cell line cannot survive the selection, set up a control for selection medium by transferring 1×10^7 donor cells lines into a 75-cm^2 flask with selection medium. (Control 1)

Because of a lower fusion efficiency, we have increased the number of donor and recipient cells (as compared to the human fusion described in the literature) in order to obtain the desired number of hybrids.

C. Irradiation of Donor Cell Line

5. The cells are exposed to irradiation produced by an industrial X-ray unit on maximum setting (150 kV, 5 mA in a Torrex X-ray machine, no filters). To ensure that donor cells are not adhered to the plastic, retrypsinize if necessary. Resuspend the cells in 10 ml of medium.

Caution: Care should be taken to ensure safe use of the X-ray unit. Ensure that the machine is warmed up before use. Check for leakage using X-ray monitor.

Keep the cells on ice before and after the irradiation process.

It is very important to ensure that the X-ray unit is calibrated regularly.

To control for the irradiation procedure, remove 1 ml of irradiated donor cells into a 75-cm^2 flask with selection medium. (Control 2)

D. Fusion of Hybrid Cells

6. Add the irradiated donor cells to an equal number of recipient cells in a conical base centrifuge tube.

A 1:1 ratio of irradiated donor cells to recipient cells was shown to be optimal in our hands. To improve low-efficiency fusions, other ratios (in excess of the donor cells) are recommended.

7. A pellet is obtained by spinning the mixed cells at 1500*g* for 15 minutes at room temperature.

8. Rinse the cells with 20 ml of phosphate-buffered saline (PBSA), and transfer to the universal tube. Spin down the cells for at 1500*g* 15 minutes at room temperature. Tap gently on the side of the universal to resuspend the pellet in 20 ml of PBSA.

To control for the fusion experiment, take 2 ml of mixed cells into a 75-cm^2 flask with selection medium. (Control 3)

9. Add 0.8 ml PEG (prewarmed to 37°C) slowly down the side of the tube, and mix gently using the pipette. Incubate the cell mixture in a 37°C waterbath for 60 seconds. Clumping of cells will be visible.

Note: The 60 seconds of incubation at 37°C is very critical for fusion efficiency. Handle the cells gently at this stage.

10. Add 5 ml of PBSA slowly down the side of the tube, gently stir with a pipette.

11. Add 5 ml of medium with serum and selective agent carefully to the cells.

The medium used for growing the radiation hybrids is the same as the one for growing the recipient cell line—except for the addition of HAT.

12. Spin the cells down at 1500*g* for 15 minutes at room temperature. Remove the medium, and add the cells to 200 ml of medium with serum and selective agent. Invert the bottle gently to distribute the clumps of cells evenly. Distribute the cells evenly among 20 wide-neck 75-cm^2 flask (i.e., 10 ml in each flask). This is optimal because it results in about 10–20 colonies/flask.

13. The medium of the cells should be changed every 3–4 days after monitoring growth by microscope.

Changes of medium is necessary in the following 3 days after the fusion. This is to remove the dead cells present in the flask. Subsequent media changes should be less frequent, and it is advisable to avoid changing of media when colonies are starting to form. This is to prevent disruption of new colonies.

Control 1 (recipient cell control) is used to look for revertants (in the case of Wg3H, this serves to detect HPRT$^+$). Control 2 (unfused irradiated donor cells) shows the effect of irradiation, and control 3 (unfused recipient and donor cells) will served as control for the PEG-mediated fusion. In our hands, control 2 usually dies off much more slowly than control 1. All the controls plates should show no growth when colonies from the fusion experiment become visible.

E. Colony Picking and Hybrid Growth

14. Colonies (25–75 cells per colony) should be visible to the naked eye in 2–4 weeks. All colonies are marked for ease of picking; keep a record of the colony and flask number.

15. Colonies are picked by using bent pipettes and transferred to 6-well tissue culture plates containing 2 ml of medium with selective agent prewarmed to 37°C.

Pipettes are bent using a Bunsen flame. Aspirate the medium from the plate containing the colonies until it is dry. Using the bent pipette, take up a drop of medium. Lower the pipette tip directly over the colony of cells premarked, and scrap the area containing the colony of cells gently. Take up the cells using the drop of medium, and resuspend it in the designated well containing the medium.

16. Each individual isolated clone is allowed to grow to confluence. Each colony is then grown up as required by splitting 1:3 into larger flasks.

Note that changes may occur in the hybrid in different passages. Hence, to extract DNA for a given hybrid, it is advisable to use the colony of cells from the same passage.

To harvest large amount of DNA (> 1 g), it is more efficient to use roller-bottle systems to grow up the colonies.

Note that the hybrid colonies will have different morphologies and growth characteristics.

F. DNA Extraction and Frozen Stock of Hybrids

17. DNA is extracted from each radiation hybrid using standard DNA extraction procedures. The quality of DNA extracted is assessed by running the DNA on agarose gell as well as the O.D. ratio at 260 nm vs. 280 nm using a UV spectrophotometer. Quantitation of DNA is calculated based on the O.D. reading at 260 nm.

18. Individual hybrid cell line is frozen down in duplicate either in liquid nitrogen or −135°C. A third of the cells from a confluent 75-cm^2 flasks worth of cells is frozen down in a cryotube.

III. Characterization of Radiation Hybrids by PCR

This protocol describes a PCR-based assay to determine the extent to which zebrafish genomic fragments are retained in different hybrid clones (i.e., the average marker retention frequency). Several zebrafish-specific markers are used for PCR amplification from DNA obtained from the hybrids. The presence or absence of PCR product in each hybrid is scored after agarose gel electrophoresis. The utility of a panel of radiation hybrids can thus be assessed. The two critical parameters are the average marker retention frequency and the size of the fragments retained. This protocol describes the general strategy; details can be adjusted according to the equipment available.

A. Materials

DNA stocks prepared from hybrid colonies as well as DNA from donor and recipient cell lines as controls

PCR primer pairs specific for zebrafish genome

PCR reaction mixture

Mineral oil

Microtiter plates (with 96 or 386 wells)

Robotic workstation or 8-/12-channel pipettor

Thermocycler (e.g., MJ, Cetus 9600)
Reagents and equipment for agarose gel electrophoresis

B. Preparation of DNA Stock Plates for PCR Amplification

1. Dilute DNA samples of the hybrids as well as zebrafish and hamster genomic DNA to a stock concentration of 15 ng/μl.

It is convenient to keep the stock DNA in 96 deep-well, sealed, microtiter plates (e.g., MARSH microtubes with plug caps), which can be stored at 4°C. Centrifugation of the DNA stock plates before usage minimizes contamination of adjacent wells.

2. Aliquot 3 μl of template DNA (i.e., total of 45 ng) for each hybrid to be tested including the appropriate controls (zebrafish and hamster genomic DNA, several H_2O samples) into microtiter plates. Proceed with preparations for PCR amplifications to prevent evaporation.

C. PCR Amplification

3. Prepare sufficient amount of PCR cocktail for the number of DNA samples to be amplified according to the following (per one sample):

$MgCl_2$ (stocks)	1 μl
dNTPs (2m*M*)	1 μl
10× PCR buffer	1 μl
Taq (2 units/μl)	0.25 μl

The best $MgCl_2$ concentration for a given primer pair has to be determined experimentally and could range from 1.5 to 5.0m*M*. Keep small aliquots of the following $MgCl_2$ stocks: 5.0, 10, 15, 20, 30, 40m*M* (corresponding to final $MgCl_2$ concentrations of 1.5, 2.0, 2.5, 3.0, 4.0, 5.0m*M*).

4. Add 3.25 μl of the PCR cocktail to each well of the microtiter plates containing the DNA samples.

5. Overlay each well with one to two drops of mineral oil.

6. Prepare the following reagents for hot-starting the PCRs in the subsequent step:

Primer 1—"forward" (22 ng/μl)	1 μl
Primer 2—"reverse" (22 ng/μl)	1 μl
Water	2 μl

7. Start the thermal cycling program with microtiter plates and program for one cycle of 5 min at 94°C; when the temperature reaches 94°C, start adding 4 μl of hot-start reagents (primer–water mixture) to each well.

8. Continue PCR according to the following profile:

94°C 1 minute
55°C 1 minute
72°C 30 seconds
for 35 PCR cycles, followed by 2 minutes at 72°C

D. Run Agarose Gels and Score Results

9. Load approximately 10 μl product from each well of the microtiter plate and run on 3% agarose gels in 1 × TBE.

A custom electrophoresis comb can be designed that allows running 48 or 96 samples in one gel and fits the microtiter plate format. Such apparatus is also commercially available from Scotlab.

Keep the area and equipment used for agarose gel electrophoresis separate from PCR preparation. In general, it is good practice to use separate pre- and post-PCR operations to avoid contamination.

10. Photograph the gel after sufficient separation of PCR product.

11. Score the PCR results for each radiation hybrid and use zebrafish and hamster amplification results as controls. Record the data in a spreadsheet or database program.

If the PCR reaction does not work try choosing a different $MgCl_2$ concentration or changing annealing temperature. The higher the $MgCl_2$ concentration, the lower the stringency of the PCR. The higher annealing temperature, the higher the stringency of the PCR.

Reagents and Solutions

10× PCR buffer
500m*M* KCl
100m*M* Tris, pH 9.0
1.0% Triton X-100
10m*M* $MgCl_2$

Store in small aliquots at −20°C for 1–2 months.

IV. Characterization of Radiation Hybrids Using FISH Analysis

This protocol describes the characterization of radiation hybrids using fluorescent *in situ* hybridization (FISH) analysis. This is performed by probing metaphase spreads of the radiation hybrids using fluorescently labeled zebrafish DNA. FISH analysis on radiation hybrids will present an overall view of the distribution of donor fragments within the nucleus of the hybrid. In addition, it can be used as a rough estimate for the retention frequency of the hybrid panel if a selection of hybrids were analyzed using FISH.

A. Materials, Reagents, and Solutions

Biotinylated-16-dUTP
Sonicated salmon sperm DNA
Zebrafish genomic DNA
Colcemid solution (BRL)
A 75-cm^2 flask of radiation hybrid cells
DNaseI
Avidin-FITC DCS: 1 in 500 dilution with TNFM (use 100 μl per slide plus 50 μl excess supplied by Vector Laboratories)
Biotinylated anti Avidin 1 in 250 dilution with TNFM (use 100 μl per slide plus 50 μl excess supplied by Vector Laboratories)
4′,6-diamidino-2-phenylindole (DAPI)
Propidium iodide
Phosphate-buffered saline
50% formamide in 2× SSC
Light and fluorescent microscope equipped with phase contrast and filters for FITC\DAPI\propidium iodide
10× nick translation solution: 0.5*M* Tris HCl pH7.5, 0.1*M* $MgSO_4$, 1m*M* Dithiothreitol, 500 μg/ml BSA
Block solution: 4× SSC, 0.05% Tween 20, 5% non-fat milk (e.g., Marvel)
TNFM/wash solution: 4× SSC, 0.05% Tween 20

Hybridization Buffer:

5 ml formamide
2 ml 50% dextran sulphate
1 ml 20× SSC
800 μl 0.5*M* hybridization buffer pH 7.0 (577 μl 0.5*M* Na_2HPO_4, 433 μl 0.5*M* NaH_2PO_4)
100 μl 10% SDS
100 μl 10× Denhardt's solution
1 ml sterile dH_2O

Fix solution: 3:1 methanol to acetic acid
Hypotonic solution: 9:1 dH_2O to FCS

B. Probe Preparation

1a. Label the zebrafish DNA with biotin-16-dUTP by nick translation using the following protocol:

10× nick translation buffer	2.5 μl
0.5m*M* dATP, dCTP and dGTP	1.9 μl
1m*M* bio-16-dUTP	0.7 μl
DNase I	1.0 μl
DNA polymerase I (5U)	0.5 μl
1 mg of DNA + dH_2O	
Total volume of reaction	25 μl

1b. Incubate at 14°C for 40 minutes.

1c. Add 1/10 vol. of 0.5*M* EDTA (pH8.0) to a 50-ng aliquot. Store the remaining reaction mix on ice.

1d. Run the aliquot on a 3% agarose micro-gel in order to check the efficiency of the DNAase. The fragments obtained should be between 100 and 500 bp. If the fragments are larger, add 1 μl of DNaseI to the samples and incubate for a further 20 minutes.

A micro-gel is prepared by pouring the agarose solution onto a microscope slide. A mini-gel comb is used to make the wells for loading.

The amount of DNase I needed to produce nicked products of the desired size needs to be determined for each new tube of DNaseI.

Care must be taken to ensure that the probe is less than 500 bp after nick translation. Failure to follow this can result in a large amount of nonspecific background signal on the slide.

1e. Add 1/10 vol. of EDTA to the remainder of the nick mix and incubate at 65°C for 2 minutes.

1f. Store the probe at −20°C.

C. Preparation of Metaphase Spread

2a. Cells should be harvested when they are 90–95% confluent (a 75-cm^2 flask of cells, i.e about 10^7 is sufficient).

A media change at 17 hours prior to harvesting will improve the mitotic index.

2b. Add 100 μl of colcemid solution (10 μg/ml) to the flask of cells, and incubate at 37°C for 2 hours.

2c. Remove media from flask and wash the cells with 0.5 ml of prewarmed trypsin.

2d. Aspirate the trypsin and harvest the cells using 1–2 ml of trypsin.

2e. Lyse the cells by resuspending the cells in a fresh tube containing 3 ml of hypotonic solution (9:1 dH_2O to FCS) and incubate at room temperature for 15 minutes.

2f. Add 10 drops of fix (3:1 methanol to acetic acid) and spin for 5 minutes at 1000 rpm.

2g. Remove supernatant, resuspend pellet, and slowly add 5 ml of fix.

2h. Spin down the cells at 1000 rpm for 5 minutes.

2i. Repeat steps 2 g and 2h twice, and store the tubes at −20°C.

The chromosome suspensions are prepared on the day and evaluated under phase contrast using a light microscope.

D. Slide Making

3a. Spin down the sample, remove the supernatent, and resuspend the pellet in approximately 1 ml of fresh fix. If the samples have been stored at −20°C, spin them down, resuspend in fresh fix, spin down again, remove supernatent, and resuspend the pellet in fix solution.

3b. Using a pasteur pipette, place one drop of cell suspension onto a clean slide and air dry.

3c. Examine the slide under the microscope. Note the number of metaphases, the number of interphase nuclei, and the amount of background (caused by the presence of cytoplasm).

3d. Adjust the fix accordingly to reduce the background. Slide-making conditions are dependent on a number of variables, such as quality of the preparations, cell density, the height from which the sample is being dropped and humidity. The addition of 1 drop of fix immediately after dropping the preparation on the slide is usually beneficial in reducing cytoplasm on the slide.

3e. Incubate the slides for 1–2 hours at 60°C prior to hybridization.

E. Hybridization and Washing

4a. Spin down 50 ng of the labeled zebrafish DNA with 10 μg of sonicated salmon sperm for 5 minute followed by incubation at 37°C from 15 minutes to 1 hour.

4b. Apply the probe to the slide, overlay the slide with a coverslip, and seal using cow gum. The slide is incubated at 70°C for 5 minutes, followed by an overnight incubation at 37°C in a humidified chamber.

4c. Remove the cow gum and wash the coverslips off using 2× SSC for 5 minutes. Wash the slide in 50% formamide in 2× SSC at 42°C for 10 minutes, followed by 0.1× SSC at 60°C for 10 minutes. This is usually performed in a glass chamber.

4d. Block the slides for 20 minutes at 37°C, followed by a 20-minute incubation with 1/500 dilution avidin-FITC at room temperature in a dark, humidified chamber.

4e. Wash the slides in 4× SSC, 0.05% Tween 20 for 15 minutes at 42°C followed by a 20 minutes incubation with biotin anti-avidin at room temperature in a dark, humidified box.

4f. Repeat the wash at step 4e, and incubate again in avidin-FITC.

4g. The slides are washed for 10 minutes at 42°C, washed in PBS for 5 minutes, and mounted in 20 μl of vector shield containing 10 μg/ml of DAPI and 4 μg/ml propidium iodide.

Analysis of 20 metaphases should be enough to give a reasonable idea of the retention rate of the hybrid. Note that fragments of up to 0.5 Mb can be detected by FISH; hence, the retention rate observed will be an underestimation of the true retention frequency.

Negative controls must be included in each FISH to check for cross species hybridization. If the background is too high, cot 1 DNA should be included in the hybridization buffer and the mixture incubated at 37°C for 15 minutes up to 2 hours after probe denaturation and before application to the slide.

V. Commentary

In 1975, a technique of irradiation and fusion gene transfer in which human donor cells were given a lethal dose of gamma irradiation and then fused to recipient hamster cells was described by Goss and Harris (1975). The human—hamster hybrid cells retained random unselected fragments of the human genome, and the coretention frequencies of syntenic markers could be used to construct maps (Goss and Harris, 1977). This technique was not used widely until 1989, when Cox and colleagues modified the protocol by using a somatic cell hybrid containing a single human chromosome as the donor cell (Cox *et al.*, 1989). This modification allowed the hybrids generated to be used for both map construction of chromosomes and as a source for ordering new markers for specific genome regions of interest (Cox *et al.*, 1989; Foster *et al.*, 1994; Schmitt *et al.*, 1996). The hybrids produced were named "radiation hybrids," and the derived genome maps were called "radiation hybrid maps." Using the Cox approach, between 100 and 200 hybrids are needed for mapping each chromosome and, as a consequence, over 4000 hybrids would be needed for mapping the whole genome. In Cambridge, the original protocols of Goss and Harris have been reinvestigated. They confirmed that by using human diploid cells as donors, it is possible to map the whole genome with a single set of 200 hybrids (Walter *et al.*, 1994). This technology provides a way of mapping any genome providing a complementary approach to recombination mapping. The advantages of this approach follow:

1. The ability to map any marker by PCR as long as it differs between the donor and recipients.
2. The integration of maps based on different marker types. This is particularly valuable when seeking to order markers as a preliminary to building contigs.
3. The number of breaks per chromosome can be controlled by altering the radiation dose. This allows hybrid panels to be tailored to different resolutions depending on the application.

4. It is suitable for high throughput mapping of different markers; hence, it allows the rapid establishment of a physical and transcript maps.

Whole genome radiation hybrid mapping technology has already been successfully applied to map the human genome (Hudson *et al.,* 1995); Gyapay *et al.,* 1996) as well as the mouse genome (Schmitt *et al.,* 1996); McCarthy *et al.,* 1997). In this unit, we have described the construction and characterization of zebrafish radiation hybrid panels. The panel will provide the necessary resource for the construction of a physical map of zebrafish, allowing the integration of markers from the genetic map and other molecular markers on a single map.

VI. Critical Parameters and Troubleshooting

Some of the most important parameters in determining the success of a radiation hybrid panel are the donor and recipient cell lines chosen for the fusion. We routinely use an early passage of the hamster cell line Wg3H. This allows us to apply the HAT selection scheme, which has worked most robustly in our hands and shows a very low rate of reversion. Other selections schemes can be employed as described by Goodfellow (1991). The growth characteristics of both donor and recipient cell lines influence the fusion frequency. Cells that are easy to propagate and show rapid growth work best. After testing several zebrafish cell lines, we have found that AB9 is superior. Again, early passages will most likely work better. We routinely check several metaphase spreads before choosing a donor cell line to make sure that no major rearrangements or chromosome losses have occurred during tissue culture.

Another parameter that can influence the success of a panel is the radiation dose. In addition, the lengths of fragments retained in the hamster background can be adjusted for different experimental purposes by varying radiation doses. However, it is best not to combine radiation hybrids generated at different doses. For other, nonroutine radiation doses (we recommend using 3000 rad), it is advised to determine a killing curve to make sure that the dose chosen is lethal to the donor cell line.

Many other factors determine the success of a fusion experiment, and it can be difficult to diagnose the exact problem if results deviate from the expected number of colonies. During the fusion process, the cells have to be handled with extreme care, especially after exposure to PEG. The age, brand, and molecular weight of PEG should be as recommended. For critical experiments, use a fresh bottle. If a careful review of all procedures does not provide any answers to the failure of an experiment, it can be useful to try another zebrafish cell line or a different stock aliquot of the same cell line. Extensive passaging of any cell line should be avoided. Another possibility can be to adjust the ratio of donor to recipient cells. Usually, an increase in the donor cell line is recommended.

The number of colonies expected after a fusion experiment is discussed later. Sometimes, many colonies can show up shortly after the fusion procedure. These arise from cells that have not been killed off yet and can present a problem. Once a small-size colony has been established, it will be difficult to kill because cells are at close density and protected from the lethal effects of HAT. If these colonies arise in the first few days after fusion, change media immediately and repeat every day to prevent further growth. However, excessive media changes should usually be avoided since it can disrupt any hybrid colonies that have started to form. As always, checking for colonies in the control plates will be important; no colonies should be visible after 1 week. Sometimes, dead cells can be mistaken for colonies, and it is important to check the progression of such clones carefully under the microscope.

Picking colonies at the right time is critical. Usually, not all colonies in a given flask will be at the same density. Picking colonies that are too small (check under the microscope) is not useful because the cell density in the destination plate is too low and will prevent survival. On the other hand, one has to watch out for the formation of satellite colonies. If this process has started, picking all the colonies in a given flask should be started immediately. Always finish picking all colonies and discard the flask. Adding new media to wait for further growth is not a good practice. For subsequent growth of hybrid colonies, avoid extensive subculturing because a loss of fragments can be expected. Freeze down two aliquots at the same time as harvesting cells for DNA preparation.

After PCR analysis, it can turn out that the retention frequency is lower that expected. Such an experiment can be rescued if an excess of 10–20% of hybrids have been prepared. Simply pick the hybrids with the highest retention rates for further use and discard the rest. Alternatively, hybrid colonies can be combined after DNA extraction to boost the retention frequencies. If this is not feasible, the experiment will have to be repeated. Pooling of hybrids from different fusion is another option to consider if the retention frequency is low. For a detailed description of the theoretical considerations for pooling, see Lunetta and Boehnke (1994).

For obtaining superior PCR results, follow the procedures in the protocol. Good choice of primer pairs and optimal PCR conditions are important. However, a small percentage of primers will not yield desirable results but can often be rescued by redesigning the primers. We always carry out PCR in duplicate and run both aliquots on a gel. If the results are discrepant, they should be scored that way because the statistical programs can use that information. Not all PCR products will yield the same intensity bands and sometimes more/less PCR product loaded onto the gel can overcome this. Always run genomic hamster and zebrafish DNA as a control and compare the expected results.

Slide preparation is one of the most critical parameters when performing FISH. The metaphases should be well spread, with little or no cytoplasm around them. For a concise description of the problems associated with slide preparation see "Current Protocols in Human Genetics," Volume 1, pp. 4.1.6–4.1.17 (1997). A nonspecific background signal is a common problem and may be caused by incomplete nick translation of the probe whereby fragments of DNA greater than 600bp are generated or by poor-quality metaphase spreads. It is essential that the temperatures of all water baths, heat blocks, and solutions are correct before use.

A. Anticipated Results

It is reasonable to expect at least 10 colonies per plate from a fusion of 1×10^7 cells in 20 plates (i.e., 5×10^5 cells per plate). It is a good practice to pick as many isolated clones as possible from the fusion to allow selection of clones that grow well. Additional clones will allow the exclusion of clones with very low retention frequency. A panel of 100–200 whole genome radiation hybrid is adequate.

The ideal retention frequency is 50%, which will provide the most power for mapping using statistical analysis. However, in practice, radiation hybrid panels with >20% retention frequency is sufficient. A lower retention frequency is expected with a higher radiation dose used. Experiments resulting in panels with an average retention <15% should be repeated.

B. Time Considerations

Allow a week before the fusion to initiate the preparation of donor and recipient cell lines. It is recommended that a few trial fusions be carried out to establish the ideal donor and recipient cell line combination. It is also good practice to perform a "killing curve" on the donor cell line, to establish the lethal dose for the cell line. The irradiation and fusion process is a full day's work. The time taken for colony picking will be dependent on the rate of colony growth. Usually this process will take about 3 to 5 days to complete. A lot of tissue culture maintenance work is involved in growing up the hybrids. This is very labor-intensive, and depending on the growth rate of the hybrids, it may take more than 3 months to obtain the desired number of cells for DNA isolation as well as a frozen stock of hybrids. DNA isolation of 200 cell lines can be completed in 4 days. Quantitation and quality check on the DNA extracted as well as dilution of DNA samples to a working concentration for PCR will require 1–2 day's work.

PCR-ready plates (containing the DNA ready for PCR) can be prepared in advance and stored in the freezer. Time taken for preparation and addition of the PCR reaction mixes to PCR-ready plates using the multichannel pipette should not take more than 15 minutes. PCR cycle time is about 2–3 hours, and the throughput of PCR is dependent upon the availability of PCR blocks. Loading of PCR products onto agarose gels using the multichannel pipette takes less than 15 minutes. The gels are run for 20 minutes at 200 V and are ready to photograph. Scoring of hybrids is labor-intensive. For high-throughput mapping, it is recommended to invest in an image-processing system (such as a gel doc system) to enable automatic scoring of hybrids, and generate data in the desired format for statistical analysis.

Slides for FISH can be made on the day or stored at room temperature for up to 1 week prior to use. Nick translation of the probe will take 1 hour including running a gel. Allow 1 hour to prepare the probe and slide for denaturation and hybridization, followed by an overnight incubation. Stringent washes and fluorescent detection should take approximately 3 hours.

References

Cox, D. R., Pritchard, C. A., Uglum, E., Casher, D., Kobori, J., and Myers, R. M. (1989). Segregation of the Huntington disease region of human chromosome 4 in a somatic cell hybrid. *Genomics* **4,** 397–407.

Foster, J. W., Dominguez-Steglich, M. A., Guioli, S., Kwok, C., Weller, P. A., Stevanovic, M., Weissenbach, J., Mansour, S., Young, I. D., Goodfellow, P. N., Brook, J. D., Schafer, A. J. (1994). Mutations in an *SRY*-related gene cause campomelic dysplasia and autosomal sex reversal. *Nature* **372,** 525–530.

Goodfellow, P. N. (1991). Irradiation and fusion gene transfer. *Methods Mol. Biol.,* 53–61.

Goss, S. J., and Harris, H. (1975). New method for mapping genes in human chromosomes. *Nature* **255,** 680–684.

Goss, S. J., and Harris, H. (1977). Gene transfer by means of cell fusion II. The mapping of 8 loci on human chromosome 1 by statistical analysis of gene assortment in somatic cell hybrids. *J. Cell Sci.* **25,** 39–57.

Gyapay, G., Schmitt, K., Fizames, C., Jones, H., Vegaczarny, N., Spillett, D., Muselet, D., Prudhomme, J. F., Dib, C., Auffray, C., Morissette, J., Weissenbach, J., and Goodfellow, P. N. (1996). A radiation hybrid map of the human genome. *Hum. Mol. Genet.* **5**(3), 339–346.

Hudson, T. J., Stein, L. D., Gerety, S. S., Ma, J. L., Castle, A. B., Silva, J., Slonim, D. K., Baptista, R., Kruglyak, L., Xu, S. H., Hu, X. T., Colbert, A. M. E., Rosenberg, C., Reevedaly, M. P., Rozen, S., Hui, L., Wu, X. Y., Vestergaard, C., Wilson, K. M., Bae, J. S., Maitra, S., Ganiatsas, S., Evans, C. A., Deangelis, M. M., Ingalls, K. A., Nahf, R. W., Horton, L. T., Anderson, M. O., Collymore, A. J., Ye, W. J., Kouyoumjian, V., Zemsteva, I. S., Tam, J., Devine, R., Courtney, D. F., Renaud, M. T., Nguyen, H., Oconnor, T. J., Fizames, C., Faure, S., Gyapay, G., Dib, C., Morissette, J., Orlin, J. B., Birren, B. W., Goodman, N., Weissenbach, J., Hawkins, T. L., Foote, S., Page, D. C., and Lander, E. S. (1995). An STS based map of the human genome. *Science,* **270,** 1945–1954.

Lunetta, K. L., and Boehnke, M. (1994). Multipoint radiation hybrid mapping comparison of methods, sample size requirements, and optimal study characteristics. *Genomics* **21,** 92–103.

McCarthy, L. C., Terret, J., Davis, M. E., Knights, C. J., Smith, A. L., Critcher, R., Schmitt, K., Hudson, J., Spurr, N. K., and Goodfellow, P. N. (1997). A first-generation whole genome-radiation hybrid map spanning the mouse genome. *Genome Res.* **7,** 1153–1161.

Schmitt, K., Foster, J. W., Feakes, R. W., Knights, C., Davis, M. E., Spillett, D. J., and Goodfellow, P. N. (1996). Construction of a mouse whole genome radiation hybrid panel and application to Mmu11. *Genomics,* **34**(2), 193–197.

Walter, M. A., and Goodfellow, P. N. (1993). Radiation hybrids: Irradiation and fusion gene transfer. *Trends Genet.* **9,** 352–356.

Key Reference

Original paper describing construction of whole genome radiation hybrid panel for human and the generation of chromosomal maps.

Walter, M. A., Spillett, D. J., Thomas, P., and Goodfellow, P. N. (1994). A method for constructing radiation hybrid maps of whole genomes. *Nat. Genet.* **7,** 22–28.

CHAPTER 17

Zebrafish/Mouse Somatic Cell Hybrids for the Characterization of the Zebrafish Genome

Marc Ekker,* Fengchun Ye,† Lucille Joly,* Patricia Tellis,† and Mario Chevrette†

* Department of Medicine and Department of Cellular and Molecular Medicine
Ottawa Civic Hospital
Loeb Institute for Medical Research
University of Ottawa
Ottawa, Ontario K1Y 4E9
Canada

† Urology Division
Department of Surgery
McGill University and Montreal General Hospital Research Institute
Montreal, Quebec
Canada

METHODS IN CELL BIOLOGY, VOL. 60

0091-679X/99 $30.00

I. Introduction

The fusion of cells from two different species in the presence of polyethylene glycol or certain inactivated viruses results in a combined cell with two separate nuclei, called heterokaryon. After the heterokaryon undergoes mitosis, a hybrid cell is produced in which the separate nuclear envelopes break up and the chromosomes from both donor species come together in a single large nucleus. Such cells are usually unstable and tend to lose chromosomes from one of the two parental species. However, presence of genes conferring a selective advantage ensure retention of at least one donor chromosome. The reduction in complexity of the donor DNA in such hybrids enables assignment of genes to chromosomes. Hybrid cells can also be used as a source of DNA from specific chromosomes.

Although whole cell fusion has been used for a few decades, characterized collections of somatic cell hybrids had mainly been made between two mammalian donor species. The successful transfer of chicken microchromosomes to rodent cells (Schwartz *et al.*, 1971; Kao, 1973; Trisler and Coon, 1973; Leung *et al.*, 1975) and their stable maintenance indicated that stable transfer of chromosomes between distantly related species was feasible, even though the chicken–rodent hybrids were not characterized in details with gene markers or techniques such as fluorescence *in situ* hybridization (FISH). Somatic cell hybrids were most useful in the assignment of genes to mammalian chromosomes, and now production of radiation hybrids (see Kwok *et al.*, this volume) has allowed a greater degree of precision in the location of genes on chromosomes. We have previously shown that it is possible to transfer zebrafish chromosomes or chromosome segments in a mouse cell and that the zebrafish chromosomal elements can be stably maintained in a mouse background, provided that selective pressure is maintained (Ekker *et al.*, 1996). We present here a summary of the techniques pertinent to the production and the characterization of zebrafish/mouse cell hybrids.

II. Production of Zebrafish/Mouse Somatic Cell Hybrids

Transfer of zebrafish chromosomes to cells of a foreign species has been accomplished successfully with three different fish cell lines—ZF4, LFF, and AB9—as donor cells, and with the mouse B78 cells or hamster Wg3H cells as recipients (Ekker *et al.,* 1996; Chevrette *et al.,* 1998; Kwok *et al.,* Ch. 16, this volume; L. Joly, M. Chevrette, and M. Ekker, unpublished observations). Use of zebrafish AB9 cells as donors is described in Kwok *et al.* (Ch. 16, this volume).

A. Parental Cell Lines

1. Zebrafish ZF4 Cells

The zebrafish embryonic fibroblast cell line ZF4 (Driever and Rangini, 1993) was one of the first cell lines from this species that was described as able to grow in culture without requirement for fish serum or fish embryo extract. ZF4 cells were characterized (Driever and Rangini, 1993) and shown to contain more than 90 chromosomes.

2. LFF Cells

Because ZF4 cells contained more chromosomes than the 50 found in a normal zebrafish cell, we decided to generate an additional embryonic fibroblast cell line, which we named LFF (Chevrette *et al.,* 1998). The resulting cell line was named LFF for "Lucille's Fish Fibroblasts."

1. Zebrafish embryos of the AB strain are collected soon after fertilization and are washed twice with sterile water, twice with Holtfreter's solution (60m*M* NaCl, 0.7m*M* KCl, 2.4m*M* $NaHCO_3$, 0.9m*M* $CaCl_2$), twice for one minute with 0.1% Javex in Holtfreter's and then kept in Holtfreter's solution at 28°C.
2. Approximately 24 hours after fertilization, the chorions are removed with pronase E (0.5 mg/ml in Holtfreter's) for about 1 hour at 28°C and rinsed in Holtfreter's.
3. The embryos are digested three times in 1 ml of a sterile solution of 0.25% trypsin + 1m*M* EDTA for a period of 30 minutes at 37°C. The embryo suspension is pipeted up and down with a Pasteur pipet a few times to facilitate dissociation.
4. At the end of each digestion period, the cell suspension is combined to 6 ml of culture media (see Section II.A.4), leaving clumps of undigested embryos behind.
5. Cells are grown at 28°C in an incubator with 5% CO_2. In our experiment, the surviving cells originated from the second trypsin digestion.
6. When a karyotype was performed on the LFF cells, the number of chromosomes varied between 46 and 50 (P. Tellis, M. Ekker, and M. Chevrette, unpublished observations).

3. The Mouse B78 Melanoma Cells

The mouse B78 melanoma cell line was chosen as recipient in whole-cell fusion experiments because it was previously shown to fuse very easily with other cell lines (Speevak *et al.,* 1995). Furthermore, the cells have a fast growth rate and are easy to maintain, characteristics that were often retained after fusion (Section II.E).

4. Cell Culture Media

LFF cells are grown in DMEM/F12 + 10% fetal bovine serum + 50 μg/ml rainbow trout embryo extract + 0.4% trout serum. Zebrafish ZF4 cells and mouse B78 cells are kept in DMEM/F12 + 10% fetal bovine serum. Zebrafish/mouse hybrids are first selected in DMEM/F12 + 10% fetal bovine serum + 800 μg/ml G418 + 6μ*M* ouabain. After expansion, hybrid cells are cultured in the same medium except that the concentration of G418 is reduced to 400 μg/ml and the ouabain is omitted.

B. Tagging Zebrafish Chromosomes with the *neo* Resistance Gene

Mouse B78 cells are not deficient in any enzyme; therefore, we introduced, into zebrafish chromosomes, the aminoglycoside phosphotransferase (*neo*) gene conferring resistance to the aminoglycoside drug G418, to select against unfused recipient mouse cells and to ensure retention of zebrafish chromosomes in hybrids. In order to obtain a high probability of having each zebrafish chromosome tagged in the cell population used for whole-cell fusion, a pool of more than 400 G418-resistant clones was produced.

1. CMV *neo* Plasmid

The cytomegalovirus (CMV) promoter/enhancer has previously been shown to have high activity in ZF4 cells (Driever and Rangini, 1993) and was therefore chosen to drive expression of the *neo* gene.

2. Transfection Procedure

Zebrafish cells (ZF4, LFF, or AB9) are transfected with the pCMV-neo plasmid using the calcium phosphate procedure essentially as described in Sambrook *et al.* (1989).

1. For 24 hours prior to transfection, 5×10^5 cells are plated in a 100 mm diameter tissue culture grade Petri dish. Typically, a total of four plates are transfected.
2. The pCMV-neo plasmid (10 μg) is mixed with 500 μl of 2× HeBSCl buffer, pH 7.05. The calcium phosphate precipitate is made by adding, dropwise, 49.6 μl of 2.5*M* $CaCl_2$ to the DNA/buffer solution.
3. After gentle mixing, the precipitate is left to form for 20–30 minutes at room temperature.
4. The suspension is pipetted up and down twice and added, dropwise, directly to the solution overlaying the cells, followed by gentle swirling of the dishes.
5. The cells are incubated overnight at 28°C.
6. The next morning, the medium is removed and 5 ml of a 15% glycerol/85% phosphate-buffered saline is added to the cells for 4 minutes. The cells are then rinsed three or four times with PBS and fed with fresh medium (for media composition see Section II.A.4).

3. Selection of G418-Resistant Colonies

1. At 48 hours after washing the cells (above section), G418 is added to the medium at a final concentration of 600 μg/ml. The stock solution of G418 is prepared at a concentration of 10 mg/ml (active) in tissue culture medium.

2. Approximately 3 weeks after the selection with G418 is applied, surviving colonies are harvested concurrently to form pools of more than 100 colonies (typically from two 100-mm diameter culture dishes). A final pool of more than 400 clones is produced.

C. Whole-Cell Fusion

Whole-cell fusion between zebrafish and mouse cells was done using polyethylene glycol (PEG) MW 1500 from NBS Biologicals. Concentrations of PEG varied from 48 to 52%, and the time of incubation varied from 60 to 90 seconds. Most hybrids produced in our laboratories were done using a 50% PEG solution for 90 seconds.

1. Plate 500,000 cells of each of the two parental lines together into a 25 cm^2 flask, 24 hours before fusion.

2. The media is removed from the cell monolayer. Add 1 ml of 50% (wt/wt) PEG solution in Dulbecco's modified Eagle medium without serum. Incubate for 90 seconds at room temperature. During incubation, keep moving the flask gently so that the whole surface of the cell monolayer is constantly in contact with the PEG solution.

3. Remove as much PEG solution as possible using aspiration (the PEG solution is quite viscous). Rinse three times with 5–10 ml PBS. The cells are then incubated at 37°C in serum-containing medium (see Section II.A.4) for a period of 24 hours after which they are split and transferred to two 162 cm^2 flasks.

4. The selection drugs G418 (800 μg/ml) and ouabain ($6 \mu M$) are added to both flasks.

5. One flask is then incubated at 37°C, the normal growth temperature for mouse cells, and the other is incubated at 28°C, the normal growth temperature for zebrafish cells. In one such experiment, colonies of cells resistant to both drugs developed only in the flask incubated at 37°C. This suggested that cells in these colonies had one or a few zebrafish chromosomes in a mouse background. Conversely, we would have anticipated that hybrid cells with one or a few mouse chromosomes in a zebrafish background would have grown better at 28°C. Because obtaining the latter kind of hybrid cells was of little interest to us, production of hybrid cells that grow at 28°C was abandoned.

D. Freezing Hybrid Cells

1. Cells are trypsinized using 0.25% trypsin/EDTA for 3 minutes at room temperature.

2. Trypsinized cells are pooled and transferred to a 15 ml centrifuge tube. They are centrifuged for about 5 minutes at 1000 rpm in a clinical centrifuge.

3. The supernatant is removed, and the cells are resuspended in a solution of 10% dymethyl sulfoxide/90% culture medium (1–2 ml for each 100 mm diameter dish).

4. Cell suspensions are distributed in 1 ml aliquots into cryovials and are cooled slowly at −80°C. They are transferred to liquid nitrogen for long-term storage.

E. Morphology and Growth Rate of Hybrid Cells

Most hybrids are morphologically similar to the B78 parental cell line (Chevrette *et al.,* 1998). Only a few hybrids have a more elongated morphology than the parental cells, and this phenomenon was not dependent on culture conditions. The majority of hybrid cells grow more slowly than the mouse B78 parent but faster than the zebrafish ZF4 or LFF parents. Approximate doubling times are: for ZF4 and LFF cells, 3.8 days; for B78 cells, 0.8 days; for hybrids, 1–2 days.

III. Analysis of Zebrafish Chromosomes in Hybrid Cells by Fluorescence *in Situ* Hybridization

A. Probes

1. Total Zebrafish Genomic DNA Probe (Painting Probe)

Total zebrafish genomic DNA was digested for at least 3 hours with a mixture of the restriction enzymes *Eco*RV, *Pst*I, and *Pvu*II in the presence of spermidine and used directly in a labeling reaction (see Section III.B.1). This probe was previously shown to label zebrafish chromosomes in their entirety (Ekker *et al.,* 1996).

2. SRI Probe

SRI, a highly repeated satellite-like DNA sequence, constitutes up to 8% of the zebrafish genome (Ekker *et al.,* 1992; He *et al.,* 1992). The monomeric SRI unit of approximately 170 bp has been subcloned and can be labeled to use as a probe in FISH experiments. SRI labels all centromeres of metaphase chromosomes from zebrafish ZF4 or LFF cells (Ekker *et al.,* 1996).

3. Other Repetitive Sequences

A second satellite-like repetitive sequence, named Alu Type Ib, has been described by He and collaborators (1992). Although of lesser abundance than SRI, it can also be used to identify zebrafish chromosomes in metaphase preparation from hybrid cells. It marks zebrafish chromosomes mainly at the centromeres but has also been shown to label preferentially one chromosome from ZF4 cells (C. C. Martin, P. Tellis, M. Chevrette, and M. Ekker, unpublished observations).

B. Probe Labeling

Two different methods are used to prepare probes for FISH: random octamer labeling and PCR.

1. Random Octamer Labeling

1. The DNA template (100 ng) is boiled for 5 minutes.
2. DNA is labeled with Biotin-14-dCTP (Life Technologies). Other reagents and reaction conditions are as defined in the Bio Prime DNA Labeling System (Life Technologies) or according to similar procedures.
3. Probe synthesis is carried out in a total volume of 50 μl using the Klenow fragment of DNA polymerase I at 37°C for 1 hour, and the reaction stopped with 5 μl of 0.2*M* EDTA pH7.5.
4. Unincorporated biotin-14-dCTP is removed by ethanol precipitation. The probe is resuspended in 10 μl of TE^{-4} (10m*M* Tris–HCl, pH 8; 0.1m*M* EDTA). Biotinylated probes are stable for at least 1 year when stored at −20°C.

2. Probe Synthesis by PCR

Biotinylation of DNA fragments of 200 bp or less was performed by PCR. In our hands, PCR synthesis of probes from short template produced stronger signals in FISH experiments than probes prepared by random octamer labeling. This method also takes out the need to excise and purify the DNA fragment to be labeled from its vector.

1. Oligonucleotide primers have to be synthesized based on selected sequences at both ends of the DNA fragment to be labeled.
2. The PCR reaction contains 10 ng of uncut plasmid containing the DNA fragment of interest; the two oligonucleotide primers, 1μ*M* each; biotin-14-dCTP, 0.1m*M*; dCTP, 0.1m*M*; dATP, dGTP, dTTP, 0.2m*M* each; the recommended buffer for the source of Taq DNA polymerase; 1.5 units of Taq DNA polymerase and water in a total volume of 50 μl.
3. The PCR conditions follow a predwell of 5 minutes at 92°C; 40 cycles of 1 min at 92°C, 1 min at 60°C (depending on the sequence and length of oligonucleotides used), and 1 min at 72°C; and a postdwell of 5 minutes at 72°C.
4. Ethanol precipitation is performed after the PCR to remove unincorporated nucleotide and other unwanted reagents. The probe is resuspended in 15 μl of TE^{-4} and stored at −20°C.

C. Chromosome and Slide Preparation

1. Chromosomes are prepared from a T75 flask at about 70–80% confluency and in exponential growth.

2. Of a 10 μg/ml solution of colcemid (Life Technologies), 75 μl are added to the cells. Ethidium bromide (0.15 ml of a 1 mg/ml solution) can also be added, optionally, if longer chromosome length is desired.

3. Cells are incubated at 37°C for 90–120 minutes (mouse cells or zebrafish/mouse hybrids) or at 28°C for 4 hours (zebrafish primary cell lines). Cultures growing slowly require longer incubation time than fast-growing cultures.

4. Media from cultures is collected in a 50-ml centrifuge tube.

5. Cells are washed once with 2 ml of PBS (zebrafish cells) or PBS + 1m*M* EDTA, and this wash is added to the centrifuge tube.

6. Cells are trypsinized as described in Section II.D and added to the centrifuge tube.

7. Cells are centrifuged (Section II.D), the supernatant is removed, and the cells are resuspended in 20 ml of hypotonic solution (1% sodium citrate for zebrafish cells; 0.075*M* potassium chloride for hybrids or mouse cells) prewarmed at 37°C.

8. The cells are incubated for 20 minutes at 37°C.

9. An optional prefixation step is performed by adding 20 drops of Carnoy's solution (methanol : acetic acid, 3 : 1) to the tube.

10. After centrifugation at 1000 rpm for 10 minutes, the cells are gently resuspended in 0.5 ml of hypotonic solution by gently tapping the tube.

11. Carnoy's solution is added to the tube, first dropwise for 20 drops, and then using a Pasteur pipet, up to the 10-ml mark of a 15-ml centrifuge tube.

12. Cells are incubated for 20 minutes at −20°C.

13. Cells are centrifuged, resuspended in 10 ml of Carnoy's solution, and incubated again for 20 min at −20°C.

14. The preceding step is repeated once more except that the cells are kept at −20°C for 30 minutes to overnight. Total fixation time should be at least 1.5 hours.

15. After the last fixation step, cells are centrifuged and resuspended in 5 ml of fresh cold Carnoy's solution, incubated at −20°C for 30 minutes, centrifuged, resuspended in 2 ml of Carnoy's, and kept on ice.

16. One or two drops of cell suspension are added to a clean microscope slide. Excess liquid is removed by tipping the slide which is then placed on top of a beaker containing steaming water. This helps prevent too rapid a drying.

17. After the slides are dry, their quality is assessed under a phase contrast microscope. Chromosomes should be observed and should be dark and well-spread. Interphase nuclei should appear flat and gray, not small and shiny. There should not be too much cytoplasmic background or debris. Many variables affect

slide quality, particularly temperature and humidity. Technique should be optimized to suit conditions.

18. The best parts of the slides for FISH are identified, and the slides are kept overnight at 50°C on a slide warmer. They are then stored at room temperature in a slide box.

D. Hybridization Procedure

FISH is performed as described (Pinkel *et al.*, 1986) with minor modifications.

1. Denaturation

1. Slides with chromosome preparations (Section III.A.3) are dehydrated successively in ice-cold 70, 90, and 100% ethanol for 2 minutes each and allowed to air dry.

2. Slides are denatured for 2 minutes at 70–72°C in 40 ml of prewarmed denaturation solution [70% formamide/2× SSC (1× SSC is 0.15*M* NaCl; 0.015*M* sodium citrate, pH 7.0)] in a glass Coplin jar. The water bath is kept at 75°C so that temperature, which is checked by putting a thermometer directly into the Coplin jar, is maintained at 70–72°C.

3. After denaturation, the slides are dehydrated, as previously, in ice-cold 70, 90, and 100% ethanol.

2. Hybridization

1. Hybridization is performed on the same day the slides are denatured.

2. The 10 μl hybridization solution contains: 100 ng of biotinylated probe in 2 μl, 1 μl of sheared salmon sperm DNA (1mg/ml) and 7 μl of a solution containing 50% formamide/10% dextran sulfate/2× SSC, pH 7.

3. The hybridization solution is denatured for 10 min at 70°C and immediately placed on ice.

4. The 10 μl of hybridization solution are placed on a chosen location on the chromosome slide. Bubbles are removed.

5. Each slide is covered with a 22 × 22 mm glass cover slip which is sealed with rubber cement.

6. Slides are placed in a humidified chamber (e.g., made of pieces of Whatman paper soaked in water plus two wooden sticks to support the slides, in a plastic box or a square Petri dish) and incubated overnight at 37°C.

3. Posthybridization Washes

1. Slides are taken out of the humidified chamber, the rubber cement is gently removed with forceps, and the slide is shaken gently in the first wash solution to loosen the glass coverslip.

2. Wash solutions are prewarmed at 37°C in a glass Coplin jar.

3. The first wash is done in 50% formamide/2× SSC for 10 min at 37°C.

4. The second is done in 2× SSC for 10 minutes, 37°C.

5. Following the two washes, the slides are kept in 1× PBD (phosphate-buffered detergent, Oncor) in a glass Coplin jar at room temperature for 2 minutes to 2 hours.

4. Fluorescence Detection and Amplification

1. Slides are removed from PBD, and 30 μl of FITC-avidin (Oncor) are added to the chosen area on the slide (see Section III.C.18).

2. Each slide is covered with a 22 × 22 mm plastic cover slip and incubated for 20 min at 37°C in the humidified chamber.

3. The plastic coverslips are removed and slides are washed three times for 2 minutes each in 40 ml of 1× PBD in a glass Coplin jar at room temperature.

4. The signal is amplified by adding 30 μl of anti-avidin antibody (Oncor) to the chosen area.

5. Slides are covered with a plastic coverslip and incubated for 5–20 minutes at 37°C in the humidified chamber.

6. Slides are washed three times, 2 minutes each wash, in 1× PBD at room temperature. This is followed by another incubation with 30 μl of FITC-avidin, and washings as in step 1.

7. Slides are counterstained with 15 μl of freshly made propidium iodide : Antifade (1:29) solution. Propidium iodide (Cambio) is first diluted 1:30 with water prior to mixing with Antifade (Oncor).

8. Slides are observed under a Leitz Diaplan incident light fluorescence microscope and photographed with a Photoautomat (Leica). Wavelengths are: for FITC, excitation 490 nm and emission 525 nm; for the propidium counterstain, excitation 520 nm and emission 610 nm. Slides are photographed with Kodak Ektachrome Elite II color slide film, 400 ASA.

9. For long-term storage, glass coverslips can be sealed with nail polish and the slides stored for up to a month in the dark at 4°C.

E. Morphology of Zebrafish Chromosomes in Hybrids

As it is often seen in hybrids resulting from whole-cell fusion, the zebrafish chromosomal elements are often translocated or inserted into a mouse chromosome (Ekker *et al.*, 1996). Many different types of chromosome rearrangements are seen in zebrafish/mouse hybrids. Thus zebrafish elements have been inserted between mouse elements, either as a single insertion (Fig. 1A, see color plate) or as multiple insertions on the same chromosome (Fig. 1B, see color plate). Many hybrids also have zebrafish insertions or translocations on more than one mouse chromosome (Fig. 1C, see color plate). In few instances, the zebrafish chromosomes are standing free from mouse chromosome or are seen as double minutes (Fig. 1D, see color plate).

F. Number of Zebrafish Chromosomes in Hybrids as Determined by FISH

FISH analysis suggested that hybrids had integrated between zero and four zebrafish chromosomal elements (Ekker *et al.,* 1996; Chevrette *et al.,* 1998), which represents a small number of transferred chromosomes for hybrids generated by whole-cell fusion. For example, in similar whole cell fusions between human and mouse cells, we have transferred more than 18 human chromosomes (Speevak *et al.,* 1995).

IV. Identification of Chromosomes in Hybrids Using PCR

A. Available Markers

The markers presently available to genotype the zebrafish/mouse cell hybrid collection include a few hundred cloned genes, close to a thousand simple sequence repeats (SSR) markers (Goff *et al.,* 1992; Knapik *et al.,* 1996, 1998), and random amplified polymorphic DNA sequences (RAPD) markers (Welsh and McClelland, 1990; Williams *et al.,* 1990; Postlethwait *et al.,* 1994). The use of RAPD markers on zebrafish/mouse hybrids has been thus far problematic, possibly because of differences in the strains of zebrafish used for the production of the RAPD map and those used in the whole-cell fusion experiments.

Primers corresponding to sequences inside the DANA/mermaid short interspersed repetitive sequence (Izsvak *et al.,* 1996; Shimoda *et al.,* 1996a, 1996b) can also be used to detect rapidly the presence of zebrafish DNA inside a newly made hybrid. However, the main use DNA/mermaid primers and primers corresponding to other zebrafish interspersed repetitive sequences will be in inter-DANA/mermaid PCR (Shimoda *et al.,* 1996a, 1996b) a technique which is the object of Section V.

B. PCR Procedure

1. Each PCR reaction contains 10 or 100 ng of hybrid cell DNA, 100 ng of DNA from each of the parental cell lines, or 100 ng of a (1 : 10) mixture of zebrafish cell (ZF4 or LFF) DNA and mouse B78 DNA; the two oligonucleotide primers, 0.2 μl each of 20m*M* stocks; dATP, dCTP, dGTP, dTTP, 0.2m*M* each; the recommended buffer for the source of Taq DNA polymerase; 1.5 units of Taq DNA polymerase (Boehringer Manheim) and water in a total volume of 25 μl.

2. To detect SSR markers in zebrafish/mouse somatic hybrid cells, two different PCR programs are used. The first one consists of 45 cycles of 1 minute at 94°C, 1.5

minutes at 55°C and 1.5 minutes at 72°C. The second one consists of 45 cycles of 1 minute at 94°C, 1.5 minutes at 50°C, and no extension step. Both programs contain a predwell of 4 minutes at 94°C and a postdwell of 7 minutes at 72°C.

The amplification conditions are first tested on the parental cell lines (both the B78 mouse cells and the zebrafish ZF4 or LFF cells) as well as on a 10:1 mixture of B78:zebrafish DNA. Conditions giving the expected amplified fragment in both the zebrafish and the mixture DNAs, but no fragment with B78 DNA, are subsequently used on the hybrids. We were able to use about 75% of the SSR that we tried using the preceding criteria. All SSR primers are commercially available (Research Genetics, Huntsville, Alabama).

C. Composition of the Hybrid Panel

Using the PCR approach described in the above section with four to thirteen markers per linkage group, we were able to assign each twenty five linkage groups (LG) to at least one of our hybrids (Table I). However, hybrids containing markers from a given LG do not contain all markers for this LG (Fig. 2). Thus it seems that during the fusion (or the subsequent culture of the cells) zebrafish chromosomes were fragmented, resulting in hybrids which may resemble radiation hybrids. Nevertheless, two closely linked markers are often found in the same hybrids. Taken together, these results suggest that although we have used whole cell fusion, an intact zebrafish chromosome was rarely transferred (or retained) in our hybrids. A detailed genotyping of the present collection of zebrafish/mouse cell hybrids is currently in progress (M. Ekker, L. Joly, P. Tellis, and M. Chevrette, unpublished data) with the currently available markers.

Comparison of chromosome numbers in individual hybrids as determined by PCR with the number of chromosome determined by FISH (Section III.F) indicates that a larger number of chromosomal elements can be detected using PCR. We never observed more than four chromosomal elements in a given hybrid using FISH, whereas PCR analysis revealed that many hybrids contain elements from as many as 12 different LGs (Chevrette *et al.,* 1998). We can assume that, using a total zebrafish painting probe, small chromosomal elements will not be detected by FISH, whereas PCR amplification using a marker from this region will detect its presence. Another, nonexclusive possibility is that the greater number of LGs detected by PCR reflects heterogeneity in the hybrid cell population at the time of harvest: all hybrids would contain the chromosome or chromosome segment with the drug selection gene but different subsets of other zebrafish chromosomes. This explanation would be consistent with the results of our experiments on the stability of zebrafish chromosomes and chromosome segments in zebrafish/mouse cell hybrids (Section VI).

D. Alternative Procedure

Although time consuming compared to PCR amplification, it is nevertheless possible to analyze zebrafish/mouse hybrids using Southern blot hybridization (Southern, 1975).

Table I

Representation of the 25 Linkage Groups of the Zebrafish Genetic Maps (Postlethwait *et al;* 1994; Knapik *et al.*, 1998) in the Zebrafish/Mouse Hybrid Collection.[a]

HYBRID	ZEBRAFISH LINKAGE GROUP																								
ZFB-71	1	2				6												18		20	21				
ZFB-213	1	2					7										17				21		23		
ZFB-212	1		3	4					9					14						20		22	23		
ZFB-57	1		3	4		6						12				16									
LFFB-17				4	5									14				18					23	24	25
ZFB-238			3			6	7													20	21		23		
LFFB-4							7							14						20					
ZFB-214			3					8								16				20					
LFFB-10	1								9					14				18			21				
ZFB-43	1		3		5	6		8		10			13	14				18	19		21				
ZFB-64		2					7				11		13			16	17								
ZFB-211												12													25
ZFB-205													13						19		21				
ZFB-228														14		16		18					23		25
LFFB-1							7							14	15								23		
ZFB-54		2	3					8				12				16				20					
ZFB-42		2			5												17								
ZFB-218														14				18					23		
ZFB-65					5	6	7					12							19	20					
ZFB-48			3					8								16				20					
ZFB-240			3				7				11			14						20	21				
ZFB-56																						22			
ZFB-50					5				9											20	21	22	23		
LFFB-24														14										24	
LFFB-15														14										24	25

[a] Selected for this table were 25 zebrafish/mouse hybrids. The presence of a linkage group in a hybrid is based on PCR detection of at least one marker from this linkage group.

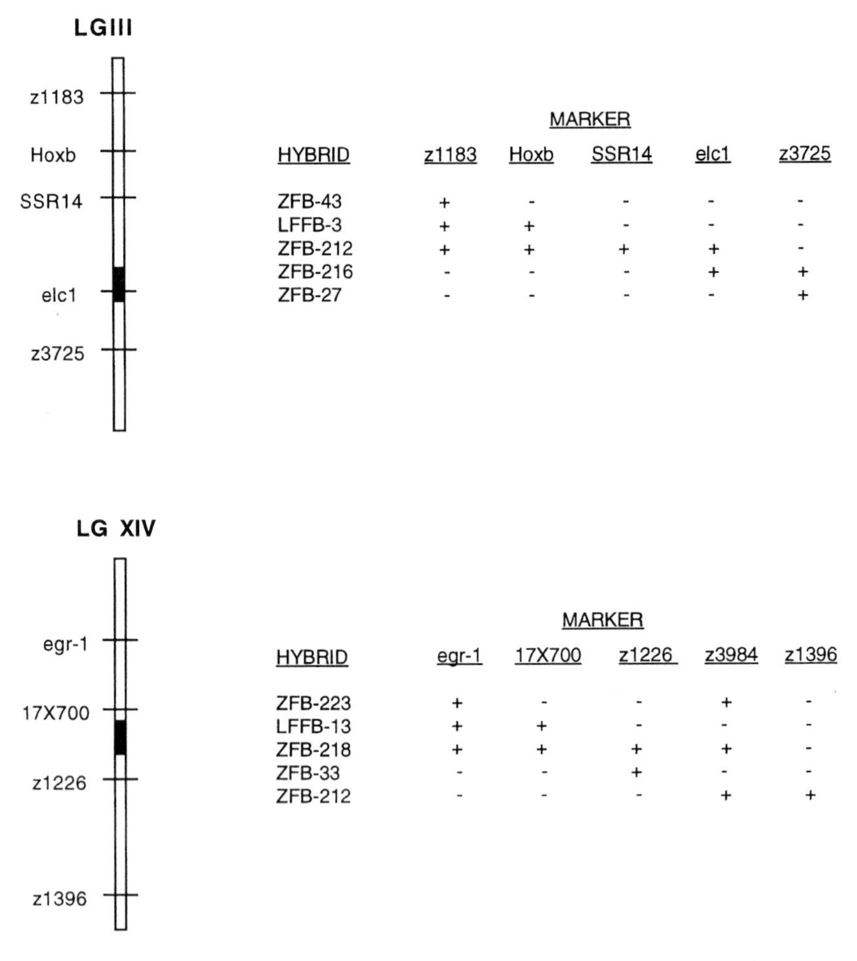

	MARKER				
HYBRID	z1183	Hoxb	SSR14	elc1	z3725
ZFB-43	+	-	-	-	-
LFFB-3	+	+	-	-	-
ZFB-212	+	+	+	+	-
ZFB-216	-	-	-	+	+
ZFB-27	-	-	-	-	+

	MARKER				
HYBRID	egr-1	17X700	z1226	z3984	z1396
ZFB-223	+	-	-	+	-
LFFB-13	+	+	-	-	-
ZFB-218	+	+	+	+	-
ZFB-33	-	-	+	-	-
ZFB-212	-	-	-	+	+

Fig. 2 Zebrafish/mouse hybrids contain distinct segments of zebrafish chromosomes. Chromosomes corresponding to LG III (top panel) and LG XIV (bottom panel) are shown. The position of the centromere is indicated in black.

1. Ten micrograms of DNA from individual hybrids are digested with restriction enzymes that have a 6-bp recognition sequence (e.g., *Eco*RI, *Hind*III, *Bam* HI).

2. Digests are subjected to electrophoresis on a 0.8% agarose gel in Tris–acetate–EDTA buffer (TAE) and transferred according to standard procedures (Sambrook *et al.,* 1989). Radiolabeled probes are synthesized using the random hexamer procedure.

Using this approach, we detected the zebrafish *bmp2* gene (Nikaido *et al.,* 1997) in a subset of the zebrafish/mouse cell hybrids. This contributed to its assignment to LG 20 of the genetic map (Postlethwait *et al.,* 1994) and its identification as the gene affected by the *swirl* mutation (Nguyen *et al.,* 1998).

V. Inter-DANA/Mermaid PCR on Hybrids

The purpose of this procedure is to amplify, using PCR, zebrafish DNA sequences from a zebrafish/mouse hybrid through the use of primers that allow amplification of DNA sequences between adjacent DANA/mermaid interspersed repetitive sequences (Izsvak *et al.,* 1996; Shimoda *et al.,* 1996a, 1996b). DANA/mermaid sequences are abundant which allows easy amplification of the DNA between two adjacent DANA sequences using PCR (Shimoda *et al.,* 1996b). Furthermore, DANA/mermaid sequences are not found in the mouse genome (Shimoda *et al.,* 1996a) which allows specific amplification of zebrafish sequences from the zebrafish/mouse hybrids (Fig. 3). Amplification is performed using combinations of four different DANA primers, in separate reactions. Following amplification, the PCR products are pooled and labeled with biotin and used as a probe for chromosome painting.

A. DANA/Mermaid PCR

1. DNA from zebrafish/mouse hybrids is diluted with TE to a final concentration of 10 ng/μl in TE.

2. Each PCR reaction contains: 5 μl zebrafish/mouse DNA (10 ng/μl); 5 μl PCR Buffer (10×)*; 5 μl dNTPs (2m*M* each of dATP, dCTP, dGTP, and dTTP);

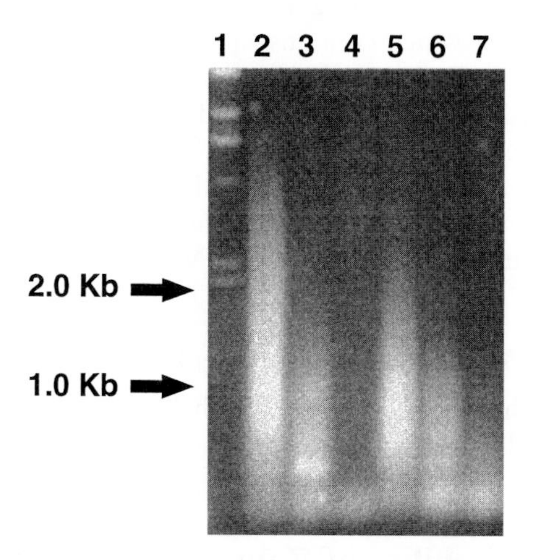

Fig. 3 DANA/mermaid PCR products from zebrafish or zebrafish/mouse hybrid DNA. Primer set B (see Section V.A.3) was used. Lane 1, *Hin*dIII digest of bacteriophage lambda DNA; lane 2, PCR amplification from ZF4 cell DNA; lanes 3–7, PCR amplification from zebrafish/mouse hybrids: 3, ZFB-8; 4, ZFB-78; 5, ZFB-209; 6, ZFB-50-1; 7, ZFB-14.

1.5 unit of Taq DNA polymerase and water to a volume of 47.5 μl. *PCR buffer (10×): 500m*M* KCl, 100m*M* Tris–HCl (pH 8.4 at 20°C), 15m*M* $MgCl_2$ and 1mg/ml of gelatin.

3. This mix (47.5 μl) is added to six eppendorf tubes (0.5 ml size) and labeled as follows:

A. primers 61C and 257 (see below)
B. primers 61C, 154C, 257, and 304 (see below)
C. primer 61C (5′-TCGAACCAGCGACCTTCTTG-3′)
D. primer 154C (5′-GTCTTTGGACTGTGGGGGAA-3′)
E. primer 257 (5′-CTGGATAAGTTGGCGGTTCA-3′)
F. primer 304 (5′-AAGGGACTAAGCCAAAAAGA-3′)

and containing 2.5 μl of the appropriate primer(s). Primer stocks are at a concentration of 100 ng/μl.

4. Parafin oil (50 μl) is added to each tube, as necessary, depending on the model of thermal cycler.

5. PCR conditions consist of 40 cycles of 2 minutes at 92°C, 1 minute at 65°C (no extension step). The program contains a predwell of 5 minutes at 94°C and a postdwell of 4 minutes at 72°C.

6. A 10 μl aliquot of the DANA/mermaid PCR products is electrophoresed on an agarose gel (0.8% in Tris–borate EDTA buffer) to verify a successful amplification. The gel should reveal a smear of DNA with some repetitive elements (seen as bands) characteristic for each primer used (Fig. 3).

7. The remaining 40 μl from tubes A, C, D, E, and F are combined whereas tube B, which contains the amplification with all 4 DANA/mermaid-primers is kept by itself. The DANA/mermaid-PCR products (240 μl) are precipitated with 24 μl of NaAcetate, 3*M* and 0.75 ml 100% EtOH (kept at −20°C). The tube is stored at −70°C for 20 minutes and then centrifuged for 15 minutes. The supernatant is removed, washed with cold 70% EtOH, and centrifuged again, then the supernatant is removed. The pellet is dryed and resuspended in 50 μl of TE. Quantification is done by UV absorbance (260 nm).

B. Use of DANA/Mermaid PCR Products in FISH Experiment

The DNA is labeled with biotin using the procedure described in Section III.B.1 and FISH is performed as described in Sections III.C and III.D.

VI. Stability of Zebrafish/Mouse Hybrids

Stability of zebrafish chromosomes in hybrids can be determined by growing cells for a large number of divisions in the presence or absence of G418, the drug that ensures their retention in a mouse background. We previously reported such an experiment with hybrids ZFB10 and ZFB37, which were both grown for over 100 generations (34 passages) either in the presence or in the absence of G418 (Ekker *et al.,* 1996). Hybrid ZFB10 contains a zebrafish chromosomal element that has inserted into a mouse chromosome, whereas translocation was not apparent for the chromosomal elements present in ZFB37. The retention of zebrafish chromosomal elements in the hybrids can be assayed by FISH analysis (Section III) of 100 interphase nuclei or by analysis of markers originally present in the hybrid cells (Section IV). For hybrid ZFB10, the zebrafish chromosomal element was detected in at least 95% of nuclei examined when the cells were maintained in G418. Furthermore, the SSR markers originally present on hybrid ZFB10 were all retained. When ZFB10 cells were grown in the absence of G418, only 68% of nuclei contained a zebrafish chromosome detectable by FISH.

More than one zebrafish chromosome can be detected in 37% nuclei of ZFB37 cells at passage 4, the starting point of the experiment. When ZFB37 cells are grown for 34 passages in the presence of G418, 95% of nuclei have kept at least one zebrafish chromosome, a result comparable to that obtained with ZFB10. However, only 13% of the cells still contain more than one zebrafish chromosome. This suggests that the chromosome tagged with the *neo* resistance gene is stably maintained in the mouse background, whereas the other(s), which presumably lack the resistance gene, are lost. In the absence of G418, a large decrease in the percentage of cells that retain zebrafish chromosomes is seen, and only 46% of the cells have retained a zebrafish chromosome.

The following conclusions can be drawn from this experiment. First, zebrafish chromosomes can be maintained for many generations in a mouse background provided that they carry a selection marker and that selective pressure is maintained. This was previously observed for other hybrid cell lines after a high number of passages (Ruddle, 1981). Second, insertion or translocation to a mouse chromosome does not seem to prevent the loss of the zebrafish chromosomal element from hybrids when selection is removed. It should be noted that for most practical purposes, like gene mapping or generation of a genomic library for a specific zebrafish chromosome, it should not be necessary to grow zebrafish/mouse hybrids for such a large number of passages.

VII. Perspectives

Using cell hybrids is the fastest route to assigning genes to chromosomes because, contrary to genetic linkage mapping, this approach does not require that any polymorphism for the gene of interest be found. Although they cannot offer the degree of precision in map position obtained with radiation hybrids, zebrafish/mouse somatic cell hybrids can nevertheless provide a means to assign genes to zebrafish chromosomes. In addition to the *bmp2/swirl* example mentioned in Section IV.D, zebrafish/mouse cell hybrids were also used to assign cloned zebrafish *Hox* genes to their respective *Hox* complexes (Prince *et al.*, 1998). Despite the finding that zebrafish chromosomes are often fragmented in zebrafish/mouse hybrids obtained by whole-cell fusion, closely linked markers will most of the time be detectable in the same hybrids. Thus, the collection of zebrafish/mouse hybrids can be used to support data obtained in genetic linkage experiments (Knapik *et al.*, 1998).

One important additional use of zebrafish/mouse somatic cell hybrids is their ability to provide a source of DNA for specific chromosome or chromosome regions. As suggested by the experiment on the stability of cell hybrids (Section VI), it should be possible to culture clones in order to retain only the chromosome or chromosome segment containing the drug resistance gene. The zebrafish donor cell population comprises more than 400 independent G418^r clones, and it is therefore reasonable to assume that most of the 25 zebrafish chromosomes exist at least once in the hybrid cell collection in the *neo*-tagged form. Monochromosomal subclones of zebrafish/mouse hybrids can be used in a DANA/mermaid PCR protocol (Section V) to generate zebrafish DNA specific to a single chromosome or chromosome segment. Such DNA can in turn be useful in the production of additional markers to specifically increase the density of the map of the zebrafish genome for a chromosome region where the present-day genetic map is poorly populated with markers or where an interesting mutation is located.

Acknowledgments

Our research is supported by grants from the Canadian Genome Analysis and Technology Program and from the Medical Research Council of Canada. We thank Ela Knapik and John Postlethwait for providing us with markers, genes, data prior to publication, and useful discussion; Marie-Andrée Akimenko for a critical reading of the manuscript; and Ted Zerucha and Stuart Joyce for help with the illustrations.

References

Chevrette, M., Joly, L., Tellis, P., and Ekker, M. (1997). Contribution of zebrafish/mouse cell hybrids in the mapping of the zebrafish genome. *Biochem. Cell Biol.* **75,** 641–649.

Driever, W., and Rangini, Z. (1993). Characterization of a cell line derived from zebrafish (*Brachydanio rerio*) embryos. *In Vitro Cell. Dev. Biol.-Animal* **29,** 749–754.

Ekker, M., Fritz, A., and Westerfield, M. (1992). Isolation of two families of satellite-like repetitive DNA sequences from the zebrafish (*Brachydanio rerio*). *Genomics* **13,** 1169–1173.

Ekker, M., Speevak, M. D., Martin, C. C., Joly, L., Giroux, G., and Chevrette, M. (1996). Stable transfer of zebrafish chromosome segments into mouse cells. *Genomics* **33**: 57–64.

Goff, D. J., Galvin, K., Katz, H., Westerfield, M., Lander, E. S., and Tabin, C. J. (1992). Identification of polymorphic simple sequence repeats in the genome of the zebrafish. *Genomics* **14,** 200–202.

He, L., Zhu, Z., Faras, A. J., Guise, K. S., Hackett, P. B., and Kapuscinski, A. R. (1992). *AluI* repeats of zebrafish (*Brachydanio rerio*) DNA: Cloning, characterization and transfer into fertilized eggs. *Mol. Marine Biol. Biotech.* **1,** 125–135.

Izsvak, Z., Ivics, Z., Garcia-Estefania, D., Fahrenkrug, S. C., and Hackett, P. B. (1996). DANA elements: A family of composite, tRNA-derived short interspersed DNA elements associated with mutational activities in zebrafish (*Danio rerio*). *Proc. Natl. Acad. Sci. U.S.A.* **93,** 1077–1081.

Kao, F.-T. (1973). Identification of chick chromosomes in cell hybrids formed between chick erythrocytes and adenine-requiring mutants of chinese hamster cells. *Proc. Natl. Acad. Sci. U.S.A.* **70,** 2893–2898.

Knapik, E. W., Goodman, A., Atkinson, O. S., Roberts, C. T., Shiozawa, M., Sim, C. U., Weksler-Zangen, S., Trolliet, M. R., Futrell, C., Innes, B. A., *et al.* (1996). A reference cross DNA panel for zebrafish (*Danio rerio*) anchored with simple sequence length polymorphisms. *Development* **123,** 451–460.

Knapik, E. W., Goodman, A., Ekker, M., Chevrette, M., Delgado, J., Neuhaus, S., Shimoda, N., Driever, W., Fishman, M. C., and Jacob, H. J. (1998). A mocrosatellite genetic linkage map for the zebrafish (*Danio rerio*). *Nat. Genet.* **18,** 338–343.

Leung, W.-C., Chen, D. R., Dubbs, D. R., and Kit, S. (1975). Identification of chick thymidine kinase determinant in somatic cell hybrids of chick erythrocytes and thymidine kinase-deficient mouse cells. *Exp. Cell Res.* **95,** 320–326.

Nguyen, V. H., Schmid, B., Trout, J., Connors, S. A., Ekker, M., and Mullins, M. C. (1998). Ventral positional information, including the neural crest, is established by a *bmp2/swirl* pathway of genes in the zebrafish. *Dev. Biol.* **199,** 93–110.

Nikaido, M., Tada, M., Saji, T. and Ueno, N. (1997). Conservation of BMP signaling in zebrafish mesoderm patterning. *Mech. Dev.* **61,** 75–88.

Pinkel, D., Straume, T., and Gray, J. W. (1986). Fluorescence *in situ* hybridization with human chromosome-specific libraries: detection of trisomy 21 and translocations of chromosome 4. *Proc. Natl. Acad. Sci. U.S.A.* **85,** 9138–9142.

Postlethwait, J. H., Johnson, S. L., Midson, C. N., Talbot, W. S., Gates, M., Ballinger, E. W., Africa, D., Andrews, R., Carl, T., Eisen, J. S., Horne, S., *et al.* (1994). A genetic linkage map for the zebrafish. *Science* **264,** 699–703.

Prince, V. E., Joly, L., Ekker, M., and Ho, R. K. (1998). Zebrafish *hox* genes: Genomic organization and modified colinear expression patterns in the trunk. *Development* **125,** 407–420.

Ruddle, F. H. (1981). A new era in mammalian gene mapping: Somatic-cell genetics and recombinant DNA methodologies. *Nature* **294,** 115–119.

Sambrook, J., Fritsch, E. F., and Maniatis, T. (1989). "Molecular cloning, a laboratory manual," 2nd ed. Cold Spring Harbor Laboratory Press, Cold Spring Harbor, NY.

Schwartz, A. G., Cook, P. R., and Harris, H. (1971). Correction of a genetic defect in a mammalian cell. *Nature New Biol.* **230,** 5–8.

Shimoda, N., Chevrette, M., Ekker, M., Kikuchi, Y., Hotta, Y., and Okamoto, H. (1996a). *Mermaid*: A family of short interspersed repetitive elements widespread in vertebrates. *Biochem. Biophys. Res. Comm.* **220,** 226–232.

Shimoda, N., Chevrette, M., Ekker, M., Kikuchi, Y., Hotta, Y., and Okamoto, H. (1996b). *Mermaid*: a family of short interspersed repetitive elements, is useful for zebrafish genome mapping. *Biochem. Biophys. Res. Comm.* **220,** 233–237.

Southern, E. M. (1975). Detection of specific sequences among DNA fragments separated by gel electrophoresis. *J. Mol. Biol.* **98,** 503–517.

Speevak, M. D., Berube, N. G., McGowan-Jordan, I. J., Bisson, C., Lupton, S. D., and Chevrette, M. (1995). Construction and analysis of microcell hybrids containing dual selectable tagged human chromosomes. *Cytogen. Cell Genet.* **69,** 63–65.

Trisler, G. D., and Coon, H. G. (1973). Somatic cell hybrids between two vertebrate classes. *J. Cell Biol.* **59,** 347.

Welsh, J., and McClelland, M. (1990). Fingerprinting genomes using PCR with arbitrary primers *Nucl. Acids Res.* **18,** 7213–7218.

Williams, J. G. K., Kubelik, A. R., Livak, K. J., Rafalski, J. A., and Tingey, S. V. (1990). DNA polymorphisms amplified by arbitrary primers are useful as genetic markers. *Nucl. Acids Res.* **18,** 6531–6535.

CHAPTER 18

Banded Chromosomes and the Zebrafish Karyotype

Angel Amores and John H. Postlethwait

Institute of Neuroscience
University of Oregon
Eugene, Oregon 97403

I. Introduction

The production and phenotypic analysis of zebrafish mutations has proceeded at a remarkable pace in the last few years (Kimmel, 1989; Driever *et al.,* 1996; Haffter *et al.,* 1996), but knowledge of zebrafish genomics, which is essential for the wholescale molecular understanding of these mutations, has advanced more slowly. Recent progress in zebrafish genomics includes the construction of genetic maps based on random amplified polymorphic DNAs (RAPDs) (Postlethwait *et al.,* 1994; Johnson *et al.,* 1996), microsatellites (Knapik *et al.,* 1996; 1998), somatic cell hybrids (Chevrette *et al.,* 1997), and cloned genes (Postlethwait *et al.,* 1998), and the isolation and characterization of chromosome rearrangements (Fritz *et al.,* 1996; Talbot *et al.,* 1997). The full utility of a detailed genetic map and a more satisfying analysis of chromosome rearrangements, however, requires, first, the unambiguous cytogenetic identification of each individual chromosome and, second, the assignment of linage groups defined by meiotic recombination to individual cytogenetically distinct chromosomes. Unfortunately, such studies

METHODS IN CELL BIOLOGY, VOL. 60

0091-679X/99 $30.00

have not yet been accomplished; consequently, there is a pressing need for molecular cytogenetics in zebrafish.

Early preparations demonstrated that the zebrafish *Danio rerio* has 25 pairs of small chromosomes, nearly all of which are metacentric or submetacentric (Endo and Ingalls, 1968), and despite more recent investigation with a variety of banding techniques (Daga *et al.,* 1996; Gornung *et al.,* 1997), sex chromosomes have not been detected. Although C banding and Ag-NOR (Nucleolar Organizer Region) banding techniques have been quite successfully applied to fish chormosomes (Garcia *et al.,* 1987; Amemiya and Gold, 1990; Amores *et al.,* 1995b), they have limited power to differentiate all chromosome pairs. Attempts to induce serial bands by G banding (Giemsa banding), replication banding (RBG), and Q banding (quinacrine banding) have not yet provided the resolution rquired to unequivocally identify all zebrafish chromosomes (Pijnacker and Ferwerda, 1995; Daga *et al.,* 1996; Gornung *et al.,* 1997). Replication banding may represent the best method to identify different chromosome pairs in fish (Giles *et al.,* 1988; Hellmer *et al.,* 1991; Amores *et al.,* 1995a; Bertollo *et al.,* 1997), and it has been successfully applied in DNA mapping studies in a number of species (Hayes *et al.,* 1992; Matsuda *et al.,* 1992), including humans (Fan *et al.,* 1990; Takahashi *et al.,* 1990). Here we report the zebrafish karyotype as defined by R banding and provide a detailed protocol for preparing such chromosomes.

II. Methods

A. Preparation of Fish Chromosomes

Zebrafish chromosomes are prepared either from embryo or adult tissues *in vivo* or from cells growing in tissue culture. Both long-term stable lines and primary cultures can provide good preparations. Embryos have the advantage that they are easy to obtain and examine, but they have relatively few mitoses. Long-term cultured lines, such as ZF4, although quite convenient, tend to be aneuploid and may have suffered rearrangements (Driever and Rangini, 1993). Primary cultures, either from adult fins or from embryos, are more time consuming to set up but display many mitoses, and probably reflect the *in vivo* situation better than long-term cultures; they appear to be the preparation of choice.

1. Fin Culture

The culture of larval or adult fins described here is according to Alvarez *et al.* (1991), with slight modifications.

1. About half of the caudal fin of a juvenile or adult fish is removed and placed in a tube containing 1 ml of L-15 mddium (Sigma), supplemented with 1% antibiotic-antimycotic solution (10,000 U penicillin/ml, 10 mg streptomycin/ml, 25 μg amphotericin B/ml).

2. Pretreat the 25 cm^2 flask by wetting the surface with medium prior to adding the tissue.

3. Rinse the fin biopsy once with 70% ethanol and twice with medium. In a small Petri dish with about 3 ml of medium, mince the fin biopsy into small pieces with forceps.

4. Transfer the pieces of one biopsy (e.g., from one fish) to the 25 cm^2 flask with forceps. (The flask may be pretreated with 1% gelatin to aid the adhesion of the tissue, but this is not essential).

5. Add 0.5 ml of complete growth medium (L-15 medium supplemented with 5% fetal calf serum, 10% zebrafish embryo extract, 1% antibiotic-antimycotic solution and 2% L-glutamine). Culture at 26–28°C.

6. After the explants are firmly attached (usually the next day), increase the total volume of medium to 2–3 ml.

7. Monitor the cultures daily for evidence of contamination and changes of pH; the medium initially has an orange tinge and will change to dark pink with increasing pH and to yellow with decreasing pH (the decreasing pH can arrise by bacterial contamination or by increased cell growth).

8. If a medium change is required (pH too high or too low), replace one half with fresh complete growth medium.

9. Fibroblasts usually begin to proliferate and migrate away from the biopsy after about 24–48 hours, but this is variable.

10. After the cells of the monolayer have reached confluency, they are ready for subculture. Alternatively, they can be harvested directly without subculture steps before they reach confluency. A culture from half a caudal fin from a single adult requires from 7–10 days to reach confluency.

2. Subculture

1. Remove the old medium and wash twice with sterile Hank's Balanced Saline Solution (HBSS) at pH 7.2 to remove any residual serum. (Alternatively use PBS without Ca, pH 7.2).

2. Add 2 ml of 1× Trypsin–EDTA (Sigma, 1×) in HBSS (or PBS minus Ca).

3. Monitor cell detachment with inverted microscope.

4. After the cells have detached, add 2 ml of medium supplemented with fetal calf serum (final concentration 5%) and $CaCl_2$ (final concentration 2mM). Transfer the suspension to a sterile, round-bottomed, 12 ml polypropylene tube (Sterilin).

5. Centrifuge the cell suspension for 5 minutes at 1200 rpm.

6. Discard the supernatant and resuspend the cells in 1 ml of complete medium. Transfer 0.5 ml of the cell suspension to a new 25 cm^2 flask and add complete medium to 3 ml after a few hours (so that the cells can sediment and attach to the flask). If desired, transfer the remaining 0.5 ml of cells into a fresh flask.

7. Monitor the cultures as before for infection or change in pH.

8. After 75% confluency is reached, the cultures are ready to be harvested (the mitotic index drops dramatically in cultures that have reached confluency).

3. Embryo Cell Culture

In general, we use 12 hour to 16 hour old embryos, and approximately 20 embryos for a 25 cm^2 flask. Older embryos can also be used, but they are generally harder to disaggregate.

1. Decorionate the embryos by hand with fine forceps.
2. Wash the embryos twice for 1 minute in 1:100 dilution of a 12% bleach (Clorox) solution.
3. Wash twice in HBSS.
4. Disaggregate in 2 ml Trypsin–EDTA 1× (in HBSS) with gentle pipetting.
5. Add medium supplemented with fetal calf serum to 5% and $CaCl_2$ to 2m*M*, to stop the action of the trypsin–EDTA
6. Transfer to a 12 ml polypropylene tube and spin down 8 minutes at 1200 rpm.
7. Discard the supernatant and resuspend the cells in 4 ml HBSS.
8. Spin down 8 minutes at 1200 rpm.
9. Resuspend cells in 4 ml of complete growth medium and plate out in 25 cm^2 flask. Culture at 26°C.
10. Monitor after the second day for cell growth and possible contamination.
11. Usually cultures are ready for harvesting after 4–5 days.
12. Harvest cells as in fin culture method.

4. Embryo Extract

This method follows the Zebrafish Book (Westerfield, 1993) and Collodi *et al.* (1992).

1. Take 100 three day old embryos.
2. Remove their chorions and chill on ice.
3. Rinse with bleach 0.1% for 1 minute.
4. Rinse two times with sterile cold PBS (ER water or HBSS) and one time with L-15 medium (4°C).
5. Transfer to a homogenizer with 0.5 ml of medium and homogenize well.
6. Transfer the homogenate to a sterile microfuge tube.
7. Centrifuge the homogenate 10 minutes at 4°C at 15,000 rpm.

8. Filter sterilize the supernatant.
9. Store at −20°C until use.

5. Harvesting Cells for Chromosome Spreads

1. Subculture primary cultures once. Fin cultures take 5 days until the cells are at the correct density for subculture.

2. Subculture in 4 ml of complete medium for 7 days (or until the culture reaches 755 confluency) during which time it is not necessary to change the medium.

3. Add colchicine to a final concentration of 0.02 μg/ml and incubate for 60 minutes (longer incubation times will increase the proportion of metaphases but also increase the contraction of the chromosomes).

4. Take off the medium and wash gently 2× with 2 ml of HBSS + glucose at 6/L (final volume about 4 ml). Incubate sample in 12 ml tube at 25°C in incubator.

5. Add 200 μl of Trypsin–EDTA solution with 1.8 ml HBSS + glucose at 6 g/L.

6. Shake to dislodge the cells; the suspension should be cloudy with the loosened cells. Withdraw all the cells and solution to a round-bottomed 12 ml polypropylene tube. Add back the original medium to stop the trypsin. (Total volume should be about 8 ml.)

7. Spin down the cells in bench top centrifuge for 8 minutes at 1500 rpm.

8. Remove the supernatant, leaving a small amount of liquid (0.5 ml) above the cell pellet.

9. Add 0.5 ml of hypotonic solution (0.5% KCl) to swell the cells. Resuspend the pellet, and then add a further 6.5 ml of hypotonic solution. Incubate at 25°C for 15 minutes.

10. Spin down the cells in bench top centrifuge for 8 minutes at 1500 rpm.

11. Fix the ice-cold 3:1 ml methanol/glacial acetic acid. Quickly resuspend the cells in 0.5 ml of fixative and then add 3.5 ml more of the fixative. Then spin down the cells in bench top centrifuge for 8 minutes at 1500 rpm.

12. Repeat resuspension and spin down again.

13. Resuspend in fix again and leave for 30 minutes at 4°C. Spin down the cells in bench top centrifuge for 8 minutes at 1500 rpm.

14. Resuspend and spin down. Remove most of the supernatant, leaving a small amount to resuspend the cells.

15. Acid wash the microscope slides and rinse them in deionized water. Store at −20°C prior to spreading. Add one to two drops of the cells to the middle of the slide using a Pasteur pipette from about 4 in. from above the slide. The cells will spread across the slide. Air dry and leave for at least 2 days before proceeding to chromosome banding.

B. Chromosome Banding

1. C-Banding

This protocol follows the one by Sumner (1972) with slight modifications.

1. Incubate slides (from step 15 in the protocol for harvesting cells for chromosome spreads, 5–6 days old) 14–18 minutes in 0.2*N* HCl at room temperature.
2. Wash the slides briefly with deionized water.
3. Incubate the slides for 5 minutes in a 5% $Ba(OH)_2$ solution at 60°C.
4. Wash the slides briefly in deionized water and 0.01*N* HCl.
5. Incubate slides in 2× SSC at 60°C for 60 minutes.
6. Rinse the slides in deionized water and stain in 10% Giemsa (in 0.025*M* phosphate buffer, pH 6.8) for 30 minutes.
7. Rinse the slides in deionized water, air dry, and view at the microscope.

2. Chromomycin A3 Banding

1. Incubate the slide for 10–15 minutes in McIlvaine citric acid–Na_2HPO_4 buffer (pH 7) (prepare solution A, 0.1*M* citric acid, and solution B, 0.2*M* Na_2HPO_4. For pH 7 mix 82 ml of solution B plus 18 ml of solution A).
2. Incubate the slide with CMA_3 solution (0.5 mg/ml in McIlvaine buffer with 10m*M* $MgCl_2$) for 15 minutes in a dark humid chamber.
3. Wash the slide with 0.15*M* NaCl/0.005*M* Hepes buffer (pH 7).
4. Incubate the slides in Methyl Green solution (0.12 mg/ml in NaCl/Hepes buffer) for 15 minutes.
5. Wash the slide with NaCl/Hepes buffer, mount the slide in glycerol, and analyze with fluorescence microscopy.

3. Trypsin G-Banding

This protocol follows Gold and Li (1991), with slight modifications.

1. Incubate slides 5 days old in 2× SSC for 2 hours at 60°C.
2. Wash the slides with 0.9% NaCl and partially air dry.
3. Treat the slides for 20 minutes with a trypsin/Giemsa solution (0.5 ml of trypsin stock solution + 1.5 ml Giemsa in 45 ml of 0.01*M* phosphate buffer pH 7.2) at room temperature. The trypsin stock solution is freshly prepared with 0.25 ml of 2.55 trypsin (Gibco) in 2.4 ml of water.
4. Rinse the slides with distilled water and air dry.

4. Replication Banding

a. Incubation with BrdU

The DNA in some portions of each chromosome replicate later in the S-phase of the cell cycle than other regions. Replication banding exploits this fact by staining late replicating DNA differently from early replicating DNA. Cells are cultured in fluorodeoxy uridine, an inhibitor of thymidylate synthetase, which blocks the production of thymidylate and arrests cells within the S-phase as stores become depleted. The addition of BrdU permits S to go to completion, but with this nucleotide analogue now incoporated into DNA instead of thymidine. Mitotic chromosomes will then be especially sensitive to breakage by ultraviolet (UV) light in regions of high BrdU, which will be the late replicating regions. UVs damaged DNA stains differently from intact DNA, giving rise to the banded chromosomes.

1. Subculture the zebrafish cells (from step 10 in the fin culture protocol or step 11 in the embryo culture protocol) in 4 ml of medium for 4–7 days (variable upon cell growth) during which time it is not necessary to change the medium.

2. When the cell culture reach about 75% confluency, add FdU (100× Fluoro-deoxyUridine, Sigma) to a final concentration of 0.123 μg/ml and incubate for 16–18 hr at 26°C.

3. Then add BrdU (1 mg/ml stock solution of BromodeoxyUridine, Sigma) to a final concentration of 30 μg/ml for 6–7 hours. Prepare chromosome slides as described before.

Note: It is essential to use a thymidine deficient medium (L-15) for the BrdU incubation.

b. Fluorochrome Photolysis Giemsa

1. Two day old slides from step 3 in the protocal for incubation with BrdU, are washed in 0.5× SSC and immersed in a Hoechst 33258 solution (0.5 μg/ml in 0.5× SSC, pH 7.05) for 30 minutes at room temperature.

2. After a rinse with deionized water, slides are covered by a thin layer of 2× SSC and exposed to long-wave (366 nm) UV products hand-held lamp at a distance of 2.5 cm from the slides (this will break DNA that had incorporated BrdU, that is, newly synthesized DNA).

3. Slides are than rinsed and treated with 0.5× SSC at 60°C for 60 minutes before staining in 4% Giemsa for 20 minutes.

Note: In general, pale chromosomes with no or faint bands result from an overtreatment, and dark chromosomes result from undertreatment.

C. Fluorescent *in Situ* Hybridization to Chromosomes

1. Probe Synthesis

a. Nick Translation

1. Mix in an eppendorf tube:

1 μg DNA

5 μl 10× nick translation buffer (0.5*M* Tris–HCl; pH 7.8, 0.5 mg/ml BSA)

5 μl 100m*M* DTT

4 μl dNTP (0.5 m*M* dATP, dGTP, dCTP, each)

1 μl 0.5 m*M* dTTP

2 μl 1m*M* DIG-11-dUTP, Biotin-16-dUTP or Fluorescein-12-dUTP

1 μl DNase I (diluted 1:1000 from 1 mg/ml stock)

10 U DNA polymerase I

Bring the volume to 50 μl with distilled water.

2. Mix and incubate 2 hours at 15°C.

3. Take an aliquot to check the probe size (probe size should be between 100 and 500 bp).

4. Stop the reaction by adding 1.5 μl of 0.5*M* EDTA, pH 7.4.

5. Purify the labeled DNA by Sephadex G-50 resin equilibrated with 10m*M* Tris–HCl, 1m*M* EDTA, 0.1% SDS.

b. Probe Labeling by DOP–PCR

For labeling of large genomic clones [P_1-derived artificial chromosomes (PACs) or bacterial artificial chromosomes (BACs)] we use the protocol of Kroisel *et al.* (1994) with slight modifications.

1. In a 10 μl reaction [50m*M* Tris–HCl (pH 7.5), 7m*M* $MgCl_2$, 200μ*M* dNTPs, 1μ*M* DOP–PCR primer (5′-CCGACTCGAGNNNNNNATGTGG-3′, N = A, C, G, or T)], denature 10–50 ng of PAC DNA for 3 minutes at 94°C (top with 30 μl of mineral oil if necessary to prevent evaporation).

2. Add 0.5 unit of Exo-Klenow DNA polymerase (Stratagene), anneal primer at 25°C for 30 seconds, followed by a 1-minute ramp to 37°C, a 2-minute extension at 37°C, and a 30-second denaturation at 94°C.

3. Add 0.5 units of DNA polymerase after each cycle and repeat three times.

4. Label 5 μl of the previous reaction by PCR in a 25 μl reaction (10m*M* Tris–HCl pH 8.3, 50m*M* KCl, 1.5m*M* $MgCl_2$, 0.001% gelatin, 200 μ*M* dNTPs, 1.3 mg/ml BSA, 2 μM DOP–PCR primer, 1 unit amplitaq DNA polymerase, Perkin Elmer (top with 50 μl of mineral oil if necessary) in the presence of DIG-11-dUTP, Biotin-16-dUTP or Fluorescein-12-dUTP (ratio of dTTP to the respective labeling nucleotide is 2:1).

5. Purify probes by ethanol precipitation in the presence of 10 μg carrier RNA and resuspend in 25 μl of TE buffer.

2. Probe Mixture

50% formamide

2× SSC

10% dextran sulfate

50m*M* sodium phosphate buffer pH 7
100 μg/ml tRNA
4 ng/μl labeled probe

For PAC/BAC probes, use 20 ng/μl of labeled probe plus 200 ng/μl of sonicated at genomic DNA.

3. Pretreatment of Slides

1. Mark the area of interest on the back of the slide with a diamond writer.
2. Wash the slide in 2× SSC at 37°C for 5 minutes.
3. Apply 150 μl RNase A (100μg/ml 2× SSC) on the slide, cover with a coverslip (22× 40 mm), and incubate for 1 hour at 37°C in a moist chamber.
4. Wash in 2× SSC for 3 × 5 minutes (Coplin jar).
5. Dehydrate in 70, 90, and 100% ethanol 5 minutes each and air dry. Use slides immediately.

4. Chromosome Denaturation

1. Prewarm the slides to 60°C in an incubator to avoid a drop in temperature when the slides are put into the denaturation solution.
2. Denature each slide by immersing for 2 minutes in 70% formamide, 2× SSC at 80°C.
3. Rinse slides in ice cold 2× SSC.
4. Dehydrate slides in ice cold 70, 90, and 100% ethanol for 3 minutes each.
5. Air dry.

5. Probe Denaturation and *in Situ* Hybridization

1. Denature probes for 10 minutes in boiling water. Keep probe in a 0.5 ml microfuge tube.
2. Chill on ice for 3 minutes and spin down condensate (for PAC/BAC probes incubate probe 30 minutes at 37°C). Apply probe to denatured chromosomes.
3. Cover with 22 × 22 mm coverslip and seal edges with rubber cement or nail polish.
4. Incubate overnight at 37°C in a moist chamber.

6. Posthybridization Washes

1. Remove coverslip and transfer slides into a Coplin jar containing 50% formamide; 2× SSC prewarmed to 42°C.

2. Incubate for 10 minutes shaking in hybridization oven at 42°C.
3. Change prewarmed solution two more times and shake 10 minutes each.
4. Incubate 3 × 5 min in 2× SSC prewarmed to 42°C.
5. Incubate 1 × 5 min in 0.1× SSC at 60°C (stringent wash).

7. Antibody Detection

1. Take slides out of the Coplin jar and apply 200 μl of 4× SSC, 3% BSA, 0.2% Tween-20.

2. Cover with 22 × 40 mm coverslip and incubate for 30 minutes at 37°C in moist chamber.

3. Apply 100 μl primary detection solution (4× SSC, 1% BSA, 0.2% Tween-20 + antibody). If detecting biotin, then use 5 μg/ml of goat anti-Biotin (Vector labs); if detecting digoxigenin, then use 2 μg/ml sheep anti-DIG-Fluo (BCL).

4. Apply coverslip and incubate for 30 minutes at 37°C in dark moist chamber.

5. Wash 3 × 5 min in 4× SSC; 0.2% Tween-20 at 42°C with shaking.

6. Takes slides out of the Coplin jar and add 200 μl of second-detection solution (4× SSC, 1% BSA, 0.2% Tween-20 + second antibody). If detecting biotin detection, then use 5 μg/ml Texas Red anti-goat IgG (Vector Labs); if detecting digoxigenin, then use 5 μg/ml Donkey anti-Sheep IgG (Sigma).

7. Cover with coverslip and incubate for 30 minutes in dark, moist chamber at 37°C.

8. Wash 3× 5 minutes with 4× SSC, 0.2% Tween-20 prewarmed to 42°C.

9. Cover with coverslip and incubate for 30 minutes in dark, moist chamber at 37°C.

10. Wash 3× 5 minutes with 4× SSC, 0.2% Tween-20 prewarmed to 42°C.

8. Counterstain

1. Take slides out of the Coplin jar and apply 200 μl of counterstain solution (1 μg/ml propidium iodide in 2× SSC for the fluorescein conjugated antibodies; 0.2 μg/ml DAPI for the Texas Red conjugated antibody).

2. Incubate 10 minutes at room temperature.

3. Rinse briefly with deionized water.

4. Drain excess fluid and apply 30 μl antifade solution [0.023% DAPCO (Sigma) in 90% glycerol, 0.1*M* Tris–HCl pH 8.0].

5. Cover with 22 × 40 mm coverslip, store into a dark container at 4°C, and examine with fluorescence microscopy using the appropriate set of filters.

III. Results and Discussion

Figure 1 shows the zebrafish karyotype produced with replication banding from cultures of adult fin cells, and Fig. 2 shows the karyogram derived from the

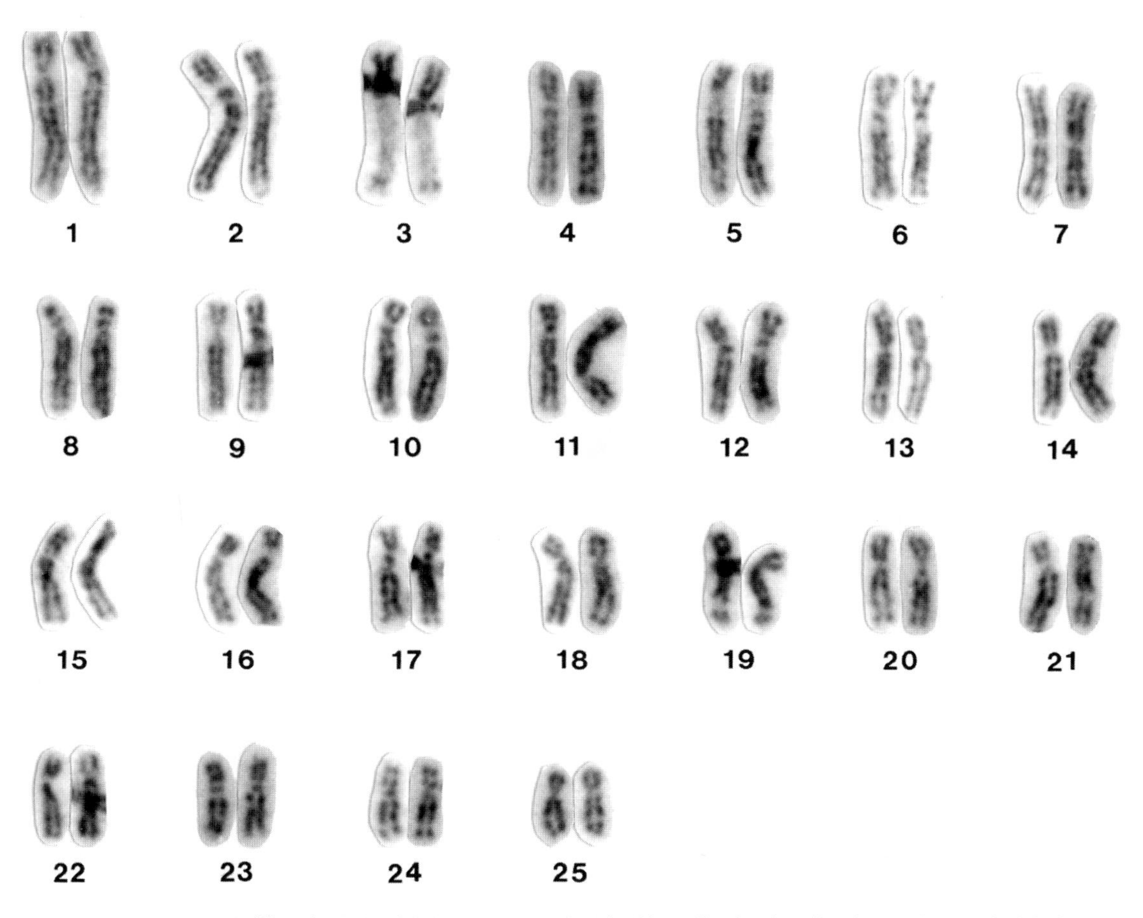

Fig. 1 Zebrafish karyotype produced with replication banding from cultures of adult fin cells.

examination of five such karyotypes. Several of the chromosomes are immediately distinguishable, confirming the results of other workers (Pijnacker and Ferwerda, 1995; Daga *et al.,* 1996; Gornung *et al.,* 1997). Chromosome 1 is substantially longer than the other chromosomes. The long arm of chromosome 3 is late replicating, and chromosomes 7 and 15 are the only relatively large chromosomes that are exactly metacentric. The other chromosomes must be distinguished by a closer examination of their replication banding pattern. Although members of a homologous pair of chromosomes usually show the same pattern, occasionally they vary in length and in the location of bands. This can be explained by technical factors (Drouin *et al.,* 1988) or due to the condensation delay of BrdU-substituted DNA (Schollmayer *et al.,* 1981).

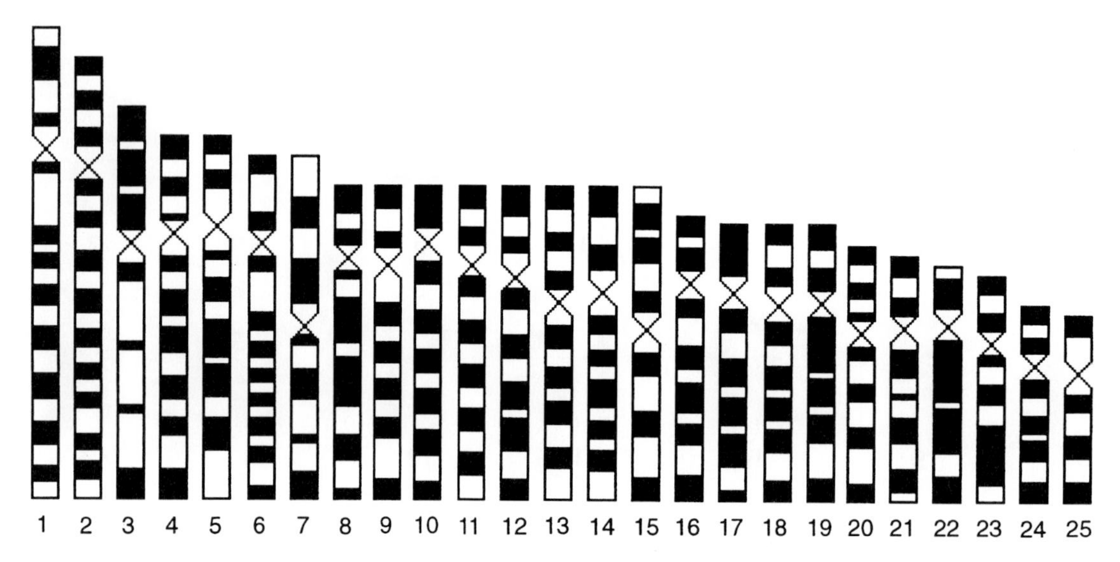

Fig. 2 Idiogram derived from the replication banding karyotype of *Danio rerio.* Black colored segments corresponds to early-replicating R-bands.

There are several differences in chromosome numbering and morphology between different authors (Schreeb *et al.,* 1993; Pijnacker and Ferwerda, 1995; Daga *et al.,* 1996; Gornung *et al.,* 1997). This is because several of the chromosomes are nearly the same length and because the degree of condensation affects different chromosomes differently. Cells with more relaxed chromosomes had 15 submetacentrics, whereas cells with more condensed chromosomes had only 10 submetacentrics, the other 5 chromosomes being subtelocentrics. This could be explained by the existence of preferential zones of condensation or by the condensation delay of BrdU substituted regions (Drouin *et al.,* 1991) that results in variations in the arm ratio. More condensed chromosomes display a smaller number of chromosomal bands (G or R bands), which does not always allow for unambiguous chromosome identification. The *in vitro* technique described here suggests that the BrdU labeling of chromosomes in cultured cells is superior to the *in vivo* technique used in most fish reports (Daga *et al.,* 1996; Pendas *et al.,* 1993a). Moreover, the cell culture technique avoids the sacrifice of specimens when studying chromosome rearrangements.

Constitutive heterochromatin as revealed by C banding is located in centromeric positions of all chromosomes (Pijnacker and Ferwerda, 1995; Daga *et al.,* 1996; Gornung *et al.,* 1997). The type I satellite probe (Ekker *et al.,* 1992) is located in the centromere of most chromosomes (Fig. 3, see color plate), and no marker chromosomes were evident. This distribution is similar to the pattern found in some fish species (Garrido-Ramos *et al.,* 1994; Oliveira and Wright, 1998) but not others (Haaf *et al.,* 1993; Reed *et al.,* 1997), where specific satellite-like sequences are not located on all chromosomes and, hence, can be used as markers for chromosome mapping or comparative purposes.

The Nucleolar Organizer Regions are important markers, and have been used in comparative studies among fish species (Amemiya and Gold, 1990; Amores *et al.,* 1995b). NOR are easily visualized by silver staining and are generally located near telomeres of the long arm of three chromosome pairs in zebrafish (Pijnacker and Ferwerda, 1995; Daga *et al.,* 1996; Gornung *et al.,* 1997). Silver staining, however, will reveal only NOR regions that are active in the previous interphase (Howell, 1977) and do not always reveal the location of all rDNA loci. *In situ* hybridization with ribosomal probes is the best alternative to detect rDNA loci that cannot be detected with alternative methods (Pendas *et al.,* 1993b). When applied to zebrafish chromosomes (Gornung *et al.,* 1997), additional rRNA loci were detected near telomeres of the long arm of chromosome 3. Because NOR are polymorphic in size, location, and number within many fish species (Fernandes-Matioli *et al.,* 1997; Phillips *et al.,* 1988; Sola *et al.,* 1992) and the zebrafish strains used by diverse authors had different origins, it is not clear if this additional NOR represents such a polymorphic situation in zebrafish. Caution should be taken when making cross-species comparisons, however, because similar looking chromosomes (as revealed by chromosome banding) can be completely different chromosomes as revealed by *in situ* hybridization with chromosome-specific probes (Stanyon *et al.,* 1995). Unfortunately, we do not yet known which zebrafish linkage group corresponds to each NOR-containing chromosome.

An important goal is the assignment of linkage groups defined genetically (Johnson *et al.,* 1996; Knapik *et al.,* 1998; Postlethwait *et al.,* 1998) to chromosomes defined cytogenetically (Fig. 1). For this purpose, it is useful to generate chromosome-specific probes that unequivocally identify each chromosome for all investigators. Such a panel exists already for human and mouse chromosomes (Shi *et al.,* 1997; Speicher *et al.,* 1996). Chromosome-specific probes are best utilized in fluorescent *in situ* hybridization (FISH) to banded chromosomes. Large clones, such as the 100 kb or so clones present in the zebrafish genomic PAC library constructed by C. Amemiya (see Amemiya *et al.,* 1996) are ideal for this purpose. We have therefore isolated two to ten PAC clones for each of the 25 linkage groups, each clone corresponding to an expressed gene. Genes where chosen at the centromere and toward each telomere of each linkage group. Using this resource in FISH experiments will identify the chromosome, and confirm the location of the physical centromere with respect to the genetically defined centromere (Johnson *et al.,* 1996; Knapik *et al.,* 1998).

In conclusion, these results provide a preliminary, complete identification and numbering of the zebrafish karyotype. They provide a basis for the assignment of linkage groups to chromosomes either by using FISH or by applying the analysis of characterized chromosome rearrangements (Fritz *et al.,* 1996). It is important to complete this work soon to place zebrafish genomics on a firm cytogenetic foundation.

References

Alvarez, M. C., Otis, J., Amores, A., and Guise, K. (1991). Short-term cell culture technique for obtaining chromosomes in marine and freshwater fish. *J. Fish Biol.* **39,** 817–824.

Amemiya, C. T., and Gold, J. R. (1990). Cytogenetics studies in North American minnows (Cyprinidae). XVII. Chromosomal NOR phenotypes of 12 species, with comments on cytosystematic relationships among 50 species. *Hereditas* **112,** 231–247.

Amemiya, C., Ota, T., and Litman, G. (1996). Construction of P1 artificial chromosome (PAC) libraries from lower vertebrates. *In* "Nonmammalian Genomic Analysis: A Practical Guide" (B. Birren and E. Lai, Eds.). San Diego: Academic Press.

Amores, A., Bejar, J., and Alvarez, M. C. (1995a). BrdU replication bands in the anguilliform fish *Echelus myrus. Chrom. Res.* **3,** 423–426.

Amores, A., Bejar, J., and Alvarez, M. C. (1995b). Replication, C and Ag-NOR chromosome-banding in two anguilliform fish species. *Mar. Biol.* **123,** 845–849.

Bertollo, L. A. C., Fontes, M. S., Fenocchio, A. S., and Cano, J. (1997). The X_1X_2Y sex chromosome system in the fish *Hoplias malabaricus.* I. G-, C- and chromosome replication banding. *Chrom. Res.* **5,** 493–499.

Chevrette, M., Joly, L., Tellis, P., and Ekker, M. (1997). Contribution of zebrafish–mouse cell hybrids to the mapping of the zebrafish genome. *Biochem. Cell Biol.* **75,** 641–649.

Collodi, P., Kamei, Y., Ernst, T., Miranda, C., Buhler, D. R., and Barnes, D. W. (1992). Culture of cells from zebrafish (*Brachydanio rerio*) embryo and adult tissues. *Cell Biol. Toxicol.* **8,** 43–61.

Daga, R. R., Thode, G., and Amores, A. (1996). Chromosome complement, C-banding, Ag-NOR and replication banding in the zebrafish *Danio rerio. Chrom. Res.* **4,** 29–32.

Driever, W., and Rangini, Z. (1993). Characterization of a cell line derived from zebrafish (*Brachydanio rerio*) embryos. *In Vitro Cell. Dev. Biol.* **28A,** 749–754.

Driever, W., Solnica-Krezel, L., Schier, A. F., Neuhauss, S. C. F., Malicki, J., Stemple, D. L., Stainier, D. Y. R., Zwartkruis, F., Abdelilah, S., Rangini, Z., Belak, J., and Boggs, C. (1996). A genetic screen for mutations affecting embryogenesis in zebrafish. *Development* **123,** 37–46.

Drouin, R., Lemieux, N., and Richer, C.-L. (1988). High-resolution R-banding at the 1250-band level. I. Technical considerations on cell synchronization and R-banding (RHG and RBG). *Cytobios* **56,** 107–125.

Drouin, R., Lemieux, N., and Richer, C.-L. (1991). Chromosome condensation from prophase to late metaphase: Relationship to chromosome bands and their replication time. *Cytogenet. Cell Genet.* **57,** 91–99.

Ekker, M., Fritz, A., and Westerfield, M. (1992). Identification of two families of satellite-like repetitive DNA sequences from the zebrafish (*Brachydanio rerio*). *Genomics* **13,** 1169–1173.

Endo, A., and Ingalls, T. H. (1968). Chromosomes of the zebra fish. A model for cytogenetic, embryologic, and ecologic study. *J. Hered.* **59,** 382–384.

Fan, Y.-S., Davis, L. M., and Shows, T. M., (1990). Mapping small DNA sequences by fluorescence *in situ* hybridization directly on banded metaphase chromosomes. *Proc. Natl. Acad. Sci. USA* **87,** 6223–6227.

Fernandes-Matioli, F. M. C., Almeida-Toledo, L. F., and Toledo-Filho, S. A. (1997). Extensive Nucleolus Organizer Region polymorphism in *Gymnotus carapo* (Gymnotoidei, Gymnotidae). *Cytogenet. Cell Genet.* **78,** 236–239.

Fritz, A., Rozowski, M., Walker, C., and Westerfield, M. (1996). Identification of selected gamma-ray induced deficiencies in zebrafish using multiplex polymerase chain reaction. *Genetics* **144,** 1735–1745.

Garcia, E., Alvarez, M. C., and Thode, G. (1987). Chromosome relationships in the genus *Blennius* (Blenniidae, Perciformes): C-banding patterns suggest two karyoevolutionary pathways. *Genetica* **72,** 27–36.

Garrido-Ramos, M. A., Jamilena, M., Lozano, R., Rejon, C. R., and Rejon, M. R. (1994). Cloning and characterization of a fish centromeric satellite DNA. *Cytogenet. Cell Genet.* **65,** 233–237.

Giles, V., Thode, G., and Alvarez, M. C. (1988). Early replication bands in two scorpion fishes, *Scorpaena porcus* and S. Notata (order Scorpaeniformes). *Cytogenet. Cell Genet.* **47,** 80–83.

Gold, J. R., and Li, Y. C. (1991). Typsin G-banding of North American cyprinid chromosomes: Phylogenetic considerations, implications for fish chromosome structure, and chromosomal polymorphism. *Cytologia* **56,** 199–208.

Gornung, E., Gabrielli, I., Cataudella, S., and Sola, L. (1997). CMA3-banding pattern and fluorescence *in situ* hybridization with 18S rRNA genes in zebrafish chromosomes. *Chrom. Res.* **5,** 40–46.

Haaf, T., Schmid, M., Steinlein, C., Galetti, P. M., and Willard, H. F. (1993). Organization and molecular cytogenetics of a satellite DNA family from *Hoplias malabaricus* (Pisces, Erythrinidae). *Chrom. Res.* **1,** 77–86.

Haffter, P., Granato, M., Brand, M., Mullins, M. C., Hammerschmidt, M., Kane, D. A., Odenthal, J., Van Eeden, F. J. M., Jiang, Y. J., Heisenberg, C. P., Kelsh, R. N., Furutani-Seiki, M., Vogelsang, E., Beuchle, D., Schach, U., Fabian, C., and Nüsslein-Volhard, C. (1996). The identification of genes with unique and essential functions in the development of the zebrafish, *Danio rerio. Development* **123,** 1–36.

Hayes, H., Petit, E., Lemieux, N., and Dutrillaux, B. (1992). Chromosomal localization of the ovine beta-casein gene by non-isotopic *in situ* hybridization and R-banding. *Cytogenet. Cell Genet.* **61,** 286–288.

Hellmer, A., Voiulescu, I., and Schempp, W. (1991). Replication banding studies in two cyprinid fishes. *Chromosoma* **100,** 524–531.

Howell, W. M. (1977). Visualization of ribosomal gene activity: Silver staining proteins associated with rRNA transcribed from oocyte chromosomes. *Chromosoma* **62,** 361–367.

Kimmel, C. B. (1989). Genetics and early development of zebrafish. *Trends Genet.* **5,** 283–288.

Knapik, E. W., Goodman, A., Atkinson, O. S., Roberts, C. T., Shiozawa, M., Sim, C. U., Weksler-Zangen, S., Trolliet, M. R., Futrell, C., Innes, B. A., Koike, G., McLaughlin, M. G., Pierre, L., Simon, J. S., Vilallonga, E., Roy, M., Chiang, P.-W., Fishman, M. C., Driever, W., and Jacob, H. J. (1996). A reference cross DNA panel for zebrafish (*Danio rerio*) anchored with simple sequence lenth polymorphisms. *Development* **123,** 451–460.

Knapik, E. W., Goodman, A., Ekker, M., Chevrette, M., Delgado, J., Neuhauss, S., Shimoda, N., Driever, W., Fishman, M. C., and Jacob, H. (1998). A microsatellite genetic linkage map for zebrafish (*Danio rerio*). *Nat. Genetic.* **18,** 338–343.

Kroisel, P. M., Ioannou, P. A., and de Jong, P. J. (1994). PCR probes for chromosome *in situ* hybridization of large-insert bacterial recombinants. *Cytogenet. Cell Genet.* **65,** 97–100.

Johnson, S. L., Gates, M. A., Johnson, M., Talbot, W. S., Horne, S., Baik, K., Rude, S., Wong, J. R., and Postlethwait, J. H. (1996). Centromere-linkage analysis and consolidation of the zebrafish genetic map. *Genetics* **142,** 1277–1288.

Matsuda, Y., Harada, Y.-N., Natsuume-Sakai, S., Lee, K., Shiomi, T., and Chapman, V. M. (1992). Location of the mouse complement factor H gene (cfh) by FISH analysis and replication R-banding. *Cytogenet. Cell Genet.* **61,** 282–285.

Oliveira, C., and Wright, J. M. (1998). Molecular cytogenetic analysis of heterochromatin in the chromosomes of tilapia, *Oreochromis niloticus* (Teleostei: Cichlidae). *Chrom. Res.* **6,** 205–211.

Pendas, A. M., Moran, P., and Garcia-Vazquez, E. (1993a). Replication banding patterns in Atlantic salmon (*Salmo salar*). *Genome* **36,** 440–444.

Pendas, A. M., Moran, P., and Garcia-Vazquez, E. (1993b). Multichromosomal location of ribosomal RNA genes and heterochromatin association in brown trout. *Chrom. Res.* **1,** 63–67.

Phillips, R. B., Pleyte, K. A., and Hartley, S. E. (1988). Stock-specific differences in the number and chromosome positions of the nucleolar organizer regions in arctic char (*Salvelinus alpinus*). *Cytogenet. Cell Genet.* **48,** 9–12.

Pijnacker, L. P., and Ferwerda, M. A. (1995). Zebrafish chromosome banding. *Genome* **38,** 1052–1055.

Postlethwait, J., Johnson, S., Midson, C., Talbot, W. S., Gates, M., Ballinger, E., Africa, D., Andrews, R., Carl, T., Eisen, J., Horne, J., Kimmel, C., Hutchinson, M., Johnson, M., and Rodriguez, A. (1994). A genetic map for the zebrafish. *Science* **264,** 699–703.

Postlethwait, J. H., Yan, Y.-L., Gates, M. A., Horne, S., Amores, A., Brownlie, A., Donovan, A., Egan, E. S., Force, A., Gong, Z., Goutel, C., Fritz, A., Kelsh, R., Knapik, E., Liao, E., Paw, B., Ransom, D., Singer, A., Thomson, M., Abduljabbar, T. S., Yelick, P., Beier, D., Joly, J.-S., Larhammar, D., Rosa, F., Westerfield, M., Zon, L. I., Johnson, S. L., and Talbot, W. S. (1998). Vertebrate genome evolution and the zebrafish gene map. *Nat. Genet.* **18,** 345–349.

Reed, K. M., Dorschner, M. O., and Phillips, R. B. (1997). Characteristics of two salmonid repetitive DNA families in rainbow trout, *Oncorhynchus mykiss. Cytogenet. Cell Genet.* **79,** 184–187.

Schreeb, K. H., Groth, G., Sachsse, W., and Freundt, K. J. (1993). The karyotype of the zebrafish (*Brachydanio rerio*). *J. Exp. Anim. Sci.* **36,** 27–31.

Schollmayer, E., Schäfer, D., Frisch, B., and Schleiermacher, E. (1981). High resolution analysis and differential condensation in RBA-banded human chromosomes. *Hum. Genet.* **59,** 187–193.

Shi, Y.-P., Mohapatra, G., Miller, J., Hanahan, D., Lander, E., Gold, P., Pinkel, D., and Gray, J. (1997). FISH probes for mouse chromosome identification. *Genomics* **45,** 42–47.

Sola, L., Rossi, A. R., Iaselli, V., Rasch, E. M., and Monaco, P. J. (1992). Cytogenetics of bisexual/unisexual species of *Poecilia.* II. Analysis of heterochromatin and nucleolar organizer regions in *Poecilia mexicana* by C-banding and DAPI, quinacrine, Chromomycin A3, and silver staining. *Cytogenet. Cell Genet.* **60,** 229–235.

Speicher, M., Ballard, G., and Ward, D. (1996). Karyotyping human chromosomes by combinatorial multi-fluor FISH. *Nat. Genet.* **12,** 368–375.

Stanyon, R., Arnold, N., Koehler, U., Bigoni, F., and Weinberg, J. (1995). Chromosomal painting shows that "marked chromosomes" in lesser apes and Old World monkeys are not homologous and evolved by convergence. *Cytogenet. Cell Genet.* **68,** 74–78.

Sumner, A. T. (1972). A simple technique for demonstrating centromeric heterochromatin. *Exp. Cell Res.* **75,** 304–306.

Takahashi, E., Hori, T., O'Connell, P., Leppert, M., and White, R. (1990). R-banding and nonisotopic *in situ* hybridization: Precise localization of the human type II collagen gene (COL2A1). *Hum. Genet.* **86,** 14–16.

Talbot, W. S., Egan, E. S., Gates M. A., Walker, C., Ullmann, B., Kimmel, C. B., and Postlethwait, J. H. (1997). Genetic analysis of chromosomal rearrangements in the *cyclops* region of the zebrafish genome. *Genetics,* **148,** 373–380.

Westerfield, M. (1993). "The Zebrafish Book." Eugene, OR: University of Oregon Press.

CHAPTER 19

Zebrafish Informatics and the ZFIN Database

Monte Westerfield

Institute of Neuroscience
University of Oregon
Eugene, Oregon 97403

Eckehard Doerry, Arthur E. Kirkpatrick, and Sarah A. Douglas

Computer and Information Science Department
University of Oregon
Eugene, Oregon 97403

I. Introduction

A. Biological Informatics

Many areas of biological research generate large amounts of information that must be organized, stored, and made accessible to researchers by computer

METHODS IN CELL BIOLOGY, VOL. 60

0091-679X/99 $30.00

database. The advent of gene cloning and sequencing vastly increased the amount of information available and motivated the development of computerized tools like GenBank for storing, searching, and comparing. The Human Genome Project has further facilitated the generation of large amounts of genomic information including genetic and physical maps, DNA and protein sequences, and linkage to genetic disorders. Parallel studies in nonhuman species are generating comparable data from a variety of organisms. Numerous databases manage these data for the genomics and biomedical research communities. Information about the research publications that describe these studies are also stored and organized in databases like MEDLINE.

The goal of biological informatics is to provide useful resources for storing, organizing, accessing, and analyzing data. Major challenges are to support the broad diversity of data types that biologists use and to develop procedures for data submission and database querying that make sense to biologists. A long-range goal for biological informatics is to develop techniques for relating information from different species. Ultimately we will want easy and useful methods to compare not only DNA sequences but also syntenic relationships, gene structure, gene expression patterns, mutant phenotypes, anatomy, physiology and developmental mechanisms in different organisms.

B. Need for the Zebrafish Database

The recent increase in the number of research studies using zebrafish to analyze vertebrate development and genetics (Eisen, 1996; Travis, 1996) has led to a significant information access problem. Although the use of zebrafish in genetic research is relatively new, the number of laboratories studying zebrafish and the amount of data generated by these laboratories are increasing at a phenomenal rate. The information generated from these studies already far exceeds the ability of individual scientists to track and organize it. For these reasons, a concerted effort has begun to establish and maintain a centralized database for the zebrafish research community. This zebrafish database project grew from our earlier World Wide Web (WWW) site, http://zfish.uoregon.edu. Because of the dramatic increase in the amount of data and the need for more sophisticated search methods, we have integrated the Web site information and other zebrafish research information into a relational database called the Zebrafish Information Network (ZFIN).

II. Database Design Process

From the outset, both biologists and computer scientists participated together to design the zebrafish database (Doerry *et al.,* 1997a; Westerfield *et al.,* 1997). The ultimate usefulness and usability of the database depend upon a careful assessment of the requirements of the users, early and continued testing of prototype interfaces by real users, and analysis of the users' interactive behavior while using the database. We followed the basic steps of user-centered (Fig. 1; Newman and Lamming, 1995) and participatory (Schuler and Namioka, 1993) design.

Step 1: Develop Database and Usability Requirements. We began by identifying how the database should be used to study biological problems and the types of data the database should contain. The computer scientists on the design team interviewed zebrafish researchers, read journal articles, participated in experiments, and attended research talks and laboratory meetings. We also used questionnaires to gather design information from zebrafish researchers around the world. We studied existing Web-accessible biological databases, and we obtained help and design suggestions from scientists who are developing databases for mouse and *Drosophila.* We used this information to formulate the functional and usability requirements for the database and an abstract data model.

Step 2: Iterate Detailed Design Process. We then used an iterative refinement process with cycles of redesign, prototype implementation, and evaluation with real users to refine the data model and develop effective interfaces. Selected zebrafish scientists evaluated the usability of each prototype using typical data submission and retrieval tasks. We videotaped and analyzed these sessions to identify problems which we solved with changes implemented in subsequent prototypes in the iterative design cycle. We then tested the prototype with a small group of zebrafish scientists (acting as beta testers) through the WWW. We used feedback from these beta testers to adjust the final form of the prototype.

Steps 3 and 4: Enter Data and Release Publicly. Step 3, data collection, was conducted in parallel with the iterative design cycle. Several laboratories collected, formatted and entered data into the database. We have released parts of the database to the public after each part passed through these various steps in the design process. We continue to obtain feedback from users which we use to update and improve the database, allowing the system to evolve to meet the changing needs of the users.

From our usability studies, we identified two major requirements of the zebrafish research community that the database must meet. First, because of the broad range of zebrafish research, the database system must support many different

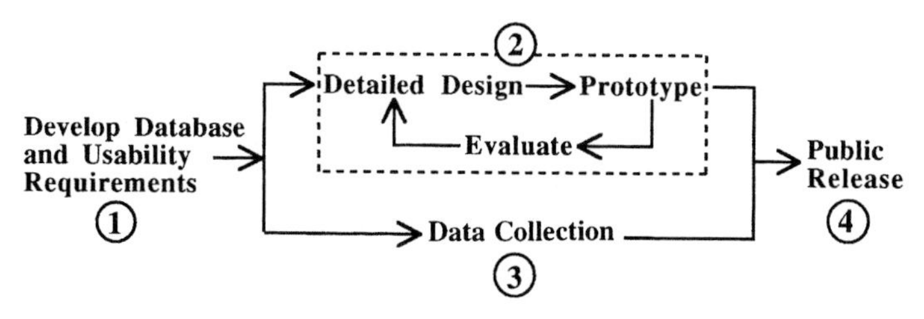

Fig. 1 The user-centered design process (reprinted with permission from Westerfield *et al.,* 1997).

types of data including text descriptions and images of wild-type and mutant fish, graphical displays of the genetic map, physiological records, and laboratory methods (Westerfield, 1995). Zebrafish researchers need to be able to search for information contained in various combinations of these data types. Thus, we chose a commercial relational database management system which supports all the required data types. The relational property of the database allows scientists to retrieve and interrelate information in many different ways. For example, one can search for and display mutations by phenotype, genetic map location, mutagen, type of chromosomal rearrangement, laboratory of origin, double mutants derived from particular single mutations, or any combinations of these or other attributes. Second, zebrafish researchers need easy access to the database with little or no knowledge of how the database operates. For this reason, we have designed and built a WWW-accessible interface which allows users to interact with the database using commonly available WWW browser software like Internet Explorer or Netscape.

III. The Data Model and the Contents of the Database

The data model is an abstract representational schema that defines the organization of information stored within the database. We have described the data model that underlies the ZFIN database in a detailed document which provides descriptions of the data categories defined for the database, how data are represented in the database, how data are entered and by whom, and how data are accessed through database queries. The data model description serves as the blueprint for database implementation and offers users a concise overview of the contents of the database. The current data model includes 25 classes of information, most of which are highly interconnected (Fig. 2). The complete and most current Data Model document is available on-line at http://zfish.uoregon.edu/zf_info/dbase/db.html.

We present, here, a summary of the data classes defined in the ZFIN data model. The classes are organized into three broad categories: Community Resources, Experimental Data, and Reference Resources (Fig. 2). For each category, we describe the abstract data classes (CAPITAL LETTERS) it contains and the attributes (*italic letters*) that describe each class. We also provide a brief description of the purpose of each category and class.

A. Community Resources

The purpose of the Community Resources data classes is to establish a searchable directory of the zebrafish research community and their publications. It provides community members with a simple method of finding who their colleagues are, where they are currently working, what work they are doing and have done in the past, and how to contact them. For example, having just read

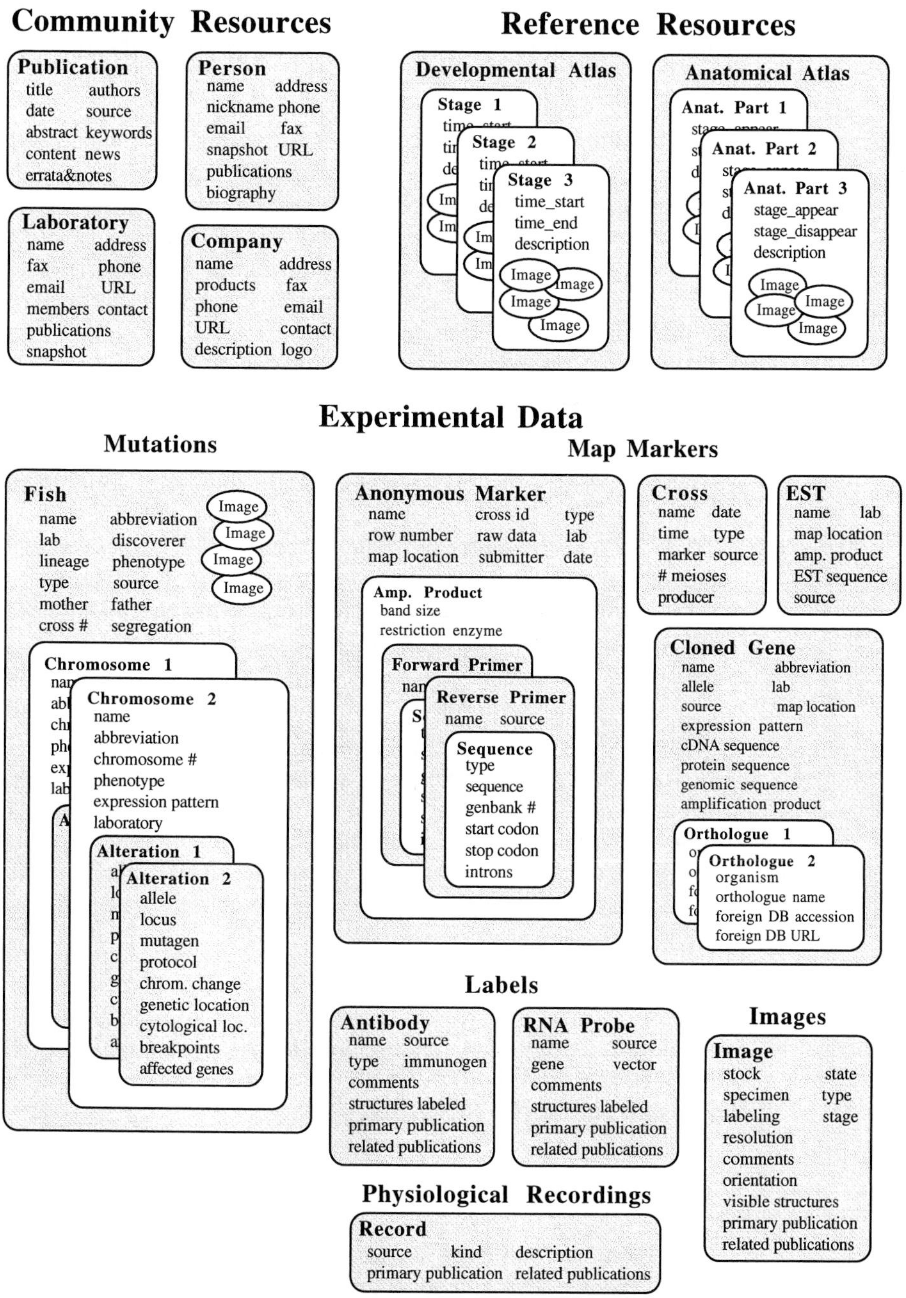

Community Resources

Publication
title authors
date source
abstract keywords
content news
errata¬es

Person
name address
nickname phone
email fax
snapshot URL
publications
biography

Laboratory
name address
fax phone
email URL
members contact
publications
snapshot

Company
name address
products fax
phone email
URL contact
description logo

Reference Resources

Developmental Atlas

Stage 1

Stage 2

Stage 3
time_start
time_end
description
Image
Image
Image
Image

Anatomical Atlas

Anat. Part 1

Anat. Part 2

Anat. Part 3
stage_appear
stage_disappear
description
Image
Image
Image
Image

Experimental Data

Mutations

Fish
name abbreviation
lab discoverer
lineage phenotype
type source
mother father
cross # segregation
Image
Image
Image
Image

Chromosome 1

Chromosome 2
name
abbreviation
chromosome #
phenotype
expression pattern
laboratory

Alteration 1

Alteration 2
allele
locus
mutagen
protocol
chrom. change
genetic location
cytological loc.
breakpoints
affected genes

Map Markers

Anonymous Marker
name cross id type
row number raw data lab
map location submitter date

Amp. Product
band size
restriction enzyme

Forward Primer

Reverse Primer
name source

Sequence
type
sequence
genbank #
start codon
stop codon
introns

Cross
name date
time type
marker source
meioses
producer

EST
name lab
map location
amp. product
EST sequence
source

Cloned Gene
name abbreviation
allele lab
source map location
expression pattern
cDNA sequence
protein sequence
genomic sequence
amplification product

Orthologue 1

Orthologue 2
organism
orthologue name
foreign DB accession
foreign DB URL

Labels

Antibody
name source
type immunogen
comments
structures labeled
primary publication
related publications

RNA Probe
name source
gene vector
comments
structures labeled
primary publication
related publications

Images

Image
stock state
specimen type
labeling stage
resolution
comments
orientation
visible structures
primary publication
related publications

Physiological Recordings

Record
source kind description
primary publication related publications

Fig. 2 Graphical overview of the data model. The data model specifies the attributes of each type of data. Data types are listed in bold lettering; the attributes that describe each data type are listed below (modified from Westerfield *et al.*, 1997).

a journal paper, a researcher might use the database to find other papers the authors have published or to locate and contact one of the authors to ask questions or request materials.

From the standpoint of the database project, a main goal of maintaining this information is to support robust, dynamic links between experimental data and researchers. For example, when a researcher submits information about a new gene or mutant to the database, it is much more useful and secure to tag the new data item with a pointer to the submitter's record in the PERSON catalog, rather than merely tagging it with the submitter's name. In this way, a researcher later viewing the gene or mutant record could follow up by asking the database for the submitter's current contact information, subsequent publications, etc.

To support these Zebrafish Community activities, two primary types of data are maintained (Fig. 2):

SOURCES: The generic SOURCE class and its subclasses, PERSON, LABORATORY, and COMPANY, catalog the research groups and sources of information and materials within the zebrafish research community.

PERSON: *name, address, phone, fax, email, URL, nicknames, laboratory, biographical sketch, publications, snapshot.* The PERSON class captures contact information on past and current members of the zebrafish research community.

LABORATORY: *name, address, phone, fax, email, URL, members, publications, contact person, snapshot.* The LABORATORY class contains contact and membership information for laboratories that are conducting zebrafish research.

COMPANY: *name, address, phone, fax, email, URL, description, contact person, logo.* The COMPANY class supports a listing of companies that produce various materials like dyes, probes, and markers commonly used in zebrafish research. The goal is to catalog the sources of materials used to generate data records contained in the database and to provide users with the information they need to contact a specific company to order these materials.

PUBLICATIONS: *authors, date, title, source, keywords, abstract, content, news & views, errata & notes.* The PUBLICATION class implements a bibliographical database for zebrafish research articles. The goal is to catalog publications that are related to other data contained in the database (i.e., listing where those data were formally published) rather than to provide a general-purpose bibliographical search engine like MEDLINE.

B. Experimental Data

The Experimental Data classes organize and catalog research results (Fig. 2) to make them readily accessible to researchers in the community. All Experimental Data classes allow records to be added or updated remotely by authorized submitters around the world. Submissions and changes to the database are automatically logged to maintain a complete historical record. Researchers can connect to the database and submit new results regardless of whether these results ever become officially known through the regular publication process. In this way, researchers in the zebrafish research community can share new and unpublished information much more rapidly than through traditional channels. Moreover, this facility provides an essentially unlimited archive for complete data sets from which only a subset of data are published.

MAP MARKERS and Related Information: The class of MAP MARKERS broadly includes descriptions of genetic material that can be used to sort the genotypes of the offspring of a cross, including cloned genes, RAPDs, SSLPs, RFLPs, STSs, ESTs, and mapped mutations. Information about specific markers is entered and displayed in standard ZFIN database forms. The spatial relationships among markers is displayed by a special-purpose graphical interface which supports dynamic analysis of the data (Fig. 3).

CROSS: *cross name, cross date, marker (reference or map cross), entry time, cross type (intercross, backcross, radiation hybrid, etc.), number of meioses, cross producer, source.* The CROSS data class catalogs information about the crosses that are used for mapping. The purpose of this class is to provide a mechanism for grouping together map markers mapped in the same genetic background. This is important because all map locations are defined relative to some background; the same marker may map at different locations in different backgrounds. When displaying mapping data, map locations are usually given with respect to a "reference cross." This is a special type of CROSS whose purpose is to provide a common "yardstick" on which other map markers can be placed. In the case of cloned genes and ESTs, this is generally done by mapping the markers directly onto the reference cross using the DNAs generated in the reference cross. This emphasizes an important characteristic of a reference cross, namely that it must generate enough genetic information to support a large number of subsequent mapping experiments. As CLONED GENES, ESTs, and ANONYMOUS MARKERS are mapped onto the genetic background of a reference cross, they effectively extend the map from that cross, increasing the accuracy of all markers previously mapped in the cross as the map becomes more dense. In the case of mutations, there is no way to map them directly into a reference cross (i.e., they can be mapped only within the background of a mutant strain). This gives rise to a second important type of CROSS, the "mapping cross." The purpose of a mapping cross is specifically to map a particular mutation. Because mutations are generated in various backgrounds, their mapped locations must be "translated" onto one of the existing reference crosses to map them with respect to previously defined markers. When a user enters a new marker that is not a mutation, the reference cross onto which it was mapped must be specified. Otherwise, if the marker is associated with a mapping cross, the user must also submit the other markers or at least the flanking markers mapped in the cross.

ANONYMOUS MARKER: *name, cross id, date, type, submitter, row number, raw data, map location, amplification product.* The ANONYMOUS MARKER data class catalogs markers, other than cloned genes, ESTs and mutations, that are used to generate the genetic map. These include RAPD, SSLP, RFLP, and STS markers.

CLONED GENE: *name, cross id, map location, amplification product, abbreviation, allele, expression pattern, cDNA sequence, genomic sequence, protein sequence, source.* The CLONED GENE data class catalogs genes cloned from zebrafish. It also identifies genes that are used as map markers. ZFIN currently stores links to sequence information in other databases (GenBank, TREMBL, etc.) rather than the complete sequences themselves.

EST: *name, cross id, map location, amplification product, EST sequence, source.* ESTs (expressed sequence tags) are used as map markers. Because ESTs may eventually be shown to be bona fide genes, they may later be reclassified as CLONED GENEs.

ORTHOLOGUE: *cloned gene id, organism, orthologue name, accession number, URL.* The ORTHOLOGUE class was established to represent, in a limited fashion, orthologies between zebrafish and other organisms. Each ORTHOLOGUE record represents one such relationship, associating a particular cloned gene in zebrafish with its orthologue in another organism.

SEQUENCE: *marker id, sequence type (genomic, protein, cDNA, EST), GenBank number, start codon, stop codon, introns.* The purpose of the SEQUENCE class is to represent sequence information associated with various map markers, with the exception of primer sequences which have their own data class.

AMPLIFICATION PRODUCT: *marker id, forward primer, reverse primer, band size, restriction enzyme.* The AMPLIFICATION PRODUCT class represents groups of primers used to generate various map markers. Amplification products are represented by a separate class, rather than listing the primers individually as map markers, because, in many cases, primers are used in pairs; pairs should be evident when searching for particular primers. The AMPLIFICATION PRODUCT class also allows representation of the band size produced by particular primer pairs.

PRIMER SEQUENCE: *name, sequence, source.* The PRIMER SEQUENCE data class stores the primers used to generate various map markers. Rather than being associated directly with such markers, the primers are identified with AMPLIFICATION PRODUCTs, which group primers into pairs (where applicable) and which are, in turn, associated with the appropriate MAP MARKER. A given AMPLIFICATION PRODUCT must have at least one and at most two PRIMER SEQUENCE records associated with it.

MUTANT AND WILD-TYPE STOCKS: FISH data records correspond directly to mutant and wild-type strains of zebrafish; when a new mutant or wild-type line is developed, a new FISH record is submitted to the database. FISH records focus primarily on the animal, its lineage, availability, who discovered it, etc., rather than on detailed genetic information.

The genetic information about a strain, on the other hand, is stored in CHROMOSOME records, which correspond directly to mutated chromosomes present in a mutant strain. Because zebrafish have 25 chromosomes, a given mutant FISH record may have between 0 and 25 CHROMOSOME records associated with it. Because most mutants have only one or two mutated chromosomes, it will be rare that a FISH record has more than one or two CHROMOSOME records associated with it. By cross-breeding, a given mutation can be transferred to a new mutant line, for example when making double mutants. In this way, a given CHROMOSOME record may ultimately be associated with many different FISH records.

Within a mutant chromosome, any number of specific genetic alterations may have occurred which, taken together, result in the mutant phenotype. It may not be known (at least not initially) which of these specific alterations is responsible for various phenotypic characteristics of the mutant. Each of these specific genetic alterations is represented by an ALTERATION record. It is at the ALTERATION level that the specific location and nature of chromosomal changes are recorded.

In sum, FISH records represent mutant and wild-type strains and serve as conceptual "containers" for CHROMOSOME records, which represent mutant chromosomes. In turn, CHROMOSOME records serve as conceptual containers for ALTERATION records, which describe specific changes to chromosomes. An advantage of this "distributed" representation of information on strains is that it automatically supports data integrity. For example, an update to a CHROMOSOME record is automatically and instantly represented in all strains that share that chromosome.

The lineage of a stock is specified in the FISH record by the *mother* and *father* used to generate the stock. This information is especially useful for entering information about stocks containing more than one mutation created by crossing parents with different single mutations. The data model associates a fish with its parents, so, for example, when information about the parent's genotype is updated, the updated information automatically appears in the description of the offspring. Thus, data entered at different times and in different contexts are interconnected by the relationships defined by the data model.

FISH: *name, abbreviation, line type (mutant, wild type), source, mother, father, discoverer, laboratory, phenotype, segregation, original cross number.* The FISH class is designed to represent both mutant and wild-type strains of fish. Note that individual generations of a given strain are not represented (i.e., no individual stock numbers); this database is not a stock management resource for a stock center. Each FISH record describes the phenotype and origins of a particular strain and provides a conceptual representation of the strain's genetic content, namely a corresponding set of CHROMOSOME records that detail the chromosomal abnormalities (if any) of the strain.

CHROMOSOME: *name, abbreviation, chromosome number, phenotype, expression pattern, laboratory.* The CHROMOSOME class stores basic genetic information on the mutants in ZFIN. The chromosome-oriented approach is based on the methods used by *Drosophila* geneticists to catalog fly mutations (i.e., the Red Book). Unlike the Red Book, however, the ZFIN database GENE class represents cloned genes which are not associated with particular CHROMOSOMEs until they are mapped).

ALTERATION: *allele, mutagen, protocol, chromosome change, locus, genetic location, cytological location, breakpoints, affected genes.* The ALTERATION class stores the specific alterations to a chromosome. One or more ALTERATION records are associated with a given CHROMOSOME record to describe the mutations present on that chromosome. Each ALTERATION record describes a specific chromosomal change (e.g., a single point mutation, insertion, translocation).

LABELS: The LABEL class is an abstract class that stores information about reagents that are used as anatomical labels (Fig. 2). This information is primarily used to annotate images by identifying the reagents used to label the pictured animals.

ANTIBODIES: *source, primary publication, related publications, comments, name, structures labeled, type, immunogen.* The ANTIBODY class catalogs the antibody probes that are used to label zebrafish.

RNA PROBES: *source, primary publication, related publications, comments, name, structures labeled, gene, vector.* The RNA PROBE class catalogs the RNA probes (primarily made from cDNAs) that are useful for labeling zebrafish.

IMAGES: *stock, specimen, state (live, fixed), type (still, movie), orientation, resolution, stage, stage hours start, stage hours end, labeling, visible structures, primary publication, related publications, comments.* The IMAGE class provides a framework for storing image data. All IMAGE data records consist of a single graphics file, along with a set of attributes that describe the image the file contains. IMAGE records can be associated with any other type of experimental data records (e.g. mutations, cloned genes).

PHYSIOLOGICAL RECORDS: *source, primary publication, related publications, comments, kind, description.* This class allows users to enter nonphotographic records (e.g., electrical or optical recordings).

C. Reference Resources

The goal of the Zebrafish Reference Resource is to provide researchers with a comprehensive and uniform reference framework for (a) naming and identifying anatomical features of the zebrafish and (b) naming and identifying the various developmental stages. The ZFIN Reference Resource contains four distinct reference facilities (Fig. 2):

1. Staging Series. The purpose of the Staging Series is to establish a common framework for identifying the developmental stage of an animal. Each develop-

mental stage, currently agreed upon by the zebrafish research community (Kimmel *et al.,* 1995), is described and documented with selected photographic IMAGE data records.

2. Developmental Atlas. The Developmental Atlas (currently still in preparation) provides a comprehensive visual atlas of the zebrafish providing a set of whole mount and section images for each developmental stage. Most of these images will be annotated to identify various anatomical structures. The goal of the atlas is to allow researchers to identify the anatomical structures observed in their own specimens.

3. Adult Atlas. The Adult Atlas is a subset of the Developmental Atlas that focuses narrowly on the anatomical structure of the adult (older than 3 months) zebrafish. Like the Developmental Atlas, it indexes annotated whole mounts and sections. Its use is similar to the Developmental Atlas.

4. Anatomical Parts Catalog. The purpose of the Anatomical Parts Catalog is to define a stable, shared taxonomy of zebrafish anatomy. It provides researchers with a standard terminology for identifying the various anatomical structures in the zebrafish. In this way, the parts catalog reduces the confusion associated with synonymous terms for the same structure. The Anatomical Parts Catalog is used to identify structures shown in images submitted to the database, thus enforcing a uniform terminology.

All four of these reference functions are closely related (Fig. 3, see color plate). The Reference Resource establishes a comprehensive, shared framework for identifying and discussing aspects of zebrafish anatomy and development at all stages in the developmental process. For each stage, the database user has access to the criteria that define that stage (the Staging Series); to a comprehensive, annotated, and detailed visual catalog of zebrafish anatomy at that stage (the Developmental/Adult Atlas); and to a detailed, cross-referenced definition of anatomical taxonomy at that stage (the Anatomical Parts Catalog).

ANATOMICAL PARTS: *name, abbreviation(s), description, stage(s).* The PART class provides a catalog of anatomical structures that can be identified at each stage during development. It contains the standard list of terms used to name anatomical structures. Although the list of PARTs has no information about structural relationships associated with it (i.e., which part is contained in which other part), the PARTs catalog can be used in conjunction with the PHYSICAL STRUCTURE class to generate dynamically the hierarchical structural relationships among the various anatomical parts. The names of PARTs can also be used to index the IMAGE database to extract images that illustrate the anatomical part, the MUTANT database to retrieve mutations that affect the part, and the CLONED GENE database to identify genes expressed by the part.

PHYSICAL STRUCTURE: The purpose of the PHYSICAL STRUCTURE class is to implement a flexible, extensible mechanism for describing the hierarchical structural relationships among the various anatomical parts of the zebrafish. In conjunction with the ANATOMICAL PARTs catalog, this data structure provides the basis for implementing vital functions including:

Enforcing uniform terminology when identifying anatomical structures visible in an image.

Constructing and displaying dynamic graphical representations (e.g., a tree) of the anatomical structure as a means for users to identify anatomical parts, to specify the region of the fish for data entry or searching, or to access other interface-related functions.

The PHYSICAL STRUCTURE class is implemented in a somewhat abstract manner. Each member of the class defines a single structural relationship between two PARTs. For example, the brain consists of a forebrain, a midbrain, and a hindbrain. This relationship is represented in the following four instances of the PHYSICAL STRUCTURE class:

Container	Containee
nervous system	brain
brain	forebrain
brain	midbrain
brain	hindbrain

The primary advantage of this approach is that hard-wired relationships between the parts do not need to be specified; it is relatively easy to revise the anatomical atlas to reflect newly identified structures or to subdivide existing structures into finer grained components. We anticipate that many modifications to the taxonomy will be made in coming years by zebrafish researchers. A disadvantage of this implementation is that multiple queries are required to extract the hierarchical structure.

D. Relational Database Searches

Although the contents of the database are divided logically into discrete data classes, the relational property of the database system allows users to relate different types of data to each other. The specification of mutant phenotypes provides a good example (Fig. 4). A mutation may have a visible phenotype in the live embryo because it alters the normal structure of an anatomical feature. The description of this type of mutant phenotype uses standard anatomical terms that are defined as ANATOMICAL PARTs by the Anatomical Parts Catalog. The affected ANATOMICAL PARTs may be present only during some developmental stages as defined by the Staging Series and the Developmental Atlas. Similarly, the mutation may affect the expression pattern of a gene. The description of this aspect of the phenotype uses the standard name of the gene, linking the mutation to information about the gene and the probe that was used to study

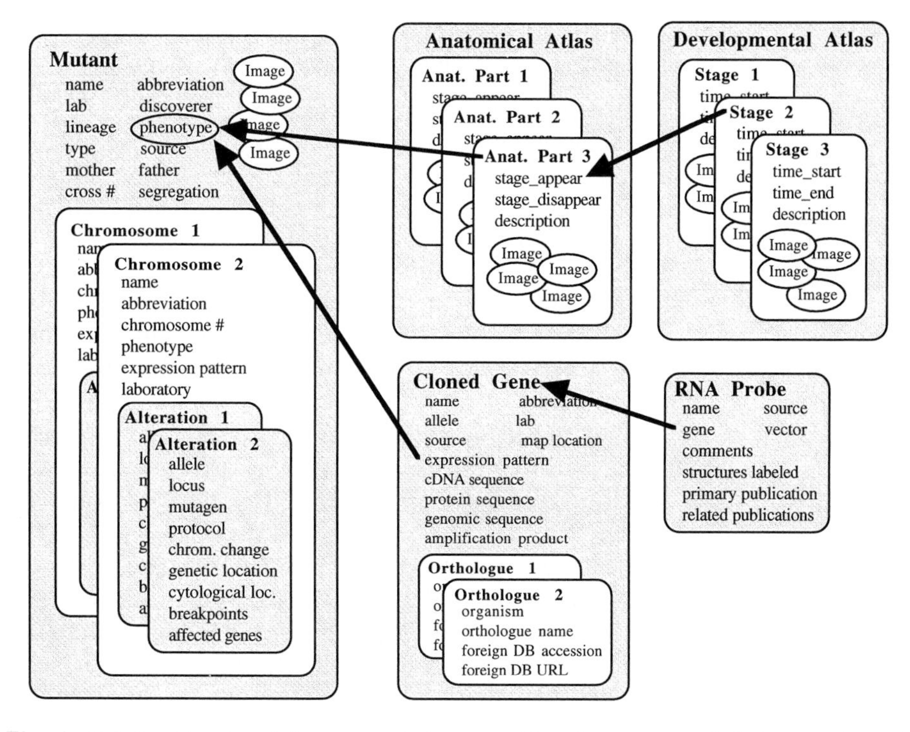

Fig. 4 Relationships that define a mutant phenotype. The phenotype may include alterations of anatomical parts at specific developmental stages and alterations of gene expression patterns studied with specific RNA probes. The relational property of the database links these different types of information together.

the altered expression pattern. Thus, a scientist can query the database and obtain information about this mutation not only by specifying its name or allele designation but also by searching for a specific phenotype, an alteration in an anatomical part or gene expression pattern, the stage in which the phenotype is visible, or any combination of these data classes. Moreover, updated information about these data items, for example new information about orthologues of the gene cloned in other species, is automatically linked to the mutation.

IV. Using the ZFIN Database

A. WWW Browser Requirements and Settings

To create a dynamic, user-friendly interface, we employed several relatively advanced WWW technologies. This means that the user's browser software must support these technologies to access ZFIN effectively. In principle, any WWW browser that supports Javascript should be able to access most of the features of the ZFIN database. Some specialized activities like image annotation and manipulation of the genetic map also require support for Java applets.

1. WWW Browser. Netscape Navigator, version 3+, is currently the only browser fully capable of displaying the ZFIN interface. Microsoft Internet Explorer and older versions of Netscape do not support all interface features and will create problems. Based on limited testing (using version 4.0), we found that Internet Explorer does not provide proper support for the advanced interface features of the ZFIN database. Some information can be accessed, but the current version of Internet Explorer limits the functionality of the interface. The latest Netscape browser software can be downloaded from Netscape (http://home.netscape.com); the software is free for people affiliated with an educational institution.

2. Language and Security Settings. Both Java and Javascript must be enabled within the Netscape browser and some security features should be disabled. To set these browser options, read the "Helpful Hints" for using ZFIN on the home page.

3. Screen Settings. Maximize screen space. Many of the ZFIN data displays and entry forms are too large to fit in a single window on an average-sized computer display. To make viewing easier, it is important to maximize the screen real estate available to ZFIN by (1) adjusting the size of the browser's window to fill the screen completely and (2) setting the browser to hide the toolbar, directory buttons, and current URL (location), which all take up space at the top of the browser window. Many of the images available in ZFIN contain more than 256 colors. To ensure data integrity, the color depth of the screen should be set to 24 or 32 bits (i.e. "thousands" or "millions" of colors) if possible.

4. Navigation. Avoid the Netscape BACK button. We have found that using the Netscape BACK button is unreliable in versions of Netscape prior to 4.01 because of a bug in the Netscape browser. Instead, ZFIN provides a task-sensitive navigation tool which appears as a set of "Path" tiles along the left side of the Netscape window. We recommend using these tiles rather than the BACK button to move within ZFIN.

B. Passwords and Security

The ZFIN database is publicly accessible via the WWW. "Guest" users are able to browse and search most types of data. To provide additional mechanisms for submitting and updating data and for limiting access to some types of unpublished data, ZFIN also employs a system of user authentication. To obtain a password, users must contact the database administrator (zfinadmn@zfish.uoregon.edu).

We use the database itself to store individual user and interface state information. By authenticating users and storing information about the state of the interface and what particular task the user is doing the database is able to customize the interface presented to each individual user. When a user logs-on with a registered user name and password, the database identifies the user and subsequently generates interface screens specifically tailored to that particular person. In this way, the database management system controls the privileges and access of each user. Individuals can thus submit and update their own data, and groups of users can share data privately among themselves, thus facilitating collaborations prior to publication.

C. Queries

In principle, users can instruct the ZFIN relational database to retrieve all possible combinations of the data records contained within the database. This facility is provided by Standard Query Language (SQL), the powerful but complex systems-level interface to the database. In practice, however, very few biologists are familiar with SQL or with the representational schema of the data model. For this reason, we have created a graphical user interface that allows biologists to express a wide variety of standard queries without specialized training. The cost of this simplicity is a reduction in expressive power; the interface allows only a subset of the queries possible with SQL. To choose which queries to support in our interface, we analyzed the tasks of zebrafish researchers as described in Section II. Our participatory design process (Doerry *et al.,* 1997b) identified the most common information needs of zebrafish researchers, and the resulting interface design allows users to express these queries efficiently. ZFIN currently provides query forms to search the database for mutants, genetic map markers, researchers, laboratories, and publications. Upon request, the database administrator provides SQL searches for other more specialized queries.

D. Data Submission

1. Data Entered by Database Staff. Some types of data generated from studies of zebrafish are currently available in other public databases. For example, most zebrafish research publications are listed in MEDLINE or other literature reference resources. The database staff searches these public sources on a regular basis, and downloads information into the zebrafish database. Similarly, DNA, gene, and protein sequences are available through GenBank and other on-line sources. The zebrafish database provides links from zebrafish gene or protein names to the raw data in these other sources.

2. Submissions by Authorized Users. A goal of the zebrafish database project is to archive as much information as possible. Although only a fraction of the data obtained in scientific studies is available to the research community in peer-reviewed journals, much of the unpublished information may be useful and important for current and future studies. If these data are not preserved in a database, much of the information will eventually be lost. For these reasons,

we have designed the zebrafish database to accept submissions directly from authorized scientists, without review. This means that some records contain preliminary or incomplete data whereas others, based on published information, are more complete and accurate. To distinguish among these sources, each record is marked with a tag which identifies whether a researcher or database staff member submitted it and whether it has been published. Thus, during a database search, the user can determine a level of confidence for each data item.

3. Updating Data. A database record can be updated at any time by the original submitter. If a change is requested by a user who did not submit the original data, the update must be performed by the database staff. The database records the previous and the updated value, retaining a history of the changes. The database allows submission of partial records, with some data missing, because information should be available in a timely manner, even if it is preliminary. The original submitter of a data record retains the privilege of updating the record, so new information can be added as it becomes available. The database staff makes all other updates. Some database records may become outdated as new studies are completed. This creates a more difficult problem because outdated records need to be identified and decisions need to be made about how to deal with them. To solve such problems, the database staff communicates with the original data submitters. An oversight committee from the zebrafish research community helps make especially difficult decisions.

E. Accessing the Zebrafish Database

The database can be accessed at http://fish.uoregon.edu. Mirror sites for the WWW server are located in France at http://www-igbmc.u-strasbg.fr/index.html and in Japan at http://www.grs.nig.ac.jp:6060/index.html. To gain authorization for data submission or to obtain additional information, contact zfinadmn@fish.uoregon.edu.

Acknowledgments

We thank Wolfgang Driever for help with design; Paul Bloch, Lauradel Collins, Don Pate, and Mike McHorse for expert technical advice; Pat Edwards for help with data entry; and our colleagues in the zebrafish research community for suggestions and help with the beta testing. Supported by the W.M. Keck Foundation and the NSF (BIR-9507401).

References

Doerry, E., Douglas, S., Kirkpatrick, A. E., and Westerfield, M. (1997a). Moving beyond html to create a multimedia database with user-centered design: A case study of a biological database. Tech. Report CIS-TR-97-02, Computer and Information Science Dept., University of Oregon.

Doerry, E., Douglas, S. A., Kirkpatrick, A. E., and Westerfield, M. (1997b). Participatory design for widely-distributed scientific communities. Proc. 3rd Conf. Human Factors and the Web.

Eisen J. S. (1996). Zebrafish make a big splash. *Cell* **87,** 969–977.

Kimmel, C. B., Ballard, W. W., Kimmel, S. R., Ullmann, B., and Schilling, T. F. (1995). Stages of embryonic development of the zebrafish. *Dev. Dyn.* **203,** 253–310.

Newman, W. M., and Lamming, M. G. (1995). "Interactive System Design." Addison-Wesley, Reading, MA.

Schuler, D., and Namioka, A. (1993). "Participatory Design: Principles and Practices." Lawrence Erlbaum Assoc., Hillsdale, NJ.

Travis, J. (1996). Gone Fishing! Scientists use mutant zebra fish to learn how vertebrate embryos develop. *Science News* **150,** 360–361.

Westerfield, M. (1995). "The Zebrafish Book. A Guide for the Laboratory Use of Zebrafish (*Danio rerio*)," 3rd ed. Univ. of Oregon Press, Eugene, OR.

Westerfield, M., Doerry, E., Kirkpatrick, A. E., Driever, W., and Douglas, S. A. (1997). An on-line database for zebrafish development and genetics research. *Sem. Cell Dev. Biol.* **8,** 477–488.

APPENDIX 1

Genetic Backgrounds and Some Standard Stocks and Strains Used in Zebrafish Developmental Biology and Genetics

Stephen L. Johnson
Department of Genetics
Washington University Medical School
St. Louis, Missouri 63110-1010

Leonard I. Zon
Howard Hughes Medical Institute
Children's Hospital
Boston, Massachusetts 02115

Several different strains and stocks of zebrafish are available for developmental analysis and genetic mapping. We may consider these different strains or stocks in two groups. The first group, or experimental group, consists of strains that are commonly used for developmental analysis, embryonic manipulation, and mutant screens. The second group, which we call the mapping group, consists of those strains that are used to generate maps when crossed to experimental strains. Strains within the mapping group were chosen because they show extensive genetic polymorphism with respect to standard experimental strains such as AB.

Summary of Genetic Backgrounds

Experimental Strains

1. AB is original strain developed by Streisinger and Walker for robust egg production. This strain was generated from a mix of different strains available in pet stores in Eugene in the early 1970s. This strain was maintained to retain maximum genetic variation. The AB strain was replaced by *AB. The Massachusetts General Hospital (MGH) microsatellite map (Knapik *et al.,* 1998) is based

METHODS IN CELL BIOLOGY, VOL. 60

0091-679X/99 $30.00

in part on AB. Many mutant screens, including those from the Kimmel, Eisen, and Westerfield labs in Oregon and the Driever and Fishman labs were performed using AB or *AB stocks.

2. *AB is Walker's lethal-free derivative of AB (personal communication) selected for quality of haploid embryos.

3. C32 is Streisinger's clonally derived strain (Streisinger *et al.,* 1981). C32 stocks were possibly contaminated by AB in the late 1980s. Subsequently, two different isolates of the C32 strain have been maintained, by Walker (at Oregon) and by Johnson (in St. Louis), under inbreeding conditions. Presumably, the Walker and Johnson isolates are now different from each other. C32 × SJD derived maps (Johnson *et al.,* 1996; Postlethwait *et al.,* 1998) are based on the Johnson isolate of C32. The C32 strain (Johnson isolate) is used for adult pigment pattern and fin regeneration studies (Johnson *et al.,* 1995a; Johnson and Weston, 1996). The Johnson isolate of C32 is 90–95% homozygous at all loci (A. Nechiporuk and S. Johnson, unpublished).

4. The Tübingen wild type is a strain of fish obtained from Tübingen pet store. This strain was used for the Tübingen mutant screens (Haffter *et al.,* 1996).

5. The Florida wild-type is drawn from pet-store fish obtained from farms in Florida. Many labs use fish purchased from pet stores for embryo generation. These are sometimes referred to as Florida wild-type. The lore is that most pet store fish come from stocks that were brought to Florida fish farms for the aquarium trade at the beginning of the twentieth century. In fact, the background of fish available in pet stores at any particular time may change, and it is not clear that there is a single genetic background or population of zebrafish raised in Florida fish farms. A convenient source for inexpensive zebrafish is EKK Will Waterlife, 7502 Symmes Rd., Gibsonton, Florida 33534. Phone: 1 (800) 237-4222. Fax: 1 (813) 677-1542. When ordering, please send a letter on institutional letterhead explaining that the fish will be used for biomedical research.

Mapping Strains

1. The India strain (also known as Darjeeling; Johnson *et al.,* 1994) is drawn from wild fish from India that were brought to Germany in the late 1980s. Isolates were initially maintained in Oregon and Tübingen and are now maintained at MGH. India is used as the complementary genetic background to AB to anchor the MGH genetic map.

2. SJD is Johnson's inbred derivative of India (or Darjeeling). SJD was generated by passing the Darjeeling line through two sequential rounds of early pressure parthenogenesis (EP). Because EP results in animals derived from a single meiotic half-tetrad, the SJD strain is homozygous and isocentromeric at all centromeres. SJD has been passed through a subsequent 7 generations of full-sib mating, to result in a stock that is 93–95% homozygous at all loci (A. Nechiporuk and S. Johnson, unpublished). SJD is one of the strains used in develop-

ment of the C32 × SJD based genetic maps (Johnson *et al.,* 1996; Postlethwait *et al.,* 1998).

3. WIK is a recent isolate from India, to Tübingen, that is used in Tübingen-based mapping projects (see http://www.eb.tuebingen.mpg.de/Abt.3/Haffter__Lab/Mapping.html).

References

Haffter, P., Granato, M., Brand, M., Mullins, M. C., Hammerschmidt, M., Kane, D. C., Odenthal, J., van Eeden, F. J. M., Jiang, Y.-J., Heisenberg, C.-P., Kelsh, R. N., Furutani-Seiki, M., Vogelsang, E., Beuchle, D., Schach, U., Fabian, C., and Nüsslein-Volhard, C. (1996). The identification of genes with unique and essential functions in the development of the zebrafish, *Danio rerio. Development* **123,** 1–36.

Johnson, S. L., Midson, C. N., Ballinger, E. W., and Postlethwait, J. (1994). Identification of RAPD primers that reveal extensive polymorphisms between laboratory strains of zebrafish. *Genomics* **19,** 152–156.

Johnson, S. L., Africa, D., Walker, C., and Weston, J. A. (1995a). Genetic control of adult pigment stripe development in zebrafish. *Dev. Biol.* **167,** 27–33.

Johnson, S. L., and Weston, J. A. (1996). Temperature-sensitive mutations that cause stage-specific defects in zebrafish fin regeneration. *Genetics* **141,** 1583–1595.

Johnson, S. L., Gates, M. A., Johnson, M., Talbot, W. S., Horne, S., Baik, K., Rude, S., Wong, J. R., and Postlethwait, J. H. (1996). Centromere-linkage analysis and consolidation of the zebrafish genetic map. *Genetics* **142,** 1277–1288.

Knapik, E. W., Goodman, A., Ekker, M., Chevrette, M., Delgado, J., Neuhauss, S., Shimoda, N., Driever, W., Fischman, M. C., and Jacob, H. J. (1998). A microsatellite genetic linkage map for zebrafish. *Nat. Genet.* **18,** 338–343.

Postlethwait, J. H., Yan, Y.-L., Gates, M. A., Horne, S., Amores, A., Brownlie, A., Donovan, A., Egan, E. S., Force, A., Gong, Z., Goutel, C., Fritz, A., Kelsh, R., Knapik, E., Liao, E., Paw, B., Ransom, D., Singer, A., Thomson, M., Abduljabbar, T. S., Yelick, P., Beier, D., Joly, J.-S., Larhammar, D., Rosa, F., Westerfield, M., Zon, L. I., Johnson, S. L., and Talbot, W. S. (1998). Vertebrate genome evolution and the zebrafish genome map. *Nat. Genet.* **18,** 345–349.

Streisinger, G., Walker, C., Dower, N., Knauber, D., and Singer, F. (1981). Production of clones of homozygous siploid zebra fish (*Brachydanio rerio*). *Nature* **291,** 293–296.

APPENDIX 2

Centromeric Markers in the Zebrafish

Donald A. Kane*, Leonard I. Zon†, and H. William Detrich III‡

* Department of Biology
University of Rochester
Rochester, New York 14627

† Howard Hughes Medical Institute
Children's Hospital
Boston, Massachusetts 02115

‡ Department of Biology
Northeastern University
Boston, Massachusetts 02115

The following simple sequence length polymorphism (SSLP) markers, each a pair of polymerase chain reaction (PCR) primers, can be used to establish centromeric linkage of mutations in the zebrafish. The primer pairs are robust and are generally polymorphic in crosses between the common zebrafish strains (Appendix 1). The markers, some described previously by Knapik *et al.* (1996, 1998) and Postlethwait *et al.* (1994) and others made available prior to publication by Drs. Nobuyoshi Shimoda, Dori Hosobuchi, and Mark C. Fishman (manuscript in preparation), can be accessed at http://zebrafish.mgh.harvard.edu/papers/unpublished/zf_map2k/. Primers are available commercially from Research Genetics (1-800-533-4363). Approximate distances to the centromere (±10 cM) are provided for some markers.

Linkage group	Marker	Distance to centromere (cM)
1	Z1154	
	Z1351	10
2	Z4300	10
	Z3430	
	Z4585	

(*continues*)

0091-679X/99 $30.00

Linkage group	Marker	Distance to centromere (cM)
3	Z4161	
	Z3725	0
	Z963	
4	Z5095	
	Z5112	
	Z374	
	Z1389	40
5	Z1167	5
	Z3516	0
6	Z880	25
	Z3011	
	GOF26	45
7	Z5467	
	Z1059	20
	Z1535	
8	Z3083	
	GOF9	
	GOF13	0
9	Z4168	5
	Z3124	
	Z1660	
10	Z3835	
	Z1145	0
11	Z3412	5
	Z4190	15
12	Z4188	10
	Z1473	20
13	Z5395	10
	Z3424	5
14	Z4896	
	Z1226	25
15	GOF18	35
	Z3760	
16	Z3104	5
	Z5022	
17	GOF12	
	Z1490	35
	Z3894	0
18	Z3838	
	Z3853	
	Z737	0
19	Z3023	
	Z3770	30
20	Z536	35
	snap25a[a]	
21	Z4425	5
	Z5503	
	Z4910	15

(*continues*)

Linkage group	Marker	Distance to centromere (cM)
22	Z230	5
	Z178	
23	Z342	
	Z4003	30
	Z4421	15
24	Z3399	
	Z249	
	Z3310	25
	Z5413	
25	Z3632	
	Z4198	
	Z3528	5

[a] Marker gene that encodes an intracellular receptor, Snap25a, that binds synaptic vesicles prior to neurotransmitter release [Risinger *et al.* (1998)]. The primer pair (ATGCACATATCACCCGG and AGTGCTCGTCGTATTAG) yields a product of 460 bp from the 3′-untranslated region of the gene and detects a polymorphic *Alu*I site.

References

Knapik, E. W., Goodman, A., Atkinson, O. S., Roberts, C. T., Shiozawa, M., Sim, C. U., Weksler-Zangen, S., Trolliet, M. R., Futrell, C., Innes, B. A., Koike, G., McLaughlin, M. G., Pierre, L., Simon, J. S., Vilallonga, E., Roy, M., Chiang, P. W., Fishman, M. C., Driever, W., and Jacob, H. J. (1996). A reference cross DNA panel for zebrafish (*Danio rerio*) anchored with simple sequence length polymorphisms. *Development* **123,** 451–460.

Knapik, E. W., Goodman, A., Ekker, M., Chevrette, M., Delgado, J., Neuhauss, S., Shimoda, N., Driever, W., Fishman, M. C., and Jacob, H. J. (1998). A microsatellite genetic linkage map for zebrafish. *Nat. Genet.* **18,** 338–343.

Postlethwait, J. H., Johnson, S. L., Midson, C. N., Talbot, W. S., Gates, M., Ballinger, E. W., Africa, D., Andrews, R., Carl, T., Eisen, J. S., Horne, S., Kimmel, C. B., Hutchinson, M., Johnson, M., and Rodriguez, A. (1994). A genetic linkage map for the zebrafish. *Science* **264,** 699–703.

Risinger, C., Salaneck, E., Sderberg, C., Gates, M., Postlethwait, J. H., and Larhammar, D. (1998). Cloning of two loci for synapse protein Snap25 in zebrafish; Comparison of paralogous linkage groups suggests loss of one locus in the mammalian lineage. *J. Neurosci. Res.,* in press.

APPENDIX 3

Collection, Storage, and Use of Zebrafish Sperm

David G. Ransom and Leonard I. Zon

Howard Hughes Medical Institute
Children's Hospital
Boston, Massachusetts 02115

I. Introduction

Zebrafish sperm can be readily collected and used for *in vitro* fertilization (IVF) of squeezed eggs. Samples of sperm can be used fresh or cryopreserved and stored indefinitely as a back up of genetic stocks. Maintenance of frozen sperm samples is essential to guard against accidental loss of important fish and to reduce the space needed to house lines of fish that are not in current use. Fresh zebrafish sperm can also be UV irradiated to inactivate the paternal genome.

METHODS IN CELL BIOLOGY, VOL. 60

0091-679X/99 $30.00

Irradiated sperm retain the ability to fertilize eggs and are used to produce haploid or gynogentic diploid zebrafish for mutant screens and genetic mapping experiments.

This appendix describes methods for zebrafish sperm collection, cryopreservation, and IVF that are based on the pioneering work of Walker and Streisinger (Westerfield, 1998) with additions and modifications from our laboratory and other investigators. Detailed methods for the production and screening of haploid zebrafish are described by Walker in this volume (Chapter 3). There are also chapters by Pelegri and Schulte-Merker (Chapter 1) and Eisen *et al.* (Chapter 4) describing the production and use of gynogenetic diploid zebrafish. These three chapters include discussions detailing how to handle females and squeeze eggs from them for IVF.

Many investigators are familiar with collecting sperm by squeezing anesthetized zebrafish. This method is effective for collecting small numbers of samples for cryopreservation and does not require sacrifice of the donors. However, collecting sperm by squeezing individual zebrafish males can take several hours if large volumes are needed to create many clutches of haploid or gynogenetic diploid embryos. We have determined that IVF methods using dissected testes work well in zebrafish and are substantially faster than collecting sperm by squeezing. Collecting sperm directly from dissected testes has long been the traditional method used for IVF of Medaka eggs (Aoki *et al.*, 1997; Yamamoto, 1967). We have observed that zebrafish also store mature sperm throughout the testis. Two dissected zebrafish testes (one male equivalent) contain approximately ten times more active sperm than can be squeezed from a healthy fish. We have also found that both testis extract and squeezed sperm can be cryopreserved. Thus, active sperm can be collected from living male zebrafish or dissected testes. The collection of sperm by either method is important for mapping mutations and conducting mutant screens as described in the preceding chapters and for maintaining frozen stocks of sperm that can be stored indefinitely.

II. Materials

A. Equipment

1. Kontes pellet pestle (Fisher #K749520-0000) and 2-ml microfuge tubes
2. No. 5 watchmaker forceps (Dummont #07379), small surgical scissors, and millipore forceps
3. 3-cm square piece of kitchen sponge with a 0.5 cm slit cut in the top
4. Drummond 5-μl and 50-μl micropipettes (Drummond #2-00-005 and 2-00-050)
5. Stereo microscope such as Nikon SMZ-U
6. 2-ml screw cap cryovials (Corning #430488)
7. 15-ml blue cap tubes (Falcon #2097)

8. 2 Styrofoam boxes, 1 filled with dry ice and 1 filled half-way with liquid nitrogen
9. Plastic spoons

B. Reagents

1. *Tricaine anesthetic* (3-aminobenzoic acid ethyl ester) (Sigma #A5040). The 10× stock solution is 0.2% Tricaine in ddH_2O adjusted to pH 7.0 with 1.0*M* Tris 9.5. Tricaine is toxic if not properly buffered. For anesthesia of zebrafish dilute stock solution 1:10 in system water.

2. *Hank's Saline:* 0.137*M* NaCl, 5.4m*M* Kcl, 0.25m*M* $Na_2H\ PO_4$, 0.44m*M* $KH_2\ PO_4$, 1.3m*M* $CaCl_2$, 1.0m*M* Mg SO_4, 4.2m*M* NaH CO_3.

Hank's (Final): 9.9 ml Hank's Premix, 0.1 ml Stock #6.

Hank's Premix: Combine the following in order: 10.0 ml Solution #1, 1.0 ml Solution #2, 1.0 ml Solution #4, 86.0 ml ddH_2O, 1.0 ml Solution #5.

Hank's Stock Solutions:

Stock #1	Stock #2	Stock #4	Stock #5	Stock #6
8.0 g NaCl 0.4 g KCl in 100 ml ddH_2O	0.358 g Na_2HPO_4 Anhydrous 0.60 g $K_2H_2PO_4$ in 100 ml ddH_2O	0.72 g $CaCl_2$ in 50 ml ddH_2O	1.23 g $MgSO_4 \times 7H_2O$ in 50 ml ddH_2O	0.35 g $NaHCO_3$ 10.0 ml ddH_2O

3. *Ginzberg's Ringers:* Made fresh from 10× stock plus buffer.

10× Ginzberg's Ringers Stock: For 500 ml final, add in order 400 ml ddH_2O, 32.5 g, NaCl 1.25 g, KCl 1.5 g $CaCl_2$. The order of reagent addition and complete mixing are essential. Add ddH_2O to 500 ml and autoclave.

1× Ginzberg's Ringers: For 50 ml, add 45 ml ddH_2O, 5 ml 10× Ginzberg's, 5 ml fresh 10× $NaHCO_3$ (0.5 g in 50 ml ddH_2O). Mix and pH to 8.0.

4. *Freezing medium,* after Walker and Streisinger (Westerfield, 1998): For 10 ml of fresh freezing medium, add the following in order: 9.0 ml 1× Ginzberg's Ringers freshly buffered, 1.0 ml methanol, 1.5 g Carnation nonfat dry milk powder. Mix by shaking for 5 minutes on a vortexer. Complete mixing is essential before chilling the solution.

5. *I buffer:* Made fresh from autoclaved stock solutions plus fresh buffer. For 50 ml fresh I buffer, add 5 ml each of the following five stock solutions in order and pH to 7.0.

10× I Part A	10× I Part B	10× I Part C	10× I Part D	10× fresh bicarb
NaCl 34 g KCl 8.5 g H_2O to 500 ml	$CaCl_2$–$2H_2O$ 4.4 g H_2O to 500 ml	$MgSO_4$–$7H_2O$ 2.46 g H_2O to 500 ml	fructose 25 g H_2O to 500 ml	1.2 g $NaHCO_3$/50 ml

Autoclave stock solutions 1–4. Make bicarb buffer stock fresh each day.

6. *E3 Medium:* 5m*M* NaCl, 0.17m*M* KCl, 0.33m*M* CaCl2*2H_2O, 0.33m*M* $MgSO_4$ *7H_2O.

60× E3 Stock: 172 g NaCl, 7.6 g KCl, 29 g $CaCl_2$ *2H_2O, 49 g $MgSO_4$ *7H_2O, add ddH_2O to 10 l.

E3: Dilute 160 ml of the 60× stock into 10 l ddH_2O.

III. Collection of Zebrafish Sperm

A. Squeezing Males

1. The night before squeezing, set up males and females in pairs for mock mating to stimulate sperm production. Separate the males the next morning when the lights turn on.

2. Place two to four males into a 250-ml beaker containing 0.02% buffered Tricaine in system water at room temperature.

3. Wait 2 to 5 minutes or until the gill movement has just slowed. Swirl water gently with a plastic spoon to test if anesthesia is complete. Do not leave zebrafish in Tricaine for longer than 5 minutes to ensure complete and rapid recovery.

4. Lift out a single male using a spoon and rinse it briefly in a mouse cage of system water; then place it belly up in the slit on a small square of damp sponge.

5. Carefully dry the area of the anal fin with a Kimwipe. It is crucial not to mix water with the sperm. The sperm are active only for a short period when diluted in water.

6. Place the fish under a stereo microscope to observe the region of the anal fin under low power.

7. Use a 5-μl glass micropipette to push aside the anal fins to expose the cloaca and then collect sperm.

8. Use Millipore forceps to squeeze the sides of the male gently just anterior to the anal fins and aspirate the sperm with the micropipette. We find that one in three healthy males aged 5 to 12 months will produce approximately 1 μl of sperm each. Good-quality sperm has a uniform milky white appearance.

9. For immediate IVF or UV irradiation, transfer the sperm to a microfuge tube containing ice-cold Hank's on ice. The Hank's should be freshly buffered that day. A single sperm sample can be diluted in 100 μl of Hank's for IVF and stored on ice for at least 3 hours. For UV irradiation, typically 50 μl of sperm are collected in 500 μl of Hank's. For cryopreservation, mix the sperm with freezing media as described later.

10. Place the male into a recovery tank and swirl the water gently with the spoon. Recovery should take less than 1 minute.

11. Males are typically rested for a month before being squeezed again.

B. Collecting Testes

1. Collect males from the appropriate system tank in the morning before they are fed.

2. Place two to four males into a 250-ml beaker containing 0.02% buffered Tricaine in system water at room temperature.

3. Wait 5 to 10 minutes until the males are fully anesthetized. Gently dry the fish with a gauze pad or paper towel.

4. Make cuts with small scissors to pith; then cut at the level of the pectoral fins to remove the head. Next cut the abdominal body wall to expose the internal organs.

5. Place the fish belly up on a plastic Petri dish and view under low power with a stereo microscope.

6. Use watchmakers forceps to open the abdomen and remove testes. The testes are remarkably large white organs that lie on either side of the ventral portion of the swim bladder. They run from just behind the pectoral fins to the anal fin. Other organs such as the stomach and intestines are more yellow to orange colored. We have found that about one in 10 fish have one hypertrophic testis. A small amount of attached fat that looks like oil droplets in the sample is not a problem.

7. For immediate IVF or UV irradiation, transfer the testes to a microfuge tube containing ice-cold Hank's on ice. The Hank's should be freshly buffered that day. A single male equivalent can be diluted in 750 μl of Hank's for IVF that day and stored on ice for at least 6 hours. For UV irradiation, we typically place ten male equivalents in 750 μl of Hank's.

8. To release sperm from the testes, grind them by hand for 30 seconds using a Kontes pellet pestle which fits snugly into the microfuge tube. Allow the tissue debris to settle for 5 minutes before using the supernatant for IVF. For cryopreservation, mix four testes with freezing media as described later.

IV. Cryopreservation of Zebrafish Sperm

A. Freezing Sperm

1. Collect sperm into a 5-μl glass micropipette by squeezing males as described earlier. Be careful to avoid mixing the sperm with water. Work quickly through the following steps.

2. Transfer 1–2 μl of sperm to a microfuge tube containing 10 μl of ice-cold sperm freezing medium that has been made fresh that morning.

3. Pipette the mixture into a 50-μl glass micropipette that has been cut to fit into a 2-ml screw cap cryovial.

4. Put the capillary into a 2-ml cryovial, cap the tube, and place it into a 15-ml Falcon tube. Close the 15-ml tube and quickly push it into crushed dry ice and leave the tube horizontal for 20 minutes.

5. After 20 minutes, remove the 2-ml cryovial from the Falcon tube and immerse it in liquid nitrogen.

6. The sperm sample can be stored for years in a liquid nitrogen freezer.

B. Freezing Testes

1. Place two male equivalents of fresh testis into 120 μl of ice-cold freezing medium. For freezing testis samples, we use an excess of sperm to be safe.

2. Use a Kontes pellet to grind the tissue gently by hand for 30 seconds. Avoid introducing bubbles into the mixture as much as possible.

3. Pipette 10-μl aliquots into 50-μl glass micropipettes cut to fit a 2-ml cryovial.

4. Quickly put capillary tubes into 2-ml cryovials and screw on the caps. Place the 2-ml cryovials into a 15-ml Falcon tubes. Screw on the caps and bury the 15-ml tubes horizontally in dry ice for 20 minutes.

5. After 20 minutes, remove the 2-ml cryovials from the Falcon tubes and immerse them in liquid nitrogen.

6. The testes samples can be stored for years in a liquid nitrogen freezer.

V. *In Vitro* Fertilization

A. IVF with Fresh Sperm

1. For normal IVF, place 1 μl of squeezed sperm directly into 100 μl of ice-cold Hank's buffer in a microfuge tube. Alternatively place four dissected testes in 750 of μl ice-cold Hank's and gently grind for 30 seconds using a Kontes pellet pestle, then add 250 μl more Hank's.

2. Add 30 μl of either sperm solution to squeezed eggs, mix gently with a pipette tip, and wait 30 seconds. Squeezing of eggs from females is described in Chapter 3.

3. Next add 800 μl of system water, wait 2 minutes, and then add 5 ml of system water.

4. Sperm solution keeps well on ice for several hours.

B. IVF with UV-Irradiated Sperm

1. For UV irradiation to make haploids or gynogentic diploids, we use an excess of sperm. Collect ten healthy male equivalents into 750 μl of ice-cold Hank's.

2. Use a Kontes pellet pestle to grind the tissue gently by hand. Six grinds for 5 seconds each is enough to make a very turbid solution.

3. Let solution settle on ice for 5 minutes. Remove 500 μl of supernatant to UV irradiate. Some tiny chunks of tissue do not block UV irradiation. This solution is stable for at least 6 hours on ice.

4. UV irradiate according to the procedures described in Chapter 3. We use the same apparatus to UV-treat testes extract for 2 minutes, 15 seconds with a set 40 cm above the watchglass of sperm.

5. Add 30 μl of UV sperm solution to squeezed eggs and wait 30 seconds.

6. Proceed with methods to generate haploids or diploids as described in Chapters 1, 3, and 4.

C. IVF with Thawed Sperm

1. Remove cryovials of sperm from storage and keep in a container of liquid nitrogen.

2. Squeeze eggs from an anesthetized female into a 60 mm dish as described in Chapter 3.

3. Open a cryovial and pull out a micropipette tube using forceps. Quickly that the sperm solution by holding the tube.

4. Pipette 90 μl of ice-cold Hank's through the tube to mix and expel the sperm into the dish next to the eggs. An alternative method uses I buffer in place of Hank's at this step (Wolfgang Driever, personal communication).

5. Quickly and gently mix the diluted sperm with the eggs using a pippetteman.

6. Wait 30 seconds; then add 900 μl of system water.

7. Wait another 2 minutes; then add 5 ml of system water.

8. After a few hours, remove unfertilized eggs and place embryos into 28°C E3 medium. Both squeezed sperm and testes produce approximately a 10–20% rate of fertilization depending, at least in part, on the quality of the eggs used.

VI. General Considerations

In this appendix, we summarize methods for the collection, storage, and use of sperm from living zebrafish based on methods developed by Walker and Streisinger (Westerfield, 1998) as well as methods for the use of dissected testes developed in our laboratory. Similar methods of sperm collection have been used successfully by medaka researchers for many years (Yamamoto, 1967). Squeezing sperm from live zebrafish males is effective and preserves the health of the donor, but this method can be time-consuming when large volumes are needed to produce stocks of UV-irradiated sperm. The dissection of testes is a simple and rapid method to collect large amounts of sperm solution that is stable for greater than 6 hours on ice. We routinely dissect 10 males in 15 minutes to yield the equivalent of 100 squeezed fish. This method allows rapid production of enough UV-inactivated sperm for a day of making many clutches of haploids or gynogenetic diploids. In addition, testes from a single male can be used to do 24 or more normal IVFs with a greater than 95% rate of fertilization. In addition, we have found that males from mutant lines or strains that are not capable of natural mating can nevertheless be easily propagated by IVF using sperm from dissected testes. Testis extract and squeezed sperm can both be frozen to store samples from important lines of zebrafish.

Whereas IVF of zebrafish eggs with fresh or UV treated sperm is greater than 90% efficient and highly reproducible, cryopreserved sperm is much less active. Typical fertilization rates with thawed sperm are below 20%. The use of I buffer in place of Hank's saline can improve the results for IVF with frozen zebrafish sperm by a further 20–30%. Aoki and coworkers have reported greater than 90% efficiency of IVF with medaka sperm squeezed from dissect testes and frozen in fetal bovine serum supplemented with 10% dimethyl sulfoxide/*N,N*-dimethyl formamide (DMF) as a cryoprotectant. (Aoki *et al.,* 1997). This suggests that further experimentation may lead to improved methods for freezing and thawing zebrafish sperm stocks.

In conclusion, active sperm can be collected from living male zebrafish or dissected testes. Sperm from either source can be frozen and stored indefinitely as a backup of important lines of zebrafish. Large volumes of sperm can also be UV irradiated and then used to generate haploids and gynogentic diploids that have no contribution of DNA from the paternal genome. These haploids and gynogenetic diploids are important for mapping mutations as well as for conducting mutant screens as described in detail in the preceding chapters of this volume.

Acknowledgments

We thank Alison Brownlie, Adriana Donovan, Barry Paw, and Stephen Pratt for helpful discussions and troubleshooting of these methods.

References

Aoki, K., Okamoto, M., Tatsumi, K., and Ishikawa, Y. (1997). Cryopreservation of medaka spermatazoa. *Zool. Sci.* **14,** 641–644.

Yamamoto, T. O. (1967). Medaka. *In* "Methods in Developmental Biology" (F. H. Wilt and N. K. Wessells, Eds.), pp. 101–111. T. Y. Crowell Co., New York.

Westerfield, M. (1998). "The Zebrafish Book," 4th ed. The University of Oregon Press, Eugene, OR.

APPENDIX 4

Zebrafish Web Site Listings

Pat Edwards

Institute of Neuroscience
University of Oregon
Eugene, Oregon 97403-1254

Zebrafish Web Servers

http://zfish.uoregon.edu/ZFIN

Database administrator can be reached at zfinadmn@zfish.uoregon.edu

Zebrafish Research Database, University of Oregon, Eugene, OR, USA

People, labs, publications, mutations, map markers, images.

http://zfish.uoregon.edu/

http://www.grs.nig.ac.jp:6060/index.html/ (Asian mirror site)

http://www-igbmc.u-strasbg.fr/index.html (European mirror site)

The Fish Net, University of Oregon WWW server, Monte Westerfield, Webmaster; Eugene, OR, USA

Provides on-line access to embryonic and larval anatomy—an annotated anatomical atlas; Genome Project—genetic map and sequence projects; Genetic Strains—zebrafish wild-type and mutant strains; Molecular Probes—DNA libraries, cloned genes, and antibodies; full text of *The Zebrafish Book,* back issues of *The Zebrafish Science Monitor,* and a listing of WWW zebrafish sites.

http://zebra.biol.sc.edu/

Zebrafish Information Server, Richard Vogt, Webmaster, University of South Carolina, Columbia, SC, USA

The Zebrafish Information Server has existed since 1994 for the Zebrafish Research Community; think of it as the friendly bulletin board at your local grocery store.

METHODS IN CELL BIOLOGY, VOL. 60

0091-679X/99 $30.00

Lab URLs

http://www.lri.ca/profiles/akimenko.htm
Akimenko Lab, Marie-Andreé Akimenko, Director; Loeb Research Institute, Ottawa, Ontario, Canada
Genetic control of fin development and regeneration in zebrafish.

http://med-humgen14.bu.edu/immunesystem.htm
Amemiya Lab, Chris T. Amemiya, Director; Boston University School of Medicine, Boston, MA, USA
Molecular genetics of the vertebrate immune system; molecular evolution; functional genomics; nonmammalian genomic models.

http://osu.orst.edu/dept/cgrb/faculty/bailey/index.html
Bailey Lab, George Bailey, Director; Oregon State University, Corvallis, OR, USA
Chemical carcinogenesis mechanisms using the rainbow trout and zebrafish models.

http://osu.orst.edu/dept/biochem/barnesresearch.html
Barnes Lab, David W. Barnes, Director; Oregon State University, Corvallis, OR, USA
The mechanisms by which interaction of the animal cell with its environment controls proliferation and differentiation.

http://edtech.cebs.wku.edu/~jbilotta/neuro.htm
Bilotta Neuroscience Laboratory, Joe Bilotta, Director; Western Kentucky University, Bowling Green, KY, USA
Examining the effects of environmental stressors on neural development.

http://wwweb.mpib-tuebingen.mpg.de/Abt.1/
Bonhoeffer Lab, Dr. Friedrich Bonhoeffer, Director; Max-Planck-Institut für Entwicklungsbiologie, Tübingen, Germany
Mutants of the retinotectal projection in zebrafish; olfactory system in zebrafish.

http://www.biop7.jussieu.fr/Labo/BiolDev/
Laboratoire de Biologie du Développement, Habib Boulekbache, Director; Université de Paris, France
Function and expression of homeobox genes during embryogenesis and fin morphogenesis.

http://www.nbio.uni-heidelberg.de/Brand.html
Brand Lab, Michael Brand, Director; Universität Heidelberg, Germany
Molecular genetics and morphogenesis of the developing mid–hindbrain and inner ear.

http://www.zoo.uni-heidelberg.de/zoo1/website/English/indexeng.htm
Heidelberg Aquatic Toxicology Lab, Thomas Braunbeck, Director; University of Heidelberg, Germany

(Eco)toxicology: early life-stage-tests, life-cycle tests, effects of endocrine disruptors, primary cultures of hepatocytes as alternative test system, *in vitro* and *in vivo* genotoxicity.

http://www.virginia.edu/~neurolab

The Brunjes Lab, Peter C. Brunjes, Director; University of Virginia, Charlottesville, VA, USA

The development of the brain with a focus on understanding the development of the ventral forebrain.

http://www.wmich.edu/bios/staff/byrd/

Byrd Lab, Christine A. Byrd, Director; Western Michigan University, Kalamazoo, MI, USA

Formation of new cells in the adult brain using the olfactory system of zebrafish as the model.

http://www.ucg.ie/bio/lucy.html

Galway Department of Biochemistry, Lucy Byrnes, Director; National University of Ireland, Galway, Ireland

Regulation of gene expression during zebrafish development.

http://www.bchs.uh.edu/People/Cahill/Cahill.html/

Cahill Lab, Greg Cahill, Director; University of Houston, TX, USA

Cellular and molecular mechanisms that generate and regulate circadian rhythmicity in vertebrates.

http://mother.biolan.uni-koeln.de/institute/ebio/Deutsch/campos.html

Campos-Ortega Lab, Jose Campos-Ortega, Director; Universität zu Köln, Germany

Cellular, genetic, and molecular bases of neurogenesis in *Drosophila melanogaster* and *Danio rerio.*

http://www.missouri.edu/~bioscwww/faculty/chandrasekhar.html

Chandrasekhar Lab, Anand Chandrasekhar, Director; University of Michigan, Ann Arbor, MI, USA

Molecular and cellular analysis of neuronal specification and axonal pathfinding.

http://home.ust.hk/~changlab/

Laboratory of Molecular Biophysics, Donald C. Chang, Director; The Hong Kong University of Science and Technology, Hong Kong

Calcium signaling in cell division.

http://www.hmc.psu.edu/biochem/faculty/cheng.htm

Cheng Lab, Keith Cheng, Director; The Jack Gittlen Cancer Research Institute, Pennsylvania State University, Hershey, PA, USA

Genetic dissection of genomic instability and cell differentiation in zebrafish.

http://www.sinica.edu.tw/imb/chung

Chung Lab, Bon-chu Chung, Laboratory Director; Academia Sinica, Taipei, Taiwan

Regulation of steroid synthesis has been under intensive studies because steroid hormones play major roles in controlling salt and sugar balance, and in the development of sex organs.

http://www.ansc.purdue.edu/grad/gradbul/gradres1.htm#collodi

Collodi Lab, Paul Collodi, Director; Purdue University, West Lafayette, IN, USA

Hormonal regulation of vertebrate development.

http://weber.u.washington.edu/~fishscop/

Cooper Lab, Mark Cooper, Director; University of Washington, Seattle, WA, USA

Axis formation, cellular mechanics of gastrulation and neurulation.

http://www.unimi.it/engl/dipart/biodip/biodipin8.html#sci

Developmental Biology Lab, Franco Cotelli, Director; Sez. Zool Citol, Milano, Italy

Maternal genes; genes involved in CNS development; analysis of the expression of genes coding for the major egg chorion components.

http://www.univ-montp2.fr/~neurodvpmt

Laboratoire de Neurogenetique du Developpement, Christine Dambly-Chaudiere and Alain Ghysen, Directors; Universite Montpellier II, France

Development of sensory organs; development of the lateral line and of its primary and secondary projections.

http://dir.nichd.nih.gov/Lmg/lmgdevb.htm

Dawid Lab, Igor Dawid, Director; NICHD, National Institutes of Health, Bethesda, MD, USA

The molecular mechanisms underlying differentiation and pattern formation in the early vertebrate embryo, in particular with respect to the establishment of the body pattern at gastrulation.

http://www.wesleyan.edu/~sdevoto/RESEARCH/resHome.html

Devoto Lab, Stephen Henri Devoto, Director; Wesleyan University, Middletown, CT, USA

The development of zebrafish muscle cell identity.

http://www.mcb.harvard.edu/faculty/dowling_bio.html

Dowling Lab, John E. Dowling, Director; Harvard University, Cambridge, MA, USA

How visual information is processed by the retina.

http://www.zebrafish.uni-freiburg.de

Driever Lab, Wolfgang Driever, Director; University of Freiburg, Germany

A genetic approach to study patterning and neuronal differentiation in the zebrafish neural plate.

http://www.umich.edu/~neurosci/faculty/easter.html

Easter Lab, Stephen S. Easter, Jr., Director; University of Michigan, Ann Arbor, MI, USA

Development of tracts in the early central nervous system and development of the retina.

http://www.neuro.uoregon.edu/faculty/eisen.html

Eisen Lab, Judith S. Eisen, Director; University of Oregon, Eugene, OR, USA

Specification and patterning of neurons and neural crest cells in embryonic zebrafish.

http://www.lri.ca/profiles/ekker.htm

M. Ekker Lab, Marc Ekker, Director; Loeb Research Institute, Ottawa, Ontario, Canada

Zebrafish developmental genetics.

http://www.med.umn.edu/dbc/ekker/?902853069

S. Ekker Laboratory, Stephen C. Ekker, Director; University of Minnesota Medical School, Minneapolis, MN, USA

Understanding patterning events in early vertebrate development.

http://www.neurobio.sunysb.edu/fetcholab/

Fetcho Lab, Joe Fetcho, Director; SUNY, Stony Brook, NY, USA

How the nervous system controls motor behavior in vertebrates.

http://cvrc.mgh.harvard.edu/

http://zebrafish.mgh.harvard.edu/

The Fishman Laboratory, Mark Fishman, Director; Massachusetts General Hospital, Charlestown, MA, USA

Understanding vertebrate morphogenesis. Genetics and embryology are used in combination to dissect organogenesis, especially of the cardiovasculature, gastrointestinal and hematopoietic systems.

http://www.nemc.org/psych/fulwiler.htm

Fulwiler Lab, Carl E. Fulwiler, Director; New England Medical Center, Boston, MA, USA

Genetic analysis of neural development in zebrafish.

http://www.biozentrum.unibas.ch/~biocomp/report9697/2nd/gerster.html

Gerster Lab, Thomas Gerster, Director; Biozentrum der Universität Basel, Switzerland

Analysis of gene expression during early zebrafish development.

http://www.med.umich.edu/mhri/res/95/goldman/goldman.html

Goldman Lab, Dan Goldman, Director; University of Michigan, Ann Arbor, MI, USA

CNS regeneration; regulation of gene expression during nervous system development.

http://spaceprojects.arc.nasa.gov/Space_Projects/aquatic/aqualab.html

Space Station Biological Research Project: Aquatic Laboratory, Edward M. Goolish, Director; NASA Ames Research Center, Moffett Field, CA, USA

Environmental physiology; adaptation to the microgravity and radiation environment of space.

http://www.med.upenn.edu/~cellbio/Granato.shtml

Granato Lab, Michael Granato, Director; University of Pennsylvania, Philadelphia, PA, USA

Genetic control of axonal guidance and neural circuit formation in the zebrafish embryo.

http://biosci.cbs.umn.edu/labs/perry

Hackett Lab, Perry Hackett, Director; University of Minnesota, St. Paul, MN, USA

Gene regulatory elements; repetitive elements in zebrafish; fish transgenics.

http://www.eb.tuebingen.mpg.de/Abt.3/

http://www.eb.tuebingen.mpg.de/Abt.3/Haffter__Lab/Mapping.html

Haffter Lab, Pascal Haffter; Max-Planck-Institut für Entwicklungsbiologie, Tübingen, Germany

Midline signaling and left–right asymmetry in the zebrafish; genetic mapping of mutations in the zebrafish.

http://www.bio.jhu.edu/faculty/Halpern/Halpern.html

Halpern Lab, Marnie E. Halpern, Director; Carnegie Institution of Washington, Baltimore, MD, USA

Genetic approaches in the zebrafish to explore how regional specializations arise in the developing vertebrate nervous system.

http://www.immunbio.mpg.de/german/main.htmlx (follow links: organisation/Hammerschmidt)

Hammerschmidt Lab, Matthias Hammerschmidt, Director; Max-Planck-Institut für Immunobiologie, Freiburg, Germany

Early dorsoventral pattern formation; development of the cardiovascular system.

http://www.med.niigata-u.ac.jp/pha/pharm-jp.html (in Japanese)

Department of Pharmacology, Hiroshi Higuchi, Director; Niigata University School of Medicine, Niigata, Japan

Gene expression of neuron-specific genes, Molecular mechanism of neural differentiation by using zebrafish.

http://www.med.osaka-u.ac.jp/pub/molonc/www/Eindex.html

Hirano Lab, Toshio Hirano, Director; Biomedical Research Center, Osaka University Medical School, Japan

Elucidation of the mechanisms by which cells of developing zebrafish respond to secreted signaling factors including how cytokines specify cell fate and cell behavior.

http://www.molbio.princeton.edu/faculty/ho.html

Ho Lab, Robert Ho, Director; Princeton University, Princeton, NJ, USA

How and when individual cells of the early embryo become "committed," that is, irreversibly restricted to expressing a particular cellular identity.

http://web.mit.edu/biology/www/Ar/hopkins.html

Hopkins Lab, Nancy Hopkins, Director; Massachusetts Institute of Technology, Cambridge, MA, USA
The genetic basis of development and simple behaviors in the zebrafish using insertional mutagenesis.

http://spike.fa.gau.hu/~tejfol/index.html
Laboratory of Fish Culture, Laszlo Horvath, Director; Godollo University, Hungary
Fish genetics and culture.

http://www.niob.knaw.nl/
Hubrecht Laboratory, S. W. deLaat, Director; Netherlands Institute for Developmental Biology, Utrecht, The Netherlands
Neurogenic signaling; 3D anatomical atlas of the zebrafish embryo.

http://hughes1.rai.kcl.ac.uk
Hughes Lab, Simon M. Hughes, Director; The Randall Institute, King's College London, UK
Formation and maintenance of fast and slow muscle fibers in vertebrates, with particular interest in the relationship of intrinsic cell programs and extrinsic signals in the early muscle patterning of limb and somite and the role of electrical activity and myogenic transcription factors of the MyoD family in the control of fiber type-specific gene expression.

http://weber.u.washington.edu/~jbhurley
Hurley Lab, James B. Hurley, Director; Howard Hughes Medical Institute, University of Washington, Seattle, WA, USA
Vision in zebrafish.

http://www.shef.ac.uk/uni/academic/A-C/biomsc/research/ingham.html
Ingham Lab, Philip W. Ingham, Director; The Krebs Institute, University of Sheffield, UK
Cell signaling in inductive interactions.

http://dbbs.wustl.edu/RIB/Johnson_Stephen.html
Johnson Lab, Stephen L. Johnson, Director; Washington University Medical School, St. Louis, MO, USA
Growth control and morphogenesis in zebrafish development.

http://www.ncl.ac.uk/~nbiochem/jowett/jowett1.html
Jowett Lab, Trevor Jowett, Director; University of Newcastle upon Tyne, UK
Zebrafish as a model of vertebrate development.

http://www.rochester.edu/College/BIO/faculty/Kane.html
Kane Lab, Donald Kane, Director; University of Rochester, Rochester, NY, USA
How early developmental genes control cell fate in the zebrafish.

http://www.uwo.ca/zoo/faculty/kelly.html
University of Western Ontario Zebrafish Pond, Greg Kelly, Director; University of Western Ontario, London, Ontario, Canada

Cytoskeletal-plasma membrane directed cell-signaling mechanisms.

http://www.bath.ac.uk/Departments/Biosciweb/kelsh.htm

Bath Zebrafish Development Lab, Robert Kelsh, Director; University of Bath, UK

How pluripotent progenitor cells become specified to one of several distinct fates; how cell migration through the embryo is controlled.

http://www.science.nus.edu.sg/~webdbs/Staff/khoohw/index.htm

Khoo Lab, Hong-Woo Khoo, Director; The National University of Singapore, Singapore

Gene transfer in aquatic organisms and crustacean maturation.

http://weber.u.washington.edu/~biowww/kimelman.html#kimelman

Kimelman Lab, David Kimelman, Director; University of Washington, Seattle, WA, USA

The intercellular signals and transcription factors that pattern the early vertebrate embryo.

http://www.neuro.uoregon.edu/~kimmelab/klab.htm

Kimmel Lab, Charles B. Kimmel, Director; University of Oregon, Eugene, OR, USA

Neuronal patterning in vertebrate embryos; embryonic cell lineage analysis; genes that control developmental patterning.

http://www.genetik.uni-koeln.de/skorsching

Molecular and Systems Neurobiology Group, Sigrun I. Korsching, Director; University of Cologne, Köln, Germany

Neuronal encoding and pattern formation in the olfactory nervous system of zebrafish.

http://www.ima.org.sg/research/fish/fish.html

Fish Developmental Biology Lab, Vladimir Korzh, PI; Institute of Molecular Agrobiology; an affiliate of The National University of Singapore

Axis formation and early neuronal differentiation in zebrafish.

http://www.science.adelaide.edu.au/genetics/lardelli/lardellilabwww

Lardelli Lab, Michael Lardelli, Director; University of Adelaide, Australia

The action of *Notch* genes/receptors in vertebrate development, especially neurogenesis; interactions between *Notch* receptors subfamilies and between *Notch* gene/receptors and other genes/proteins/discovery and functional analysis of genes controlling neural development.

http://www.science.nus.edu.sg/~webdbs/Labs/deve-lab/index.html

Singapore Developmental Biology Laboratory, Tit-Meng Lim, Director; The National University of Singapore

The developmental expression of intermediate filament genes in the zebrafish.

http://mouse.mc.duke.edu

Linney Lab, Elwood Linney, Director; Duke University Medical Center, Durham, NC, USA

Retinoic acid and estrogen signaling during embryogenesis in zebrafish.

http://www.oldcity.com/wml/linser.html#Paul

Whitney Marine Laboratory, Paul J. Linser, Director; University of Florida, St. Augustine, FL, USA

Molecular and cell biology of visual development.

http://www.ag.auburn.edu/genemap/molgen.html

Fish Molecular Genetics and Biotechnology Lab, Zhanjiang (John) Liu, Director; Auburn University, Auburn, AL, USA

Transgenic fish and gene mapping.

http://www.wmdc.edu/ (follow links: academics/biology/Wilbur Long)

Long Lab, Wilbur Long, Director; Western Maryland College, Westminster, MD, USA

Early development and pattern formation.

http://vigyan.nsu.edu/~genelab/

Molecular Genetics Research Unit, Debabrata Majumdar, Director; Norfolk State University, Norfolk, VA, USA

Cloning and expression of developmentally important genes.

http://www.hms.harvard.edu/dms/bbs/fac/malicki.html

Malicki Lab, Jarema Malicki, Director; Massachusetts Eye and Ear Infirmary, Boston, MA, USA

Genetic basis of neuronal identity and pattern in the zebrafish retina.

http://home.ust.hk/~aequorin/calcium.htm

HKUST Calcium-Aequorin Imaging Laboratory, Andrew L. Miller, Director; HKUST, Hong Kong

Use of aequorins to visualize patterns of calcium signaling during the development of zebrafish embryos.

http://info.med.yale.edu/pharm/moczydl1.htm

Moczydlowski Lab, Edward Moczydlowski, Director; Yale Medical School, New Haven, CT, USA

Transferrin-related genes in zebrafish, in particular a transferrin homolog called saxiphilin.

http://weber.u.washington.edu/~rtmoon

Moon Lab, Randall Moon, Director; University of Washington, Seattle, WA, USA

The biochemical mechanisms of signal transduction by the Wnt family of secreted ligands, and the functions of the Wnt family in vertebrate embryos.

http://molly.hsc.unt.edu/~smoorman/

Moorman Lab, Stephen Moorman, Director; University of North Texas Health Science Center, Fort Worth, TX, USA

Developmental neurobiology and spinal cord regeneration.

http://www.healthsci.tufts.edu/labs/lgmoss/home.htm

Moss Lab, Larry Gene Moss, Director; Tufts New England Medical Center, Boston, MA, USA

Development biology of the pancreas and molecular physiology of insulin.

http://fishnet.bio.temple.edu/

Myers Lab, Paul Z. Myers, Director; Temple University, Philadelphia, PA, USA

Neurodevelopment, early embryonic behavior and circuitry, and teratogenesis.

http://pharma1.med.osaka-u.ac.jp/Welcome.html (in Japanese)

Osaka Department of Pharmacology 1, Osamu Muraoka, Director; Osaka University Medical School, Japan

Gene expression in the nervous system (including drug dependence); molecular mechanisms of neurogenesis by cell adhesion molecules; signal transduction in the nervous system.

http://wwweb.mpib-tuebingen.mpg.de/Abt.3/

Nüsslein-Volhard Lab, Christiane Nüsslein-Volhard, Director; Max-Planck-Institut für Entwicklungsbiologie, Tübingen, Germany

Somite formation in the zebrafish; neural crest development; behavior and sensory systems in the zebrafish.

http://www.neuro.uoregon.edu/~postlela/mydoc.htm

Postlethwait Zebrafish Genetics Laboratory, John H. Postlethwait, Director; University of Oregon, Eugene, OR, USA

Zebrafish genetic mapping.

http://femur.uchicago.edu/Faculty/Prince.cgi

Prince Lab, Victoria Prince, Director; University of Chicago, Chicago, IL, USA

Patterning of the primary axis in vertebrate embryos.

http://weber.u.washington.edu/~draible/

Raible Lab, David Raible, Director; University of Washington, Seattle, WA, USA

Zebrafish neural crest development.

http://www.umich.edu/~praymond

Raymond Lab, Pamela Raymond, Director; University of Michigan, Ann Arbor, MI, USA

The development and regeneration of the vertebrate retina using both goldfish and zebrafish as model systems.

http://som1.ab.umd.edu/AquaticPath/fac_res.html

Aquatic Pathobiology Center, Renate Reimschuessel, Director; University of Maryland School of Medicine, Baltimore, MD, USA

Pathogenesis of cell injury, carcinogenesis, environmental and comparative pathology, molecular biology, immunology and aquaculture science; aquatic toxicology and developing nonmammalian models for human diseases.

http://cscwww.cats.ohiou.edu/~neuro/NEUROFAC/ross.HTM

Ross Lab, Linda S. Ross, Director; Ohio University, Athens, OH, USA

Understanding the functional organization of the nervous system by exploring the developmental origin of sensory systems.

http://www.science.nus.edu.sg/~webdbs/Labs/card-lab/index.html

Singapore Cardiovascular Molecular Biology Lab, Ge Ruowen, Director; National University of Singapore

The molecular mechanism of blood vessel formation (angiogenesis) both in embryonic development (zebrafish) as well as in tumor.

http://www.ummed.edu/dept/pharmacology/sagerst.htm

Sagerström Lab, Charles Sagerström, Director; University of Massachusetts Medical Center, Worcester, MA, USA

Cellular and molecular mechanisms regulating development of the vertebrate central nervous system (CNS), with a particular emphasis on early development of the caudal CNS (i.e., the hindbrain and the spinal cord).

http://129.49.19.42/neuro/sun95fin.html (scroll down to Schechter)

Schechter Lab, Nisson Schechter, Director; SUNY, Stony Brook, NY, USA

Molecular correlates of visual pathway development and regeneration in goldfish and zebrafish.

http://saturn.med.nyu.edu/~schier/

Schier Lab, Alex Schier, Director; Skirball Institute, New York University New York, NY, USA

Head patterning in the zebrafish.

http://www.albany.edu/~js213/

Schmidt Lab, John T. Schmidt, Director; SUNY, Albany, NY, USA

The activity-driven refinement of retinotopic connections in the visual system and the role of myosin light chain kinase in regulating actin–myosin-based growth cone motility.

http://www.ncl.ac.uk/~nbiochem/sharrock/sharrock.html

Sharrocks Lab, Andrew D. Sharrocks, Director; The Medical School, Newcastle University, Newcastle upon Tyne, United Kingdom

Eukaryotic transcriptional regulation; studies on the ETS-domain and MADS-box transcription factor families.

http://www.wi.mit.edu/sive/home.html

Sive Lab, Hazel Sive, Director; Whitehead Institute for Biomedical Research, Cambridge, MA, USA

Analyses of the mechanisms of neural determination and patterning in zebrafish and frog, using explant, molecular, and genetic assays.

http://itsa.ucsf.edu/~shorne/

Stainier Lab, Didier Stainier, Director; University of California, San Francisco, CA, USA

Study of heart, vessel, and blood formation, as well as the role of the yolk syncytial layer (YSL) in early embryonic development.

http://web.mit.edu/biology/www/Ar/steiner.html

Steiner Lab, Lisa Steiner, Director; Massachusetts Institute of Technology, Cambridge, MA, USA

The study of the development of the immune system in two model organisms: the amphibian, *Xenopus laevis,* and the zebrafish, *Danio rerio.*

http://www.uidaho.edu/LS/BioSc/faculty/dstenkamp.html

Stenkamp Lab, Deborah L. Stenkamp, Director; University of Idaho, Moscow, ID, USA

Examination of the cellular and molecular mechanisms of vertebrate retinal development, with a specific focus on the patterning of photoreceptor mosaics, using goldfish and zebrafish as experimental models; specific interests in (1) the patterning of photoreceptor mosaics and (2) the role of hedgehog genes in retinal development.

http://mama.indstate.edu/dls/facstaff/stuart.html

Stuart Lab, Gary W. Stuart, Director; Indiana State University, Terre Haute, IN, USA

The molecular mechanisms by which vertebrate organisms develop from single-celled embryos into complex organisms.

http://www.uni-konstanz.de/FuF/Bio/forsch/zoology/stuermerlab

Stürmer Lab, Claudia Stürmer, Director; University of Konstanz, Germany

The function of specific cell surface proteins in the central nervous system of fish during axon growth in the embryo and axon regeneration in the adult.

http://www.med.nyu.edu/people/W.Talbot.html

Talbot Lab, Will Talbot, Director; Skirball Institute, New York University School of Medicine, New York, NY, USA

The genetic mechanisms that govern cell fate specification and morphogenesis in the zebrafish.

http://darkwing.uoregon.edu/~zfish/

University of Oregon Zebrafish Facility, Bill Trevarrow, Director; University of Oregon, Eugene, OR, USA

Zebrafish Facility research focused on developing more efficient, higher-quality methods for zebrafish husbandry, breeding, and feeding.

http://zebra.biol.sc.edu/biology/faculty/vogt.html

Vogt Lab, Richard Vogt, Director; University of South Carolina, Columbia, SC, USA

Molecular biology, function and development of olfactory systems in insects and fish.

http://www.uwm.edu/Dept/MFB

Marine & Freshwater Biomedical Science Center, Daniel Weber, Director; University of Wisconsin, Milwaukee, WI, USA

Effects of xenoestrogens on gene expression of sex hormones (Reinhold Hutz); chromosomal mapping (Ruth Phillips); developmental abberations from dioxin exposure and effects on ARNT and AHR genes (Richard Peterson).

http://www.sas.upenn.edu/biology/faculty/research/weinberg.research.html

Weinberg Lab, Eric Weinberg, Director; University of Pennsylvania, Philadelphia, PA, USA

Genetic control of central nervous system and muscle development in the zebrafish embryo.

http://www.neuro.uoregon.edu/faculty/westerf.html

Westerfield Lab, Monte Westerfield, Director; University of Oregon, Eugene, OR, USA

Mechanisms that regulate the differentiation of neurons and muscles.

http://www.neuro.uoregon.edu/faculty/weston.html

Weston Lab, James A. Weston, Director; University of Oregon, Eugene, OR, USA

Analysis of mutations that affect the development of neural crest derivatives.

http://www.shef.ac.uk/uni/academic/A-C/biomsc/research/whitfiel.html

Whitfield Lab, Tanya Whitfield, Director; The Krebs Institute, University of Sheffield, UK

Genetic analysis of the development of sensory epithelium in the zebrafish inner ear and lateral line.

http://lenti.med.umn.edu/zebrafish/zfish_top_page.html

Zebrafish Sequence Analysis Project, University of Minnesota, St. Paul, MN, USA

http://hhmi.med.harvard.edu/~zonlab

The Zon Lab, Leonard Zon, Director; Howard Hughes Medical Institute's Children's Hospital, Boston, MA, USA

The regulation of hematopoiesis during vertebrate embryonic development.

INDEX

ISBN 0-12-544162-2

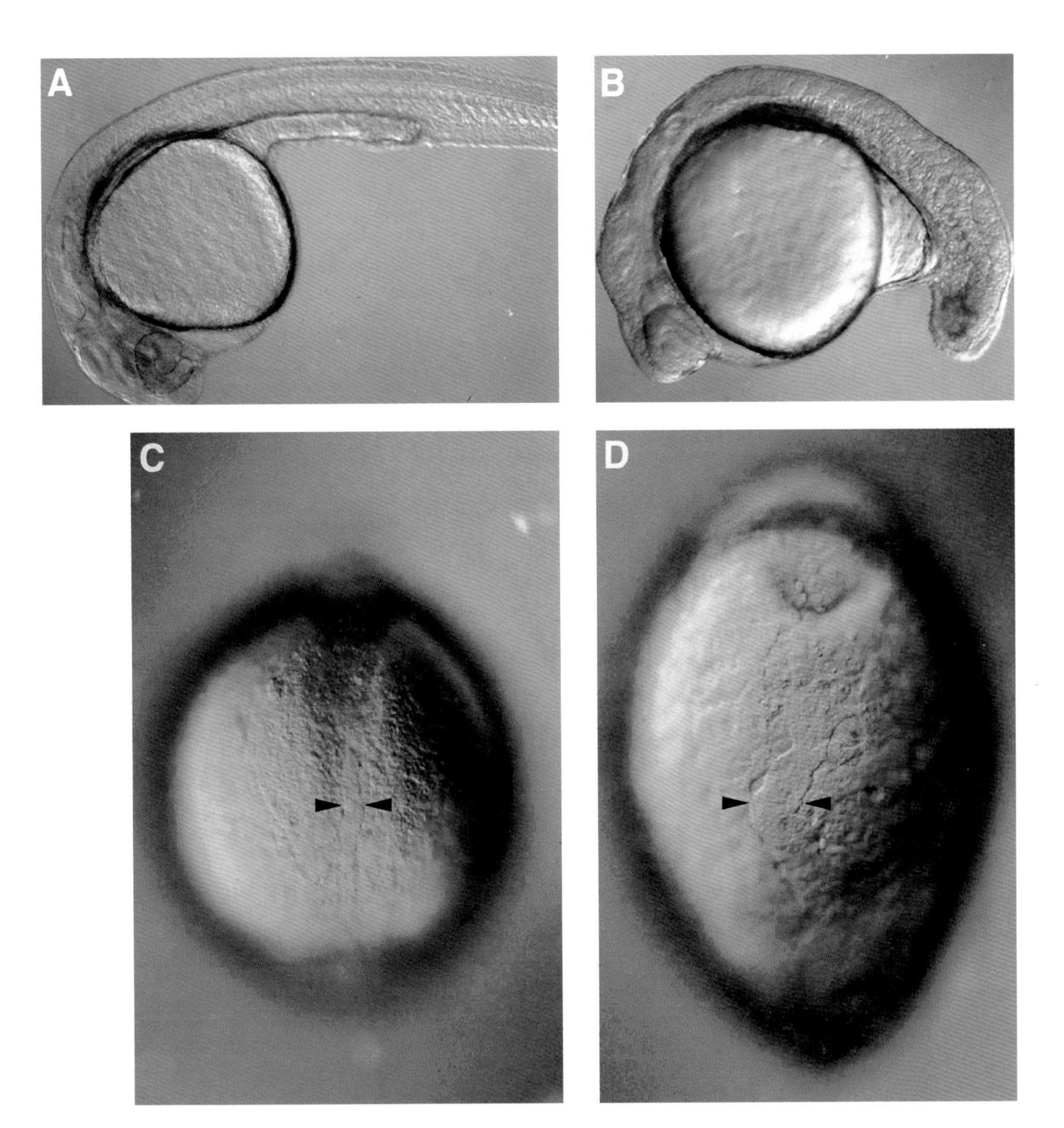

Ch. 1, Fig. 1. Embryos derived from homozygous *yob* and heterozygous *sbn* females exhibit maternal-effect phenotypes. Lateral view of a wild-type embryo (A) and an embryo derived from a yob^{d} homozygous female (B) at the pharyngula period. Note the reduced body length in (B). Dorsal view of a wild-type embryo (C) and an embryo derived from a heterozygous sbn^{C24} female (D) at the tailbud stage. As sbn^{C24} is a dominant maternal mutant, all embryos show the dorsalized phenotype with a broadened notochord anlagen (flanked by arrowheads).

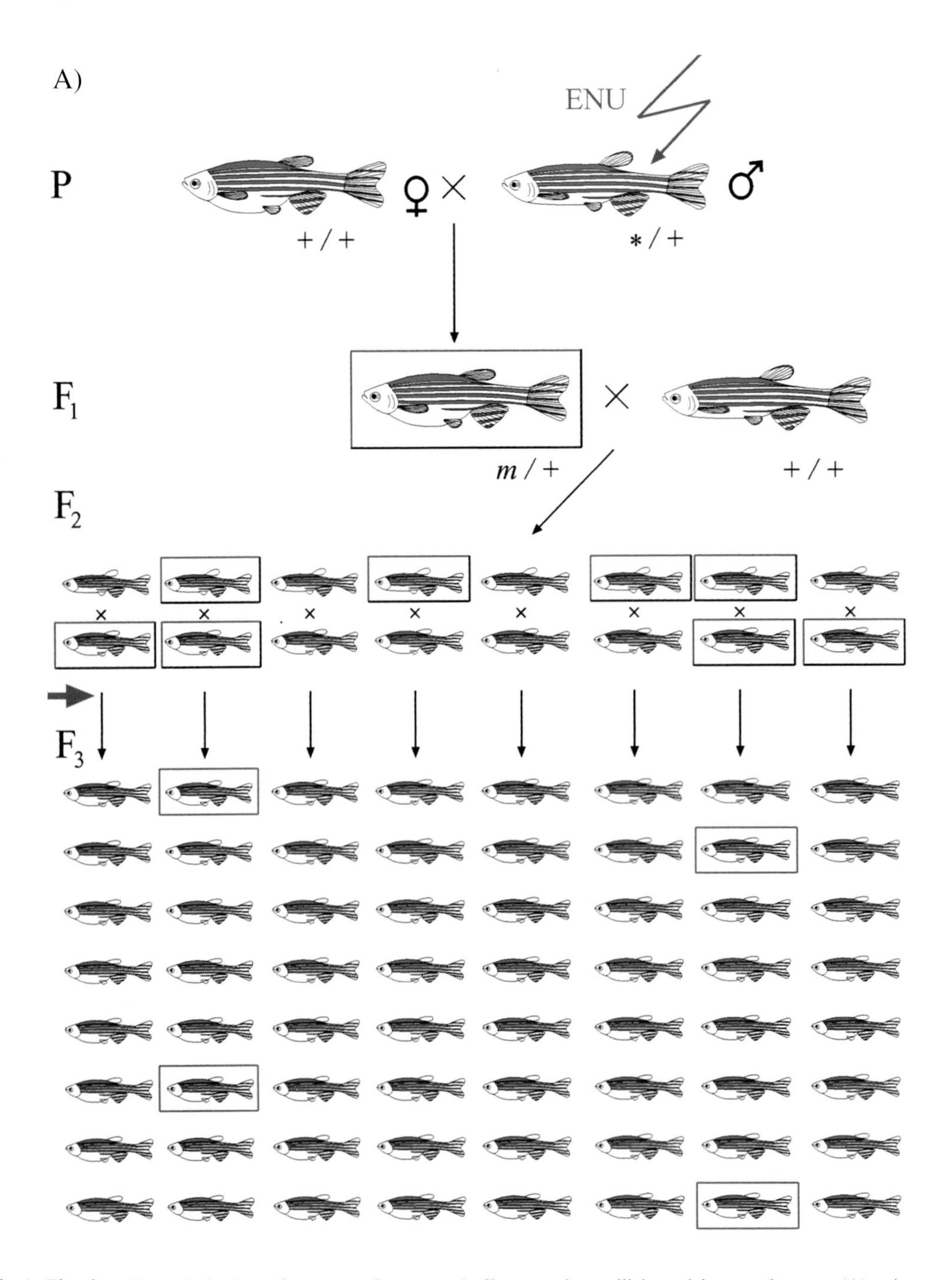

Ch. 1, Fig. 2. Theoretical scheme for a screen for maternal-effect mutations utilizing solely natural crosses (A) or incorporating gynogenesis (B). In both schemes, mutations are induced in males of the parental (P) generation using the chemical mutagen ENU. P males are crossed to wild-type females to produce, in the next (F_1) generation, fish heterozygous for the newly induced mutations. In A, F_1 individuals are used to start F_2 families. The sex of the F_1 fish is irrelevant. Random intercrosses between F_2 siblings result in crosses between heterozygous carriers (in red/black boxes) in one quarter of cases, which in turn produce, in the next (F_3) generation, one quarter of homozygous mutant individuals (in red boxes). For comparison, a red arrow in A indicates the

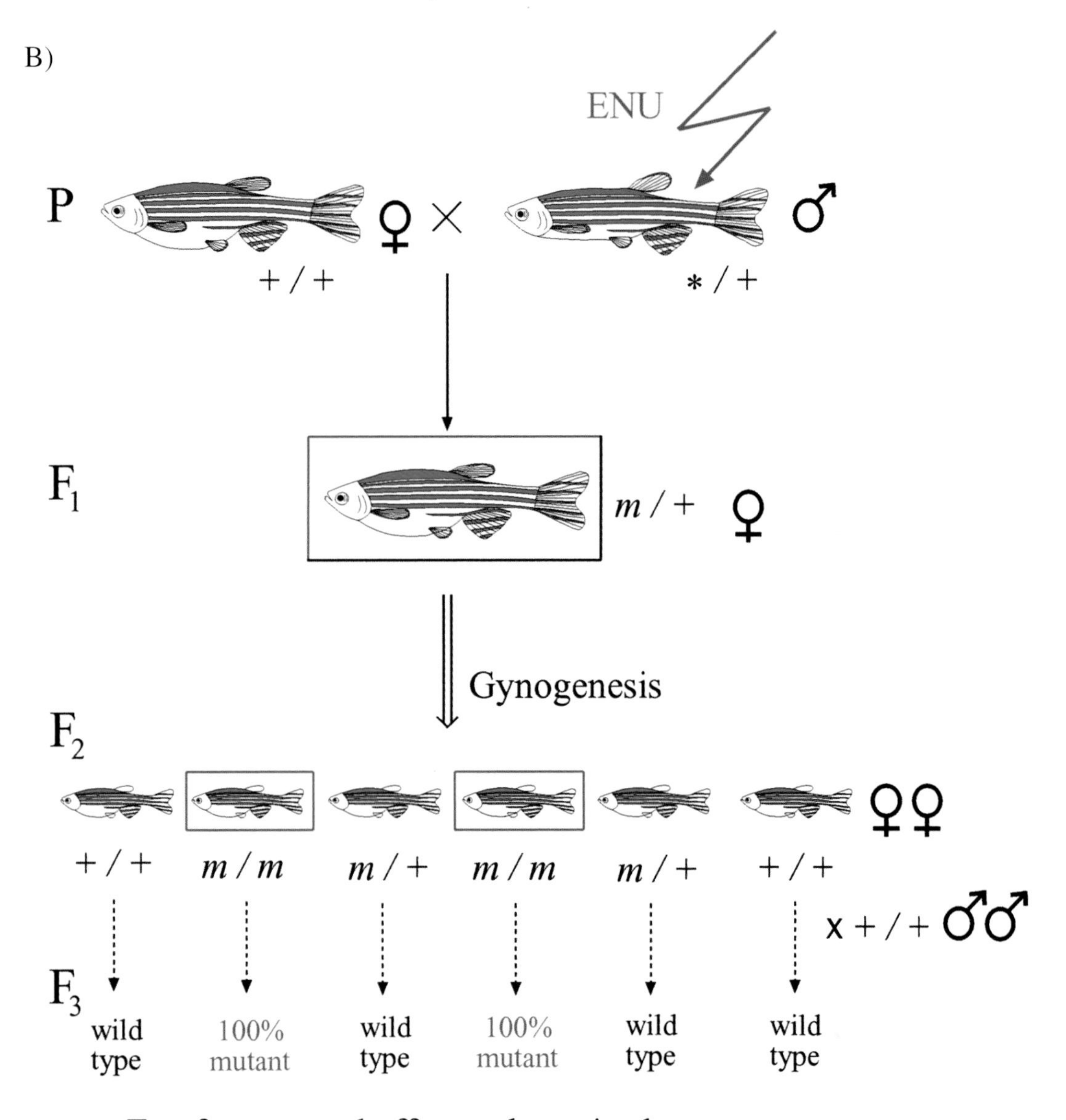

point in which embryos were screened for mutations in zygotic genes in the Boston and Tübingen screens (see text). In B, F_1 heterozygous females (in a red/black box) are treated to induce gynogenesis, producing F_2 homozygous mutants (in red boxes). Hypothetical results are shown using Early Pressure and a gene with an average centromere–locus distance (see text and Fig. 3). Adult females homozygous for newly induced mutations (F_3 in A and F_2 in B) can be tested for maternal-effect phenotypes by fertilizing their eggs with sperm of any genotype, either naturally or *in vitro* (dotted arrows, shown in B only). Females mutant for recessive maternal-effect mutations are expected to produce eggs and embryos exhibiting highly penetrant phenotypes, regardless of the paternal genotype. Both schemes have been drawn to represent the approximate number of crosses and maternal-effect tests needed to identify with 80% certainty mutations present in the F_1 carriers (maternal-effect tests shown only in B). The number of crosses in A can in principle be reduced by a factor of 2 if the F_1 carrier is crossed to another F_1 carrying a different set of mutations (not shown). These numbers are: in A, 1 (or 0.5 if both F_1 are carriers of different mutations) F_2 crosses, 8 (or 4) F_3 crosses, and 64 (or 32) F_4 maternal-effect tests per mutagenized genome; and in B, 1 gynogenetically derived F_2 clutch and 6 maternal-effect tests per mutagenized genome.

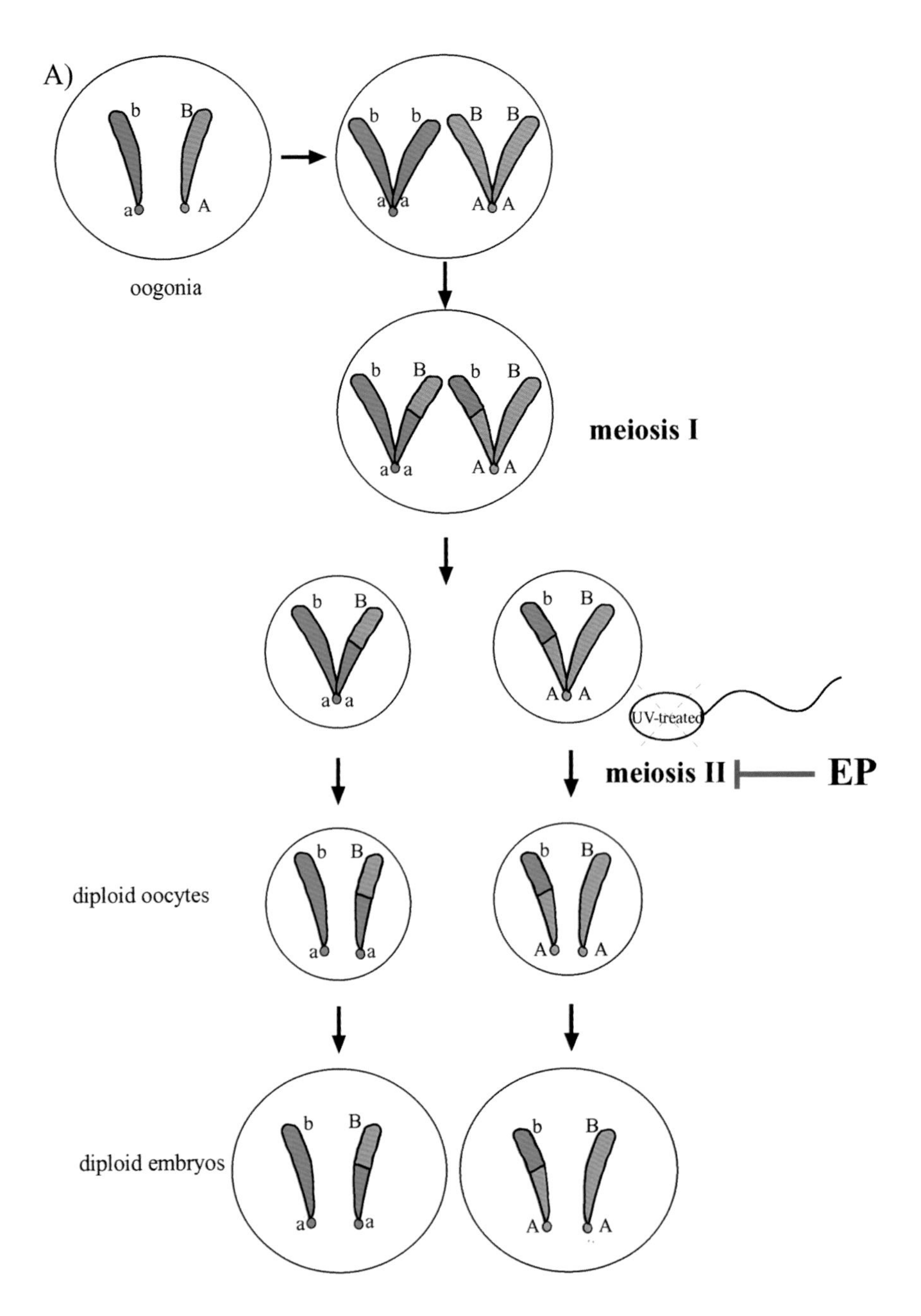

Ch. 1, Fig. 3. Methods to induce gynogenesis. Early Pressure (EP, in A) and Heat Shock (HS, in B). Oocytes are arrested at metaphase of the second meiotic division (Selman and Wallace, 1989). Upon egg activation and fertilization, meiosis resumes and produces haploid oocytes. Fertilization with sperm whose genetic content has been inactivated with UV light results in haploid embryos (not shown), which are inviable. EP inhibits cytokinesis of meiosis II, thus producing diploid oocytes that can be fertilized with UV-treated sperm to produce viable diploid embryos. HS inhibits the first embryonic cell division, thus transforming haploid embryos into diploid ones. Segregation patterns for a proximal, or centromere-linked, gene (with alleles A and a) and a distal gene (with alleles B and b), are shown in each panel. In EP-derived oocytes, recombination events between the centromere and the locus produce heterozygosity at loci distal to the recombination event, thus preferentially reducing their degree of homozygosity. HS-derived embryos are homozygous at every locus, regardless of chromosomal position or recombination. See text for more details.

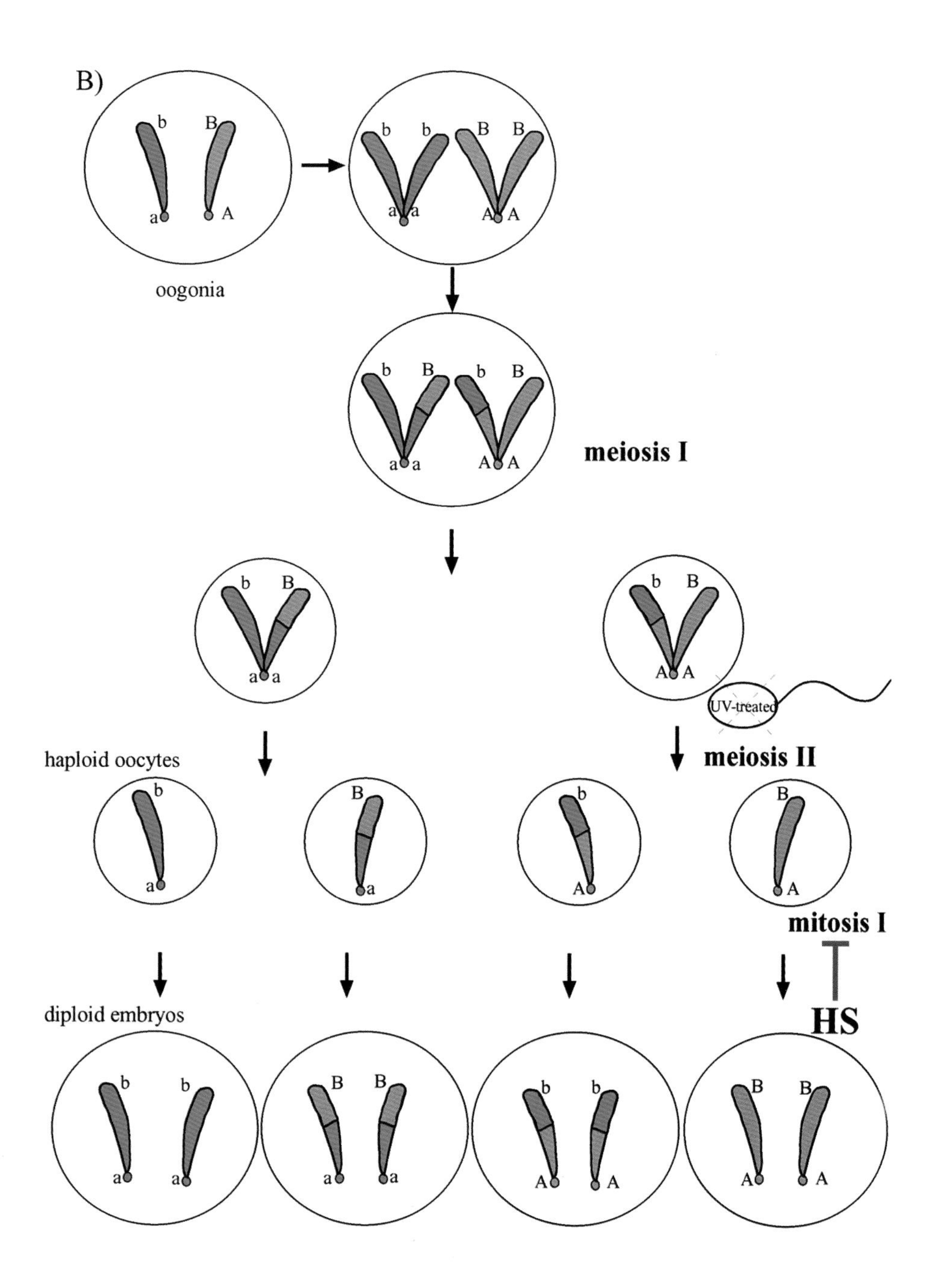
B)
oogonia
meiosis I
UV-treated
haploid oocytes
meiosis II
mitosis I
HS
diploid embryos

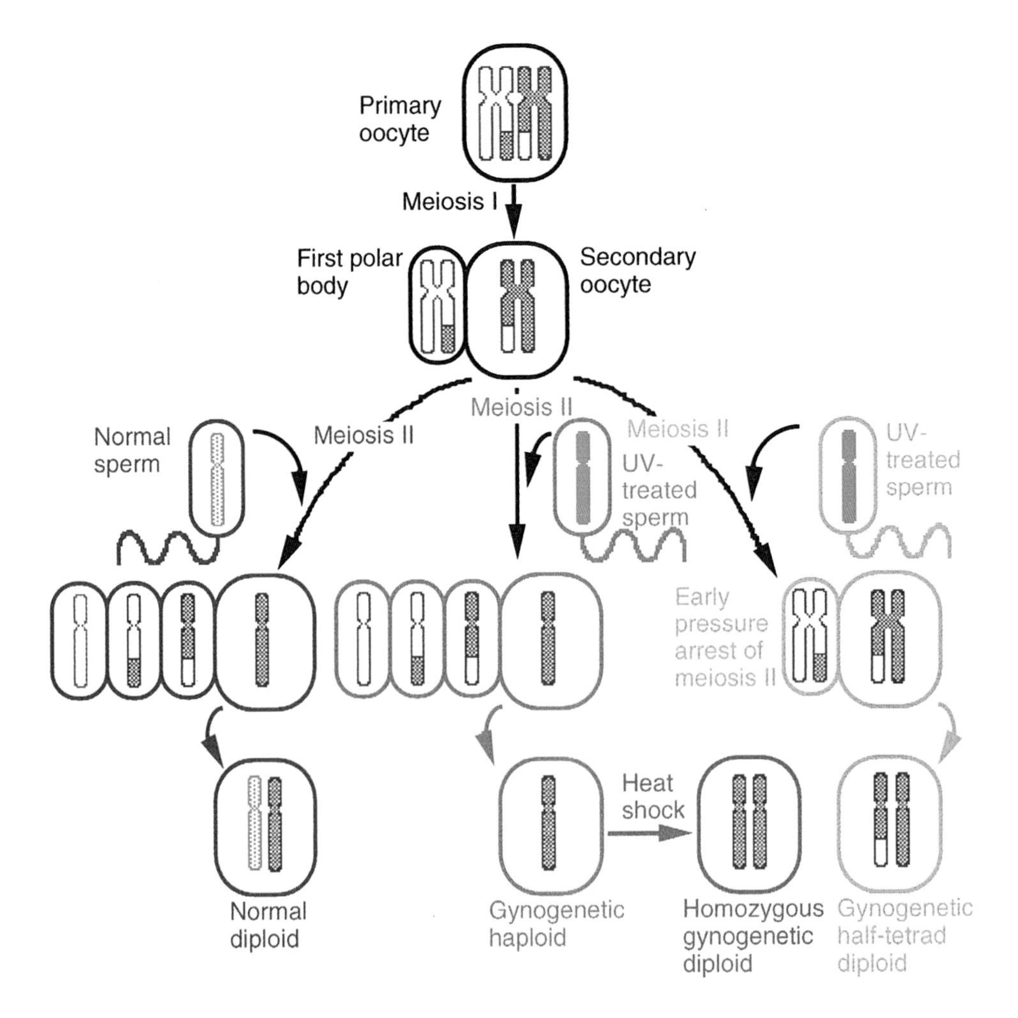

Ch. 4, Fig. 1. Ploidy can be manipulated to produce haploid and diploid zebrafish embryos. Normal diploid embryos result from fertilization of secondary oocytes by potent sperm. Gynogenetic haploid embryos are produced by activating secondary oocytes with sperm rendered genetically impotent by UV irradiation. These embryos can be made homozygous diploid at all maternal loci using heat shock to suppress the first mitotic division. Gynogenetic half-tetrad diploid embryos are produced by activating secondary oocytes with UV-irradiated sperm and then arresting the second meiotic division with early pressure. Details of these methods can be found in *The Zebrafish Book* (Westerfield, 1993) or on the World Wide Web at http://zfish.uoregon.edu/zf_info/zfbook/zfbk.html. Adapted from Postethwait, J. H., and Talbot, W. S., Zebrafish genomics: From mutants to genes, Copyright 1996, pp. 183–190, with kind permission from Elsevier Science Ltd., The Boulevard, Langford Lane, Kidlington 0X5 1GB, UK.

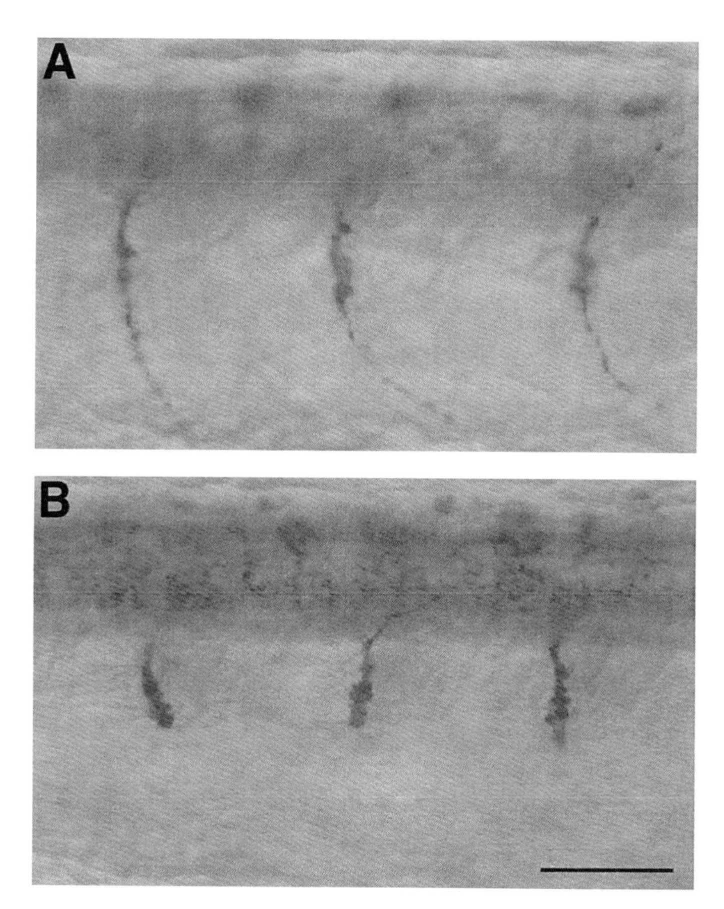

Ch. 4, Fig. 3. *stumpy* mutant embryos have truncated CaP axons. Side views of 24 h wild-type (A) and *sty* mutant (B) embryos in which motor axons have been labeled with the zn-1 and znp-1 monoclonal antibodies. Axons in *sty* mutants terminate at the level of the nascent horizontal myoseptum. Scale bar, 25 μm.

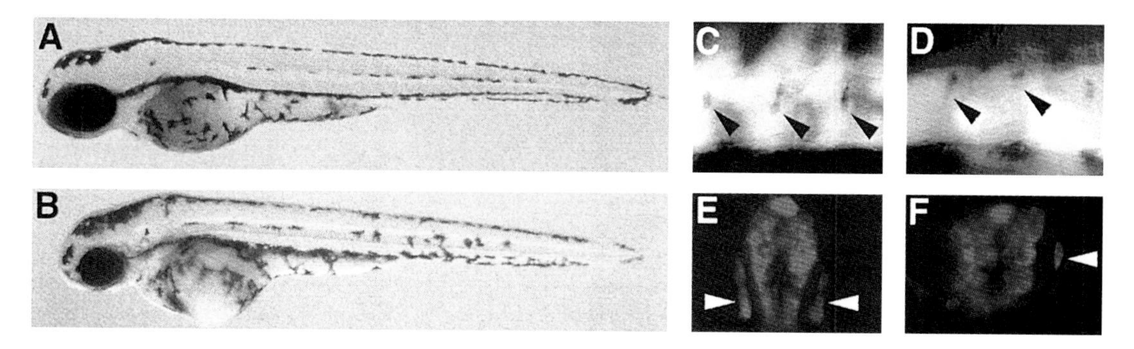

Ch. 4, Fig. 4. *laughing man* mutant embryos have defective DRGs. Side views of living wild-type (A) and *laughing man* (B) 3 day embryos. In whole-mount (C, D) and frozen transverse sections (E, F) of wild-type embryos labeled with the anti-Hu antibody, DRG neurons (arrowheads) are located adjacent to the ventrolateral portion of the spinal cord. In *laughing man* mutant embryos, DRG neurons are displaced dorsally (E, F) and are not always bilaterally symmetric. Adapted from: Screen for mutations affecting development of zebrafish neural crest, P. D. Henion, D. W. Raible, C. E. Beattie, K. L. Stoesser, J. A. Weston, and J. S. Eisen, In *Development Genetics*, Copyright © 1996 Wiley-Liss, Inc.

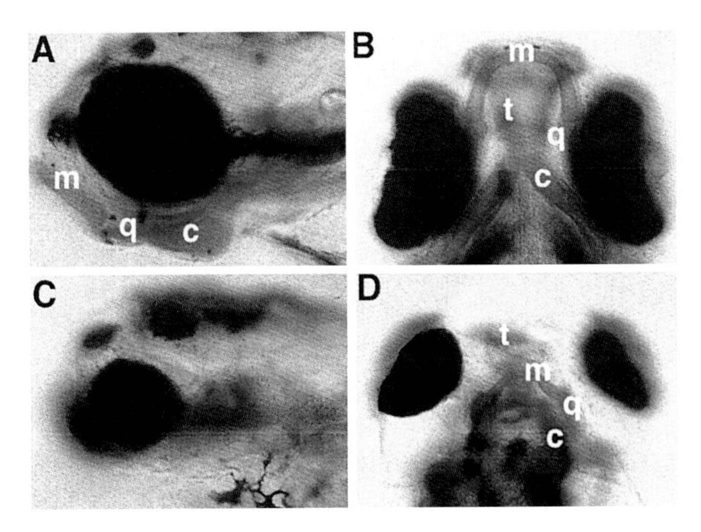

Ch. 4, Fig. 5. *gobbler* mutant embryos have defective cranial cartilage development. Lateral (A,C) and ventral (B,D) views of the heads of 5 day wild-type and mutant embryos stained with alcian blue to reveal cartilaginous elements. In wild-type embryos (A,B), both neurocranial (trabeculae, t) and visceral (Meckel's cartilage, m; quadrate, q; ceratohyal, c) cartilage elements are visible. In *gobbler* mutant embryos, although all cartilage elements appear to be present in (C,D), they are abnormally shaped. Adapted from: Screen for mutations affecting development of zebrafish neural crest, P. D. Henion, D. W. Raible, C. E. Beattie, K. L. Stoesser, J. A. Weston, and J. S. Eisen, *Developmental Genetics*, Copyright © 1996 Wiley-Liss, Inc.

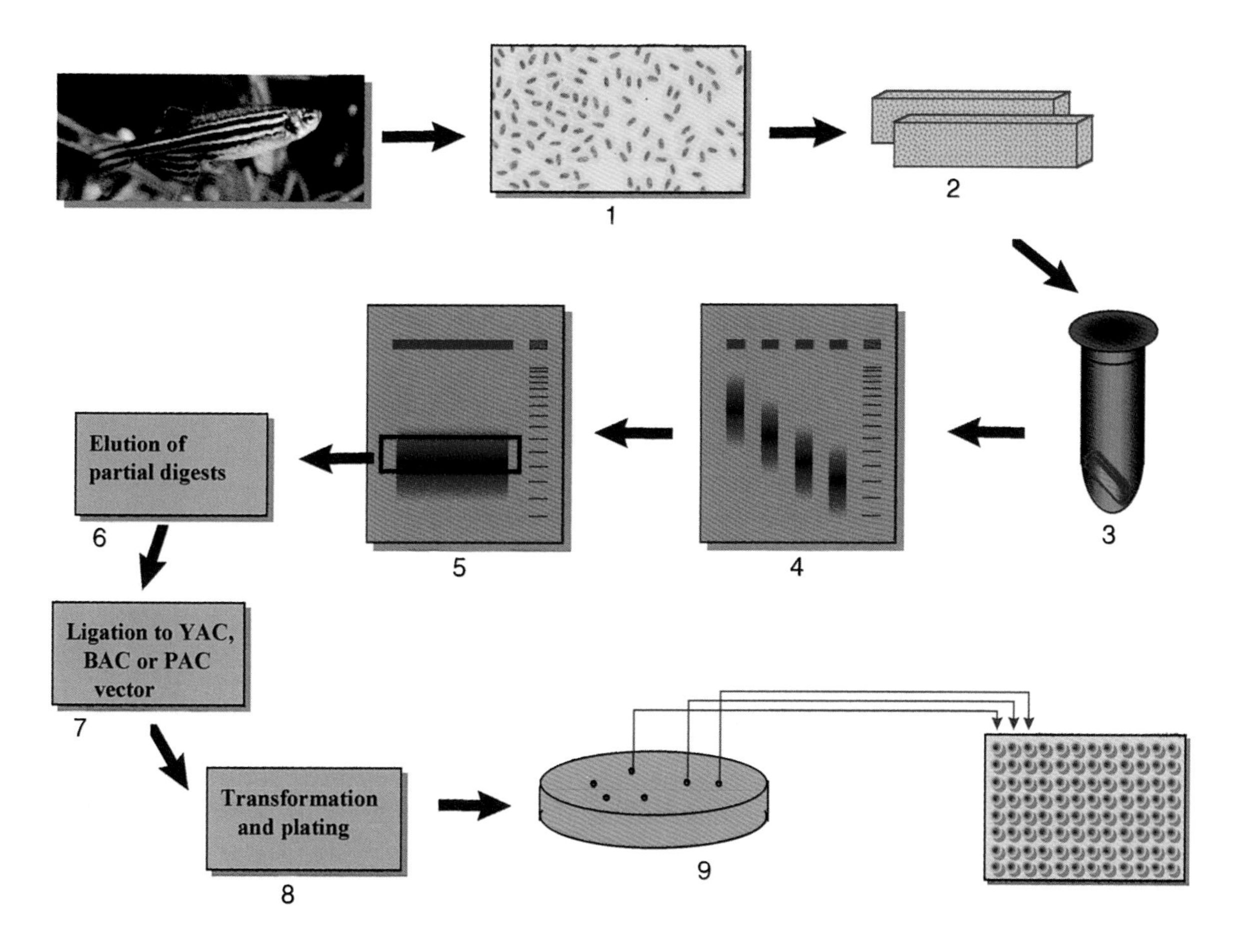

Ch. 14, Fig. 1. Flow chart for the preparation of a large-insert genomic library *Danio rerio* and other vertebrates. This figure is meant to serve only as a basic overview; it does not encompass slight modifications of the general method, especially when generating YAC libraries. (1) Cells (or nuclei) are isolated, washed, and counted using a hemacytometer. (2) The cells (or nuclei) are embedded at a fixed density (roughly 100–200 μg DNA per ml) in low melting point agarose blocks. The blocks are then treated *in situ* with proteolytic agents to lyse the cells (nuclei). The agarose embedding prevents shearing of the high molecular weight (HMW) DNA during handling. (3) The blocks are extensively dialyzed against TE buffer to remove salts and other small molecules and are partially digested *in situ* with suitable restriction enzymes for YAC, BAC, or PAC cloning. The partial digestion reactions are carried out using either a time-course experiment or methylase competition (Larin *et al.*, 1991; Amemiya *et al.*, 1996). (4) Aliquots of the digests are assessed on an analytical pulsed field gel in order to determine optimal conditions for partial digestion. (5) A preparative pulsed field gel is then run on the selected partial digests, and the regions of interest are excised without staining of the DNA with ethidium bromide and exposure to UV irradiation. (6) DNA is isolated from the gel using an agarose-digesting enzyme (e.g., agarase) or by electroelution (Strong *et al.*, 1997). (7) The DNA is ligated to dephosphorylated YAC, BAC, or PAC vectors. In the case of the BAC and PAC systems, the ligations are carried out using a very low DNA-to-volume ratio, which fosters the formation of large, circular molecules. For YAC cloning, the ligations are often electrophoresed on a second preparative pulsed field gel (to eliminate smaller ligation products). (8) Ligation products for YACs are transformed into *S. cerevisiae* via spheroplast transformation ligation products for BACs and PACs are transformed into the cell wall-deficient DH10B™ (Life Technologies) *E. coli* via electroporation (Shizuya *et al.*, 1992; Ioannou *et al.*, 1994). Transformants are plated on agar-based selective media (see Table 1) and allowed to grow. (9) Recombinant colonies are robotically picked (arrayed) into individual wells of 96-well or 384-well microtiter dishes that contain media supplemented with a cryoprotectant (for archiving the library at –70°C). Notably, the arraying of these libraries (as opposed to conventional λ or cosmid libraries) individualizes each clone with a unique address in the library (Shepherd *et al.*, 1994).

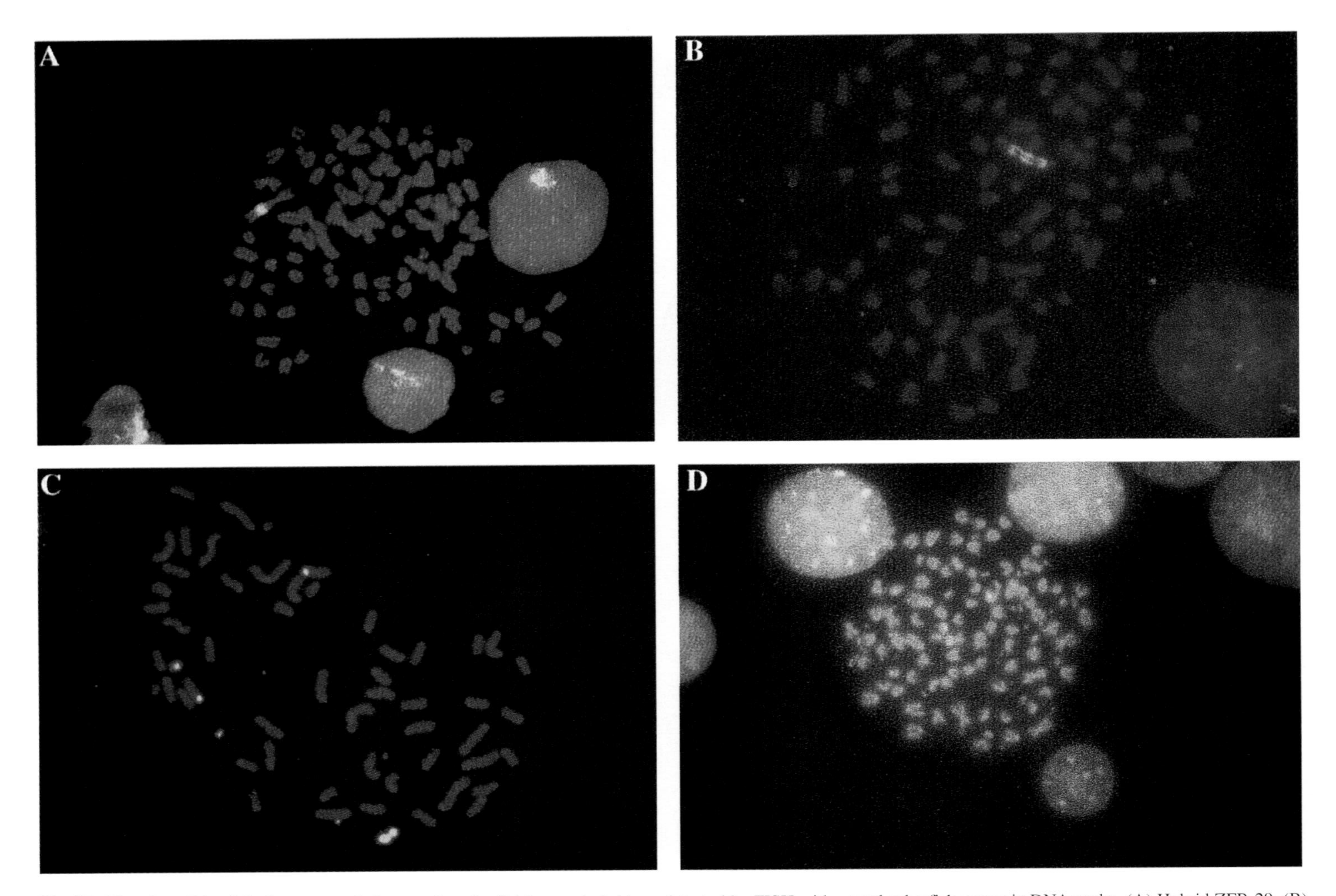

Ch. 17, Fig. 1. Zebrafish chromosomal elements in zebrafish/mouse hybrids as detected by FISH with a total zebrafish genomic DNA probe. (A) Hybrid ZFB-29, (B) hybrid LFFB-13, (C) hybrid LFFB-10, (D) hybrid ZFB-226.

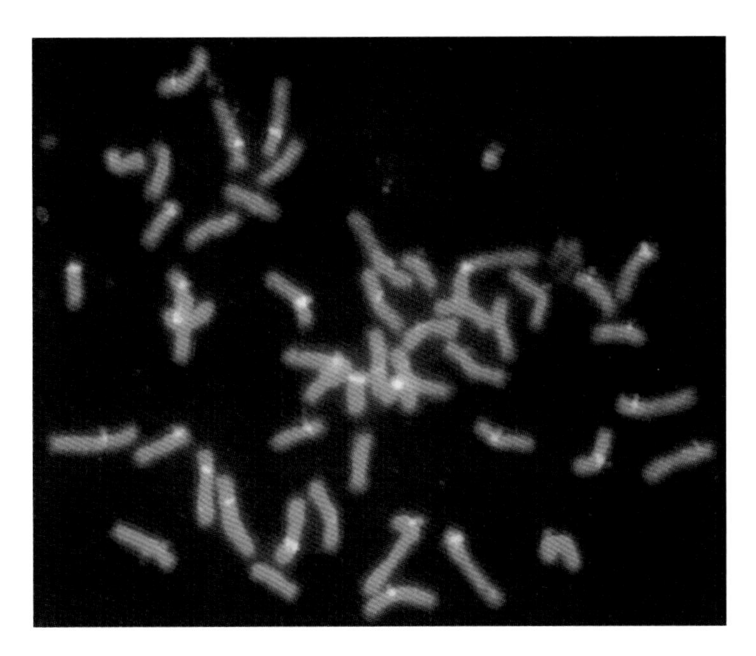

Ch. 18, Fig. 3. The location of the type I satellite probe to zebrafish chromosomes by fluorescent *in situ* hybridization. The fluorescein-labeled probe (yellow) is located on chromosomes counterstained with propidium iodide (red).

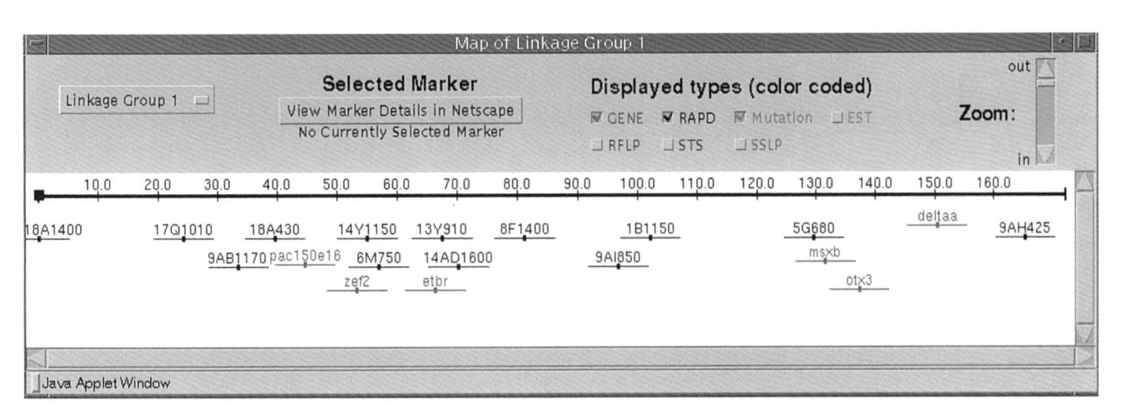

Ch. 19, Fig. 3. Graphical interface for viewing the genetic map. This implementation is derived from a mapview bioWidget developed by Gregg Helt at the Berkeley *Drosophila* Genome Project.